Thomas P. Wangler
RF Linear Accelerators

Related Titles

Padamsee, H.

RF Superconductivity

Science, Technology & Applications

2008
ISBN: 978-3-527-40572-5

Reiser, M.

Theory and Design of Charged Particle Beams

1994
ISBN: 978-0-471-30616-0

Brown, I. G. (ed.)

The Physics and Technology of Ion Sources

2004
ISBN: 978-3-527-40410-0

Edwards, D. A., Syphers, M. J.

An Introduction to the Physics of High Energy Accelerators

1993
ISBN: 978-0-471-55163-8

Thomas P. Wangler

RF Linear Accelerators

2nd completely revised and enlarged edition

WILEY-VCH Verlag GmbH & Co. KGaA

The Author

Thomas P. Wangler
Michigan State University
USA

Cover
Peter Hesse Design

All books published by **Wiley-VCH** are carefully produced. Nevertheless, authors, editors, and publisher do not warrant the information contained in these books, including this book, to be free of errors. Readers are advised to keep in mind that statements, data, illustrations, procedural details or other items may inadvertently be inaccurate.

Library of Congress Card No.: applied for

British Library Cataloguing-in-Publication Data
A catalogue record for this book is available from the British Library.

Bibliographic information published by the Deutsche Nationalbibliothek
Die Deutsche Nationalbibliothek lists this publication in the Deutsche Nationalbibliografie; detailed bibliographic data are available on the Internet at <http://dnb.d-nb.de>.

© 2008 WILEY-VCH Verlag GmbH & Co. KGaA, Weinheim

All rights reserved (including those of translation into other languages). No part of this book may be reproduced in any form – by photoprinting, microfilm, or any other means – nor transmitted or translated into a machine language without written permission from the publishers. Registered names, trademarks, etc. used in this book, even when not specifically marked as such, are not to be considered unprotected by law.

Printed on acid-free paper

Composition Laserwords Private Ltd, Chennai, India
Printing betz-druck GmbH, Darmstadt
Bookbinding Litges & Dopf GmbH, Heppenheim

ISBN: 978-3-527-40680-7

Contents

Preface to the Second Edition *XI*

Preface to the First Edition *XIII*

1 Introduction *1*
1.1 Linear Accelerators: Historical Perspective *2*
1.2 Linac Structures *6*
1.3 Linac Beam Dynamics *10*
1.4 Multiparticle Effects *12*
1.5 Applications of Modern RF Linacs *13*
1.6 Accelerator-Physics Units, Unit Conversions, and Physical Constants *15*
1.7 Useful Relativistic Mechanics Relationships *16*
1.8 Maxwell's Equations *17*
1.9 Conducting Walls *19*
1.10 Group Velocity and Energy Velocity *20*
1.11 Coaxial Resonator *22*
1.12 Transverse-Magnetic Mode of a Circular Cylindrical Cavity *24*
1.13 Cylindrical Resonator Transverse-Magnetic Modes *26*
1.14 Cylindrical Resonator Transverse Electric Modes *27*
References *30*

2 RF Acceleration in Linacs *32*
2.1 Particle Acceleration in an RF Field *32*
2.2 Energy Gain on Axis in an RF Gap *33*
2.3 Longitudinal Electric Field as a Fourier Integral *36*
2.4 Transit-Time-Factor Models *39*
2.5 Power and Acceleration Efficiency Figures of Merit *42*
2.6 Cavity Design Issues *44*
2.7 Frequency Scaling of Cavity Parameters *46*
2.8 Linac Economics *47*
References *52*

RF Linear Accelerators. 2nd, completely revised and enlarged edition.
Thomas P. Wangler
Copyright © 2008 Wiley-VCH Verlag GmbH & Co. KGaA, Weinheim
ISBN: 978-3-527-40680-7

3 Periodic Accelerating Structures 53
3.1 Synchronous Acceleration and Periodic Structures 53
3.2 Floquet Theorem and Space Harmonics 54
3.3 General Description of Periodic Structures 57
3.4 Equivalent Circuit Model for Periodic Structures 59
3.5 Periodic Array of Low-Pass Filters 61
3.6 Periodic Array of Electrically Coupled Circuits 62
3.7 Periodic Array of Magnetically Coupled Circuits 63
3.8 Periodic Array of Cavities with Resonant Coupling Element 64
3.9 Measurement of Dispersion Curves in Periodic Structures 65
3.10 Traveling-Wave Linac Structures 68
3.11 Analysis of the Periodic Iris-Loaded Waveguide 69
3.12 Constant-Impedance Traveling-Wave Structure 72
3.13 Constant-Gradient Structure 74
3.14 Characteristics of Normal Modes for Particle Acceleration 76
3.15 Physics
Regimes of Traveling-Wave and Standing-Wave Structures 79
References 81

4 Standard Linac Structures 83
4.1 Independent-Cavity Linacs 83
4.2 Wideröe Linac 87
4.3 H-Mode Structures 89
4.4 Alvarez Drift-Tube Linac 91
4.5 Design of Drift-Tube Linacs 96
4.6 Coupled-Cavity Linacs 98
4.7 Three Coupled Oscillators 99
4.8 Perturbation
Theory and Effects of Resonant-Frequency Errors 101
4.9 Effects from Ohmic Power Dissipation 103
4.10 General Problem of $N+1$ Coupled Oscillators 105
4.11 Biperiodic Structures for Linacs 108
4.12 Design of Coupled-Cavity Linacs 111
4.13 Intercell Coupling Constant 114
4.14 Decoupling of Cavities Connected by a Beam Pipe 116
4.15 Resonant Coupling 117
4.16 Accelerating Structures for Superconducting Linacs 121
$\lambda/4$ Superconducting Structures 121
$\lambda/2$ Superconducting Structures 121
TM Superconducting Structures 122
RF Properties and
Scaling Laws for TM and $\lambda/2$ Superconducting Structures 125
Shunt
Impedance for TM and $\lambda/2$ Superconducting Structures 127
Stored Energy for TM and $\lambda/2$ Superconducting Structures 129

Scaling Formulas for λ/4 Superconducting Structures *131*
References *133*

5   Microwave Topics for Linacs *135*
5.1   Shunt Resonant Circuit Model *135*
5.2   Theory of Resonant Cavities *137*
5.3   Coupling to Cavities *138*
5.4   Equivalent Circuit for a Resonant-Cavity System *139*
5.5   Equivalent Circuit for a Cavity Coupled to two Waveguides *144*
5.6   Transient Behavior of a Resonant-Cavity System *146*
5.7   Wave Description of a Waveguide-to-Cavity Coupling *148*
5.8   Microwave Power Systems for Linacs *156*
5.9   Multipacting *159*
5.10 Electron Field Emission *162*
5.11 RF Electric Breakdown: Kilpatrick Criterion *163*
5.12 Adiabatic Invariant of an Oscillator *164*
5.13 Slater Perturbation Theorem *165*
5.14 Quasistatic Approximation *167*
5.15 Panofsky–Wenzel Theorem *168*
      References *173*

6   Longitudinal Particle Dynamics *175*
6.1   Longitudinal Focusing *175*
6.2   Difference
      Equations of Longitudinal Motion for Standing-Wave Linacs *177*
6.3   Differential Equations of Longitudinal Motion *178*
6.4   Longitudinal Motion when Acceleration Rate is Small *178*
6.5   Hamiltonian and Liouville's Theorem *182*
6.6   Small Amplitude Oscillations *186*
6.7   Adiabatic Phase Damping *187*
6.8   Longitudinal
      Dynamics of Ion Beams in Coupled-Cavity Linacs *189*
6.9   Longitudinal Dynamics in Independent-Cavity Ion Linacs *190*
6.10 Longitudinal
      Dynamics of Low-Energy Beams Injected into a $v = c$ Linac *192*
6.11 Rf Bunching *194*
6.12 Longitudinal Beam Dynamics in H-Mode Linac Structures *196*
      References *199*

7   Transverse Particle Dynamics *201*
7.1   Transverse RF Focusing and Defocusing *201*
7.2   Radial Impulse from a Synchronous Traveling Wave *203*
7.3   Radial Impulse near the Axis in an Accelerating Gap *204*
7.4   Including Electrostatic Focusing in the Gap *207*
7.5   Coordinate Transformation through an Accelerating Gap *208*

7.6	Quadrupole Focusing in a Linac	*209*
7.7	Transfer-Matrix Solution of Hill's Equation	*211*
7.8	Phase-Amplitude Form of Solution to Hill's Equation	*213*
7.9	Transfer Matrix through One Period	*214*
7.10	Thin-Lens FODO Periodic Lattice	*215*
7.11	Transverse Stability Plot in a Linac	*217*
7.12	Effects of Random Quadrupole Misalignment Errors	*218*
7.13	Ellipse Transformations	*221*
7.14	Beam Matching	*222*
7.15	Current-Independent Beam Matching	*224*
7.16	Solenoid Focusing	*225*
7.17	Smooth Approximation to Linac Periodic Focusing	*226*
7.18	Radial Motion for Unfocused Relativistic Beams	*227*
	References	*230*

8	**Radiofrequency Quadrupole Linac**	*232*
8.1	Principles of Operation	*232*
8.2	General Potential Function	*236*
8.3	Two-Term Potential Function Description	*238*
8.4	Electric Fields	*240*
8.5	Synchronous Acceleration	*241*
8.6	Longitudinal Dynamics	*242*
8.7	Transverse Dynamics	*243*
8.8	Adiabatic Bunching in the RFQ	*245*
8.9	Four-Vane Cavity	*248*
8.10	Lumped-Circuit Model of Four-Vane Cavity	*249*
8.11	Four-Vane Cavity Eigenmodes	*251*
8.12	Transmission-Line Model of Quadrupole Spectrum	*254*
8.13	Radial-Matching Section	*260*
8.14	RFQ Transition Cell	*265*
8.15	Beam Ellipses in an RFQ	*271*
8.16	Tuning for the Desired Field Distribution in an RFQ	*273*
8.17	Four-Rod Cavity	*274*
8.18	Four Vane with Windows RFQ	*276*
	References	*280*

9	**Multiparticle Dynamics with Space Charge**	*282*
9.1	Beam Quality, Phase Space, and Emittance	*283*
9.2	RMS Emittance	*285*
9.3	Transverse and Longitudinal Emittance	*287*
9.4	Emittance Conventions	*288*
9.5	Space-Charge Dynamics	*289*
9.6	Practical Methods for Numerical Space-Charge Calculations	*292*
9.7	RMS Envelope Equation with Space Charge	*296*
9.8	Continuous Elliptical Beams	*297*

9.9	Three-Dimensional Ellipsoidal Bunched Beams	*299*
9.10	Beam Dynamics Including Linear Space-Charge Field	*300*
9.11	Beam-Current Limits from Space Charge	*302*
9.12	Overview of Emittance Growth from Space Charge	*303*
9.13	Emittance Growth for rms Matched Beams	*306*
9.14	Model of Space-Charge-Induced Emittance Growth in a Linac	*314*
9.15	Emittance Growth for rms Mismatched Beams	*316*
9.16	Space-Charge Instabilities in RF Linacs from Periodic Focusing: Structure Resonances	*318*
9.17	Longitudinal-Transverse Coupling and Space-Charge Instabilities for Anisotropic Linac Beams	*319*
9.18	Beam Loss and Beam Halo	*325*
9.19	Los Alamos Beam Halo Experiment	*329*
9.20	Scaling of Emittance Growth and Halo	*331*
9.21	Longitudinal Beam Dynamics Constraint on the Accelerating Gradient	*332*
	References	*338*
10	**Beam Loading** *341*	
10.1	Fundamental Beam-Loading Theorem	*342*
10.2	The Single-Bunch Loss Parameter	*344*
10.3	Energy Loss to Higher-Order Cavity Modes	*344*
10.4	Beam Loading in the Accelerating Mode	*345*
10.5	Equations Describing a Beam-Loaded Cavity	*347*
	General Results	*348*
	Optimum Detuning	*350*
	Extreme Beam-Loaded Limit	*351*
	Numerical Example of a Beam-Loaded Cavity	*351*
	Example of a Heavily Beam - Loaded Superconducting Cavity with Bunches Injected on the Crest of the Accelerating Wave	*352*
10.6	Generator Power when the Beam Current is Less than Design Value	*352*
10.7	Transient Turn-On of a Beam-Loaded Cavity	*354*
	References	*360*
11	**Wakefields** *361*	
11.1	Image Force for Line Charge in Round Pipe	*362*
11.2	Fields from a Relativistic Point Charge and Introduction to Wakefields	*364*
11.3	Wake Potential from a Relativistic Point Charge	*367*
11.4	Wake Potentials in Cylindrically Symmetric Structures	*368*
11.5	Scaling of Wake Potentials with Frequency	*370*
11.6	Bunch Wake Potentials for an Arbitrary Charge Distribution	*371*
11.7	Loss Parameters for a Particular Charge Distribution	*376*
11.8	Bunch Loss Parameters for a Gaussian Distribution	*377*
11.9	Beam-Coupling Impedance	*378*

11.10	Longitudinal- and Transverse-Impedance Definitions *380*
11.11	Impedance and Wake Potential for a Single Cavity Mode *381*
11.12	Short-Range Wakefields-Parasitic Losses *383*
11.13	Short-Range Wakefields: Energy Spread *383*
11.14	Short-Range Wakefields: Compensation of Longitudinal Wake Effect *384*
11.15	Short-Range Wakefields: Single-Bunch Beam Breakup *384*
11.16	Short-Range Wakefields: BNS Damping of Beam Breakup *386*
11.17	Long-Range Wakefields and Multibunch Beam Breakup *389*
11.18	Multipass BBU in Recirculating Electron Linacs *397*
	References *402*

12 Special Structures and Techniques *405*

12.1	Alternating-Phase Focusing *405*
12.2	Accelerating Structures Using Electric Focusing *406*
12.3	Coupled-Cavity Drift-Tube Linac *410*
12.4	Beam Funneling *411*
12.5	RF Pulse Compression *413*
12.6	Superconducting RF Linacs *414*
	Brief History *415*
	Introduction to the Physics and Technology of RF Superconductivity *416*
12.7	Examples of Operating Superconducting Linacs *419*
	Atlas *419*
	CEBAF *419*
	Spallation Neutron Source *421*
12.8	Future Superconducting Linac Facilities *423*
	International Linear Collider *423*
	Next-Generation Rare Isotope Facility *426*
	Free-Electron Lasers *427*
	References *430*

Index *433*

Preface to the Second Edition

In the nine years since this book was published there has been continuing progress in many areas of the RF-linac field. This is not surprising because linacs are a critical enabling technology for high-energy physics, nuclear physics, neutron spallation sources, and free electron lasers (FELs). My motivation for the first edition was that the RF-linac field needed a textbook. The second edition is motivated by two considerations: 1) the need to add additional material that is important for those who are designing today's new linacs, and 2) to include new developments and new linacs that have been built during the past decade, as well as proposed new linacs for which major design efforts have been underway. This is still a textbook, but addition of the new material makes the second edition more useful as a reference book.

Several chapters have been expanded to address many developments during the past decade. New developments in the linac field during this time have included increased development of H-mode structures (Chapter 4), accelerating cavities for superconducting linacs (Chapter 4), multipass BBU in recirculating linacs and energy-recovery linacs (Chapter 11), and two new promising electric-focused structures under development (Chapter 12), the RF-DTL or RFD and the RF-Interdigital or RFI structure. The last paragraph of Section 4.15 in this new addition describes a simple basic principle for resonant coupling of two cavities.

In addition, Chapter 6 includes a simple model for RF bunching and longitudinal beam dynamics for long H-mode structures. Chapter 7 includes a new section on current-independent matching. Chapter 8 on the Radiofrequency Quadrupole Linac is expanded to include treatment of the RFQ radial matching section, beam ellipses in the RFQ, RFQ tuning, RFQ transition cells, and the Four-Vane with windows RFQ. Chapter 9, which covers multiparticle beam dynamics with space charge, is considerably expanded to include emittance growth for both rms-matched and rms-mismatched beams, space-charge instabilities from periodic focusing in RF-linacs, longitudinal-transverse coupling, and space-charge instabilities for anisotropic beams and how to avoid them. The Los Alamos beam-halo experiment which tested the two main analytic models of beam-halo formation is described in Chapter 9;

RF Linear Accelerators. 2nd, completely revised and enlarged edition.
Thomas P. Wangler
Copyright © 2008 Wiley-VCH Verlag GmbH & Co. KGaA, Weinheim
ISBN: 978-3-527-40680-7

also included in Chapter 9 is the longitudinal beam-dynamics constraint on accelerating gradient.

Discussed in Chapter 12 are the new linacs that have been built and operated within the past decade. Included is the spallation neutron source (SNS), an accelerator based neutron source that was designed by six National Laboratories and constructed at Oak Ridge National Laboratory for the U.S. Department of Energy. The majority of the SNS linac uses superconducting cavities for beam acceleration. At its full beam power of 1.4 MW, SNS will provide the worlds most intense pulsed neutron beams for scientific research and industrial development. In February, 2007 the SNS accelerated beam to its design energy of 1 GeV, a new world record for a proton linear accelerator.

Another important machine built and operated during the past decade is the Jefferson Laboratory FEL, which operates as an RF-superconducting energy-recovery linac (ERL). JLAB has demonstrated ERL operation in continuous wave (CW) mode at high power. In July, 2004 10 kW of radiated power was produced at 6 micron wavelength, and in October, 2006, 14.2 kW CW was achieved at 1.6 microns wavelength.

Included among proposed future linacs are the International Linear Collider (ILC), a proposed facility based on RF-superconducting cavities that would produce high-energy collisions between electron and positron beams at center-of-mass energies from 200 to 500 GeV. Included for future nuclear physics is a next-generation radioactive ion beam facility.

As was remarked in the preface to the first edition, superconducting RF linacs will become increasingly important in the RF-linac field for many applications. This is clearly illustrated by the examples in Chapter 12 of both new operating linacs and proposed future linacs. In all cases superconducting linacs are an important part of these machines. The discussion of superconducting RF technology in Chapter 12 is intended to provide a brief and systematic presentation of the physics and technology for the new student. A more comprehensive treatment is available in the excellent textbook *RF Superconductivity for Accelerators*, by Hasan Padamsee, Jens Knobloch, and Thomas Hays.

I wish to acknowledge a very productive collaboration with Jim Billen in teaching the USPAS linac courses. Jim's contribution included developing and documenting the problem solutions, and working with the students. In putting together this new edition, I have benefited from many colleagues. I especially want to acknowledge Jim Billen, Ken Crandall, Jean Delayen, Bob Gluckstern, Ingo Hofmann, Andy Jason, Martin Reiser, Uli Ratzinger, Alwin Schempp, Don Swenson, and Peter Walstrom. I want also to acknowledge the support and helpful suggestions of my wife Julie.

Los Alamos, 2007   *Thomas P. Wangler*

Preface to the First Edition

This book is based on my RF linear accelerator course presented through the US Particle Accelerator School (USPAS), sponsored by the Department of Energy; and organized by Mel Month. The course has been presented through the USPAS at the University of Texas, Harvard University, Duke University, and at University of New Mexico, Los Alamos. The material has been written assuming that the student has a basic knowledge of classical and relativistic mechanics, as well as electromagnetism. Thus, the material should be suitable for engineering students as well as advanced undergraduate or graduate students of physics.

The field of charged-particle accelerators is an impressive product of twentieth-century science and technology. It is an interdisciplinary field using state-of-the-art technologies, and draws on the skills of experts from different science and engineering fields. The modern era of radio-frequency linear accelerators began just after the end of World War II. Since that time, the RF linac has emerged as a device of major scientific and technical importance, addressing a broad range of important applications in basic research, energy, medicine, and defense. The field of RF linacs is undergoing rapid development, stimulated particularly by the interest in linear colliders for high-energy physics, high-power proton linacs for advanced neutron sources, and smaller linacs for medical and commerial applications. In many respects the RF linac is well suited to these new applications, especially those which emphasize high-intensity or high-brightness output beams. Continuing developments, especially in RF superconducting technology, are contributing significant advances in the linac field. Because of the diversity of the future applications, and the steady advances of accelerator technology, we can expect the field of RF linacs to continue to grow as we move into the twenty-first century.

My main motivation for writing this book has been that the RF-linac field needs a textbook, and until now there has been none. A textbook is invaluable to new students, who need a systematic and comprehensive introduction to basic principles. Additionally, the textbook can also be of value to accelerator experts. The accumulating body of knowledge in the field continues to be communicated mostly in laboratory technical reports and in acclerator conference proceedings. There is an international linac conference,

RF Linear Accelerators. 2nd, completely revised and enlarged edition.
Thomas P. Wangler
Copyright © 2008 Wiley-VCH Verlag GmbH & Co. KGaA, Weinheim
ISBN: 978-3-527-40680-7

as well as international particle-acclerator conferences in the U.S. and Europe, each of which takes place every two years. Nevertheless, this published material is generally very condensed, and is written primarily for the specialist in a particular sub-discipline. The nature of the RF linac field is such that at the major laboratories, experts can be found in the main sub-disciplines, including beam dynamics, accelerating structures, mechanical engineering, microwave engineering, radio-frequency power systems, vacuum technology, and cryogencis. To contribute more effectively to an RF-linac project, an expert often needs information from sources outside his or her area of expertise. Without a textbook, both the new student and the working scientist or engineer, often struggle to find the material needed to broaden their knowledge. The objective of this textbook is to provide a systematic presentation of both the science principles and the technology aspects that are the basis of the RF-linac field.

I have organized the book in the way that I have found most useful for teaching an introductory course on the principles of RF linacs. The core material consists of the basic introductory linac course. Those sections marked with an asterisk (*) are presented as optional material, which can be omitted without disrupting the overall continuity of the book. Selected optional sections can be added to the basic course to meet the specific needs of the students. Detailed presentations of specific applications have been omitted in the book, not because I think these are unimportant, but because it is relatively easy to find good and up-to-date review articles on the applications.

For all of the major topics, the basic analytic theory is presented, because I believe that understanding the analytic theory is the best way to obtain the insight required for producing a modern linac design. Computer codes and simulation methods are the next step, and I have included introductory presentations of the physics algorithms for some of the most commonly-used beam-dynamics codes that are used to predict the performance of modern linacs. I believe that analytic theory and the computer codes are both important tools; however, in these exciting times of rapid development in the computer field, there can be a temptation for overreliance on the computer. Even with such powerful computer codes, there is still no substitute for understanding the basic physics principles.

If the trends of the past decade continue, superconducting RF linacs will become increasingly important in the RF linac field for many applications. In this text, the superconducting RF technology topics are distributed throughout the chapters; technical information is provided as is needed for each specific discussion. The discussion of superconducting RF technology in Chapter 12 is indended to provide a brief and systematic presentation for the new student, including the physics and the technology together with some examples of superconducting RF linacs. I recommend to the student that a reading of this introductory section should be followed by the more detailed treatment in the excellent textbook in this same Wiley series, *RF Superconductivity for Accelerators*, by Hasan Padamsee, Jens Knobloch, and Thomas Hays.

The problems at back of each chatere have been chosen to develop additional insight to the material presented, to illustrate further a particular point, or to provide the student with a better feeling for typical numerical magnitudes of important quantities. I believe it is important that the latter capability be developed by anyone who needs to apply the material in this book to the design or real accelerators. The references provided are not claimed to be complete, but represent those that I have found most useful. Throughout the book, I have tried to use conventional notation in the accelerator field, even thoug this results in an unfortunate duplication of symbols. Such is the nature of the accelerator field, much of whose language has been adopted from the notation of older fields of physics and engineering that form its basis.

Although no previous text book on RF linacs exists, books and articles have been published over the years that can be valuable resources for the serious student. Most notable is the 1970 red bible title *Linear Accelerators*, a compilation of articles edited by P. Lapostolle and A. Septier, which has served over the years as an important reference source for many in the linac field. Also abailable are the 1959 *Handbuch der Physik* article on "Linear Accelerators" by L. Smith, the 1965 book *Theory of Linear Accelerators* by A. D. Vlasov, the 1982 lectures thorugh the USPAS by G. Loew and R. Talman in A.I.P Conference Proceedings No. 105, and the 1985 book *Theory of Resonance Linear Accelerators* by I. M. Kapchinskiy. The books *Theory and Design of Charged Particle Beams* by M. Reiser, and *The Physics of Charged Particle Beams* by J. D. Lawson also contain much useful material related to linac beam dynamics.

I wish to acknowledge a very productive collaboration with Jim Billen in teaching the USPAS linac courses and his substantial efforts in developing and documenting solutions to the problems. I am grateful to Jim Billen, Dick Cooper, and Pierre Lapostolle, who read much of the manuscript and made valuable suggestions, and to Ken Crandall for helpful suggestions on the beam dynamics material. I am grateful to a long list of colleagues in the accelerator field from whom I have learned, and who have made important indirect contributions to this book. Thanks to my daughter, Anne, who helped me with many of the figures. Thanks to Jim Billen, Frank Krawczyk, and Peter Kneisel, who provided some key figures. Thanks to Mel Month, who asked me to develop and teach a linac course for the USPAS and encouraged me to write this book.

Los Alamos   *Tom Wangler*

1
Introduction

> My little machine was a primitive precursor of this type of accelerator which today is called a 'linac' for short. However, I must now emphasize one important detail. The drift tube was the first accelerating system which had earthed potential on both sides, i.e. at both the particles' entry and exit, and was still able to accelerate the particles exactly as if a strong electric field was present. This fact is not trivial. In all naiveté one may well expect that, when the voltage on the drift tube was reversed, the particles flying within would be decelerated, which is clearly not the case. After I had proven that such structures, earthed at both ends, were effectively possible, many other such systems were invented. – Rolf Wideröe
> [From *The Infancy of Particle Accelerators*, edited by Pedro Waloschek, see ref. [4]]

During the second half of the twentieth century, the linear accelerator has undergone a remarkable development. Its technological base is a consequence of the science of both the nineteenth and twentieth centuries, including the discoveries of electromagnetism by Faraday, Maxwell, and Hertz in the nineteenth century and the discovery of superconductivity in the twentieth century. The design of a linear accelerator requires an understanding of the major areas of classical physics, especially classical mechanics, and electromagnetism, as well as relativity theory. The linear accelerator has developed as a great tool for learning about the world of subatomic particles. The linear accelerator provides beams of high quality and high energy, sufficient to resolve the internal structure of the nucleus and of its constituent subnuclear particles. Like a microscope, it has probed the internal structure of the atomic nucleus and of the nuclear constituents, the proton and neutron. Measurements made using the beams from an electron linear accelerator have given us our present picture of the proton, that it is made of pointlike particles called *quarks*. Furthermore, the linear accelerator has been applied in hospitals throughout the world as a source of X rays for radiation therapy to treat cancer. This application may represent the most significant spin-off of high-energy and nuclear physics research for the benefit of mankind. The linear accelerator is truly one of the most significant examples of high-technological development in the postindustrial era. The sizes of linacs range from a few meters to a few kilometers, and the costs range from a few million to a billion dollars, depending on the final energy.

As a research tool alone, we can expect that the linear accelerator will have a great future in the twenty-first century. The straight-line trajectory avoids power losses caused by synchrotron radiation that accompanies circular radio frequency (RF) accelerators. The capability for providing strong focusing allows high-quality and high-intensity beams that enable precision measurements to be made, and provides high-power beams for many applications. We can anticipate continuing progress in areas such as radio-frequency quadrupole (RFQ) linacs, colliding beams, high-power beams, high-frequency RF power and microwave technology, and RF superconductivity. Further developments in these areas will lead the linac to new performance levels with higher currents, better beam quality, and lower power requirements. We can confidently expect an expansion to new applications in the medical and industrial areas. The purpose of this book is to present the scientific and technical foundations of the linear accelerator, how it works, and why it will continue to serve as a powerful tool for the study of nature, and for many other practical applications.

1.1
Linear Accelerators: Historical Perspective

It might be expected that the term *linear accelerator* should refer to any device in which particles are accelerated along a straight line. However, through common usage in the accelerator field the term *linear accelerator* has been reserved for an accelerator in which charged particles move on a linear path, and are accelerated by time-dependent electromagnetic fields. The abbreviation *linac* is commonly used for the term *linear accelerator*. In a RF linac, the beam is accelerated by RF electromagnetic fields with a harmonic time dependence [1]. The first formal proposal and experimental test for a linac was by Rolf Wideröe in 1928, but linear accelerators that were useful for research in nuclear and elementary particle research did not appear until after the developments of microwave technology in World War II, stimulated by radar programs. Since then, the progress has been rapid, and today, the linac is not only a useful research tool, but is also being developed for many other important applications.

A particle accelerator delivers energy to a charged-particle beam by application of an electric field. The first particle accelerators were electrostatic accelerators in which the beam gains energy from a constant electric field. Each particle acquires an energy equal to the product of its electric charge times the potential drop, and the use of electrostatic fields led to a unit of energy called the electron volt (eV), equal to the product of the charge times the voltage drop. The main limitation of electrostatic accelerators is that the maximum energy obtainable cannot exceed the product of the charge times the potential difference that can be maintained, and in practice this potential difference is limited by electric breakdown to no more than a few tens of megavolts. RF accelerators bypass this limitation by applying a harmonic

1.1 Linear Accelerators: Historical Perspective

time-varying electric field to the beam, which is localized into bunches, such that the bunches always arrive when the field has the correct polarity for acceleration. The time variation of the field removes the restriction that the energy gain be limited by a fixed potential drop. The beam is accelerated within electromagnetic-cavity structures, in which a particular electromagnetic mode is excited from a high-frequency external power source. For acceleration, the beam particles must be properly phased with respect to the fields, and for sustained energy gain they must maintain synchronism with those fields. The latter requirement has led to the name *resonance accelerators*, which includes the linac, cyclotron, and synchrotron. The ideal particle orbit in an RF accelerator may be either a straight line for a linac, a spiral for a cyclotron, or a circle for a synchrotron.

In 1924, Gustav Ising of Stockholm proposed the first accelerator that used time-dependent fields, consisting of a straight vacuum tube, and a sequence of metallic drift tubes with holes for the beam [2]. The particles were to be accelerated from the pulsed voltages that were generated by a spark discharge and applied across adjacent drift tubes. Synchronism of the applied voltage pulses with the beam particles was to be obtained by introducing transmission lines, chosen to delay the pulse from the voltage source to each of the drift tubes. The concept proposed by Ising was not tested at that time, but the publication was very important because it influenced the young Norwegian student, Rolf Wideröe.

The first RF linear accelerator was conceived and demonstrated experimentally by Wideröe in 1927 at Aachen, Germany. It was reported in a paper [3] that is one of the most significant in the history of particle accelerators,[4] and which inspired E. O. Lawrence to the invention of the cyclotron [5]. The linac built by Wideröe was the forerunner of all modern RF accelerators. The Wideröe linac concept, shown in Fig. 1.1, was to apply a time-alternating voltage to a sequence of drift tubes, whose lengths increased with increasing particle velocity, so that the particles would arrive in every gap at the right time to be accelerated. In the figure, D are drift tubes connected to an alternating voltage source V that applies equal and opposite voltages to sequential drift tubes, G are the gaps between the drift tubes in which the electric force acts to accelerate the particles, and S is the source of a continuous ion beam. For efficient acceleration the particles must be grouped into bunches (shown by

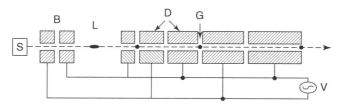

Figure 1.1 Concept of the Wideroe drift-tube linac.

the black dots), which are injected into the linac at the time when the polarity of the drift tubes is correct for acceleration. The bunching can be accomplished by using an RF gap B between the dc source and the linac. This gap impresses a velocity modulation on the incoming beam that produces spatial bunching at the end of a suitable drift space L. The net effect of the sequence of voltage kicks is to deliver a total voltage gain to the beam, which is greater than the impressed voltage V in any single gap.

In Wideröe's experiment, an RF voltage of 25 kV from a 1-MHz oscillator was applied to a single drift tube between two grounded electrodes, and a beam of singly charged potassium ions gained the maximum energy in each gap. A final beam energy of 50 keV was measured, which is twice that obtainable from a single application of the applied voltage. This was also the first accelerator that had ground potential at both the entrance and the exit ends, and was still able to deliver a net energy gain to the beam, using the electric fields within. The experiment established the principle that, unlike an electrostatic accelerator, the voltage gain of an RF accelerator could exceed the maximum applied voltage. There was no reason to doubt that the method could be repeated as often as desired to obtain unlimited higher energies. In 1931 Sloan and Lawrence [6] built a Wideröe-type linac with 30 drift tubes, and by applying 42 kV at a frequency of 10 MHz, they accelerated mercury ions to an energy of 1.26 MeV at a beam current of 1 μA. By 1934 the output energy had been raised to 2.85 MeV [7] using 36 drift tubes.

The original Wideröe linac concept was not suitable for acceleration to high energies of beams of lighter protons and electrons, which was of greater interest for fundamental physics research. These beam velocities are much larger, approaching the speed of light, and the drift-tube lengths and distances between accelerating gaps would be impractically large, resulting in very small acceleration rates, unless the frequency could be increased to near a gigahertz. In this frequency range, the wavelengths are comparable to the ac circuit dimensions, and electromagnetic-wave propagation and electromagnetic radiation effects must be included for a practical accelerator system. For example, for an electron linac the lengths of the drift tubes and supporting stems would equal nearly a half a wavelength, and instead of isopotential electrodes they would function more like resonant antennas with high power losses. Thus, linac development required higher-power microwave generators, and accelerating structures better adapted for high frequencies and for acceleration requirements of high-velocity beams. High-frequency power generators, developed for radar applications, became available after World War II. At this time, a new and more efficient high-frequency proton-accelerating structure, based on a linear array of drift tubes enclosed in a high-Q cylindrical cavity, was proposed by Luis Alvarez [8] and coworkers at the University of California. The drift-tube linac (DTL) concept was to excite a mode with a uniform electric field in the gaps and zero field inside the drift tubes to avoid deceleration when the field was reversed. A 1-m diameter, 12-m DTL with a resonant frequency of 200 MHz was built,[9] which accelerated protons from 4 to 32 MeV. At about the same time at Stanford a new, efficient accelerating structure for relativistic electrons was proposed, consisting of an array of pillbox-cavity resonators with a central hole in each end wall for propagation of both the beam and the electromagnetic energy. The structure was called the *disk-loaded* or *iris-loaded waveguide*,[10] and this development led eventually to the 3-km Stanford Linear Accelerating Center (SLAC) linac. From these two projects the first modern proton and electron linacs were born. [11]

The RF linear accelerator is classified as a resonance accelerator. Because both ends of the structure are grounded, a linac can easily be constructed as a modular array of accelerating structures. The modern linac typically consists of sections of specially designed waveguides or high-Q resonant cavities that are excited by RF electromagnetic fields, usually in the VHF and UHF microwave frequency ranges. The accelerating structures are tuned to resonance and are driven by external, high-power RF-power tubes, such as klystrons, or various types of gridded vacuum tubes. The ac (wall plug) to RF efficiencies of these tubes typically range from about 40 to 60%. The output electromagnetic energy from the tubes is transported in conventional transmission lines or waveguides to the linac structure. The accelerating structures must efficiently transfer the electromagnetic energy to the beam, and this is accomplished in two important ways. First, the resonant buildup of the fields in the high-Q structure transforms the low field levels of the input waveguide into high fields within the structure and produces a large ratio of stored electromagnetic energy relative to the ohmic energy dissipated per cycle. Second, through an optimized configuration of the internal geometry, the structure can concentrate the electric field along the trajectory of the beam promoting maximal energy transfer. The most useful figure of merit for high field concentration on the beam axis and low ohmic power loss is the shunt impedance.

One of the main advantages of the linear accelerator is its capability for producing high-energy, and high-intensity charged-particle beams of high beam quality, where high beam quality can be related to a capacity for producing a small beam diameter and small energy spread. Other attractive characteristics of the linac include the following: (1) strong focusing can be easily provided to confine a high-intensity beam; (2) the beam traverses the structure in a single pass, and therefore repetitive error conditions causing destructive beam resonances are avoided; (3) because the beam travels in a straight line, there is no power loss from synchrotron radiation, which is a limitation for high-energy electron beams in circular accelerators; (4) injection and extraction are simpler than in circular accelerators, since the natural orbit of the linac is open at each end; special techniques for efficient beam injection and extraction are unnecessary; (5) the linac can operate at any duty factor, all the way to 100% duty or a continuous wave (CW), which results in acceleration of beams with high average current.

1.2
Linac Structures

A simplified block diagram of a linac in Fig. 1.2 shows a linac structure with accelerating cavities and focusing magnets, and supplied with electromagnetic energy by an RF-power system. Beam is injected from a dc injector system. A vacuum system is required for good beam transmission. Electric power is used primarily by the RF-power system. A cooling system (water for normal-conducting linacs and liquid helium for superconducting linacs) removes the heat generated by the resistive wall losses. Because the linac uses a sinusoidally varying electric field for acceleration, particles can either gain or lose energy depending on the beam phase relative to the crest of the wave. To provide efficient acceleration for all the particles, the beam must be bunched as shown in Fig. 1.3. The bunches may be separated longitudinally by one or more RF periods.

Figure 1.4 shows the electric- and magnetic-field patterns in a simple cylindrical cavity operated in a transverse-magnetic resonant mode. Such a

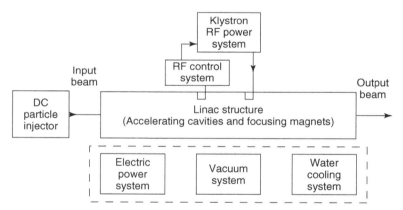

Figure 1.2 Simplified block diagram of a linac.

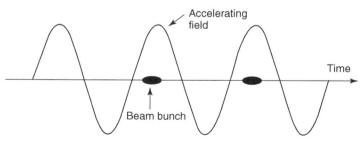

Figure 1.3 Beam bunches in an RF linac.

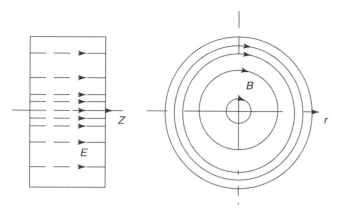

Figure 1.4 Electric (E) and magnetic (B) fields for the transverse-magnetic resonant mode in a cylindrical cavity.

mode is characterized by a longitudinal electric field on axis, which is ideal for acceleration of a charged-particle beam. An important practical consideration is how to construct an efficient linac using these cavities. There have been several solutions. First, an array of independent cavities can be used, each driven by its own RF generator, and each phased independently to provide acceleration along the entire length. This solution is used for superconducting linacs, where its main advantage is operational flexibility.

Another solution is to launch an electromagnetic traveling wave in a long structure consisting of many electromagnetically coupled cells. The traveling-wave structure was used for the 50-GeV electron linac at the SLAC. Although the simplest accelerating structure might appear to be a uniform cylindrical waveguide, it cannot provide continuous acceleration of electrons, because the phase velocity of an electromagnetic wave in a uniform waveguide always exceeds the velocity of light, so that synchronism with the beam is not possible. A structure with modified geometry is required to lower the phase velocity to that of the beam. At SLAC, the linac structure consists of a cylindrical waveguide that contains a periodic array of conducting disks with axial holes, as shown in Fig. 1.5. Each individual cell within a pair of disks is essentially identical to the basic cavity of Fig. 1.4, and the whole structure is equivalent to an array of coupled cylindrical cavities. It can be shown that for this structure the phase velocity can be reduced below the speed of light, as required for particle acceleration. The electrical characteristics of the disk-loaded waveguide structure will be described in more detail in Chapter 3.

The other common method of producing acceleration in a linac is to excite a standing wave in a multicell or coupled-cavity array. Several types of multicell structures have been invented for optimum application over specific ranges of beam velocity. One type of structure is the Alvarez DTL, discussed earlier and shown in Fig. 1.6, which is used to accelerate protons and other ions in the

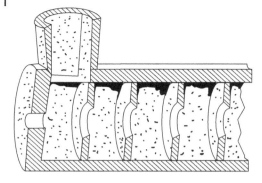

Figure 1.5 The disk-loaded traveling-wave structure, also showing the input waveguide through which the electromagnetic wave is injected into the structure at the end cell. The beam propagates along the central axis and is accelerated by the electric field of the traveling wave.

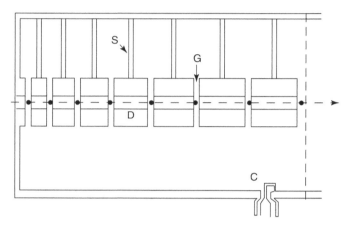

Figure 1.6 Alvarez drift-tube linac structure used for acceleration of medium-velocity ions. The beam particles are bunched before injection into the drift-tube linac. The beam bunches being accelerated in the gaps G are shown. They are shielded from the field by the drift tubes, when the field has the wrong polarity for acceleration. The drift tubes D are supported by the stems S. The cavity is excited by the RF current flowing on a coaxial line into the loop coupler C.

velocity range from about 0.04 to about 0.4 times the speed of light. Unlike the Wideröe structure, in the DTL the fields in adjacent gaps are in phase, and the spacing of the accelerating gaps is nominally equal to the distance the beam travels in one RF period. The DTL structure is not used for electrons, because electrons are so light that their velocity is already above the applicable velocity region at injection from the dc electron gun. Other coupled-cavity linac structures are used for both electrons and protons in the velocity range above about 0.4 times the speed of light. This velocity corresponds to kinetic energies near 50 keV for electrons, the typical injection energy from an electron gun, and near 100 MeV for protons. For example, a coupled-cavity structure called the side-coupled linac (SCL) is used at the Los Alamos Neutron Science Center

(LANSCE) linac at Los Alamos to accelerate the proton beam from 100 to 800 MeV. The transverse-focusing requirements are provided by magnetic-quadrupole lenses mounted within the drift tubes of the DTL, and between structures in a coupled-cavity linac.

The newest accelerating structure for the very-low-velocity range from about 0.01 to 0.06 times the velocity of light is the (RFQ), shown in Fig. 1.7. An electric-quadrupole mode is excited in a cavity resonator loaded with four conducting rods or vanes, placed symmetrically about the beam axis. The RFQ electric field provides strong transverse electric focusing, which is an important requirement for low-velocity protons and heavy ions. Acceleration in the RFQ is obtained by machining a longitudinal-modulation pattern on these four elements to create an array of effective accelerating cells and a longitudinal accelerating field. The RFQ bunches and captures a dc beam injected from the ion source, and then accelerates the beam to high-enough energies for injection into the DTL. The overall result is a significant increase in the focusing strength at low velocities, which enables acceleration of higher-current beams in linacs.

In pulsed linacs, one must distinguish between micropulses and macropulses. We will see later that, within each RF cycle, the longitudinal electric field produces a stable region (the bucket) for the beam. Consequently, the linac fields form a sequence of stable RF buckets separated by one RF period. Each bucket may contain a stable bunch of particles called a *micropulse*.

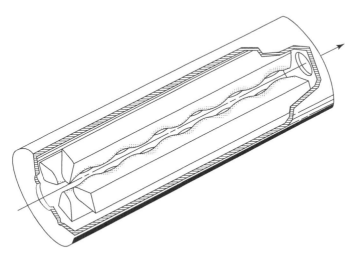

Figure 1.7 The radio-frequency quadrupole (RFQ), used for acceleration of low-velocity ions, consists of four vanes mounted within a cylindrical cavity. The cavity is excited in an electric-quadrupole mode in which the RF electric field is concentrated near the vane tips to produce a transverse RF electric-restoring force for particles that are off-axis. The modulation of the vane tips produces a longitudinal electric-field component that accelerates the beam along the axis.

When the RF generator itself is pulsed, with a period that is generally very long compared with the RF period, the generator pulses are called *macropulses*. The linac may be operated continuously, which is called *continuous-wave* operation. The choice to operate pulsed or continuously depends on several issues. One important issue is the total RF efficiency. If the accelerated beam current is small, most of the power in CW operation is not delivered to the beam, but is dissipated in the structure walls. Instead, if the accelerator is operated pulsed, and the current per RF bucket is increased while maintaining the same average beam current, then a larger fractional power is delivered to the beam, and the efficiency is improved. Another important advantage for pulsed operation is that the peak surface electric field attainable is generally larger for shorter pulses. Thus, if high accelerating fields are required, pulsed operation may be preferred. The main advantage for either longer pulse or CW operation is to reduce the space-charge forces or other beam-current-dependent effects associated with acceleration of beam with high average currents. These effects can be reduced by spreading the total beam charge over more RF buckets, as is done in longer pulse or CW operation.

Because the linac is a single-pass device, the linac length and the ohmic power consumption in the cavity walls may be large compared with circular accelerators, which use the same accelerating cavities over and over. To shorten the accelerator for a given energy gain, it is necessary to raise the longitudinal electric field, but this increases the power dissipation and increases the risk of RF electric breakdown. For high-duty-factor operation, the average power density from RF losses on the cavity walls can produce challenging cooling requirements for the conventional copper-cavity technology. Another approach to these problems that has become increasingly successful in recent years is the use of superconducting niobium cavities.

1.3
Linac Beam Dynamics

Multicell ion linacs are designed to produce a given velocity gain per cell. Particles with the correct initial velocity will gain the right amount of energy to maintain synchronism with the field. For a field amplitude above a certain threshold, there will be two phases for which the velocity gain is equal to the design value, one earlier and the other later than the crest, as shown in Fig. 1.8.

The earlier phase is called the *synchronous phase* and is the stable operating point. It is a stable point because nearby particles that arrive earlier than the synchronous phase experience a smaller accelerating field, and particles that arrive later will experience a larger field. This provides a mechanism that keeps the nearby particles oscillating about the stable phase, and therefore provides phase focusing or phase stability. The particle with the correct velocity at exactly the stable phase is called the *synchronous particle*, and it maintains exact synchronism with the accelerating fields. As the particles approach relativistic

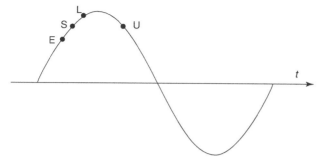

Figure 1.8 Stable (S) and unstable (U) phases, early (E), and late (L) phases.

velocities, the phase oscillations slow down, and the particles maintain a nearly constant phase relative to the traveling wave. After beam injection into electron linacs, the velocities approach the speed of light so rapidly that hardly any phase oscillations take place. With the electromagnetic wave traveling at the speed of light, electrons initially slip relative to the wave and rapidly approach a final phase, which is maintained all the way to high energy. The final energy of each electron with a fixed phase depends on the accelerating field and the value of the phase. In contrast, the final energy of an ion that undergoes phase oscillations about a synchronous particle is approximately determined not by the field, but by the geometry of the structure, which is tailored to produce a specific final synchronous energy. For an ion linac built from an array of short independent cavities, each capable of operating over a wide velocity range, the final energy depends on the field and the phasing of the cavities, and can be changed by changing the field, as in an electron linac.

Longitudinal focusing, obtained by injecting the beam on the leading edge of the wave, is essential for nonrelativistic beams of high intensity. However, RF transverse electric fields also act on the beam as shown by the radial field lines near the edges of the gap in Fig. 1.9, and except for some special cases, the particles that are focused longitudinally experience transverse defocusing forces. Furthermore, additional defocusing effects arise because the injected

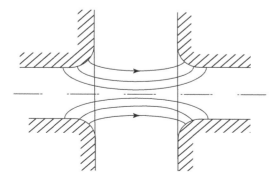

Figure 1.9 Electric-field lines in an accelerating gap.

beam particles always have finite transverse velocities, and the beam particles also exert mutually repulsive Coulomb forces. Thus, provision for transverse focusing must be provided. The most successful solutions for transverse focusing have been either to include separate magnetic-quadrupole focusing lenses or to invent accelerating structures that can provide focusing from the RF transverse electric fields, such as the RFQ.

1.4
Multiparticle Effects

Some applications require beams of high quality that occupy a small volume of phase space, called the *emittance*. Small beam phase volume is necessary if a small output focal spot or small output energy spread is required. As the beam intensity increases, several effects begin to increase the phase volume occupied by the beam, and these may eventually lead to loss of the beam. The most serious intensity limitation in ion linacs is caused by the repulsive space-charge forces, which are usually the most important at lower velocities, where the beam density is highest. The repulsive space-charge forces cause additional defocusing, and because these forces are nonlinear, they distort the particle distribution. Space-charge forces can also produce an extended halo of large-amplitude particles surrounding the main core of the beam. The halo particles can strike the walls and contribute to beam loss that causes radioactivity along the accelerating structure. The radioactivity increases the difficulty of providing routine maintenance of the linac, and thereby reduces the overall operational availability of the linac. For applications requiring high average beam current, control of the halo and beam losses through strong focusing, adequate aperture radius, and proper matching of the beam distribution to the focusing system becomes an important design requirement.

For relativistic electron linacs, the electric (space charge) and magnetic self-fields from the beam tend to cancel, nearly eliminating the total effective space-charge effect. But, short bunches of relativistic particles produce a highly Lorentz-compressed field distribution, and these fields from the beam interact with conducting-boundary discontinuities, producing scattered radiation, called *wakefields*, that act on trailing charges in both the same and later bunches. The wakefields can also increase the beam emittance. Wakefield effects can be reduced by damping the higher-order modes that are the major contributors and reducing discontinuities whenever possible. Certain cavity modes, called *deflecting modes*, can be excited by an off-axis beam, and are the most dangerous. These modes can cause further deflection of trailing particles and under certain conditions lead to an instability known as the *beam-breakup instability*, which results in loss of the beam.

Finally, beam loading occurs, as the beam itself excites the accelerating mode in the cavities. The beam-induced field adds vectorially to the contribution from the generator to produce a modified amplitude and phase. From another viewpoint, energy is transferred from the beam to each cavity, which unless otherwise corrected, reduces the cavity field and may shift the phase. Beam-loading compensation methods are used successfully to maintain the correct amplitudes and phases in the presence of the beam. Solutions for controlling all these high-intensity effects can significantly influence the design choices for the main accelerator parameters, including frequency, aperture radius, focusing characteristics, cavity tuning, and RF system operation.

1.5
Applications of Modern RF Linacs

For electron linacs, applications of recent interest include (1) electron-positron colliders for elementary-particle-physics research, (2) high-quality electron beams for free-electron lasers, (3) pulsed neutron sources for nuclear physics and material sciences, and (4) X-ray sources for radiotherapy. Electron-positron linear colliders are preferred over circular colliders because synchrotron radiation losses, experienced by relativistic electrons in circular accelerators, are avoided. Furthermore, because of the strong focusing in a linac, high beam quality is achieved, which is required for high luminosity and a high collision rate. Design studies, and research and development for linear colliders in the tetraelectron volt range, are being carried out within the framework of an international collaboration [12]. The most successful commercial application of RF accelerators is the small 10–20-MeV electron linacs for cancer therapy. A few thousand electron linacs are used worldwide for medical irradiations, and this number is growing. Small electron linacs are also used for industrial radiography and radiation processing, including radiation sterilization.

For proton linacs, modern applications include (1) injectors to high-energy synchrotrons for elementary-particle-physics research; (2) high-energy linacs for CW spallation neutron sources used for condensed matter and materials research, production of nuclear fuel, transmutation of nuclear wastes, and accelerator-driven fission-reactor concepts; (3) CW neutron sources for materials irradiation studies related to fusion reactors; and (4) low-energy neutron sources for medical applications such as boron–neutron capture therapy. Design studies for large proton linacs have been carried out for the Accelerator Production of Tritium (APT) project [13] and the neutron spallation source projects in Europe, the European Spallation Source (ESS) [14] and in the United States, where the Spallation Neutron Source (SNS) was recently constructed [15]. There are also linac applications for heavy ions, including (1) linacs for nuclear physics research, (2) ion implantation for semiconductor fabrication, and (3) multigigaelectron volt linacs for heavy-ion-driven inertial-confinement fusion. The most recently commissioned heavy-ion linac is the lead-ion linac at CERN [16].

A recent worldwide compendium of existing and planned scientific linacs [17] listed 174 linacs distributed over the Americas, Europe, and Asia.

Historically, two significant large linac projects are the SLC electron–positron linear collider at SLAC, shown in Fig. 1.10, and the LANSCE linac at Los Alamos, shown in Fig. 1.11. The main parameters of these two linacs are summarized in Table 1.1. The SLC at SLAC is the first linear collider, built to produce the Z^0 vector boson near a center-of-mass energy of 91 GeV. It used the 2-mile SLAC electron linac, which was built in the 1960s [18]. Positrons were produced by bombarding a target with a 30-GeV e^- beam. More details of this unique facility are summarized elsewhere [19].

The LANSCE linac,[21] formerly known as *LAMPF*, began operation in the early 1970s as a pion factory for research in nuclear and high-energy physics. It delivers the highest average proton beam power of any existing accelerator. It can deliver 1-mA, 800-MeV proton beams to a fixed target, or an H beam for multiturn injection into the proton storage ring, where the accumulated beam is extracted in short pulses for neutron scattering research.

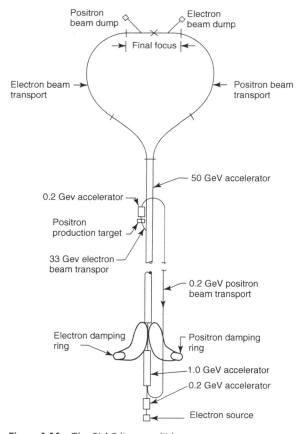

Figure 1.10 The SLAC linear collider.

1.6 Accelerator-Physics Units, Unit Conversions, and Physical Constants

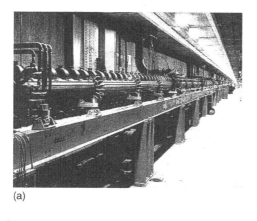

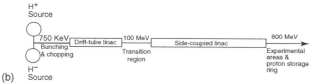

Figure 1.11 The 800-MeV LANSCE proton linac.

Table 1.1 Parameters of SLC and LANSCE linacs [20].

Facility	SLC at SLAC	LANSCE at Los Alamos (H⁻ data)
Application	e^-/e^+ collider for high-energy physics research	Linac for high-intensity beams of H^- and H^+
First beam	1967	1972
Species	Electrons and positrons	H^+ and H^-
Beam intensity	$2-3.5 \times 10^{10}$ particles per pulse	11 mA peak H^- (average over 825 μs macropulse)
Beam pulse	120 Hz, 0.06–3 μs	120 Hz, 825 μs
Output energy	46.6 GeV	800 MeV
Accelerating structure data	960 3-m structures for traveling-wave acceleration at 2856 MHz; 60–130 MW peak power, 25 kW average power	201.25-MHz DTL from 0.75 to 100 MeV 805-MHz SCL from 100 to 800 MeV
Length	3000 m	62-m DTL, 731-m SCL

1.6
Accelerator-Physics Units, Unit Conversions, and Physical Constants

In this book we will use the SI or MKS units, with two notable exceptions. The magnetic flux density will sometimes be expressed in gauss. The conversion factor between the SI unit tesla and gauss is 1 Gauss = 10^{-4} tesla. Beam-particle energy will be given in electron volts or eV units, rather than joules. The electron volt is defined as the energy acquired by a particle with charge equal to the electron charge that has been accelerated through a potential difference of 1 V. The conversion factor between electron volts and joules is approximately $1.602 \times 10^{-19} = 1$ eV. Finally, instead of the particle mass, we will usually give the rest energy, mc^2 in megaelectron volt units. Some frequently used physical constants are given in Table 1.2 [22].

1.7
Useful Relativistic Mechanics Relationships

We assume that the reader has a basic knowledge of classical and relativistic mechanics. In this section we present a brief review of basic formulas from relativistic mechanics that will be useful. Consider a particle of mass m and speed v. If c is the speed of light, it is customary to define a normalized velocity β as $\beta = v/c$, and a relativistic mass factor γ, defined as $\gamma = 1/\sqrt{1-\beta^2}$. Some other important definitions include the relativistic momentum of a particle, $\mathbf{p} = \gamma m\mathbf{v}$, the kinetic energy, $W = (\gamma - 1)mc^2$, the rest energy, mc^2, and the total energy, $U = W + mc^2 = \gamma mc^2$. The nonrelativistic limit applies when $\beta \ll 1$. It is often convenient to convert between velocity, energy, and momentum, and the following relationships are helpful. The conversion from velocity β to kinetic energy W is

$$\gamma = 1/\sqrt{1-\beta^2}, \; W = (\gamma - 1)mc^2 \tag{1.1}$$

The inverse conversion is

$$\gamma = (W + mc^2)/mc^2, \; \beta = \sqrt{1 - 1/\gamma^2} \tag{1.2}$$

The following relationships between small differences are sometimes useful: $\delta\gamma = \gamma^3 \beta \delta\beta$, $\delta\gamma = \beta\delta(\beta\gamma)$, $\delta W = mc^2 \delta\gamma$, $\delta p = mc\delta(\beta\gamma)$. Particle dynamics

Table 1.2 Physical constants.

Speed of light	c	2.99792458×10^8 m/s
Elementary charge	e	$1.60217733 \times 10^{-19}$ C
Electron mass	m_e	0.510 999 06 MeV/c^2
Proton mass	m_p	938.272 31 MeV/c^2
Atomic mass unit[a]	m_u	931.494 32 MeV/c^2
Permeability of free space	μ_0	$4\pi \times 10^{-7}$ T-m/A
Permittivity of free space	ε_0	$1/\mu_0 c^2 = 8.854187817 \ldots \times 10^{-12}$ F/m
DC resistivity of copper (293 K)	$1/\sigma$	1.7×10^{-8} Ω-m (nominal)

[a] Mass of (^{12}C)/12.

is obtained from Newton's law relating the force and the rate of change of momentum:

$$\mathbf{F} = \frac{d\mathbf{p}}{dt} = m\frac{d(\gamma v)}{dt} \tag{1.3}$$

For a particle of charge q in an electromagnetic field, the Lorentz force on particle with charge q and velocity v in an electric field $\mathbf{E}$ and a magnetic field $\mathbf{B}$, is given by

$$\mathbf{F} = q(\mathbf{E} + v \times \mathbf{B}) \tag{1.4}$$

1.8
Maxwell's Equations

The laws describing all classical electromagnetic phenomena are known as *Maxwell's equations*. These equations relate the electric and magnetic fields, and the charge and current sources. Maxwell's four equations in vacuum (where $\mathbf{D} = \varepsilon_0 \mathbf{E}$ and $\mathbf{B} = \mu_0 \mathbf{H}$), expressed in differential form, are as follows:

$$\nabla \cdot \mathbf{E} = \rho/\varepsilon_0, \text{ Gauss's law} \tag{1.5}$$

$$\nabla \cdot \mathbf{B} = 0 \tag{1.6}$$

$$\nabla \times \mathbf{E} = -\partial \mathbf{B}/\partial t, \text{ Faraday's law} \tag{1.7}$$

$$\nabla \times \mathbf{B} = \mu_0 \mathbf{J} + \mu_0 \varepsilon_0 \partial \mathbf{E}/\partial t, \text{ Ampére's law} \tag{1.8}$$

where ρ is the charge density and $\mathbf{J}$ is the current density. A charge-continuity equation $\nabla \cdot \mathbf{J} = -\partial \rho/\partial t$ is derived from these equations. Maxwell's equations can also be expressed in what is often a more convenient integral form:

$$\int \mathbf{E} \cdot d\mathbf{S} = \frac{1}{\varepsilon_0} \int \rho \, dV, \text{ Gauss's law} \tag{1.9}$$

$$\int \mathbf{B} \cdot d\mathbf{S} = 0 \tag{1.10}$$

$$\oint \mathbf{E} \cdot d\mathbf{l} = -\int \frac{\partial \mathbf{B}}{\partial t} \cdot d\mathbf{S}, \text{ Faradays law} \tag{1.11}$$

$$\oint \mathbf{B} \cdot d\mathbf{l} = \mu_0 \int \left(\mathbf{J} + \varepsilon_0 \frac{\partial \mathbf{E}}{\partial t} \right) \cdot d\mathbf{S}, \text{ Ampére's law} \tag{1.12}$$

It can be shown that the electric and magnetic fields can propagate as electromagnetic waves. When the charges and currents are zero, the wave equations in Cartesian coordinates are as follows:

$$\nabla^2 \mathbf{E} - \frac{1}{c^2} \frac{\partial^2 \mathbf{E}}{\partial t^2} = 0; \quad \nabla^2 \mathbf{B} - \frac{1}{c^2} \frac{\partial^2 \mathbf{B}}{\partial t^2} = 0 \tag{1.13}$$

where $c = 1/\sqrt{\mu_0 \varepsilon_0}$ is the speed of light in vacuum.

Maxwell's equations are composed of four coupled first-order partial differential equations. Two of the equations, Eqs. (1.6) and (1.7), have no charge or current source terms, and may be called the *homogeneous equations*. The other two equations, Eqs. (1.5) and (1.8), do contain source terms and may be called the *inhomogeneous equations*. Although in principle the four equations may be solved for any given problem, it is often convenient to solve a problem using potentials from which the fields may be derived. It is common to define the scalar potential ϕ, and the vector potential $\mathbf{A}$, which are functions of space and time, such that

$$\mathbf{B} = \nabla \times \mathbf{A}, \quad \mathbf{E} = -\nabla \phi - \frac{\partial \mathbf{A}}{\partial t} \quad (1.14)$$

With these definitions it can be shown [23] that the two homogeneous equations are automatically satisfied. The potentials are not uniquely specified from Eq. (1.14), and uncoupled source equations may be obtained by substituting Eq. (1.14) into Eqs. (1.5) to (1.8), and by imposing what is called the *Lorentz condition*,[24]

$$\nabla \cdot \mathbf{A} + \frac{1}{c^2}\frac{\partial \phi}{\partial t} = 0 \quad (1.15)$$

The resulting equations for the potentials have the symmetric, decoupled form of inhomogeneous wave equations [25]

$$\nabla^2 \phi - \frac{1}{c^2}\frac{\partial^2 \phi}{\partial t^2} = -\rho/\varepsilon_0 \quad (1.16)$$

and

$$\nabla^2 \mathbf{A} - \frac{1}{c^2}\frac{\partial^2 \mathbf{A}}{\partial t^2} = -\mu_0 \mathbf{J} \quad (1.17)$$

In the course of some of our discussions on RF cavities, we will consider the solution of Eqs. (1.15) to (1.17) within a closed region of space containing no free charges, surrounded by an equipotential surface. In this case Eq. (1.6) can be satisfied by choosing $\phi = 0$ [26]. Then, from Eq. (1.15) we have

$$\nabla \cdot \mathbf{A} = 0 \quad (1.18)$$

and Eq. (1.14) reduces to [27]

$$\mathbf{B} = \nabla \times \mathbf{A}, \quad \mathbf{E} = -\frac{\partial \mathbf{A}}{\partial t} \quad (1.19)$$

For this case, the electric and magnetic fields are obtained from the vector potential alone, which must satisfy Eqs. (1.17) and (1.18).

1.9
Conducting Walls

The boundary conditions at the interface between vacuum and an ideal perfect conductor can be derived by applying the integral forms of Maxwell's equations to small pillbox-shaped volumes at the interface. One finds that only the normal electric-field component and the tangential magnetic-field component can be nonzero just outside the conductor surface. If $\hat{n}$ is a vector normal to the interface, Σ is the surface charge density on the conductor, and **K** is the surface-current density, the boundary conditions that must be satisfied by the fields just outside the conductor are

$$\hat{n} \cdot \mathbf{E} = \Sigma/\varepsilon_0$$
$$\hat{n} \times \mathbf{H} = \mathbf{K}$$
$$\hat{n} \cdot \mathbf{B} = 0$$
$$\hat{n} \times \mathbf{E} = 0 \qquad (1.20)$$

No perfect conductors exist, but certain metals are very good conductors. Copper, with a room-temperature resistivity of $\rho = 1/\sigma = 1.7 \times 10^{-8}$ Ω-m, is the most commonly used metal for accelerator applications. For a good but not perfect conductor, fields and currents are not exactly zero inside the conductor, but are confined to within a small finite layer at the surface, called the *skin depth*. In a real conductor, the electric and magnetic fields, and the current decay exponentially with distance from the surface of the conductor, a phenomenon known as the *skin effect*. The skin depth is given by

$$\delta = \sqrt{\frac{2}{\sigma \mu_0 \omega}} \qquad (1.21)$$

Because of the skin effect, the ac and dc resistances are not equal. It is convenient to define the ac or RF surface resistance $R_s = 1/\sigma\delta$, and using Eq. (1.21), we find $R_s = \sqrt{\mu_0 \omega/2\sigma}$, which shows that the ac surface resistance is proportional to the square root of the frequency. If dS is the area element on the cavity walls, the average power dissipation per cycle is

$$P = \frac{R_s}{2} \int H^2 dS \qquad (1.22)$$

Physically, the skin effect is explained by the fact that RF electric and magnetic fields applied at the surface of a conductor induce a current, which shields the interior of the conductor from those fields. For frequencies in the 100 MHz range and for a good conductor like copper, the skin depth δ is of the order 10^{-6} m, and R_s is in the milliohm range. The use of superconducting materials dramatically reduces the surface resistance. For the RF surface resistance of superconducting niobium, we will use an approximate formula

$$R_s(\Omega) = 9 \times 10^{-5} \frac{f^2(GHz)}{T(°K)} \exp\left[-\alpha \frac{T_c}{T}\right] + R_{res} \quad (1.23)$$

where $\alpha = 1.83$, and $T_c = 9.2$ K is the critical temperature. R_{res} is known as the *residual resistance*; it is determined by imperfections in the surface, and typically is approximately 10^{-9} to 10^{-8} Ω. The superconducting surface resistance is roughly 10^{-5} lower than that of copper.

1.10
Group Velocity and Energy Velocity

Linac technology requires the propagation of electromagnetic waves in transmission lines, waveguides, and cavities. There are no truly monochromatic waves in nature. A real wave exists in the form of a wave group, which consists of a superposition of waves of different frequencies and wave numbers. If the spread in the phase velocities of the individual waves is small, the envelope of the wave pattern will tend to maintain its shape as it moves with a velocity that is called the *group velocity*. The simplest example of a wave group, shown in Fig. 1.12, consists of two equal-amplitude waves, propagating in the $+z$ direction, with frequencies ω_1 and ω_2, and wave numbers k_1 and k_2, which we can express in complex exponential form as

$$V(z,t) = e^{j(\omega_1 t - k_1 z)} + e^{j(\omega_2 t - k_2 z)}$$
$$= 2\cos\left[\frac{(\omega_1 - \omega_2)t - (k_1 - k_2)z}{2}\right] e^{j[(\omega_1 + \omega_2)t - (k_1 + k_2)z]/2} \quad (1.24)$$

The exponential factor describes a traveling wave with the mean frequency and mean wave number, and the first factor represents a slowly varying

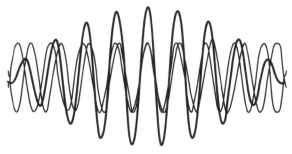

Figure 1.12 Wave composed of two components with different frequencies and wave numbers.

modulation of the wave amplitude. The phase velocities of component waves are ω_1/k_1 and ω_2/k_2, and the mean phase velocity is

$$v_p = \frac{\omega_1 + \omega_2}{k_1 + k_2} = \frac{\bar{\omega}}{\bar{k}} \qquad (1.25)$$

The group velocity is defined as the velocity of the amplitude-modulation envelope, which is

$$v_g = \frac{\omega_1 - \omega_2}{k_1 - k_2} \rightarrow \frac{d\omega}{dk} \qquad (1.26)$$

Generally, the mean phase velocity and the group velocity are not necessarily equal; they are equal when there is a linear relation between frequency and wave number, as for the ideal transmission line. The waveguide dispersion curve is a plot of ω versus k. Figure 1.13 shows an example of a dispersion curve for a uniform waveguide. The phase velocity at any point on the curve is the slope of the line from the origin to that point, and the group velocity is the slope of the dispersion curve, or tangent at that point. For the uniform guide, one finds $v_p v_g = c^2$, where $v_g < c$ and $v_p > c$.

A more general example of wave group is the wave packet, which describes a spatially localized wave group, as shown in Fig. 1.14. Again, the group velocity, rather than the phase velocity, must be used to characterize the motion of a wave packet. For example, the transient filling of a waveguide with

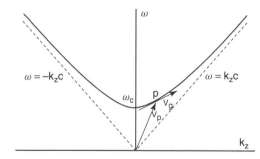

Figure 1.13 Example of dispersion curve for uniform waveguide, $\omega^2 = \omega_c^2 + (k_z c)^2$, showing graphically the meaning of phase and group velocity at the point p on the curve. The group velocity at point p is the tangent to the curve at that point. The phase velocity is the slope of the line from the origin to the point p.

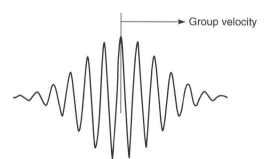

Figure 1.14 Wave packet.

electromagnetic energy must be described in terms of the motion of a wave packet, which will have a leading edge that moves at approximately the group velocity for practical cases where dispersion is not too large. The phase velocity does not appear in the description, because it describes the speed of individual waves that make up the wave packet, rather than the wave packet itself, which really consists of an interference pattern of these waves.

The energy velocity is defined as the velocity of electromagnetic energy flow, which, for a traveling wave moving in the $+z$ direction, is

$$v_E = \frac{P_+}{U_\ell} \quad (1.27)$$

where P_+ is the wave power, the electromagnetic energy per unit time crossing a transverse plane at fixed z, and U_ℓ is the stored electromagnetic energy per unit length. For cases of practical interest, the energy velocity is equal to the group velocity [28]. This result is useful for evaluating the energy velocity because the group velocity at the operating point is easy to determine from the slope of the dispersion curve.

1.11
Coaxial Resonator

Some accelerating cavities, especially for relatively low frequencies below about 100 MHz, are variants of the simple coaxial resonator. Without worrying now about the specific modifications needed to produce a practical accelerating cavity, we consider the properties of a coaxial resonator. A coaxial resonator, shown in Fig. 1.15, is formed by placing conducting end walls on a section of coaxial line formed by an inner conductor of radius a and an outer conductor with radius b. When the enclosed length is an integer multiple of half wavelengths, transverse electromagnetic (TEM) resonant standing-wave modes exist, where both the electric and magnetic fields have only transverse components. Resonance occurs when the boundary condition on the end walls, $E_r = 0$, is satisfied. This condition occurs when the conducting walls are separated by a distance $\ell = p\lambda/2$, $p = 1, 2, 3, \ldots$. To obtain the solution, first imagine a current wave on the inner conductor traveling in the $+z$ direction, $I_0 e^{j(\omega t - kz)}$. From the integral form of Ampere's law, the current produces an azimuthal magnetic field given by $B_\theta = I_0 e^{j(\omega t - kz)} \mu_0 / 2\pi r$. Given the magnetic field, the radial electric field can be obtained from the differential form of Faraday's law as $E_r = I_0 e^{j(\omega t - kz)} \mu_0 c / 2\pi r$. Likewise, we find that a wave traveling in the $-z$ direction has components $I_0 e^{j(\omega t + kz)}$, $B_\theta = I_0 e^{j(\omega t + kz)} \mu_0 / 2\pi r$, and $E_r = -I_0 e^{j(\omega t + kz)} \mu_0 c / 2\pi r$. Adding these two waves produces a standing wave satisfying the boundary condition that the tangential electric field E_r vanishes on the end walls at $z = 0$ and ℓ. The nonzero field components are

$$B_\theta = \frac{\mu_0 I_0}{\pi r} \cos(p\pi z/\ell) \exp[j\omega t] \quad (1.28)$$

$$E_r = -2j\sqrt{\frac{\mu_0}{\varepsilon_0}} \frac{I_0}{2\pi r} \sin(p\pi z/\ell) \exp[j\omega t] \qquad (1.29)$$

where $\omega = k_z c = \dfrac{p\pi c}{\ell}$, $p = 1, 2, 3, \ldots$. We note that the complex j factor in Eq. (1.29) denotes a 90° phase shift in time between the left and right sides of the equation, which can be obtained explicitly by substituting the identity $j = e^{j\pi/2}$. The electromagnetic stored energy is

$$U = \frac{\mu_0 \ell I_0^2 \ln(b/a)}{2\pi} \qquad (1.30)$$

and the quality factor or Q, including the losses on the end walls, is

$$Q_0 = \frac{p\pi}{R_s}\sqrt{\frac{\mu_0}{\varepsilon_0}} \frac{\ln(b/a)}{\left[\ell\left(\dfrac{1}{a} + \dfrac{1}{b}\right) + 4\ln\dfrac{b}{a}\right]} \qquad (1.31)$$

The lowest mode corresponds to $p = 1$, the half-wave resonator. Figure 1.15 shows the peak current and voltage distributions for $p = 1$, where the voltage is $V = \int_a^b E_r dr$. The cavity in Fig. 1.15 could be modified to make it suitable for acceleration by introducing beam holes in the inner and outer conductors at $z = \ell/2$ where the voltage is maximum. The beam, moving along a radial path, will see no field when it is within the inner conductor, and can see an accelerating field in the region on both sides between the inner and outer conductors. The injection phase could be chosen so that the beam travels across the inner conductor while the field reverses sign, so that the beam can be accelerated on both the entrance and exit sides of the inner conductor. The conductor radii could be chosen so that the beam receives the maximum energy gain on each side.

Another widely used type of resonator for accelerator applications is the coaxial line terminated at one end by a short and at the other end by a capacitance, as shown in Fig. 1.16. The capacitive termination can be

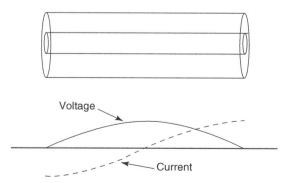

Figure 1.15 Coaxial resonator with voltage and current standing waves for $p = 1$.

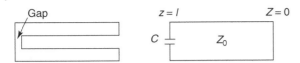

Figure 1.16 Coaxial resonator.

accomplished in practice with a coaxial line that has a gap at one end between the center conductor and the conducting end wall. An electric field suitable for acceleration may exist between the inner conductor and either the end wall or the cylindrical wall. Thus, beam holes can be introduced near the gap, allowing either a radial or an axial trajectory. Resonant modes correspond approximately to the length λ equal to an odd multiple of a quarter wavelength, the lowest mode being a quarter-wave resonator. Design formulas for the quarter-wave resonator, including the contribution to the capacitance from fringe fields, are given by Moreno [29].

1.12
Transverse-Magnetic Mode of a Circular Cylindrical Cavity

Most cavity resonators used in electron and proton linacs are derived from the simple cylindrical or pillbox cavity. Fortunately, an analytic solution exists for the fields in a pillbox cavity. Beginning with a cylinder of radius R_c, we place conducting end plates at the axial coordinates $z = 0$ and ℓ. In the pillbox cavity, the holes on the end plates that must be provided for the beam are ignored. We assume a simple azimuthally symmetric trial solution of the form $E_z(r, z, t) = R(r)e^{j\omega t}$. This solution must satisfy the wave equation with the condition that E_z vanishes at the cylindrical boundary, $r = R_c$, where it is tangential. The wave equation in cylindrical coordinates is

$$\frac{\partial^2 E_z}{\partial z^2} + \frac{1}{r}\frac{\partial E_z}{\partial r} + \frac{\partial^2 E_z}{\partial r^2} - \frac{1}{c^2}\frac{\partial^2 E_z}{\partial t^2} = 0 \tag{1.32}$$

Substituting the trial solution into Eq. (1.32), we obtain a differential equation for the radial function $R(r)$, which is the well-known Bessel's equation of order zero. The magnetic field is obtained from Ampére's law from Section 1.8. The nonzero field components of the complete solution are given by

$$E_z = E_0 J_0(k_r r) \cos(\omega t)$$
$$B_\theta = -\frac{E_0}{c} J_1(k_r r) \sin(\omega t) \tag{1.33}$$

The radial field distributions are shown graphically in Figures 1.4 and 1.17. To satisfy the boundary condition, the resonant frequency of this mode must be $\omega_c = k_r c = 2.405 c/R_c$, which is independent of the cavity length. The mode

1.12 Transverse-Magnetic Mode of a Circular Cylindrical Cavity

is called a *transverse-magnetic mode* because the z component of the magnetic field is zero, and in the conventional nomenclature the mode is called a TM_{010} mode for reasons that will be explained shortly. The total electromagnetic stored energy can be calculated from the peak electric stored energy, and the result is

$$U = \frac{\pi \varepsilon_0 \ell R_c^2}{2} E_0^2 J_1^2(2.405) \tag{1.34}$$

The average power dissipated on the cylindrical walls and the end walls is

$$P = \pi R_c R_s E_0^2 \left(\frac{\varepsilon_0}{\mu_0}\right) J_1^2(2.405)(\ell + R_c) \tag{1.35}$$

The quality factor is

$$Q = \frac{\omega_c U}{P} = \frac{2.405 \sqrt{\mu_0/\varepsilon_0}}{2R_s} \frac{1}{1 + R_c/\ell} \tag{1.36}$$

The electric field is maximum at $r = 0$, where J_0 is maximum. Two useful values of J_1 are the maximum value, which is $J_1(1.841) = 0.5819$, and the value of J_1 at the cylindrical wall, which is $J_1(2.405) = 0.5191$. The magnetic field is maximum at $k_r R = 0.5819$, where $J_1(k_r R)$ is maximum. Therefore, $B_{max}/E_{max} = 0.5819/c = 19.4 \ G/MV/m$.

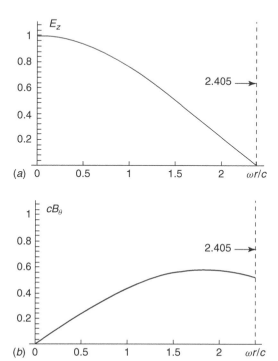

Figure 1.17 Fields for a TM_{010} mode of a cylindrical (pillbox) cavity resonator.

1.13
Cylindrical Resonator Transverse-Magnetic Modes

There are other transverse-magnetic modes with the same radial Bessel-function solution, corresponding to fitting a half-integer number of guide wavelengths within the length ℓ. We label the different longitudinal modes with the index p, and adopt the conventional nomenclature TM_{01p}, $p = 1, 2, 3, \ldots$. The dispersion relation is the same as for a uniform waveguide, except that the longitudinal wave number is restricted to those discrete values required to satisfy the boundary conditions at the two ends. The modes lie on the curve given by $\omega^2/c^2 = k_r^2 + k_z^2$, where $k_r = 2.405/R_c$ and $k_z = 2\pi/\lambda_g = \pi p/\ell$. Then, the dispersion relation becomes a discrete spectrum of points that are sprinkled on a hyperbolic curve, as shown in Fig. 1.18. The TM_{010} mode, discussed in Section 1.12, is the lowest mode with $p = 0$. The dispersion relation gives the resonant frequency of this mode as the cutoff frequency $\omega_c = k_r c = 2.405 c/R_c$.

There exist additional transverse-magnetic modes of a cylindrical cavity, corresponding to different radial and azimuthal solutions. The general expressions for the field components are as follows:

$$E_z = E_0 J_m(k_{mn}r) \cos m\theta \cos(p\pi z/\ell) \exp[j\omega t]$$

$$E_r = -\frac{p\pi}{\ell}\frac{a}{x_{mn}} E_0 J'_m(k_{mn}r) \cos m\theta \sin(p\pi z/\ell) \exp[j\omega t]$$

$$E_\theta = -\frac{p\pi}{\ell}\frac{ma^2}{x_{mn}^2 r} E_0 J_m(k_{mn}r) \sin m\theta \sin(p\pi z/\ell) \exp[j\omega t]$$

$$B_z = 0$$

$$B_r = -j\omega \frac{ma^2}{x_{mn}^2 r c^2} E_0 J_m(k_{mn}r) \sin m\theta \cos(p\pi z/\ell) \exp[j\omega t]$$

$$B_\theta = -j\omega \frac{a}{x_{mn} c^2} E_0 J'_m(k_{mn}r) \cos m\theta \cos(p\pi z/\ell) \exp[j\omega t] \tag{1.37}$$

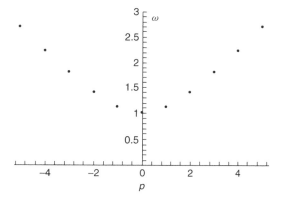

Figure 1.18 Dispersion curve for the TM_{01p} family of modes of a circular cylindrical cavity.

The general dispersion relation is $\omega^2/c^2 = k_{mn}^2 + k_z^2$, where $k_{mn} = x_{mn}/R_c$ and $k_z = 2\pi/\lambda_{guide} = p\pi/\ell$, $p = 0, 1, 2, \ldots$. Some values of the zeros of the Bessel functions, x_{mn}, are given in Table 1.3. The nomenclature of the TM$_{mnp}$ modes is defined as follows. The subscript m ($m = 0, 1, 2, \ldots$) is the number of full period variations in θ of the field components. The subscript n ($n = 1, 2, 3, \ldots$) is the number of zeros of the axial field component in the radial direction in the range $0 < r \leq R_c$, excluding $r = 0$. The subscript p ($p = 0, 1, 2, \ldots$) is the number of half period variations in z of the fields.

1.14
Cylindrical Resonator Transverse Electric Modes

Similarly, there exist additional transverse electric modes of a cylindrical cavity, corresponding to solutions with the zero axial component of the electric field. The general field-component expressions for the transverse electric modes are as follows:

$$B_z = B_0 J_m(k_{mn}r) \cos m\theta \sin(p\pi z/\ell) \exp[j\omega t]$$

$$B_r = \frac{p\pi}{\ell} \frac{a}{x'_{mn}} B_0 J'_m(k_{mn}r) \cos m\theta \cos(p\pi z/\ell) \exp[j\omega t]$$

$$B_\theta = -\frac{p\pi}{\ell} \frac{ma^2}{x'^2_{mn} r} B_0 J_m(k_{mn}r) \sin m\theta \cos(p\pi z/\ell) \exp[j\omega t]$$

$$E_z = 0$$

$$E_r = j\omega \frac{ma^2}{x'^2_{mn} r} B_0 J_m(k_{mn}r) \sin m\theta \sin(p\pi z/\ell) \exp[j\omega t]$$

$$E_\theta = j\omega \frac{a}{x'_{mn}} B_0 J'_m(k_{mn}r) \cos m\theta \sin(p\pi z/\ell) \exp[j\omega t] \quad (1.38)$$

The general dispersion relation is $\omega^2/c^2 = k_{mn}^2 + k_z^2$, where $k_{mn} = x'_{mn}/R_c$ and $k_z = 2\pi/\lambda_{guide} = p\pi/\ell$, $p = 0, 1, 2, \ldots$. The x'_{mn} are the zeros of the derivatives of the Bessel functions and are given in Table 1.4. The nomenclature of the TE$_{mnp}$ modes is defined as follows. The subscript m ($m = 0, 1, 2, \ldots$) is the number of full period variations in θ of the field components. The subscript n ($n = 1, 2, 3, \ldots$) is the number of zeros of the axial field component in

Table 1.3 Zeros of $J_m(x)$ or x_{mn}.

m	x_{m1}	x_{m2}	x_{m3}
0	2.405	5.520	8.654
1	3.832	7.016	10.173
2	5.136	8.417	11.620

1 Introduction

Table 1.4 Zeros of J'_m or x'_{mn}.

m	x'_{m_1}	x'_{m_2}	x'_{m_3}
0	3.832	7.016	10.174
1	1.841	5.331	8.536
2	3.054	6.706	9.970

the radial direction in the range $0 < r \le R_c$, excluding $r = 0$. The subscript p ($p = 0, 1, 2, \ldots$) is the number of half period variations in z of the fields.

Problems

1.1. What is the kinetic energy in units of both joules and electron volts for an electron accelerated through a dc potential of 1 MV?

1.2. Find an expression for the fractional error when the nonrelativistic approximation for kinetic energy as a function of β is used. (a) At what values of β and γ does the error in kinetic energy equal 1%? (b) To what kinetic energy does this correspond, for electrons and for protons?

1.3. If the only nonzero components of the electromagnetic field in cylindrical coordinates are E_r, E_z, and B_θ, write the nonzero components of the Lorentz force for a particle of mass m and charge q moving along the z direction with velocity v.

1.4. The rate of work done by a force $\mathbf{F}$ acting on a particle with velocity $\mathbf{v}$ is $\mathbf{F} \cdot \mathbf{v}$. Using the definition of the Lorentz force and the appropriate vector relationship, derive the expression for the rate of kinetic energy gain for a particle of charge q, and show that the magnetic force does not contribute.

1.5. A cylindrical resonator has a diameter of 1.5 in. (3.81 cm) and length ℓ of 1 in. (2.54 cm). (a) Calculate the resonant frequency of the TM_{010}, TM_{110}, TE_{011}, TE_{111}, and TE_{211} modes, and list in order of increasing frequency. (b) For the two lowest-frequency modes, plot the dispersion relation, $f(=\omega/2\pi)$ versus $k_z(=p\pi/\ell)$, both on the same graph. For simplicity, label the abscissa with the longitudinal mode index p (i.e., units of π/ℓ for $p = 0$ to $p = 5$). (Recall that the TE modes have no resonance at $p = 0$.)

1.6. Repeat the exercise of Problem 1.5 for the same diameter resonator but with different lengths. (a) $\ell = 7.725$ cm. Note the frequency of the TE_{112} mode compared to the TM_{010} mode. How did it change relative to the result of Problem 1.5? (b) $\ell = 25.4$ cm. Note that all of the first five TE_{11p} modes now lie below the TM_{010} frequency.

1.7. Calculate the RF surface resistance and skin depth of room-temperature copper at 400 MHz. Use a dc resistivity $\sigma^{-1} = 1.7 \times 10^{-8}$ Ω-m.

1.8. Calculate the RF surface resistance of superconducting niobium at 400 MHz. Assume a residual resistance $R_{\text{residual}} = 100 \times 10^{-9}$ Ω. What is the ratio of the RF surface resistance of superconducting niobium to that of room-temperature copper? (a) Assume $T = 4.2$ K; (b) assume $T = 2.0$ K.

1.9. Design a room-temperature cylindrical cavity that operates in the TM$_{010}$ mode at 400 MHz with an axial electric field $E_0 = 1$ MV/m, and a length $\ell = \lambda/2$, where λ is the RF wavelength in free space. (a) Calculate the length and diameter of the cavity. (b) Calculate the maximum B and H fields on the cavity wall. Where does this occur? (c) Calculate the B and H fields on the cylindrical wall. (d) Calculate the electromagnetic stored energy in the cavity. (e) Use the value of R_s from Problem 1.7 for a room-temperature copper surface to calculate the power loss P, the quality factor Q_0, and the decay time τ. (f) Repeat part (e) using R_s from Problem 1.8 for a 4.2-K niobium surface.

1.10. Design a half-wave coaxial cavity to be used as a 100-kW resonant load at 400 MHz. To absorb the RF power, use a 20-cm-diameter stainless steel pipe as the center conductor inside a 60-cm-diameter copper cylinder with copper end walls. This type of cavity is easily cooled by flowing water through the center conductor. (a) Ignoring any effects of the coupling loop and probe, calculate the length of the cavity. (b) Assume that the room-temperature surface resistance of stainless steel is 6.5 times that of copper. (From Problem 1.7 the copper surface resistance at 400 MHz is 0.0052 Ω.) Calculate the power dissipated on the center conductor, end walls, and outer wall. (c) Calculate the cavity stored energy and the unloaded Q. (d) What is the peak power density in watts per square centimeter on the inner and outer conductors?

1.11. A 25-MHz quarter-wave coaxial-cavity resonator with characteristic impedance 50 Ω is designed as a buncher for heavy-ion beams. (a) If the impedance at the open end of the cavity (where the electric field is maximum) could really be made infinite, what would be the length of the inner conductor? (b) If we want to reduce the size by restricting the length of the inner conductor to $\ell = 1$ m, what lumped capacitance would be required at the open end?

1.12. Accelerator cavities require ports through the cavity walls, not only for RF drive and RF pickup probes, but also for the beam and for vacuum pumping. Cylindrical pipes are commonly used, and such pipes will support waveguide modes. Consider a cylindrical cavity operating in a TM$_{010}$ mode with beam pipes connected at the center of each end wall. Assume each beam pipe has the same inner radius b, which is much less than the cavity radius. (a) Why do you expect the fields from the cavity to attenuate in the pipes? (b) Show that if the cavity excites a TM$_{01}$ waveguide mode in the beam pipes at a frequency well below cutoff, the wave power and the E and B fields attenuate with distance x along the pipe according to the approximate formula dB $= -20.9x/b$. Note that it is convenient to describe the attenuation of a wave with power P and field E in decibels, or dB, where dB $= 10\log_{10} P/P_0 = 20\log_{10} E/E_0$, where P_0 and E_0 are the input reference values. (c) The TE_{11} mode has the lowest cutoff frequency of the modes in a cylindrical pipe, and below the cutoff frequency the attenuation will be the slowest. If this mode is excited in the pipe, show that at a frequency well below cutoff, the attenuation is described by the approximate formula dB $= -16.0x/b$. (d) Assuming a pipe with radius $b = 0.5$ inches that is excited in a TE_{10} mode by the cavity, calculate the total attenuation in decibels if the pipe length is 2 in. Also express the answer as the fractional attenuation of the field. (e) The attenuation in a waveguide below cutoff frequency was derived ignoring ohmic losses. What do you think has happened to the energy in the wave?

References

1 The betatron and the related induction linac accelerate beams using nonharmonic time-dependent fields. These machines produce pulsed electric fields by induction from a magnetic pulse in accordance with Faraday's law. The orbit of the betatron is circular, and that of the induction linac is straight.
2 Ising G., (1924) *Ark. Mat. Fys.* **18**, (No. 30), 1–4.
3 Wideröe R., (1928) *Arch. Electrotech.* **21**, 387.
4 (1994) For a discussion of this and other work by Wideröe, see P. Waloschek, *The Infancy of Particle Accelerators-Life and Work of Rolf Wideröe*, DESY 94–039, Deutsches Elektronen-Synchrotron, Notkestrasse 85–22603, Hamburg, March.
5 Lawrence E.O. and Edlefsen N.E., (1930) *Science* **72**, 376.
6 Sloan D.H. and Lawrence E.O., (1931) *Phys. Rev.* **38**, 2021.
7 Sloan D.H. and Coate W.M., (1934) *Phys. Rev.* **46**, 539.
8 Alvarez L.W., (1946) *Phys. Rev.* **70**, 799.
9 Alvarez L.W., Bradner H., Franck J.V., Gordon H., Gow J.D., Marshall L.C., Oppenheimer F., Panofsky W.K.H., Richmond C., and Woodward J.R., (1955) *Rev. Sci. Instr.* **26**, 111, and (1955); *Rev. Sci. Instr.* **26**, 210.
10 Hansen W.W., Kyhl R.L., Neal R.B., Panofsky W.K.H., Chodorow M., and Ginzton E.L., (1955) *Rev. Sci. Instrum.* **26**, 134.
11 For a more complete early history of linacs, see Blewett J.P., in *Linear Accelerators*, Lapostolle P.M., and Septier A.L., North Holland Publishing, 1970 pp. 1–16.
12 For an overview of the issues and of some very interesting advanced electron-linac concepts for linear-collider applications, see *International Linear Collider Technical Review Committee Report*, prepared for the Interlaboratory Collaboration for R&D Towards TeV-scale Electron-Positron Linear Colliders, Committee Chairman, G. A. Loew (SLAC), December, 1995; see also *Design Issues of TeV Linear Colliders*, CERN/PS 96–30 (LP), 1996 European Part. Accel. Conf., ed. J. P. Delahaye(CERN).
13 Lawrence G.P., *et al.*, (1996) *Conventional and Superconducting RF Linac Designs for the APT Project*, Proc. 1996 Int. Linac Conf., Geneva, August 26–30.
14 *Outline Design of the European Spallation Neutron Source*, ed. Gardner I.S.K., Lengeler H., and Rees, G.H., ESS 95–30-M, September, 1995.
15 Holtkamp, N., *Status of the SNS Linac: an Overview*, Proc. of EPAC 2006, Edinburgh, Scotland, June 26–30, pp. 29–33.

16 Haseroth H.R., *The Lead Ion Injector at CERN*, Proc. 1996 Int. Linac Conf., Geneva, August 26–30, 1996.

17 *Compendium of Scientific Linacs*, Provisional Edition, 18th Int. Co., CERN-PS Division, Geneva, August, 26–30, 1996.

18 *The Stanford Two-Mile Accelerator*, Neal R.B., Benjamin W.A., New York, (1968); Panofsky W.K.H., *The Creation of SLAC Leading to 30 Years of Operation*, Proc. 18th Int. Linear Accel. Conf., August, 26–30, 1996, Geneva, ed. Hill C. and Vretenar M., CERN 96-07, Geneva, November 15, 1996 p. 3.

19 *SLC Design Handbook*, Erickson R., Stanford 1984; also Seeman J., (1991) *Ann. Rev. Nucl. Part. Sci.* **41**, 3891; also Seeman J.T., *Advances of Accelerator Physics and Technologies*, Advanced Series on Directions in High Energy Physics, Vol. 12, Schopper H., 1993 p. 219.

20 Erickson R.E.O., Hughes V.W., and Nagel D.E., *The Meson Factories*, University of California Press, Los Angeles, 1991; also Livingston M.S., *LAMPF, A Nuclear Research Facility*, LA-6878-MS, September 1977; also Livingston M.S., *Origins and History of the Los Alamos Meson Physics Facility*, LA-5000.

21 *Compendium of Superconducting Linacs*, Provisional Edition, 18th Int. Linac Conf. Geneva, August, 26–30, 1996.

22 (1987) *Rev. Mod. Phys.* **57**, 1121.

23 Jackson J.D., 1975 *Classical Electrodynamics*, 2nd ed., John Wiley & Sons, New York, pp. 219–223.

24 Potentials that satisfy Eq. (11) are said to belong to the Lorentz gauge.

25 The beauty of this result has been previously noted. See Feynman R.P., Leighton R.B., and Sands M., *The Feynman Lectures on Physics*, Addison Wesley, 1963 Vol. 2, p. 18–11.

26 See Condon E.U., (1940) *J. App. Phys.* **11**, 502; Condon E.U., (1941) *J. App. Phys.* **12**, 129.

27 Potentials that satisfy Eq. (14) are said to belong to the Coulomb gauge. Note that for this case, because $\phi = 0$, Eq. (11) is also satisfied and the potentials belong simultaneously to the Lorentz and Coulomb gauges.

28 Watkins D.A., 1958 *Topics in Electromagnetic Theory*, John Wiley & Sons, New York, pp. 12–14; Bevensee R.M., 1964 *Electromagnetic Slow Wave Systems*, John Wiley & Sons, New York, pp. 17–19.

29 Moreno T., 1948 *Microwave Transmission Design Data*, Dover, New York, pp. 227–230.

2
RF Acceleration in Linacs

In all RF accelerators energy is delivered to the beam from an RF electric field, which must be synchronous with the beam for a sustained energy transfer. In this chapter we introduce some important parameters that characterize the energy-gain process.

2.1
Particle Acceleration in an RF Field

First, consider an electromagnetic wave propagating along the $+z$ direction in a waveguide, whose electric field along the axis is given by

$$E_z(z, t) = E(z) \cos\left(\omega t - \int_0^z dz\, k(z) + \phi\right) \quad (2.1)$$

where the wave number k is expressed in terms of the phase velocity v_p by $k(z) = \omega/v_p(z)$. For efficient particle acceleration, the phase velocity of the wave must be closely matched to the beam velocity. If we consider a particle of charge q moving along the $+z$ direction, whose velocity $v(z)$ at each instant of time equals the phase velocity of the traveling wave, the particle arrives at position z at time $t(z) = \int_0^z dz/v(z)$, and the electric force on the particle is given by $F_z = qE(z)\cos\phi$. This particle is called a *synchronous particle*, and the phase ϕ is called the *synchronous phase*.

Energy can also be transferred to a particle from an electromagnetic standing wave in an RF cavity. The electric field along the axis is given by

$$E_z(z, t) = E(z) \cos(\omega t + \phi) \quad (2.2)$$

In general, the cavity may be constructed as a periodic or quasiperiodic array of cells, each with a single accelerating gap, where the parameters in the quasiperiodic structure vary monotonically along the structure to maintain synchronism with the beam particles, whose velocities are increasing. The field acting on a particle of charge q and velocity v at position z is obtained by

RF Linear Accelerators. 2nd, completely revised and enlarged edition.
Thomas P. Wangler
Copyright © 2008 Wiley-VCH Verlag GmbH & Co. KGaA, Weinheim
ISBN: 978-3-527-40680-7

substituting $t(z) = \int_0^z dz/v(z)$ into Eq. (2.2). It is simplest to consider cavities with cells that are similar to the pillbox cavity that we have studied in Chapter 1. Each cell can be excited in a TM$_{010}$-like standing-wave mode. For this mode, a particle is acted upon by the nonzero E_z and B_θ fields, and when holes are included in the end walls for the beam, a nonzero E_r component is also present. The general equations of motion are

$$\frac{dp_z}{dt} = q\left[E_z(r,z)\cos(\omega t + \phi) + v_r B_\theta(r,z)\sin(\omega t + \phi)\right] \quad (2.3)$$

and

$$\frac{dp_r}{dt} = q\left[E_r(r,z)\cos(\omega t + \phi) - v_z B_\theta(r,z)\sin(\omega t + \phi)\right] + F_{\text{ex}} \quad (2.4)$$

where F_{ex} includes any external radial field that might be applied for focusing. Particle trajectories can be calculated by numerical integration of Eqs. (2.3) and (2.4), using measured fields or fields calculated from an electromagnetic field-solver code.

2.2
Energy Gain on Axis in an RF Gap

Consider the accelerating gap shown in Fig. 2.1, and express the electric field on the axis experienced by a particle with a velocity v, as

$$E_z(r=0, z, t) = E(0, z)\cos[\omega t(z) + \phi] \quad (2.5)$$

where $t(z) = \int_0^z dz\, v(z)$ is the time the particle is at position z. We choose time $t = 0$ when the particle is at the origin, which can be arbitrarily chosen within the gap. At $t = 0$, the phase of the field relative to the crest is ϕ.

Suppose that the field is confined within an axial distance L containing the gap.

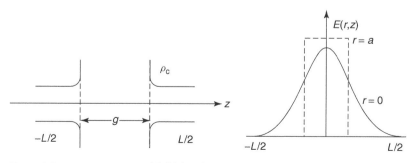

Figure 2.1 Gap geometry and field distribution.

The energy gain of an arbitrary particle with charge q traveling through the gap is

$$\Delta W = q \int_{-L/2}^{L/2} E(0, z) \cos(\omega t(z) + \phi)\, dz \qquad (2.6)$$

The use of a common trigonometric identity allows us to write the energy gain as

$$\Delta W = q \int_{-L/2}^{L/2} E(0, z) [\cos \omega t \cos \phi - \sin \omega t \sin \phi]\, dz \qquad (2.7)$$

Next, we write Eq. (2.7) in the form

$$\Delta W = q V_0 T \cos \phi \qquad (2.8)$$

where we have introduced an axial RF voltage, defined by

$$V_0 \equiv \int_{-L/2}^{L/2} E(0, z)\, dz \qquad (2.9)$$

which is assumed to be nonzero, and the transit-time factor, defined by [1]

$$T \equiv \frac{\int_{-L/2}^{L/2} E(0, z) \cos \omega t(z)\, dz}{\int_{-L/2}^{L/2} E(0, z)\, dz} - \tan \phi \frac{\int_{-L/2}^{L/2} E(0, z) \sin \omega t(z)\, dz}{\int_{-L/2}^{L/2} E(0, z)\, dz} \qquad (2.10)$$

Equation (2.8) is an important result that we will use frequently. The phase $\phi = 0$ if the particle arrives at the origin when the field is at a crest. It is negative if the particle arrives at the origin earlier than the crest, and positive if it arrives later. Maximum energy gain occurs when $\phi = 0$, which is often the choice for relativistic electrons. But we will see later that for nonrelativistic particles, where longitudinal focusing is important, it is necessary to operate with negative phases for the particles in the bunch.

It is useful to define an average axial electric-field amplitude $E_0 = V_0/L$, where V_0 is the axial RF voltage given by Eq. (2.9). V_0 is the voltage gain that would be experienced by a particle passing through a constant dc field equal to the field in the gap at time $t = 0$. The value of E_0, which is an average field over the length L, does depend on the choice of L. Therefore, when a value of E_0 is quoted for a cavity, it is important to specify the corresponding length L. For a multicell cavity, the natural choice for L is the geometric cell length. We will call the quantity $E_0 T$ the *accelerating gradient*.

In terms of E_0, the energy gain may be expressed as

$$\Delta W = q E_0 T \cos \phi L \qquad (2.11)$$

This result is sometimes called the *Panofsky equation* [2]. It is deceptively simple in appearance, but much of the physics is still contained within the transit-time factor. Regardless of the phase ϕ, the energy gain of a particle in a harmonically time-varying field is always less than the energy gain in a constant dc field equal to that seen by the particle at the center of the gap. This is known as the transit-time effect, and the transit-time factor T is the ratio of the energy gained in the time-varying RF field to that in a dc field of voltage $V_0 \cos(\phi)$. Thus, T is a measure of the reduction in the energy gain caused by the sinusoidal time variation of the field in the gap. The phase and the transit-time factor depend on the choice of the origin. It is convenient to simplify the transit-time factor, and remove its dependence on the phase, by choosing the origin at the electrical center.

$E(z)$ is usually at least approximately an even function about a geometric center of the gap. We will choose the origin at the electrical center of the gap, defined to give

$$0 = \int_{-L/2}^{L/2} E(0, z) \sin \omega t(z) \, dz \tag{2.12}$$

When $E(z)$ is an even function about the geometric center of the gap, the electrical center and the geometric center coincide. When Eq. (2.12) applies, the transit-time factor simplifies to

$$T = \frac{\int_{-L/2}^{L/2} E(0, z) \cos \omega t(z) \, dz}{\int_{-L/2}^{L/2} E(0, z) \, dz} \tag{2.13}$$

Then the transit-time-factor expression in Eq. (2.13) is the average of the cosine factor weighted by the field. The transit-time factor increases when the field is more concentrated longitudinally near the origin, where the cosine factor is largest. Thus, the larger the gap between drift tubes and the more the field can penetrate into the drift tubes, the smaller the transit-time factor. The practice of constructing drift tubes with nose cones that extend further into the gap forces a concentration of the field near the center of the gap, and raises the transit-time factor.

In most practical cases the change of particle velocity in the gap is small compared with the initial velocity. If we ignore the velocity change, we have $\omega t \approx \omega z/v = 2\pi z/\beta\lambda$, where $\beta = v/c$ and $\beta\lambda$ is the distance the particle travels in an RF period. The transit-time factor simplifies to the form most often seen in the literature,

$$T = \frac{\int_{-L/2}^{L/2} E(0, z) \cos(2\pi z/\beta\lambda) \, dz}{\int_{-L/2}^{L/2} E(0, z) \, dz} \tag{2.14}$$

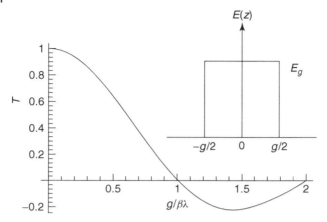

Figure 2.2 Transit-time factor for square-wave electric-field distribution.

We note that the transit-time factor is finite only if the denominator of Eq. (2.14) is nonzero. For most practical cases this is true, but exceptional cases arise. A TM_{011} cavity mode, which has a node in E_z at the origin, is an example for which the integral of the axial field is zero, and a transit-time factor for the whole cell would be infinite. When this happens, the energy gain can still be calculated directly from Eq. (2.6). We will usually consider TM_{010}-like modes where this is not a problem.

A first approximation for T assumes that the electric field has a square profile, as shown in Fig. 2.2. Thus, $E(0, z) = E_g$ is constant over a gap of length g, and falls immediately to zero outside the gap. This would be the exact result for a simple TM_{010} pillbox cavity of length g, and would be an approximate result for a simple gap with beam holes at the ends in the limit that the beam-aperture radius is very small. Then we find $E_0 = E_g$ if $L = g$, and Eq. (2.14) becomes

$$T = \frac{\sin \pi g/\beta\lambda}{\pi g/\beta\lambda} \tag{2.15}$$

The result for T versus $g/\beta\lambda$ is plotted in Fig. 2.2. To achieve maximum energy gain for a given V_0, it is clear that we want $T = 1$, which corresponds to $g = 0$. However, other considerations must be taken into account to determine the optimum gap geometry, such as the risk of RF electric breakdown, and RF power efficiency.

2.3
Longitudinal Electric Field as a Fourier Integral

A description of the fields in an accelerating gap in terms of a Fourier integral allows us to deduce some useful relationships for the transit-time factor. For a single gap, where the fields vanish at some finite distance outside the gap, the value of L in Eq. (2.14) is arbitrary as long as it is large enough to include the nonzero field region. Then, we are free to let $L \to \infty$, without affecting any of the physics. We may write

$$V_0 T(k) = \int_{-\infty}^{\infty} E(0, z) \cos(kz) \, dz \quad (2.16)$$

where $k = 2\pi/\beta\lambda$, depends on the particle velocity. The integral has the form of the Fourier cosine integral, and its Fourier transform is

$$E(0, z) = \frac{V_0}{2\pi} \int_{-\infty}^{\infty} T(k) \cos(kz) \, dk \quad (2.17)$$

We can extend these results to include the transit-time factor and energy gain for off-axis particles. For the field at a displacement r from the axis, assuming azimuthal symmetry, we write

$$E_z(r, z, t) = E(r, z) \cos(\omega t + \phi) \quad (2.18)$$

and

$$V_0 T(r, k) = \int_{-L/2}^{L/2} E(r, z) \cos(kz) \, dz \quad (2.19)$$

where in a single gap we can let $L \to \infty$. The Fourier transform of Eq. (2.19) is

$$E(r, z) = \frac{V_0}{2\pi} \int_{-\infty}^{\infty} T(r, k) \cos(kz) \, dk \quad (2.20)$$

Now, we substitute this solution for $E_z(r, z, t)$ into the wave equation in cylindrical coordinates:

$$\frac{\partial^2 E_z}{\partial z^2} + \frac{1}{r} \frac{\partial E_z}{\partial r} + \frac{\partial^2 E_z}{\partial r^2} - \frac{1}{c^2} \frac{\partial^2 E_z}{\partial t^2} = 0 \quad (2.21)$$

Integrating over k and using the orthogonality of the cosine functions, we obtain

$$\frac{\partial^2 T}{\partial x^2} + \frac{1}{x} \frac{\partial T}{\partial x} + T = 0 \quad (2.22)$$

where $x^2 \equiv K^2 r^2$ and

$$K^2 = \left\{ \left(\frac{2\pi}{\lambda}\right)^2 - k^2 \right\} \quad (2.23)$$

Equation (2.22) is the Bessel equation. Thus, the sign of K^2 can be either positive or negative, depending on whether k is larger or smaller than $2\pi/\lambda$.

2 RF Acceleration in Linacs

Substituting the physical result $k = 2\pi/\beta\lambda$, we have

$$K^2 = \left(\frac{2\pi}{\lambda}\right)^2 \left[1 - \frac{1}{\beta^2}\right] = \left(\frac{2\pi}{\lambda}\right)^2 \left[\frac{\beta^2 - 1}{\beta^2}\right] = -\left(\frac{2\pi}{\gamma\beta\lambda}\right)^2 \quad (2.24)$$

and we find $K^2 < 1$. But, for the nonphysical velocities and k values, which also are included mathematically in Eqs. (2.17) and (2.20), we can have $K^2 \geq 1$. To simplify the notation we will we will always write K as a positive number given by $K = \sqrt{|K^2|}$. The solution is

$$T(r,k) = \begin{cases} T(k)J_0(Kr), & k \leq 2\pi/\lambda \\ T(k)I_0(Kr), & k \geq 2\pi/\lambda \end{cases} \quad (2.25)$$

where J_0 is the Bessel function and I_0 is the modified Bessel function of order zero. We use the notation $T(k) = T(0, k)$.

For most beams in linacs, the paraxial approximation applies, which means that the particle radius or displacement from axis is nearly constant through the gap. We can calculate the energy gain of a particle at constant radius r. It is convenient to express the Fourier integrals in exponential form. Thus

$$E(r,z) = \frac{V_0}{2\pi} \int_{-\infty}^{\infty} dk\, T(r,k)\, e^{jkz} \quad (2.26)$$

$$V_0 T(r,k) = \int_{-\infty}^{\infty} dz\, E(r,z)\, e^{-jkz} \quad (2.27)$$

The cosine forms are obtained by taking the real part. The energy gain for a particle with radius r and velocity β_0 is obtained by calculating

$$\Delta W = q \int_{-\infty}^{\infty} dz\, E_z(r,z,t = z/\beta_0 c) \quad (2.28)$$

We find

$$\Delta W = \frac{qV_0}{2\pi} \int_{-\infty}^{\infty} dk\, T(r,k) \int_{-\infty}^{\infty} dz \left[\frac{e^{j[(k+k_0)z+\phi]} + e^{j[(k-k_0)z-\phi]}}{2}\right] \quad (2.29)$$

where $k_0 = 2\pi/\beta_0\lambda$. We apply a familiar result for the δ function,

$$\delta(k) = \frac{1}{2\pi} \int_{-\infty}^{\infty} dx\, e^{jkx} \quad (2.30)$$

and

$$\Delta W = \frac{qV_0}{2} \left[T(r,-k_0)\, e^{j\phi} + T(r,k_0)\, e^{-j\phi}\right] \quad (2.31)$$

We assume $T(r, k)$ is an even function of k, so $T(r, -k) = T(r, k)$. Also, for k corresponding to particle velocities $\beta < 1$, $T(r,k) = T(k)I_0(Kr)$. Substituting these results into Eq. (2.31), and taking the real part, we obtain

$$\Delta W = qV_0 T(k_0) I_0(K_0 r) \cos\phi \quad (2.32)$$

where $K_0 = 2\pi/\gamma_0\beta_0\lambda$. This equation is like the axial energy gain result of Eq. (2.8), except now we have included the radial dependence of the energy gain through the modified Bessel function.

For additional insight to the energy-gain result, using a familiar trigonometric identity, we can write

$$E_z(r, z, t) = \frac{V_0}{2\pi} \int_{-\infty}^{\infty} dk\, T(r, k) \cos(kz) \cos(\omega t + \phi)$$

$$= \frac{V_0}{4\pi} \int_{-\infty}^{\infty} dk\, T(r, k)\{\cos(\omega t - kz + \phi) + \cos(\omega t + kz + \phi)\}$$

(2.33)

This equation describes the standing-wave field pattern as resulting from an infinite sum of forward and backward traveling waves with phase velocities given by $\beta = \pm\omega/kc$. If we imagine a particle traveling from $z = -\infty$ to $+\infty$, only the single wave with the same phase velocity as the particle will be synchronous with the particle, and will deliver net energy to it. The effects of all other waves average to zero, because they do not maintain a constant phase relationship with the particle. From this we conclude that the radial dependence of the energy gain is the same as that of the synchronous wave.

The z component of the electric field can also be written as follows:

$$E_z(r, z, t) = \frac{V_0}{2\pi} \int_{-\infty}^{\infty} dk\, T(k) \left\{ \begin{array}{c} I_0(Kr) \\ J_0(Kr) \end{array} \right\} \cos(kz) \cos(\omega t + \phi) \quad (2.34)$$

where we use $I_0(Kr)$ when $|k| \geq 2\pi/\lambda$, and $J_0(Kr)$ when $|k| \leq 2\pi/\lambda$. Then, the radial electric-field component can be derived from Gauss's law as

$$E_r(r, z, t) = \frac{V_0}{2\pi} \int_{-\infty}^{\infty} dk\, \frac{k}{K} T(k) \left\{ \begin{array}{c} I_1(Kr) \\ J_1(Kr) \end{array} \right\} \sin(kz) \cos(\omega t + \phi) \quad (2.35)$$

and using Ampére's law, we obtain,

$$B_\theta(r, z, t) = \frac{-V_0}{c\lambda} \int_{-\infty}^{\infty} dk\, \frac{T(k)}{K} \left\{ \begin{array}{c} I_1(Kr) \\ J_1(Kr) \end{array} \right\} \cos(kz) \sin(\omega t + \phi) \quad (2.36)$$

Equations (2.34) to (2.36) comprise a complete solution to Maxwell's equations together with the remaining components B_z, B_r, and E_θ, which are zero, as can be shown by direct substitution.

2.4
Transit-Time-Factor Models

We can improve on the approximation for T in Eq. (2.15) by accounting for penetration of the field into the axial bore holes of drift tubes. Field penetration into the drift tubes reduces T, because it reduces the concentration of the field near the center of the gap. We assume that the electric field at the drift-tube bore radius ($r = a$) is constant within the gap and zero outside the gap within the metallic walls, as shown in Fig. 2.1. This is a more realistic approximation, because it constrains the field to vanish only at the bore radius inside the conductor, rather than on the axis at the end of the gap. Thus, we require that

$$E(r = a, z) = \begin{cases} E_g, & 0 \leq |z| \leq g/2 \\ 0, & g/2 < |z| \end{cases} \quad (2.37)$$

From Eq. (2.19) this leads to

$$V_0 T(a, k) = \int_{-\infty}^{\infty} E(a, z) \cos(kz) \, dz = \frac{2E_g}{k} \sin(kg/2) \quad (2.38)$$

From Eq. (2.25), $T(a, k) = T(k) I_0(Ka)$, where $K = 2\pi/\gamma\beta\lambda$, and we obtain

$$V_0 T(k) = \frac{E_g g}{I_0(Ka)} \frac{\sin(kg/2)}{kg/2} \quad (2.39)$$

Also, $V_0 T(0, 0) = \int_{-\infty}^{\infty} dz \, E(0, z) = V_0$, or $T(0) = T(0, 0) = 1$, and at $k = 0$ and $r = a$, Eq. (2.25) gives $T(a, 0) = J_0(K_0 a)$, where $K_0 = 2\pi/\lambda$. Then

$$V_0 T(a, 0) = \int_{-\infty}^{\infty} dz \, E(a, z) = E_g g \quad (2.40)$$

or

$$V_0 = \frac{E_g g}{J_0(2\pi a/\lambda)} \quad (2.41)$$

For $\beta \leq 1$ we obtain from Eqs. (2.25) and (2.39),

$$T(r, k) = J_0(2\pi a/\lambda) \frac{I_0(Kr)}{I_0(Ka)} \frac{\sin(kg/2)}{kg/2} \quad (2.42)$$

and the axial transit-time factor is

$$T(k) = \frac{T(r, k)}{I_0(Kr)} = \frac{J_0(2\pi a/\lambda)}{I_0(Ka)} \frac{\sin(\pi g/\beta\lambda)}{\pi g/\beta\lambda} \quad (2.43)$$

It is usually a good approximation that $J_0(2\pi a/\lambda) \approx 1$. In Eq. (2.43) the axial transit-time factor has been expressed as a product of an aperture-dependent factor and a gap-dependent factor, where the aperture factor takes into account the penetration of the field into the borehole.

The result of Eq. (2.43) applies for a single gap. For a multigap periodic structure, the transit-time factor depends on both the synchronous velocity β_s and the velocity β of the particle. One finds that in the argument of the Bessel functions, the quantity K becomes K_s, where $K_s = 2\pi/\gamma_s\beta_s\lambda$. Instead of Eq. (2.43), the result becomes [3]

$$T(k) = \frac{J_0(2\pi a/\lambda)}{I_0(K_s a)} \frac{\sin(\pi g/\beta\lambda)}{\pi g/\beta\lambda} \quad (2.44)$$

2.4 Transit-Time-Factor Models

An empirical correction to the gap factor can be made [4] to account for the finite chamber radius ρ_c that is put on a drift tube to remove the sharp edge, and avoid electric breakdown. The effective gap is obtained by replacing g in the formula for the transit-time factor by

$$g_c = g + 0.85\rho_c \tag{2.45}$$

For π-mode structures an additional effect occurs. The axial electric field changes sign from cell to cell and must cross zero at the cell boundaries, as shown in Fig. 2.3. For a large gap and a large borehole, the axial field distribution will be more strongly influenced by the boundary condition that the field vanishes at the beginning and end of the cell. An example is the typical multicell superconducting elliptical cavity structure with a large aperture and without any nose cone that extends into the gap. To construct a simple model for this situation, we assume that in a large-bore, π-mode structure, the axial field in the gap of a cell of length L has a cosine dependence, given by

$$E(r = 0, z) = E_g \cos(k_s z) \tag{2.46}$$

where $k_s = \pi/L$. The field in Eq. (2.46) vanishes when $z = \pm L/2$, satisfying the π-mode constraint. Generally, the cell length will have been designed to give synchronous acceleration for a particle with velocity β_s, by choosing $L = \beta_s \lambda/2$. Substituting Eq. (2.46) into Eq. (2.19) evaluated on axis, we find

$$\begin{aligned} V_0 T(0, k) &= \int_{-L/2}^{L/2} dz\, E(0, k) \cos(kz) \\ &= E_g \int_{-L/2}^{L/2} dz\, \cos(kz) \cos(kz) \end{aligned} \tag{2.47}$$

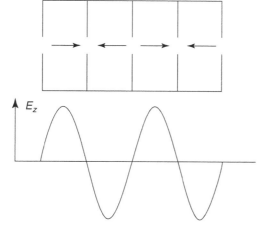

Figure 2.3 Axial electric-field distribution showing the effect of the π-mode boundary conditions, which causes the field to cross the axis at the boundaries of each cell.

The axial transit-time factor for a particle with $\beta = \beta_s$, or $k = k_s$ is

$$T(k_s) = T(0, k_s) = \frac{E_g}{V_0} \int_{-L/2}^{L/2} dz \, \cos^2(k_s z) = \frac{\pi E_g}{2 k_s V_0} \qquad (2.48)$$

Also we have

$$V_0 = \int_{-L/2}^{L/2} dz \, E(0, z) = \frac{E_g}{k_0} \int_{-L/2}^{L/2} dz \, \cos(k_s z) = \frac{2 E_g}{k_s} \qquad (2.49)$$

Substituting Eq. (2.49) into Eq. (2.48), we obtain the axial transit-time factor for the synchronous particle

$$T(k_s) = \frac{\pi}{4} \approx 0.785 \qquad (2.50)$$

Unlike the result of Eq. (2.44), the axial transit-time factor is independent of both the gap, and the radial aperture of the borehole. For completeness the radial dependence of the transit-time factor from the π-mode boundary condition is given from Eq. (2.25), by

$$T(r, k_s) = T(k_s) I_0(K_s r) = \frac{\pi}{4} I_0(K_s r) \qquad (2.51)$$

where $K_s = 2\pi/\gamma_s \beta_s \lambda$. For the case of a particle with velocity β which is not equal to β_s, a general result analogous to Eq. (2.44) is

$$T(r, k) = I_0(K_s r) \frac{\pi}{4} \left[\frac{\sin[(k/k_s - 1)\pi/2]}{(k/k_s - 1)\pi/2} + \frac{\sin[(k/k_s + 1)\pi/2]}{(k/k_s + 1)\pi} \right] \qquad (2.52)$$

2.5
Power and Acceleration Efficiency Figures of Merit

There are several figures of merit that are commonly used to characterize accelerating cavities, and we will define them in this section. Some of these depend on the power, which is dissipated because of electrical resistance in the walls of the cavities. The well-known quality factor of a resonator is defined in terms of the average power loss P as

$$Q = \omega U / P \qquad (2.53)$$

It is also convenient to define the *shunt impedance*, a figure of merit that is independent of the excitation level of the cavity and measures the effectiveness of producing an axial voltage V_0 for a given power dissipated. The shunt impedance r_s of a cavity is usually expressed in megohms, and is defined by [5]

$$r_s = \frac{V_0^2}{P} \qquad (2.54)$$

2.5 Power and Acceleration Efficiency Figures of Merit

In an accelerating cavity we are really more interested in maximizing the particle energy gain per unit power dissipation. The peak energy gain of a particle occurs when $\phi = 0$, and is $\Delta W_{\phi=0} = qV_0 T$. Consequently, we define an *effective shunt impedance* of a cavity as

$$r = \left[\frac{\Delta W_{\phi=0}}{q}\right]^2 \frac{1}{P} = \frac{[V_0 T]^2}{P} = r_s T^2 \quad (2.55)$$

This parameter in megohms measures the effectiveness per unit power loss for delivering energy to a particle. For a given field both $V_0 = E_0 L$ and P increase linearly with cavity length, as do both r and r_s. For long cavities we often prefer to use a figure of merit that is independent of both the field level and the cavity length. Thus, it is also convenient to introduce *shunt impedances per unit length*. The shunt impedance per unit length, Z, is simply

$$Z \equiv \frac{r_s}{L} = \frac{E_0^2}{P/L} \quad (2.56)$$

Similarly, the effective shunt impedance per unit length is

$$ZT^2 = \frac{r}{L} = \frac{(E_0 T)^2}{P/L} \quad (2.57)$$

The units of shunt impedance per unit length and effective shunt impedance per unit length are usually megohms per meter. Especially for normal-conducting cavities, one of the main objectives in cavity design is to choose the geometry to maximize effective shunt impedance per unit length. This is equivalent to maximizing the energy gain in a given length for a given power loss. Another useful parameter is the ratio of effective shunt impedance to Q, often called *r over Q*,

$$\frac{r}{Q} = \frac{(V_0 T)^2}{\omega U} \quad (2.58)$$

We see that r/Q measures the efficiency of acceleration per unit stored energy at a given frequency. One may wish instead to quote the ZT^2/Q. Either of these ratios is useful, because they are a function only of the cavity geometry, and are independent of the surface properties that determine the power losses.

Some fraction of the RF power that goes into the cavity is also transferred to the beam as a result of the interaction of the beam with the cavity fields. The power P_B delivered to the beam is easily calculated from

$$P_B = I \Delta W / q \quad (2.59)$$

where I is the beam current and ΔW is the energy gain. The sum of the dissipated power plus that delivered to the beam is $P_T = P + P_B$, and the efficiency of the structure is measured by the beam-loading ratio $\varepsilon_s = P_B/P_T$. The overall system efficiency must also include the power losses in the input waveguide or transmission line, reflected power from the input coupler, which ends up in a load, and power dissipation in the RF generator. The latter is usually the largest contributor. The RF generator efficiency, which is the ratio of the output RF power to the total ac input power, typically lies in the range from about 40 to 60%. The overall efficiency increases with increasing beam current.

2.6
Cavity Design Issues

Cavity design generally requires electromagnetic field–solver codes that numerically solve Maxwell's equations for the specified boundary conditions. The optimization procedures depend on the constraints of the problem. Suppose the objective is to maximize the effective shunt impedance of a single cavity. If we begin with a simple pillbox cavity with beam holes on the end plates, adding nose cones to create a region of more concentrated axial electric field, as shown in Fig. 2.4, reduces the gap and raises the transit-time factor. The nose cone is part of the drift tube, and its shape may be constrained by the requirement that it must hold a focusing quadrupole magnet. If the gap between drift tubes is too small, then the voltage gain for a given peak surface electric field becomes small. Also a capacitance between the opposite nose cones increases with decreasing gap size, increasing the ratio of wall current to axial voltage. Some designers prefer to optimize in steps. First, the gap size and the shape of the nose cone can be chosen to maximize the r/Q parameter to increase the energy gain per unit stored energy. Then, the geometry of the outer walls can be adjusted to maximize Q, thus minimizing the power dissipation per unit stored energy. This latter procedure usually leads to a spherical shape for the outer wall, which corresponds to the smallest surface area for a given volume.

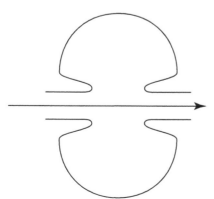

Figure 2.4 Cross section of cavity with nose cones and spherical outer wall.

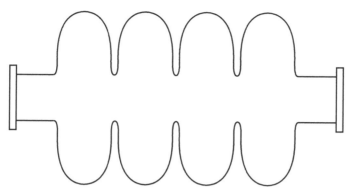

Figure 2.5 Cross section of elliptical cavity used for a superconducting proton linac. The cross section for each cell consists of an outer circular arc, an ellipse at the iris, and a connecting straight line.

In superconducting cavities, the RF power efficiency may not be as important as other requirements such as minimizing the peak surface fields, which limit the performance. The spherical or elliptical cavity shape without nose cones, shown in Fig. 2.5, is the universally accepted choice for a superconducting cavity shape for high-velocity particles, because of its low ratio of peak surface field to E_0, its lack of strong multipacting levels, and practical advantages for drainage of fluids during chemical processing.

The peak surface electric field and magnetic field are important constraints in cavity design. In normal-conducting cavities, too large a peak surface electric field can result in electric breakdown. For a cavity with nose cones or drift tubes, a typical ratio of peak surface electric field E_s to average axial field E_0, is $E_s/E_0 = 6$, whereas the ratio is unity for a pillbox cavity. The Kilpatrick criterion, discussed later, is often used as the basis for the peak surface electric-field limit in normal-conducting cavities. In superconducting cavities high electric fields cause field emission, which produces electrons in the cavity volume that absorb RF energy and create additional power loss. The surface magnetic fields correspond to surface currents that produce resistive heating. For normal-conducting cavities, cooling requirements restrict the peak magnetic field, and average power densities of about 20 W/cm² may begin to produce cooling difficulties in drift-tube linacs. Superconducting cavities can quench and make a transition to the normal-conducting state, when a critical magnetic field is exceeded. This field depends on temperature, and is near $B = 0.2$ T in the range of operating temperatures from about 2 to 4.2 K. The presence of normal-conducting impurities lowers this limit. Few designers of superconducting cavities would exceed about $B = 0.05$ to 0.1 T. The ratio of peak surface magnetic field to average axial field E_0 is 19.4 G/MV/m for a pillbox cavity. This value can increase by a factor of 2 to 3, especially in drift tubes.

The aperture radius is usually chosen to satisfy the beam-dynamics requirements for high transmission. Large aperture results in a reduced transit-time factor. As the aperture radius increases, the axial electric field decreases for a given peak surface field near the aperture. Introducing nose cones or drift tubes increases the capacitance, and introducing stems that carry a net RF current to the drift tubes increases the local magnetic flux. Loading a cavity with lumped elements produces more localized fields. This approach reduces the fields at the outer cavity walls, and the diameter of those walls can be reduced without producing much change in the frequency. The size reduction can make the structures easier to build and handle. This approach is especially useful for very low-frequency structures and is widely used for heavy-ion accelerating structures.

2.7
Frequency Scaling of Cavity Parameters

One of the most important parameters to choose when designing a linac is the operating frequency. To make this choice, it is important to know how the cavity parameters vary with frequency. Consider a cavity with fixed accelerating field E_0 and fixed total energy gain ΔW, so that the total length is fixed, independent of frequency. Suppose we scale all other cavity dimensions with wavelength or as f^{-1}. How do the other parameters behave? The transit-time factor and the fields are independent of frequency. At a fixed total length, the surface area is inversely proportional to the frequency, and the total cavity volume and stored energy are inversely proportional to the frequency squared. Surface-resistance and power loss scaling depend on whether the linac is normal conducting or superconducting. In what follows we will ignore the residual-resistance term in the superconducting RF resistance, which is a good approximation at high frequency. One finds that the RF surface resistance scales as

$$R_s \propto \begin{cases} f^{1/2} & \text{normal conducting} \\ f^2 & \text{super conducting} \end{cases} \quad (2.60)$$

The RF power losses are not zero in the superconducting case. They are caused by normal-conducting electrons that are present at any finite temperature. Although these losses are small in terms of RF power, they affect the requirements for cryogenic refrigeration and must be taken into account. The RF power dissipation scales as

$$P = \frac{R_s}{2} \left|\frac{B}{\mu_0}\right|^2 dA \propto \begin{cases} f^{-1/2} & \text{normal conducting} \\ f & \text{super conducting} \end{cases} \quad (2.61)$$

Thus, higher frequency gives reduced power loss for normal-conducting structures, but increased power losses for the superconducting case. The value of Q scales as

$$Q = \frac{\omega U}{P} \propto \begin{cases} f^{-1/2} & \text{normal conducting} \\ f^{-2} & \text{super conducting} \end{cases} \quad (2.62)$$

The effective shunt impedance per unit length scales as

$$ZT^2 = \frac{(E_0 T)^2 L}{P} \propto \begin{cases} f^{+1/2} & \text{normal conducting} \\ f^{-1} & \text{super conducting} \end{cases} \quad (2.63)$$

The shunt impedance increases at higher frequencies for normal-conducting structures, and at lower frequencies for superconducting structures. The ratio ZT^2/Q scales as

$$\frac{ZT^2}{Q} \propto \begin{cases} f^1 & \text{normal conducting} \\ f^1 & \text{super conducting} \end{cases} \quad (2.64)$$

This is the same for normal and superconducting, as it should be since this is a figure of merit that is independent of the surface properties.

2.8
Linac Economics

Accelerator designers must understand the impact of their design choices on the total costs. Included may be the capital costs of construction, and the costs of operation over some period of time, perhaps several decades. To illustrate the issues, we present a simplified linac cost model, which includes two main cost factors. First, we introduce a capital cost per meter, C_L, which includes the costs of fabrication of the accelerating structure, and any other costs that scale in proportion to the length of the linac. We introduce a capital cost per watt of RF power, C_P. We define the total cost C, the total length L, and the total RF power P, which is expressed as the sum of the resistive losses in the accelerating structure P_S, and the power P_B delivered to the beam. We can express the total cost for the design as

$$C = C_L L + C_P(P_S + P_B) \quad (2.65)$$

Suppose we want to choose the quantities L and P_S to minimize C. Before we can do so, we must include some constraints. First, we must ensure that the accelerator design produces the correct final energy. Assuming that the energy gain of the linac must be ΔW, we write

$$\Delta W = qEL\cos\phi \quad (2.66)$$

where we have introduced $E = E_0 T$, the charge q, and the RF phase ϕ. We introduce two other relationships. From the definition of effective shunt impedance, we write $P_S = E^2 L/ZT^2$, where ZT^2 is the effective shunt impedance, and the beam power $P_B = I\Delta W/q$, where I is the beam current. Expressing L and P_S as functions of E, we substitute $L = \Delta W/qE\cos\phi$, and $P_S = E\Delta W/qZT^2\cos\phi$ into Eq. (2.65) to obtain

$$C(E) = \frac{\Delta W}{q}\left[\frac{C_L}{E\cos\phi} + \frac{C_P E}{ZT^2\cos\phi} + C_P I\right] \quad (2.67)$$

An example is shown in Fig. 2.6 for the three separate cost terms and the total, as a function of the accelerating field E. The first term is the structure length cost, the second is the RF structure power cost, and the third term is the RF beam power cost. It can be seen that there is an optimum value of E that minimizes the cost. We can calculate the minimum cost by differentiating C with respect to E, which leads to

$$\frac{C_L}{E} = \frac{C_P E}{ZT^2} \quad (2.68)$$

The left side of Eq. (2.68) is proportional to the structure length cost, and the right side to the RF structure power cost. The result is that the minimum total cost corresponds to equal costs for these two contributions. The RF beam power gives a fixed cost that does not affect the minimum. Solving Eq. (2.68) for the optimum field, we find

$$E = \sqrt{\frac{C_L ZT^2}{C_P}} \quad (2.69)$$

From Eqs. (2.66) and (2.69), we find that the corresponding optimum length L is

$$L = \frac{\Delta W}{q\cos\phi}\sqrt{\frac{C_P}{C_L ZT^2}} \quad (2.70)$$

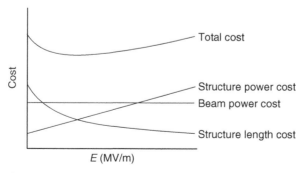

Figure 2.6 Costs versus accelerating field. The structure power cost, the beam power cost, structure length cost, and the total cost are shown.

The structure power for the optimum solution is

$$P_S = \frac{C_L L}{C_P} \tag{2.71}$$

and the minimum cost can be written as

$$C_{MIN} = 2C_L L + C_P P_B = C_P(2P_S + P_B) \tag{2.72}$$

These results show that if E is too large, L is small but this pushes the RF power costs too high. Likewise, if E is too small, L is large and the length cost will be too high. Cost parameters vary depending on many factors, but for rough cost numbers, one might choose $C_p = \$2$ per watt for the installed cost of an RF system, and $C_L \approx \$250{,}000$ per meter to include the installed cost of accelerating structure, support stand, vacuum system, diagnostics, and tunnel. Other constraints, especially technological ones, may enter that prevent achieving the optimum choices. For example the optimum field E may be too large for reliable operation. An accurate cost estimate requires a detailed accounting of the cost of all the components with careful attention to many details. Furthermore, complete estimates for large projects include additional costs such as engineering and design, project management, and allowance for contingency. These costs are far from negligible, and may increase the total cost by a factor of more than two. In recent years, operational costs are also being included, summed over a few decades of operation.

Problems

2.1. Consider a cylindrical pillbox cavity of length L operated in a TM_{014} mode. Complete the following steps to determine whether one can use this mode to achieve synchronous acceleration at each antinode. (a) Sketch the wave pattern for E_z versus z. Express the spacing between adjacent antinodes as a function of the cavity length. How many antinodes are there? (b) Express a synchronism requirement by equating the spacing between adjacent antinodes to the distance the particle travels in one half an RF period. This should relate the cavity length L to $\beta\lambda$. (c) But the cavity length and λ are already related by the dispersion relation. Write the dispersion relation and calculate the frequency f for the TM_{014} mode. Substitute the synchronism condition from part (b). Solve for β, and determine the range of allowed β values. Can one use this mode to achieve synchronous acceleration at each antinode?

2.2. Consider a 2.5-cm-long TM_{010} pillbox cavity that is resonant at 400 MHz and is excited with an axial electric field of $E_0 = 2$ MV/m. Suppose a proton with an initial kinetic energy of 5 MeV is accelerated along the axis of the cavity and arrives at the center when the phase of the field is 30° before the crest. (a) Ignoring the effects of the aperture, and assuming the velocity β remains constant at its initial value, calculate the transit-time factor. **(b)** Calculate the final energy of the particle as it leaves the cavity. Also calculate the fractional change in the velocity.

2.3. Consider a room-temperature copper TM_{010} pillbox cavity that is resonant at 400 MHz with an axial electric field $E_0 = 1$ MV/m and a length $\lambda/2$, where λ is the RF wavelength. Assume the cavity accelerates relativistic electrons with velocity $\beta = 1$. **(a)** Use the power loss formula for the pillbox cavity from Section 1.12 to calculate the RF power dissipated in the walls. **(b)** Ignoring the effects of the apertures calculate the transit-time factor. **(c)** Calculate the shunt impedance, the effective shunt impedance, and the effective shunt impedance per unit length. **(d)** Calculate the energy gain of an electron with synchronous phase $\phi = 0$.

2.4. Consider a room-temperature 10-m-long electron linac with synchronous phase $\phi = 0$, that accelerates the beam from nearly zero kinetic energy to 100 MeV. Suppose that the effective shunt impedance per unit length is $ZT^2 = 50$ MΩ/m. **(a)** Assuming the power delivered to the beam is negligible, how much RF power is required? **(b)** Repeat part (a) assuming the accelerator length is 100 m.

2.5. Use the definition of effective shunt impedance per unit length and the expression for energy gain in a cavity to show that the resistive power loss for fixed energy gain and fixed ZT^2 is proportional to $E_0 T$. For a 100-MeV room-temperature electron linac with synchronous phase $\phi = 0$, and $ZT^2 = 50$ MΩ/m, plot the curve of power (MW) required versus $E_0 T$ (MV/m).

2.6. Suppose that you have to design a TM_{010} pillbox cavity with a constant axial electric field, but you are free to choose the length. If the cavity is too short the voltage gain across it is small; if it is too long the transit-time factor is small. **(a)** Ignoring the effect of the beam aperture, find the ratio of length to $\beta\lambda$ that maximizes the energy gain for the cavity. **(b)** At this length, what is the transit-time factor? **(c)** Show that a zero-length cavity is required to maximize the energy gain per unit length. (In practice you would be limited by peak surface electric fields.)

2.7. Use the analytic expressions for the pillbox cavity given in Section 1.12 to derive expressions for r, r/Q, and ZT^2 for a cylindrical pillbox cavity in a TM_{010} mode. Express the results as the product of a constant and algebraic combinations such as length, radius, and transit-time factor.

2.8. Calculate numerical values for r, r/Q, and ZT^2 for a cylindrical pillbox cavity in a TM_{010} mode, assuming a room-temperature cavity for acceleration of relativistic electrons. Assume a cavity resonant frequency of 400 MHz and a cavity length of $\lambda/2$. Use the formula for transit-time factor and ignore the aperture factor.

2.9. Suppose that we want to design a continuous-wave (CW) room-temperature drift-tube linac to accelerate a 100-mA proton beam from 2.5 to 20 MeV. Assume we can purchase 350-MHz klystron tubes of 1-MW capacity each for the RF power. Suppose that we run the SUPERFISH cavity code and obtain the following results for all β values: $T = 0.8$, $ZT^2 = 50$ MΩ/m, and $E_s/E_0 = 6$. Let us choose to restrict the peak surface electric field at a bravery factor $b = E_s/E_K = 1$. For adequate longitudinal acceptance we choose the synchronous phase $\phi = -30°$. (a) Calculate the average axial field E_0. (b) Calculate the length of the linac assuming it consists of a single tank. (c) Calculate the structure power (power dissipated in the cavity walls), the beam power, and the total RF power required (assuming the klystron generator is matched to the cavity structure with the beam present). (d) What is the structure efficiency (ratio of beam power to total RF power)? Assuming a generator efficiency of 60%, what is the overall RF efficiency (ac power to beam power)? (e) How many klystrons do we need?

2.10. For largest effective shunt impedance (highest power efficiency), we prefer high frequencies for room-temperature linacs. But if the frequency is too large, the radial variation of beam parameters may become intolerable. Choose the linac frequency for the cases below, by using the criterion that at the injection energy the accelerating-field variation with radius over the beam aperture must not exceed 10%. Assume for all cases the radial aperture $a = 1$ cm for acceptable transmission through the entire accelerator. Assume that $E_z(r) \propto I_0(2\pi r/\gamma\beta\lambda)$ and use the approximation $I_0(x) = 1 + x^2/4$. Choose the linac frequency for the following injection conditions for the structure phase velocity: (a) electron linac near 200 keV (use $\beta\gamma = 1$, or $\beta = 0.707$ instead of $\beta = 1$ which is really used), (b) proton linac at 750 keV ($\beta = 0.04$), (c) proton linac at 3 MeV ($\beta = 0.08$), and (d) proton linac near 100 MeV ($\beta = 0.4$).

2.11. To illustrate the choice of the gap length to optimize the cavity shunt impedance, suppose we represent an accelerating cavity by a shunt LRC circuit, where the gap capacitance has the form $C = \alpha/\sqrt{g}$, α is a constant, and g is the gap length. Assuming ω is the angular resonant frequency, and V is the required gap voltage, the magnitude of the peak current charging the capacitance is $I = |\omega C V|$. Suppose that the gap length is varied, and the inductance L is varied to keep the resonant frequency constant. (a) If β is the particle velocity and $\lambda = 2\pi c/\omega$ is the RF wavelength, use the simple expression for transit-time factor T to determine what gap length $g/\beta\lambda$ maximizes T. (b) If R is the resistance of the circuit, show that the average power dissipated over an RF cycle is $P = R(\alpha\omega V)^2/2g$, which is infinite at $g = 0$. (c) If g is very small the power dissipation is too large, and if g is very large the transit-time factor is too small. There exists some value of g that maximizes the effective shunt impedance ZT^2. Using the expression for the effective shunt impedance, given in the form $ZT^2 = (VT)^2/(P\ell)$, where l is the cell length, and using the simple expression for T versus g, find the dependence of ZT^2 versus $g/\beta\lambda$, and plot the result. What approximate value of $g/\beta\lambda$ gives the maximum ZT^2.

2.12. Prove that the quantity

$$\frac{\left|\int e^{jkz} E(z)\, dz\right|}{\int E(z)\, dz} = \sqrt{\left\{\int \cos(kz) E(z)\, dz\right\}^2 + \left\{\int \sin(kz) E(z)\, dz\right\}^2}$$

is independent of the choice of origin, and is equal to the transit-time factor when the origin is at the electrical center of the gap. This provides a convenient way of calculating the transit-time factor corresponding to the origin is at the electrical center, without having to find the exact location of the electrical center. (Hint: First, consider the integrand in the numerator on the left side as a phasor with a phase kz and magnitude $E(z)$. The integral represents a sum of all these vectors, and the numerator gives the magnitude of this sum. Then consider an arbitrary coordinate transformation to a new origin displaced by Δ. Consider what happens to the magnitudes and phases of each of the transformed vectors. Does the magnitude of the new sum change? Finally, does the denominator on the left side depend on the choice of origin?)

References

1 The case $\phi = \pm \pi/2$ can cause problems with this definition. When in doubt the reader can return to Eq. (2) to avoid singularities.
2 Panofsky, W.K.H. *Linear Accelerator Beam Dynamics*, University of California Radiation Laboratory Report UCRL-1216, University of California, Berkeley, Calif., February, 1951, available from the Technical Information Service, Oak Ridge, Tenn.
3 Carne, A., Lapostolle, P., Schnitzer, B. and Promé, M. Numerical methods, Acceleration in a Gap, in *Linear Accelerators*, P.M. Lapostolle and A.L. Septier, John Wiley & Sons, New York (1970), p. 755.
4 Lapostolle, P.M. *Proton Linear Accelerators: A Theoretical and Historical Introduction*, Los Alamos Report LA-11601-MS, (1989) p. 74.
5 Sometimes one finds a factor of two in the denominator for this definition, so be careful.

3
Periodic Accelerating Structures

For an electromagnetic wave to deliver a continuous energy gain to a moving charged particle, two conditions must be satisfied: (1) the wave must have an electric field component along the direction of particle motion and (2) the particle and wave must have the same velocity to maintain synchronism. The first condition is not satisfied by electromagnetic waves in free space, but can be satisfied by a transverse magnetic wave propagating in a uniform waveguide. However, the second condition is not satisfied for a uniform waveguide, because the phase velocity $v_p > c$. The most widely used solution for obtaining phase velocity $v_p < c$ in linacs has been the use of accelerating structures with periodic geometries. A periodic structure has the property that its modes are composed of a Fourier sum of waves, some of which are suitable for synchronous-particle acceleration.

3.1
Synchronous Acceleration and Periodic Structures

Two pictures can be used to describe periodic accelerating structures: (1) considering the structure as a periodically loaded waveguide in which reflections from the periodic loading elements reduce the phase velocity compared with the uniform guide and (2) considering the structure as a periodic array of coupled resonant cavities. The periodic waveguide picture leads to a qualitative description of how the dispersion curve of a uniform waveguide is modified by the presence of the periodic loading elements [1]. However, the coupled-cavity picture of a periodic accelerating structure provides a more quantitative description of the dispersion curve. The iris-loaded structure can be viewed as an array of pillbox cavities that are coupled through the irises. An analytic treatment based on Bethe's theory of coupling of cavities through apertures, the Slater perturbation theorem, and the Floquet theorem leads to an analytic description for the dispersion curve and a simple formula for the cavity coupling constant. In the remainder of this chapter we will look more closely at periodic structures and their properties.

RF Linear Accelerators. 2nd, completely revised and enlarged edition.
Thomas P. Wangler
Copyright © 2008 Wiley-VCH Verlag GmbH & Co. KGaA, Weinheim
ISBN: 978-3-527-40680-7

3.2
Floquet Theorem and Space Harmonics

In a lossless uniform waveguide with azimuthal symmetry, the axial electric field for the lowest transverse-magnetic mode, the TM_{01} mode, is

$$E_z(r, z, t) = EJ_0(Kr)e^{j(\omega t - k_0 z)} \tag{3.1}$$

This describes a wave propagating in the $+z$ direction, with wavenumber $k_0 = 2\pi/\lambda_g$, where λ_g is the guide wavelength. The uniform waveguide has a dispersion relation, shown in Fig. 3.1, given by

$$\omega^2 = (Kc)^2 + (k_0 c)^2 \tag{3.2}$$

where K is the cutoff wavenumber for the TM_{01} mode, related to the cutoff angular frequency by $\omega_c = Kc$. Then the phase velocity is expressed as

$$v_p = \frac{\omega}{k_0} = \frac{c}{\sqrt{1 - (Kc)^2/\omega^2}} > c \tag{3.3}$$

Because the phase velocity is always larger than c, the uniform guide is unsuitable for synchronous-particle acceleration, and we must modify the structure to obtain a lower phase velocity. The solution that has been uniformly adopted is the periodic structure, Fig. 3.2.

One might expect that converting the uniform guide to a periodic structure might perturb the field distribution by introducing a z-periodic modulation of the amplitude of the wave, giving a TM_{01} propagating-wave solution of the form

$$\mathbf{E}(r, z, t) = \mathbf{E}_d(r, z)e^{j(\omega t - k_0 z)} \tag{3.4}$$

where $\mathbf{E}_d(r, z)$ is a periodic function with the same period d as the structure. Equation (3.4) is indeed found to be a correct solution, and it satisfies the

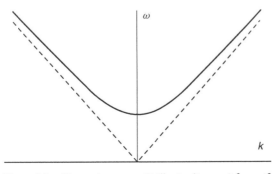

Figure 3.1 Dispersion curve (Brillouin diagram) for uniform waveguide.

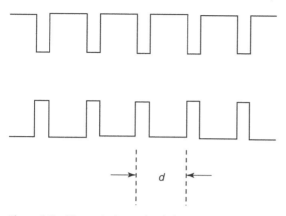

Figure 3.2 The periodic iris-loaded structure.

Floquet theorem stated in the following form. *In a given mode of an infinite periodic structure, the fields at two different cross sections that are separated by one period differ only by a constant factor, which in general is a complex number.* We will find that there are intervals of ω called *stopbands* where the constant is real with magnitude less than 1. For these regions the modes are evanescent – hence the name *stopband*. There are also passbands in which waves will propagate, within which in the loss-free case the complex constant is $\exp(jk_0 d)$, which physically represents a cell-to-cell phase shift $k_0 d$ of the field. To include losses, the constant factor can be written as $e^{-\gamma z}$, where $\gamma = -\alpha + jk_0$ corresponds to propagation with attenuation. At the ends of the passband solutions are found where the constant is ± 1, representing the $k_0 d = 0$ and π modes. The Floquet theorem [2] in a passband is expressed as

$$\mathbf{E}(r, z + d) = \mathbf{E}(r, z)e^{\pm jk_0 d} \tag{3.5}$$

where the sign depends on the sign of the wave propagation.

The plausibility of the Floquet theorem for a propagating mode can be outlined for the loss-free periodic structure. Physically, one expects that the form of the time-independent amplitude function $E_d(r, z)$ depends only on the details of the cell geometry. If this function is a solution for any one cell, it must also be a solution for any other cell in the structure, since all cells are identical. Therefore, a solution should exist using a periodic function for $E_d(r, z)$. The exponential factor accounts for differences in the initial conditions that result in different phases of the oscillation in different cells.

Because $E_d(r, z)$ is periodic, it can be expanded in a Fourier series as

$$E_d(r, z) = \sum_{n=-\infty}^{\infty} a_n(r) e^{-j 2\pi n z/d} \tag{3.6}$$

We obtain the radial solution by requiring Eq. (3.4) to satisfy the wave equation. We find

$$a_n''(r) + \frac{a_n'(r)}{r} - K_n^2 a_n(r) = 0 \qquad (3.7)$$

for all n, where the prime notation refers to differentiation with respect to r, and where $K_n^2 = (\omega/c)^2 - (k_0 + 2\pi n/d)^2$. For $K_n^2 > 0$, the solution for a propagating wave is

$$E_z(r,z,t) = \sum_{n=-\infty}^{\infty} E_n J_0(K_n r) e^{j(\omega t - k_n z)} \qquad (3.8)$$

where

$$k_n = k_0 + \frac{2\pi n}{d} \qquad (3.9)$$

We can interpret the solution of Eq. (3.8) as representing an infinite number of traveling waves, which are called *space harmonics*, each of which is denoted by the index n. We refer to the principle wave as the one with $n = 0$. The space harmonics have the same frequency but different wavenumbers, and each has a constant amplitude E_n independent of z. The waves for $n > 0$ travel in the $+z$ direction, and those for $n < 0$ travel in the $-z$ direction. The wavenumber for the nth space harmonic is shifted from the wavenumber k_0 of the principle wave by $2\pi n/d$. The phase velocity for the nth space harmonic is

$$\beta_n = \frac{\omega}{k_n c} = \frac{\beta_0}{1 + (n\beta_0 \lambda/d)} \qquad (3.10)$$

and by choosing n sufficiently large, one can obtain an arbitrarily low phase velocity. Not only has the introduction of periodicity led to the generation of space harmonics, but the phase velocity of the principle wave $\beta_0 = \omega/ck_0$ is also generally modified. Physically, this can be attributed to the addition of reflections from the periodic elements to the principle wave. Computationally and experimentally, the resulting phase velocity for the principle wave at a given frequency is obtained from the slope of the dispersion curve. Generally, it is found that the value of β_0 generally decreases as the perturbation of the periodic elements increases, as is discussed by Slater.

When $K_n^2 > 0$, the other nonzero components of the TM$_{01}$ solution are

$$E_r(r,z,t) = j \sum_{n=-\infty}^{\infty} E_n \frac{k_n}{K_n} J_1(K_n r) e^{j(\omega t - k_n z)} \qquad (3.11)$$

$$B_\theta(r,z,t) = j \sum_{n=-\infty}^{\infty} E_n \frac{\omega}{K_n c^2} J_1(K_n r) e^{j(\omega t - k_n z)} \qquad (3.12)$$

When $K_n^2 < 0$, the Bessel functions J_0 and J_1 are replaced with modified Bessel functions I_0 and I_1. The integrated effect on a beam particle that is synchronous with one of the space-harmonic waves is obtained by assuming that only the synchronous space harmonic waves act on the beam. The effects of the nonsynchronous waves are assumed to average to zero and are ignored. To see this for the case where the synchronous-particle velocity matches that of the $n = 0$ principle wave, we use Eq. (3.8) and calculate the energy gain of the synchronous particle as it experiences the force from the wave over one structure period d. We assume that the synchronous particle is at an arbitrarily chosen origin $z = 0$ at time $t = t_0$, when the phase of the field relative to the wave crest is $\phi_0 = \omega t_0$. Substituting $t = z/\beta_0 c + t_0$, the energy gain of the synchronous particle is

$$\Delta W = q \int_0^d dz \, Re[E_z(r, z, t = z/\beta_0 c + t_0)] = q E_0 I_0(K_0 r) \cos(\phi_0) d \quad (3.13)$$

where after substituting Eq. (3.8) only the $n = 0$ term survives. In Eq. (3.13), we have assumed that $K_n^2 < 0$, resulting in the modified Bessel function.

The principle wave usually has the largest Fourier amplitude and consequently most periodic accelerating structures are designed so that the principle wave $n = 0$ is synchronous with the beam (i.e., $\beta_0 = \beta_s$). The effective field components experienced by the synchronous particle can be identified from the $n = 0$ terms in Eqs. (3.8), (3.11), and (3.12). From the definition of the phase velocity, for $n = 0$ we have $k_0 = \omega/\beta_0 c$, and we find $K_0^2 = (\omega/c)^2 - k_0^2 = -\omega^2/\gamma_s^2 \beta_s^2 c^2$. The effective field components seen by the beam are the components of the synchronous wave, given by

$$E_z = E_0 I_0(K_0 r) \cos(\phi) \quad (3.14)$$

$$E_r = -\gamma_s E_0 I_1(K_0 r) \sin(\phi) \quad (3.15)$$

$$B_\theta = -\frac{\gamma_s \beta_s}{c} E_0 I_1(K_0 r) \sin(\phi) \quad (3.16)$$

where we have written $K_0 = \sqrt{|K_0^2|}$, and

$$\phi = \omega t - \frac{2\pi}{\lambda} \int \frac{dz}{\beta_s(z)} \quad (3.17)$$

For a structure with phase velocity equal to c, we have $K_0 = 0$ and there is no radial dependence of the fields. These effective field components will be used in later chapters when we discuss linac beam dynamics.

3.3
General Description of Periodic Structures

In Chapter 1 we have seen that a single cavity such as the pillbox cavity has an infinite number of resonant modes, which we will call *cavity modes*. For the pillbox cavity, these modes were labeled as *transverse-electric* (TE$_{mnp}$)

and *transverse-magnetic* (TM$_{mnp}$) cavity modes. Typical accelerator cavities are constructed as periodic or almost periodic arrays of coupled cavities, and we must understand the characteristics of coupled-cavity periodic arrays. It is well known that for any system of coupled oscillators, there exists a family of so-called normal modes, each mode behaving like an independent harmonic oscillator with its own characteristic resonant frequency. In general, when any normal mode is excited by a suitable driving force at the right frequency, each of the individual oscillators participates in the motion, and each oscillates at the same frequency with a characteristic phase difference from one oscillator to the next. For an array of coupled cavities, each of the individual cavity modes generates its own family of normal modes. Each such family lies within a definite frequency band called a *passband*, which is centered near the resonant frequency of the uncoupled cavity mode. Each passband includes all the normal modes associated with a single cavity mode, such as the familiar TM$_{010}$ mode of the pillbox cavity. One can describe each normal-mode solution in terms of a characteristic wave that can propagate through the cavity array with a characteristic frequency, and a characteristic wavelength or wavenumber. For a periodic array, the wave solutions are expressed graphically in the form of a so-called Brillouin diagram or dispersion relation, which is a plot of resonant frequency versus wavenumber. Each normal mode is represented by a single point on this plot. For an infinite periodic array, there are an infinite number of modes, and the passband is a continuous curve. For a finite length array there are a finite number of modes, equal to the number of coupled oscillators.

The dispersion relation for an infinite periodic structure typically shows passbands, frequency bands within which waves associated with a given family of normal modes can propagate with little attenuation, and stop bands, frequency bands where waves do not propagate. Figure 3.3 shows an example of the lowest passband corresponding to the TM$_{010}$ resonant mode of the individual cavities. At any frequency within a passband, there are an infinite number of waves, each corresponding to a different space harmonic, as described in Section 3.2. The $n = 0$ waves propagating in the $+z$ direction correspond to the range from $0 < k_z < \pi/d$, where d is the spatial period, the $n = 1$ waves correspond to the range from $2\pi/d < k_z < 3\pi/d$, and so on. For a periodic structure, the dispersion curves express the frequency as a periodic function of wavenumber k_z. Each cycle or zone in k_z space represents the behavior of ω versus k_z for the nth space harmonic. Because of the symmetry of the curve, all the information is presented if the plot is simply restricted to the range $0 \leq k_z \leq \pi/d$. At any given frequency, each space-harmonic component of a normal mode has a unique phase velocity, corresponding to the slope of the line from the origin to that point on the dispersion curve. All space-harmonic components have the same group velocity, corresponding to the same tangent on the dispersion curve.

3.4 Equivalent Circuit Model for Periodic Structures

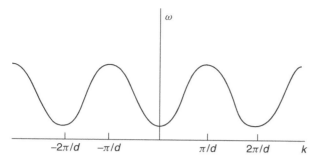

Figure 3.3 Dispersion curve of the lowest passband of an infinite periodic structure.

3.4
Equivalent Circuit Model for Periodic Structures

More insight into the characteristics of periodic structures is obtained by considering some examples of periodic electrical circuits. Consider an infinite chain of identical cells, each with a series impedance Z and a shunt admittance Y as shown in Fig. 3.4.

Given a voltage V_n and current I_n for the nth cell, as shown in the figure, we can use Ohm's law to write

$$I_{n-1} - I_n = YV_n$$
$$I_n - I_{n+1} = YV_{n+1}$$
$$V_n - V_{n+1} = ZI_n \tag{3.18}$$

Substituting the first two equations into the third to eliminate the voltage, we find

$$\frac{I_{n-1}}{Y} - I_n\left(Z + \frac{2}{Y}\right) + \frac{I_{n+1}}{Y} = 0 \tag{3.19}$$

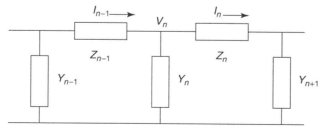

Figure 3.4 Periodic chain of identical electrical cells.

Suppose the period of the structure is ℓ, and consider a propagating wave with wavenumber $k_z = 2\pi/\lambda_g$. Then, the phase advance of the wave per period ℓ is $\phi = k_z \ell$.

Applying the Floquet theorem, we have for a wave traveling in the $+z$ direction $I_n = I_{n-1} e^{-j\phi}$, and $I_{n+1} = I_n e^{-j\phi}$. Substituting this result into Eq. (3.19), we have

$$I_n \left[\frac{e^{j\phi}}{Y} - \left(Z + \frac{2}{Y} \right) + \frac{e^{-j\phi}}{Y} \right] = 0 \qquad (3.20)$$

The solution is obtained when the square bracket is zero, or

$$\cos\phi = 1 + \frac{YZ}{2} \qquad (3.21)$$

Particular choices of Z and Y introduce a frequency dependence, and then Eq. (3.21) relates frequency ω to phase advance per period ϕ. Equation (3.21) is an important result, because it yields the dispersion relation, and defines the passband from the requirement that $-1 \leq \cos\phi \leq 1$. The limits of the passband are at $\phi = 0$ and π.

Equation (3.21) can also be expressed in a compact form using 2×2 matrix notation. The effect of the series impedance Z on an input voltage and current (V_i, I_i) can be written as

$$\begin{pmatrix} V_f \\ I_f \end{pmatrix} = \begin{bmatrix} 1 & -Z \\ 0 & 1 \end{bmatrix} \begin{pmatrix} V_i \\ I_i \end{pmatrix} \qquad (3.22)$$

where (V_f, I_f) are the voltage and current at the output end of the impedance. Likewise, the effect of the shunt admittance Y can be written in terms of a 2×2 transfer matrix, as

$$\begin{pmatrix} V_f \\ I_f \end{pmatrix} = \begin{bmatrix} 1 & 0 \\ -Y & 1 \end{bmatrix} \begin{pmatrix} V_i \\ I_i \end{pmatrix} \qquad (3.23)$$

The transformation of the voltage and current through the whole cell can be written in terms of the product transfer matrix M, as

$$\begin{pmatrix} V_{n+1} \\ I_{n+1} \end{pmatrix} = M \begin{pmatrix} V_n \\ I_n \end{pmatrix} \qquad (3.24)$$

where M is the transfer matrix through one period.

$$M = \begin{bmatrix} 1 & 0 \\ -Y & 1 \end{bmatrix} \begin{bmatrix} 1 & -Z \\ 0 & 1 \end{bmatrix} = \begin{bmatrix} 1 & -Z \\ -Y & 1+YZ \end{bmatrix} \qquad (3.25)$$

The trace of the matrix through one period is $Tr(M) = 2 + YZ$. Thus, Eq. (3.21) can be written as

$$\cos\phi = 1 + ZY/2 = Tr(M)/2 \qquad (3.26)$$

from which the dispersion relation is obtained for any particular choice of impedance and admittance. Next we apply the equivalent circuit model to four examples.

3.5
Periodic Array of Low-Pass Filters

As the first example, we study the simple low-pass filter periodic array, consisting of an array of series inductances coupled by shunt capacitances, as shown in Fig. 3.5. We identify $Z = j\omega L_1$, and $Y = j\omega C_2$.

From Eq. (3.26)

$$\cos\phi = 1 + ZY/2 = 1 - \omega^2/2\,\omega_0^2 \tag{3.27}$$

where $\omega_0^2 = 1/L_1 C_2$. The dispersion relation can be written as

$$\omega = \omega_0 \sqrt{2(1 - \cos\phi)} \tag{3.28}$$

and is plotted in Fig. 3.6. The limits of the passband are $\omega = 0$ at $\phi = 0$ (0 mode), and $\omega = 2\omega_0$ at $\phi = \pi$ (π mode). The π mode lies higher in frequency than the 0 mode, and the bandwidth of the passband is $\omega_\pi - \omega_0 \cong 2\,\omega_0$.

As shown in Fig. 3.5, the low-pass filter looks like an ideal transmission line with finite rather than infinitesimal cell lengths. This suggests that the dispersion relation should reduce to that of an ideal transmission line, when the phase advance per period becomes small. When $\phi \ll 1$ we

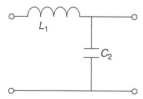

Figure 3.5 Basic cell of a periodic low-pass filter.

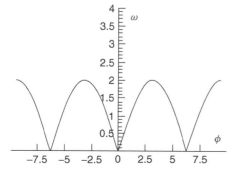

Figure 3.6 Dispersion curve for periodic low-pass filter.

find $\omega \cong \omega_0 \phi = \omega_0 k_z \ell$, or $\omega \cong k_z \ell / \sqrt{L_1 C_2}$. Identifying L_1/ℓ and C_2/ℓ as the inductance and capacitance per unit length, we find that the dispersion relation does reduce to that of the ideal transmission line.

3.6
Periodic Array of Electrically Coupled Circuits

We consider in Fig. 3.7 an example that may serve as an equivalent circuit for a periodic array of electrically coupled cavities, such as the iris-loaded waveguide structure. The figure shows a single period of a periodic circuit in which an inductance and capacitance in series form a resonator that is coupled to the next identical period by a shunt capacitance. We have $Z = j\omega L_1 + 1/j\omega C_1 = j\omega L_1(1 - \omega_0^2/\omega^2)$, where $\omega_0^2 = 1/L_1 C_1$, and $Y = j\omega C_2$. Then,

$$\cos \phi = 1 + ZY/2 = 1 - \frac{C_2}{2C_1}(\omega^2/\omega_0^2 - 1) \tag{3.29}$$

Solving for ω^2, we find the dispersion curve

$$\omega = \omega_0 \sqrt{2(C_1/C_2)(1 - \cos \phi) + 1} \tag{3.30}$$

The dispersion curve of Eq. (3.30) is plotted in Fig. 3.8. The limits of the passband correspond to $\phi = 0$ and π. At $\phi = 0$ (0 mode) $\omega = \omega_0$. At $\phi = \pi$ (π mode), $\omega = \omega_\pi \cong \omega_0 \sqrt{4(C_1/C_2) + 1}$, which for $C_1/C_2 \ll 1$ becomes $\omega_\pi \cong \omega_0(1 + 2C_1/C_2)$. The π mode lies higher in frequency than the 0 mode.

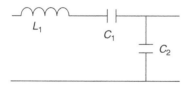

Figure 3.7 Basic cell of a periodic array of electrically coupled TM circuits.

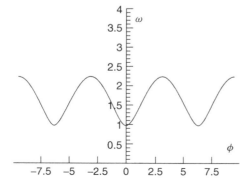

Figure 3.8 Dispersion relation for an electrically coupled periodic TM cell array.

It is sometimes convenient to normalize frequency differences to the frequency of the $\phi = \pi/2$ mode, which is in the center of the passband. We have $\omega_{\pi/2}^2 = \omega_0^2[1 + 2C_1/C_2]$, and can write

$$\omega^2 = \omega_{\pi/2}^2[1 - k\cos\phi] \qquad (3.31)$$

where $k = 2C_1/(C_2 + 2C_1)$ is defined as the intercell coupling constant, a measure of the strength of the coupling. The parameter k is not to be confused with the wavenumber. For $k \ll 1$, the fractional bandwidth of the passband is $(\omega_\pi - \omega_0)/\omega_{\pi/2} = k$. Therefore the coupling constant is also the fractional width of the passband. If the capacitance $C_2 \to \infty$, the intercell coupling vanishes. In this limit the cells become independent, and the bandwidth of the passband is zero.

3.7
Periodic Array of Magnetically Coupled Circuits

We now modify the circuit of Section 3.6 by changing the shunt admittance, which served as the coupling element between adjacent cells, from a capacitance to an inductance as shown in Fig. 3.9. This circuit represents a periodic array of coupled cavities with magnetic rather than electrical coupling, representing a model for a periodic array of magnetically coupled cavities that are coupled through slots cut in the iris walls. We have $Z = j\omega L_1 + 1/j\omega C_1 = j\omega L_1(1 - \omega_0^2/\omega^2)$, where $\omega_0^2 = 1/L_1 C_1$ and $Y = 1/j\omega L_2$. Then,

$$\cos\phi = 1 + ZY/2 = 1 + \frac{L_1}{2L_2}(1 - \omega_0^2/\omega^2) \qquad (3.32)$$

Solving for ω^2, we find the dispersion curve

$$\omega = \frac{\omega_0}{\sqrt{2(L_2/L_1)(1 - \cos\phi) + 1}} \qquad (3.33)$$

The dispersion curve is plotted in Fig. 3.10. At $\phi = 0$, the 0 mode, $\omega = \omega_0$. At $\phi = \pi$, the π mode, $\omega = \omega_\pi \cong \omega_0/\sqrt{4(L_2/L_1) + 1}$, which for $L_2/L_1 \ll 1$ becomes $\omega_\pi \cong \omega_0(1 - 2L_2/L_1)$. Now, the π mode lies lower in frequency than the 0 mode. By comparison in Sections 3.5 and 3.6, electric coupling gave a

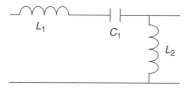

Figure 3.9 Basic cell of a periodic array of magnetically coupled TM circuits.

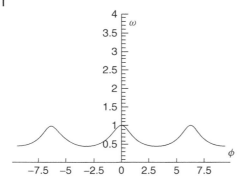

Figure 3.10 Dispersion relation for circuit model of magnetically coupled periodic TM cavity array.

π mode higher in frequency than the 0 mode. Comparing phase velocity, represented by the slope of the line from the origin to any point on the dispersion curve, with the group velocity, represented by the tangent to the curve at that point, we see that these two velocities have opposite sign, which, by definition, is called a *backward-wave mode*. We have $\omega_{\pi/2}^2 = \omega_0^2/[1 + 2(L_2/L_1)]$, and Eq. (2) can be written

$$\omega^2 = \frac{\omega_{\pi/2}^2}{1 - k \cos \phi} \tag{3.34}$$

where $k = 2L_2/(L_1 + 2L_2)$ is defined as the cavity coupling constant. For $k \ll 1$, the fractional bandwidth of the passband is $(\omega_\pi - \omega_0)/\omega_{\pi/2} = -k$. The minus sign indicates only that the π mode lies lower in frequency than the 0 mode. If the inductance $L_2 \to 0$, the intercell coupling vanishes. In this limit the cells become independent and the bandwidth of the passband is zero.

3.8
Periodic Array of Cavities with Resonant Coupling Element

As a final example, we now consider the case of a shunt resonant admittance as the coupling element, as shown in Fig. 3.11.

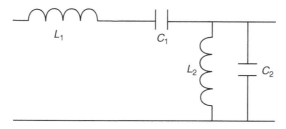

Figure 3.11 Basic cell of a periodic array of circuits with resonant coupling element.

We now have two resonating elements within the same basic cell, and we will see that this leads to an interesting and important effect, when the series and shunt resonant frequencies are equal. We begin by writing $Z = j\omega L_1 + 1/j\omega C_1 = j\omega L_1(1 - \omega_1^2/\omega^2)$, where $\omega_1^2 = 1/L_1 C_1$ and $Y = j\omega C_2 + 1/j\omega L_2 = j\omega C_2(1 - \omega_2^2/\omega^2)$, where $\omega_2^2 = 1/L_2 C_2$. Then,

$$\cos\phi = 1 + ZY/2 = 1 - \frac{\omega^2 L_1 C_2}{2}(1 - \omega_1^2/\omega^2)(1 - \omega_2^2/\omega^2) \qquad (3.35)$$

Solving this as a quadratic equation for ω^2, we find

$$\omega = \sqrt{b \pm \sqrt{b^2 - \omega_1^2 \omega_2^2}} \qquad (3.36)$$

where

$$b = \left(\frac{\omega_1^2 + \omega_2^2}{2}\right)(1 + k^2(1 - \cos(\phi))) \qquad (3.37)$$

and we have defined $k^2 = 2/[L_1 C_2(\omega_1^2 + \omega_2^2)]$

The dispersion relation has two branches, depending on the choice of the sign in Eq. (3.36). Figure 3.12 shows the dispersion relation for several choices of the parameters with $\omega_2 \geq \omega_1$. Because Eq. (3.35) is symmetric with respect to interchange of ω_1 and ω_2, there is again a stopband when $\omega_1 > \omega_2$.

Figure 3.12 shows an important general effect that occurs as the frequencies ω_1 and ω_2 approach each other; one can observe a steepening of the slopes near the 0 mode. When $\omega_1 = \omega_2$, the two curves intersect, an effect known as *confluence*. Each curve continues into the other branch with the same finite slope. As a consequence, the group velocity near the 0 mode increases from zero to a finite value at confluence. This phenomenon has been applied to standing-wave accelerating structures to produce increased mode spacing at confluence, giving significantly greater field stability in a long array of coupled cavities. Resonantly coupled accelerating structures are described in Chapter 4.

3.9
Measurement of Dispersion Curves in Periodic Structures

The dispersion relation of a structure contains the detailed information needed to determine its suitability for particle acceleration, such as phase and group velocities as a function of frequency. Analytic calculations or models, such as the ones given earlier, can be helpful for obtaining physical insight, but are generally not adequate for design purposes. To determine a dispersion relation for a periodic structure, designers rely on either laboratory measurements or numerical calculations using electromagnetic field–solving codes. In either case the usual method for determining the dispersion

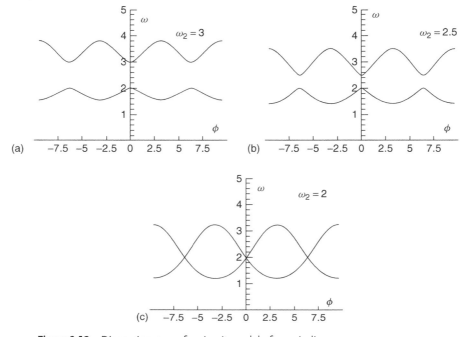

Figure 3.12 Dispersion curve for circuit model of a periodic array of cavities with a resonant coupling element. Defining $\omega_0 = 1/\sqrt{L_1 C_2}$, the curves are plotted for the following parameters $\omega_0 = 1$, $\omega_1 = 2$, and the three values $\omega_2 = 3, 2.5,$ and 2.

curve is to measure the resonant frequencies for different sections of the structure, defined by the placement of two shorting planes, as shown in Fig. 3.13.

Recall that each standing-wave mode is the superposition of two traveling waves of equal amplitude moving in opposite directions as described by the equation

$$\text{Re}\{V_0 e^{j(\omega t - k_z z)} + V_0 e^{j(\omega t + k_z z)}\} = 2V_0 \cos(k_z z)\cos(\omega t) \quad (3.38)$$

where k_z is the wavenumber. In the laboratory, by inserting weakly coupled radio-frequency (RF) drive probes, the resonant standing-wave modes can be excited, and their frequencies can be measured for any position of the shorting planes. To excite the correct cavity mode, the shorting planes are generally placed at a symmetry plane, where the electric field of the mode is normal to the plane. For example, in the electrically coupled iris-loaded structure with periodic iris spacing ℓ, the longitudinal electric field is maximum at the midpoints between the irises. If the shorting planes are located at these midpoints and include just one iris, or one full cell, two normal modes, based on the TM_{010} cavity mode can be excited, the lower frequency mode

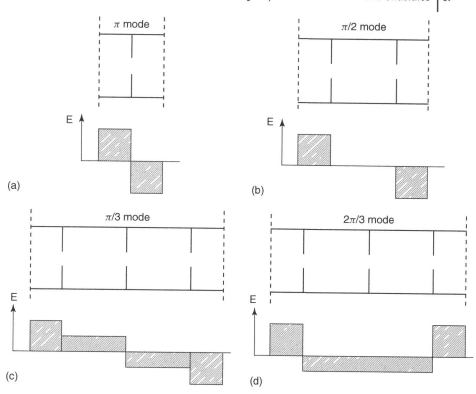

Figure 3.13 Determining the dispersion curve by measurement of the resonant frequencies for different sections of the structure, defined by the placement of two shorting planes. (a) π mode, (b) $\pi/2$ mode, (c) $\pi/3$ mode, and (d) $2\pi/3$ mode.

is the 0 mode, where the fields are in phase on each side of the iris and the higher frequency mode is the π mode, where the fields reverse sign at the location of the iris [3]. From this measurement alone, the bandwidth is determined, which, as was shown in the previous sections, gives a measure of the intercell coupling strength k. Next, if the shorting planes are positioned to include two irises, or two cells, the two previous modes will be excited, and also there will be a new normal mode, which is the $k_z \ell = \pi/2$ mode. Likewise, when three cells are included, one can excite the $k_z \ell = \pi/3$ and $k_z \ell = 2\pi/3$ normal modes. As more cells are included, more normal modes can be excited, and more points on the dispersion curve can be determined. The same procedure can also be followed using the computer, when designing a periodic structure. This procedure allows the effect of design changes in the cell geometry to be evaluated in terms of their effects on the dispersion curve.

3.10
Traveling-Wave Linac Structures

Next we describe the disk-loaded or iris-loaded waveguide structure (Fig. 3.14) used to produce traveling-wave acceleration of relativistic electrons [4].

The basic principles of operation of a traveling-wave accelerator are simple conceptually. The linac consists of a sequence of identical tanks, each consisting of an array of accelerating cavities or cells separated by the irises. The electromagnetic wave is launched at the input cell of each tank; the wave propagates along the beam direction, and beam bunches are injected along the axis for acceleration by the wave. The electromagnetic energy is absorbed by the conductor walls, and by the beam, and the field amplitude attenuates along each tank. At the end of each tank the remaining energy is delivered to an external resistive load. The structural parameters should be generally chosen to achieve the acceleration with high power efficiency. It is desirable to maximize the energy gain of the beam over a given distance and to minimize the power lost to the walls and to the external load. Perhaps the most important design parameter and usually the first parameter chosen is the frequency. Some basic considerations for the choice include: (1) higher power efficiency at higher frequencies, because of the $\omega^{1/2}$ dependence of shunt impedance and (2) tighter beam-positioning tolerances at higher frequencies, because of smaller apertures. Other considerations are often equally important. For applications requiring acceleration of very short intense bunches of electrons, it is desirable to provide large stored energy per unit length, which scales as ω^{-2}, favoring lower frequencies. For linacs used in linear collider applications, where the beam emittance of high-intensity bunches must be controlled, the undesirable effects of wakefields must be considered. These effects scale as ω^2 for the longitudinal case, and as ω^3 for the transverse case. Thus, the choice of frequency is inevitably a compromise, because the frequency seems to have a pervasive impact effecting nearly every linac-design issue. The most common frequency chosen has been 2856 MHz, which is the frequency of the SLAC linac.

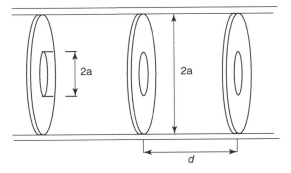

Figure 3.14 Iris(disk)-loaded traveling-wave structure.

Another important parameter to choose is the group velocity v_g that can be controlled by the choice of the aperture radius. The group velocity is important because (1) it affects the electromagnetic fill time $t_F = L/v_g$, where L is the length of the accelerator section and (2) it affects the field amplitude and the power dissipation for any given input RF power. Thus, for any traveling-wave power P_w, one can express the basic relation $P_w = Uv_g$, where U is the electromagnetic stored energy per unit length in the wave. For the same wave power one can consider two extreme cases. By choosing a low group velocity, one has a large stored energy and high fields. This results in a high initial acceleration rate and can also lead to rapid decay of the wave power, because of the large resistive power dissipation in the structure walls. Alternatively, by choosing a high group velocity, the stored energy and the fields may be too low for efficient transfer of power to the beam. We will see that the group velocity must be chosen to balance these considerations, and we will consider two ways of doing this in the following two sections. Other parameters must also be optimized. For example, the disk separation must be optimized for maximum shunt impedance per unit length. If the disks are too far apart, the amplitude of the synchronous harmonic waves is reduced relative to the other harmonics. If they are too close, the power loss on the disks becomes significant. The disk thickness is not a critical parameter, and is chosen large enough for good mechanical strength and good tolerance for electrical breakdown. The radius b of the waveguide and the aperture radius a determine the phase velocity of the synchronous wave. As was mentioned earlier, the radius a also strongly influences the group velocity.

3.11
Analysis of the Periodic Iris-Loaded Waveguide

In this section, we obtain an approximate formula for the dispersion curve for the iris-loaded structure, shown in Fig. 3.15. The results presented in this section [5] are useful for providing scaling formulas and physical insight, rather than to replace the accurate numerical results, which could be obtained from the methods introduced in Section 3.9. We describe the iris-loaded periodic structure as an infinite array of pillbox cavities, each excited in a TM_{010} mode, coupled through small apertures on the axis. The Slater perturbation theorem (discussed later), the Bethe theory [6] of the coupling of cavities through small holes, and the Floquet theorem can be used to obtain an expression for the dispersion curve and other useful results.

The Slater perturbation theorem relates the resonant-frequency change, resulting from a cavity perturbation, to the change in the electric and magnetic stored energies of the cavity. The theorem states that the perturbed resonant frequency ω is given by

$$\frac{\omega - \omega_r}{\omega_r} = \frac{1}{U}(\Delta U_m - \Delta U_e) \tag{3.39}$$

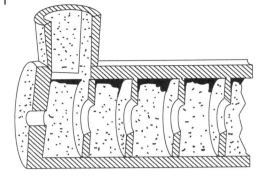

Figure 3.15 The iris-loaded traveling-wave accelerating structure including the input waveguide through which the electromagnetic wave is injected. The beam moves along the central horizontal axis.

where ω_r is the unperturbed frequency, U is the unperturbed electromagnetic stored energy, and ΔU_m and ΔU_e are the time-averaged magnetic and electric stored energies removed as a result of the perturbation. Bethe's theory describes the perturbing effect on the cavity fields of a hole in the cavity wall. It can be shown that, when the hole size is small compared with the wavelength, the effect of a hole is equivalent to adding electric and/or magnetic dipoles located at the center of the hole, whose dipole moments are proportional to the unperturbed electric and/or magnetic fields at the center of the hole. For a small circular aperture, the equivalent electric-dipole moment representing the field inside the cavity caused by the perturbation of the hole is $P = -2a^3 \varepsilon_0 E_0/3$, where a is the aperture radius, and E_0 is the unperturbed electric field at the location of the coupling aperture. The electric-dipole direction is normal to the plane of the hole; the sign of the dipole moment is negative because the induced dipole moment is opposite the direction of the unperturbed field [7]. For a wall of infinitesimal thickness, the field outside the cavity resulting from the hole is also represented by an electric-dipole moment located at the hole, which has the same magnitude but opposite sign, compared with the dipole moment describing the field within the cavity. [8] The finite thickness of the wall can be represented by a factor that represents the decay of the field below cutoff of a waveguide, whose length equals the wall thickness. The basic assumption of the model is that the interaction energy of the dipoles will change the stored energy, and as a consequence of Eq. (3.39), there will be a resonant frequency shift of each normal mode. For an array of identical TM_{010}-mode cavities separated by small circular apertures of radius a, one obtains $\Delta U_m = 0$, and

$$\Delta U_e = -a^3 \varepsilon_0 E_0^2 (1 - e^{-\alpha h} \cos \psi)/6 \qquad (3.40)$$

where

$$\alpha = \sqrt{(2.405/a)^2 - \omega^2/c^2} \approx 2.405/a \qquad (3.41)$$

is the attenuation per unit length of the field for the TM_{01} waveguide mode [9] through a hole in a wall of thickness h, and where ψ is the phase advance per

3.11 Analysis of the Periodic Iris-Loaded Waveguide

cavity of the traveling wave. For a finite array of coupled cavities only discrete values of ψ will be possible. The cavities are indistinguishable and the total frequency shift for each is the same; from Eq. (3.39) it is given by

$$\frac{\omega - \omega_r}{\omega_r} = \frac{-N\Delta U_e}{U} \tag{3.42}$$

where N is the number of apertures per cavity. For two cavities separated by a single wall, $N = 1$. For an array of coupled cavities (for proper termination the cells on the ends should be half cells), each cavity sees two apertures, so this is $N = 2$. Substituting Eq. (3.40) and $N = 2$ into Eq. (3.42), the expression for the frequency change is

$$\frac{\omega - \omega_r}{\omega_r} = \frac{a^3 \varepsilon_0 E_0^2 (1 - e^{-\alpha h} \cos \psi)}{3U} \tag{3.43}$$

Recall that the stored energy for the TM_{010} mode of an unperturbed pillbox cavity is

$$U = \frac{\pi b^2 \ell}{2} \varepsilon_0 E_0^2 J_1^2(2.405) \tag{3.44}$$

where b is the cavity radius, ℓ is the axial length of the cavity, and $J_1(2.405) = 0.5191$. Also for the TM_{010} mode, b is related to the unperturbed resonant frequency by $\omega_r = 2.405c/b$. Substituting Eq. (3.44) into Eq. (3.43), we obtain the basic dispersion relation

$$\omega = \frac{2.405c}{b} \sqrt{1 + \kappa(1 - \cos(\psi)e^{-\alpha h})} \tag{3.45}$$

where

$$\kappa \equiv \frac{4a^3}{3\pi J_1^2(2.405) b^2 \ell} \tag{3.46}$$

The phase advance per cavity can be written as $\psi = k_z \ell$, where $k_z = \omega/v_p$ and v_p is the phase velocity. The phase velocity can be expressed as

$$v_p \equiv \frac{\omega}{k_z} = \frac{2.405 c \ell}{b \psi} \sqrt{1 + \kappa(1 - \cos(\psi)e^{-\alpha h})} \tag{3.47}$$

Generally $\kappa \ll 1$ so that v_p is influenced mostly by ψ/ℓ and the radius b. The group velocity is obtained from Eq. (3.45) by substituting $\psi = k_z \ell$ and differentiating is

$$v_g = \frac{d\omega}{dk_z} = \frac{2(2.405)c}{3\pi J_1^2(2.405)} \left(\frac{a}{b}\right)^2 \sin(\psi) e^{-\alpha h} \tag{3.48}$$

The group velocity is a sensitive function of the radii a and b. If b is used to define the phase velocity according to Eq. (3.47), then a can be used to define the group velocity. In the approximation that $\kappa \ll 1$, we find that the fractional bandwidth of the TM_{010} passband is

$$\frac{\omega(\psi=\pi)-\omega(\psi=0)}{\omega(\psi=\pi/2)} \cong \frac{\kappa e^{-\alpha h}}{1+\kappa/2} \cong \kappa e^{-\alpha h} = k \qquad (3.49)$$

where we define $k = \kappa e^{-\alpha h}$. The quantity k should not be confused with the wavenumber k_z. On the basis of the result discussed earlier in this chapter that the fractional width of the passband is equal to the intercell coupling constant, we identify k as the intercell coupling constant and express it as

$$k \equiv \frac{4a^3 e^{-\alpha h}}{3\pi J_1^2(2.405) b^2 \ell} \qquad (3.50)$$

If we specify the operating mode parameters as ω_0, ψ_0, k_{z0}, and v_0 is the velocity of the synchronous particle, the required cavity length is

$$\ell = \frac{\psi_0 v_0}{\omega_0} = \frac{\psi_0 \lambda_0}{2\pi} \frac{v_0}{c} \qquad (3.51)$$

Typical parameter values are $\omega_0/2\pi = 3$ GHz, $v_0 = c$, $\psi_0 = 2\pi/3$, $b/\lambda_0 = 0.4$, $a/\lambda_0 = 0.1$, $h/\lambda_0 = 0.05$, which leads to $\ell/\lambda_0 = 1/3$, a coupling constant $k = 0.009$ and a group velocity $v_g/c = 0.008$.

3.12
Constant-Impedance Traveling-Wave Structure

It is convenient to introduce some important relationships between the longitudinal accelerating field amplitude E_a for the traveling wave, the stored energy per unit length U, and the traveling wave power P_w. The traveling-wave power is obtained by integrating the Poynting vector over the aperture, and is $P_w = \int_0^a E_r H_\theta 2\pi r dr$, where a is the radial aperture. The resistive power dissipation per unit length in the walls of the structure is $-dP_w/dz$. The quality factor is $Q = \omega U/(-dP_w/dz)$, and the shunt impedance per unit length is $r_L = E_a^2/(-dP_w/dz)$. The group velocity, v_g, is also the energy-flow velocity, and relates the traveling-wave power to the stored energy per unit length according to $P_w = v_g U$. Eliminating U from this expression and from the expression for Q yields a differential equation for traveling-wave power

$$\frac{dP_w}{dz} = -\frac{\omega P_w}{Q v_g} \qquad (3.52)$$

We define the field attenuation per unit length as $\alpha_0 = \omega/2Qv_g$, so Eq. (3.52) becomes

$$\frac{dP_w}{dz} = -2\alpha_0 P_w \qquad (3.53)$$

3.12 Constant-Impedance Traveling-Wave Structure

Now we consider the simplest case of an iris-loaded traveling-wave structure with uniform cell geometry independent of z, and identical parameters for each cell including Q, v_g, r_L, and α_0. This is called a *constant-impedance* structure in the literature. The attenuation per unit length is constant, so the solution to Eq. (3.53) is

$$P_w(z) = P_0 e^{-2\alpha_0 z} \tag{3.54}$$

which shows that the wave power is exponentially damped. We can obtain a similar expression for the accelerating field amplitude. Thus, from the basic definitions given above, it is straightforward to show that the accelerating field and the traveling-wave power are related by $E_a^2 = \omega r_L P_w / Q v_g$ and from Eq. (3.53)

$$\frac{dE_a}{dz} = -\frac{\omega E_a}{2Q v_g} = -\alpha_0 E_a \tag{3.55}$$

The solution is $E_a(z) = E_0 e^{-\alpha_0 z}$. At the end of a tank of length L, we have $P_w(L) = P_0 e^{-2\tau_0}$, and $E_a(L) = E_0 e^{-\tau_0}$, where

$$\tau_0 = \alpha_0 L = \frac{\omega L}{2Q v_g} \tag{3.56}$$

is the total power attenuation parameter for that tank. The energy gain of a synchronous particle riding at a phase ϕ relative to the crest of the wave is

$$\Delta W = q \cos\phi \int_0^L E_a(z)\,dz = q E_0 L \frac{(1 - e^{-\tau_0})}{\tau_0} \cos\phi \tag{3.57}$$

Using the relation between the input power P_0 and input field E_0 evaluated at $z = 0$, we have $E_0^2 = 2 r_L \alpha_0 P_0$, which leads to

$$\Delta W = q\sqrt{2 r_L P_0 L} \frac{(1 - e^{-\tau_0})}{\sqrt{\tau_0}} \cos\phi \tag{3.58}$$

If the input power and shunt impedance are fixed, the energy gain over a tank of length L depends on the total attenuation parameter τ_0. The total attenuation parameter can be controlled by choosing the group velocity, which is a strong function of the aperture radius a. If the value of τ_0 is to be chosen to maximize the energy gain ΔW in the length L, we find that the maximum occurs when $\tau_0 = (e^{\tau_0} - 1)/2$, which has for its solution $\tau_0 \cong 1.26$, and Eq. (3.58) yields the maximum energy gain per tank

$$\Delta W_{\max} = 0.903 q \sqrt{r_L P_0 L} \cos\phi \tag{3.59}$$

Also, from the definition of τ_0, if L is fixed and Q is known, the value of τ_0 determines the optimum group velocity. If v_g is too small, the attenuation of the wave is too great, and the field in the latter part of the structure is very small. In this case, most of the power that is traveling relatively slowly through the structure is dissipated in the walls of the structure, resulting in poor transfer of power to the beam. If v_g is too large, the initial accelerating field for a given input power is too small, and therefore the accelerating field throughout the structure is also too small for efficient energy transfer to the beam. In this case, most of the beam power passes rapidly through the structure and is delivered to the external resistive load.

The value of τ_0 also affects the filling time of the waveguide, which is calculated as the time for the energy to propagate at the group velocity from the input to the output end of the guide. The fill time is

$$t_F = \frac{L}{v_g} = \tau_0 \frac{2Q}{\omega} \tag{3.60}$$

which is $t_F = 2.52 Q/\omega$ for the optimum value of τ_0. One may need to choose τ_0 to be less than the optimum value, if the filling time needs to be reduced.

3.13
Constant-Gradient Structure

In the constant-impedance structure with uniform cell geometry and uniform parameters, we found that the RF power and the electric field decay exponentially as the wave propagates away from the input. This raises the question of whether we could do better by varying the transverse geometry to keep the accelerating field constant along the structure. This design approach is called the *constant-gradient* structure in the literature, where the term *gradient* refers to the voltage gradient, that is, accelerating field. The group velocity and the attenuation per unit length α_0, which depends on the group velocity, are very sensitive to the aperture radius, and their variation will be included explicitly. In the approximate treatment that follows, the quantities Q and r_L, which are not as sensitive to the transverse geometry, will be assumed constant. Beginning with

$$\frac{dP_w}{dz} = -2\alpha_0(z) P_w \tag{3.61}$$

we can write

$$\int_{P_0}^{P_L} \frac{dP_w}{P_w} = -2 \int_0^L \alpha(z) dz \tag{3.62}$$

where P_L is the traveling-wave power at the end of the section, which is delivered to a resistive load. Integrating Eq. (3.62), we obtain

$$P_L = P_0 e^{-2\tau_0} \tag{3.63}$$

where P_0 is the input power, and the total power attenuation is $\tau_0 \equiv \int_0^L \alpha_0(z) dz$. From the definition of r_L, which is assumed to be essentially constant, we

3.13 Constant-Gradient Structure

conclude that for E_a to be constant, so must be dP_w/dz. Then $P_w(z)$ must be linear in z, or

$$P_w(z) = P_0 + \frac{P_L - P_0}{L} z \qquad (3.64)$$

Substituting Eq. (3.63), we find

$$P_w(z) = P_0 \left[1 - \frac{z}{L}(1 - e^{-2\tau_0}) \right] \qquad (3.65)$$

Therefore, for a constant field, the traveling-wave power decreases linearly with z, rather than exponentially.

Next we want an expression for $\alpha_0(z)$. We first differentiate Eq. (3.65)

$$\frac{dP_w}{dz} = -\frac{P_0}{L}(1 - e^{-2\tau_0}) \qquad (3.66)$$

Using Eqs. (3.61) and (3.66), and eliminating dP_w/dz, we find

$$\alpha_0(z) = \frac{1}{2L} \frac{(1 - e^{-2\tau_0})}{[1 - (z/L)(1 - e^{-2\tau_0})]} \qquad (3.67)$$

The group velocity is

$$v_g(z) = \frac{\omega}{2Q\alpha_0(z)} = \frac{\omega L}{Q} \frac{[1 - (z/L)(1 - e^{-2\tau_0})]}{(1 - e^{-2\tau_0})} \qquad (3.68)$$

The group velocity required for a constant accelerating field also decreases linearly with z. From Eqs. (3.65) and (3.68), the stored energy per unit length, like the accelerating field, is constant, given by $U = P_0 Q (1 - e^{-2\tau_0})/\omega L$. The energy gain in a section of length L for a particle riding at a phase ϕ relative to the crest of the wave is

$$\Delta W = q \cos\phi \int_0^L E_a(z)\, dz = q E_0 L \cos\phi \qquad (3.69)$$

The relation between the constant accelerating field amplitude and the input power is

$$E_0^2 = -r_L \frac{dP_L}{dz} = \frac{r_L P_0}{L}(1 - e^{-2\tau_0}) \qquad (3.70)$$

Using Eq. (3.70) to eliminate E_0 from Eq. (3.69) gives

$$\Delta W = q\sqrt{r_L P_0 L (1 - e^{-2\tau_0})} \cos\phi \qquad (3.71)$$

Comparing Eq. (3.71) with Eq. (3.58), one finds that for a given τ_0, the energy gain of the constant-gradient structure is larger than for the constant-impedance structure, although for $\tau_0 < 1$ the differences are not large. In the constant-gradient case the optimum τ_0 is infinite, corresponding to transferring all the energy in the structure. In practice, we also need to consider the effect of τ_0 on the filling time of the waveguide. The filling time is

$$t_F = \int_0^L \frac{dz}{v_g(z)} = \frac{Q}{\omega L}(1 - e^{-2\tau_0}) \int_0^L \frac{dz}{1 - (z/L)(1 - e^{-2\tau_0})} \qquad (3.72)$$

Carrying out the integration gives

$$t_F = \tau_0 \frac{2Q}{\omega} \qquad (3.73)$$

which is equal to the result for the constant-impedance structure, given in Eq. (3.60).

The constant-gradient design was chosen over the constant-impedance design at SLAC for reasons which include the uniformity of the power dissipation, and the lower value for the peak surface electric field for the same energy gain. To achieve the constant-gradient design, the structure radius was tapered from about $b = 4.2$ to 4.1 cm, the iris radius was tapered from about 1.3 to 1.0 cm, and $\tau_0 = 0.57$ was chosen as a compromise between maximizing the energy gain and minimizing the fill time [10].

3.14
Characteristics of Normal Modes for Particle Acceleration

In this section, we explore the question of which normal modes are most suitable for efficient particle acceleration [11]. For simplicity, suppose that an accelerating structure consists of an array of $N+1$ identical coupled cavities, each with a short accelerating gap, so that we can approximate the transit-time factor $T \cong 1$. We assume that the field is zero outside the gaps, and that the gaps are spaced at a distance ℓ apart. As will be discussed in Chapter 4, the standing-wave field in each cavity of a periodic structure with $N+1$ cells is described by

$$E_n = E_0 \cos\left(\frac{n\pi q}{N}\right) \cos \omega t \qquad (3.74)$$

where $n = 0, 1, 2, \ldots, N$, is the cavity number and $q = 0, 1, \ldots, N$ is the normal-mode number. The usual nomenclature for each mode is to identify it by the quantity $\pi q/N$. Thus, $q = 0$ corresponds to the 0 mode, and $q = N$ corresponds to the π mode. A standing wave can be expressed as a sum of two traveling waves moving in opposite directions; the forward wave is proportional to $e^{j(\omega t - n\pi q/N)}$, and for the backward wave we have $e^{j(\omega t + n\pi q/N)}$. The wavenumbers of these two waves are identified as $k_z = \pm \pi q/N\ell$, and the corresponding phase velocities are given by $v_p = \omega/k_z = \pm \omega N\ell/\pi q$. If v_s is the synchronous-particle velocity, synchronism with the forward wave requires

$$v_s = \frac{\omega N \ell}{\pi q} \qquad (3.75)$$

3.14 Characteristics of Normal Modes for Particle Acceleration

Assuming that v_s is constant and that the synchronous particle starts at the center of gap $n = 0$ at time $t = 0$, the time for the particle to travel to gap n is given by $t = n\pi q/\omega N$. Then, from Eq. (3.74) the field in the nth gap seen by a synchronous particle is

$$E_n = E_0 \cos^2(n\psi) \tag{3.76}$$

where $\psi = \pi q/N$ is the phase advance of the wave per cavity.

It is instructive to look at this result from the equivalent point of view of two traveling waves. Expressed as the sum of traveling waves, we have

$$E_n = E_0 \cos(n\psi)e^{j\omega t} = E_0 \left(\frac{e^{jn\psi} + e^{-jn\psi}}{2} \right) e^{j\omega t} \tag{3.77}$$

Taking the real part and introducing the synchronism condition, again we find the field in the nth gap seen by the synchronous particle, where $\omega t = n\psi$,

$$E_n = \frac{E_0}{2}[\cos(\omega t - n\psi) + \cos(\omega t + n\psi)] = \frac{E_0}{2}[1 + \cos(2n\psi)] \tag{3.78}$$

This result is identical to the standing-wave result of Eq. (3.76) as can be seen by applying a simple trigonometric identity for the half angle. The first term in the last bracket of Eq. (3.78) corresponds to the contribution from the forward wave, and the cosine term is the contribution from the backward wave.

Next, the total voltage gain of the synchronous particle, assumed for simplicity to arrive at each gap at the crest of the wave, is

$$\Delta V = \sum_{n=0}^{N} E_n \ell = \sum_{n=0}^{N} \frac{E_0 \ell}{2}[1 + \cos(2n\psi)] = \begin{cases} \dfrac{(N+2)}{2} E_0 \ell, & 0 < q < N \\ (N+1)E_0 \ell, & q = 0, N \end{cases} \tag{3.79}$$

Note that if the backward wave contributed zero energy gain, the cosine term in Eq. (3.79) would not be present and we would have

$$\Delta V = (N+1)E_0 \ell/2 \tag{3.80}$$

Comparison of Eqs. (3.79) and (3.80) shows that the backward wave does contribute to the voltage gain. For the case $0 < q < N$, the fractional contribution of the backward wave is small, and vanishes in the limit $N \to \infty$, since in that limit the result of Eq. (3.79) equals that of Eq. (3.80). For the case $q = 0$ (the 0 mode), or $q = N$ (the π mode), the energy gain is twice that of the forward wave alone, which means that the backward wave contributes the same amount to particle acceleration as the forward wave. This result is less surprising when one realizes that for the 0 mode and π modes, both the backward and forward waves move (in opposite directions) from one gap to the next at the same time as the particle moves that distance. Then, both the backward and the forward waves can be said to be synchronous with the beam, arriving at the gaps at the right time for acceleration. That the backward wave can deliver a net energy gain to the forward-moving beam is possible, because the beam only sees the field from the backward wave when it is in the gap. From the point of space harmonics, one finds that both the forward and backward waves contribute space harmonics of equal strength that are synchronous with the beam [12].

Standing-wave operation in the 0 and π modes and traveling-wave operation in any mode are equally efficient, as can be confirmed by calculation of the shunt impedances. To show this, we rewrite the time dependent voltage for the nth cavity as $V_n = V_{n0}\cos(\omega t)$, where the nth amplitude is $V_{n0} = V_0\cos(n\psi)$, and $V_0 = E_0\ell$. The effective shunt impedance per cavity is $r = V_{n0}^2/P_n$, where P_n is the power dissipated in the nth cavity. We assume that all the cavities are identical and have the same shunt impedance. The total power dissipation for the entire array of $N+1$ cavities is

$$P = \sum_{n=0}^{N} P_n = \sum_{n=0}^{N} \frac{V_{n0}^2}{r} = \sum_{n=0}^{N} \frac{V_0^2 \cos^2(n\psi)}{r} = \frac{V_0^2}{r} \begin{cases} \frac{N+2}{2}, & 0 < q < N \\ N+1, & q = 0, N \end{cases} \tag{3.81}$$

The effective shunt impedance for the array of $N+1$ cavities is

$$ZT^2 = \frac{(\Delta V)^2}{P} = \begin{cases} \left(\frac{N+2}{2}\right)r, & 0 < q < N \\ (N+1)r, & q = 0, N \end{cases} \tag{3.82}$$

Finally, we calculate the effective shunt impedance per unit length, by dividing Eq. (3.82) by the total length $(N+1)\ell$, and we find

$$\frac{ZT^2}{(N+1)\ell} = \begin{cases} \left(\frac{N+2}{N+1}\right)\frac{r}{2\ell}, & 0 < q < N \\ \frac{r}{\ell}, & q = 0, N \end{cases} \tag{3.83}$$

Equation (3.83) shows the efficiency advantage, as measured by the shunt impedance per unit length, of the 0 and π modes over all other modes for standing-wave operation. As $N \to \infty$, the efficiency of the 0 and π modes are a factor of 2 better than for any other mode. The reason for this result is easiest to see, when comparing the 0 and π modes with the $\pi/2$ standing-wave mode. In the later case, half the cavities contain no field, and for the same total voltage gain, the excited cavities must have twice the voltage needed for 0 and π mode operation, where all cavities have equal field. As a consequence, the total power dissipation is twice as much in the $\pi/2$ mode. (In Chapter 4 we will learn that a $\pi/2$-like mode of a biperiodic structure can be made nearly as efficient as the 0 and π modes of a periodic structure.) For modes other than the 0, $\pi/2$, and π modes, the efficiency is low because (1) the field amplitudes at the gaps are not equal and (2) the particles do not arrive at the right time to see the peak field.

3.15 Physics Regimes of Traveling-Wave and Standing-Wave S

It is instructive to calculate the shunt impedance per unit length for traveling-wave operation. Since the single traveling wave has voltage V_0 total voltage gain for the synchronous particle riding the crest of the wave is $\Delta V = (N+1)V_0/2$. The power dissipated per cavity is $(V_0/2)^2/r$, and the total power dissipated in the $N+1$ cavity array is $P = (N+1)(V_0/2)^2/r$. The effective shunt impedance per unit length for traveling-wave operation in all modes is

$$\frac{ZT^2}{(N+1)\ell} = \frac{(\Delta V)^2}{P(N+1)\ell} = \frac{r}{\ell} \qquad (3.84)$$

that agrees with the 0 and π-mode standing-wave result in Eq. (3.83).

To obtain a high group velocity the operating mode for traveling-wave operation is generally chosen somewhere in the middle of the passband, where the slope of the dispersion curve is large; for example, the SLAC linac operates in the $\psi = 2\pi/3$ mode. All standing-wave accelerators are operated so that the structures that accelerate beam are excited in a 0 mode or a π mode. A difficulty with the 0 and π modes in a strictly periodic structure is that the group velocity is zero, which is unattractive for traveling-wave structures, and the mode spacing (related to the zero group velocity) is small, which is unattractive for standing-wave structures. Because of error-induced mode mixing, the standing-wave field distribution becomes sensitive to fabrication errors. This problem, however, has been solved by the use of biperiodic structures that operate in a $\pi/2$-like mode, as will be discussed further in Chapter 4.

3.15
Physics Regimes of Traveling-Wave and Standing-Wave Structures

Given that traveling and standing-wave structures may both be described as coupled-cavity arrays, it is natural to ask whether the traveling and standing-wave structure parameters should be chosen differently if one seeks to achieve optimal performance. To answer this question, first consider a structure designed for traveling-wave operation. In Section 3.12 we defined the attenuation per unit length as $\alpha_0 = \omega/2Qv_g$. If α_0 is constant from cell to cell along the structure, the total attenuation of the field in a single transit of the wave through a structure of length L is given by $E = E_0 e^{-\alpha_0 L}$, and a total attenuation constant is defined by the quantity $\tau_0 = \alpha_0 L$. We found that the energy gain of a synchronous particle riding the crest of a traveling wave in a structure of length L is maximum when the structure is designed so that $\tau_0 = 1.26$. Thus, efficient energy transfer from the traveling wave to the beam implies that $\tau_0 = \alpha_0 L \approx 1$.

By contrast, one expects that for acceleration by a standing wave the minimum resistive power dissipation in the walls occurs when the field is distributed uniformly throughout the structure, or $\tau_0 = \alpha_0 L \ll 1$. Therefore, for efficient operation the traveling-wave and standing-wave structures can be expected to have very different values of the total attenuation constant τ_0; the traveling-wave structure operates with large attenuation, and the standing-wave structure has small attenuation.

The total attenuation constant may also be written as

$$\tau_0 = \alpha_0 L = \frac{\omega L}{2 Q v_g} = \frac{t_F}{\tau} \tag{3.85}$$

where $t_F = L/v_g$ is the electromagnetic filling time of the structure in a single transit of the wave, and $\tau = 2Q/\omega$ is the cavity time constant for build up of the fields from multiple reflections. From the arguments given above, we expect that the most efficient traveling-wave operation occurs in a regime where $\tau < t_F$, and efficient standing-wave structures satisfy $\tau \gg t_F$. This provides us with the following approximate picture. In traveling-wave operation the individual cavities tend to fill roughly sequentially with electromagnetic energy. In standing-wave operation the field fills the whole structure before the individual cavity fields build up very much; then, the fields build up from multiple reflections in all the cavities, almost simultaneously. In other words, in traveling-wave operation the fields appear to build up in space from cavity to cavity along the structure, whereas in standing-wave operation the fields appear to build up in time, almost simultaneously in all cavities.

Problems

3.1. Consider an iris-loaded structure, designed as a traveling-wave accelerator for relativistic electrons. For the purpose of calculating the structural properties, we consider the structure as an array of coupled, pillbox cavities each excited in a TM_{010} cavity mode. Suppose the parameters are near to those of the SLAC linac structures: frequency $\omega_0/2\pi = c/\lambda_0 = 2856$ MHz at the $2\pi/3$ structure mode (phase advance per cavity $\phi_0 = 2\pi/3$), cavity radius $R/\lambda_0 = 0.39$, iris thickness $h/\lambda_0 = 0.055$, and iris radial aperture $a/\lambda_0 = 0.1024$. Use the coupled, pillbox-cavity formulas to carry out the following steps: (a) calculate the coupling constant k, (b) calculate the frequencies of the $\phi = 0, \pi/3, \pi/2, 2\pi/3$, and π modes, and plot a dispersion relation ($\omega/2\pi$ versus ϕ), (c) calculate and plot the phase velocity v_p/c versus ϕ for the same modes as in part (b), and (d) calculate and plot the group velocity v_g/c versus ϕ for the same modes as in part (b).

3.2. Suppose you want to increase the group velocity of the structure in Problem 3.1 by increasing the radial aperture a, while adjusting the cavity radius R to maintain the same value of the frequency for the $2\pi/3$ structure mode. As an example, consider the same linac as in problem 3.1, except that we increase a/λ_0 to 0.2, and increase R/λ_0 to 0.430 to maintain the same frequency. Use the coupled, pillbox-cavity formulas to calculate: (a) the new coupling constant k, (b) the phase velocity of the $2\pi/3$ mode, and (c) the group velocity of the $2\pi/3$ mode. Did the group velocity increase? Why does keeping the frequency the same for the operating mode keep the phase velocity the same? (Hint: Note that the slope of the line from the origin to the operating point on the dispersion curve is the same).

3.3. Two superconducting elliptical cavities, each operating in a TM_{010}-like cavity mode, are coupled through an axial hole in the wall separating the cavities. Assume that these cavities can be approximated by two coupled pillbox cavities with pillbox-cavity radius $R = 2.405\lambda/2\pi$, aperture radius a, and length ℓ. Use the coupled pillbox-cavity result from Section 3.11 to calculate the coupling constant k, ignoring the thickness of the wall at the radius of the aperture, assuming $a/R = 1/3$, and assuming the following conditions: (a) the cavity lengths are $\ell = \lambda/4$ and (b) the cavity lengths are $\ell = \lambda/2$. Does the coupling constant depend on the length of the cavity?

3.4. Suppose that the two superconducting elliptical cavities from problem 3.3 are connected through an axial beam pipe of length h, and assume that the beam pipe radius a/R is the same as in problem 3.3. Use the coupled pillbox-cavity formula to calculate the coupling constant k assuming $h = R$. Why does separating the cavities by a long beam pipe reduce the coupling constant?

3.5. Assume the following parameters for SLAC, which is a constant-gradient traveling-wave linac: $f = 2856$ MHz, $Q = 13{,}000$, $r_L = 57$ MΩ/m, $\tau_0 = 0.57$, and 932 waveguide sections of length $= 3.05$ m each. (a) Show that the energy gain per section of an electron at the crest of the wave is $\Delta W(MeV) = 10.9\sqrt{P_0(MW)}$, P_0 is the peak input RF power per waveguide section. (b) Calculate the fill time of the waveguide and compare this with a pulse length of 1.6 μs. (c) Calculate the group velocity relative to the speed of light at both the input and output ends of a section. (d) If $P_0 = 16$ MW, what is the peak power at the end of a section. (This will be delivered to an external load.) (e) If $P_0 = 16$ MW, calculate the energy gain per section using the formula of part (a), and calculate the total energy gain in the SLAC linac. What is the accelerating field E_0?

References

1. Slater, J.D., *Microwave Electronics*, Third Printing, D. Van Nostrand, New York, 1954, pp. 17–177.
2. Mathews, Jon. and Walker, R.L., *Mathematical Methods of Physics*, 2nd ed., W. A. Benjamin, Menlo Park, Calif., 1970, pp. 198–199.
3. For magnetic coupling, the π mode would have a lower frequency than the 0 mode.
4. For a comprehensive article on this subject see Loew, G.A. and Neal, R.B., Accelerating structures, in *Linear Accelerators*, ed. Lapostolle, P.M. and Septier,

A.L., North-Holland Publishing Company, Amsterdam, and John Wiley & Sons, New York, 1970 pp. 39–113.

5 This treatment is based on the work of Gao, J. Analytic formulas for the resonant frequency changes due to opening apertures on cavity walls, *Nucl. Instrum. Methods* **A311**, 437–443 1992; and Gao, J., Analytical approach and scaling laws in the design of disk-loaded traveling wave accelerating structures, *Part. Accel.* **43**, 235–257 (1994).

6 Bethe H.A. *Phys. Rev.* **66**, 163 (1944).

7 An effective magnetic dipole moment is also induced, given by $M = 4a^3 \mu_0 H_0/3$, where H_0 is the unperturbed magnetic field at the location of the hole. The magnetic dipole moment for an elliptical slot is discussed in Section 4.12, where the coupling constant for magnetic coupling is derived.

8 Jackson, J.D. *Classical Electrodynamics*, 2nd ed., John Wiley & Sons, 1975, p. 410.

9 The TM_{01} mode is used because it has the lowest cutoff frequency, and therefore the lowest attenuation per meter of any TM waveguide mode. Only TM modes can be excited in the pipe on the axis by the TM_{010} cavity mode.

10 Loew, G. and Talman, R., Elementary principles of linear accelerators, *AIP Conf. Proc.* **105**, 1–91 (1983).

11 Miller, R.H., Proc. of the 1986 Linear Accel. Conf., SLAC, Stanford, CA 1986, p. 200.

12 Loew, G.A. and Neal, R.B., Accelerating Structures, in *Linear Accelerators*, ed. Lapostolle, P.M. and Septier, A.L., North-Holland Publishing Company, Amsterdam, and John Wiley&Sons, New York, 1970, p. 68.

4
Standard Linac Structures

In Chapter 3 we identified the periodic structure as a practical device for radio frequency (RF) acceleration. The periodic disk-loaded waveguide was discussed as an accelerating structure for relativistic particles. This was our first example, and in this chapter we continue what might be called the *classical* or *standard* approaches to linac accelerating structures. In practice, except for the iris-loaded structure, all of the structures have been operated as standing-wave devices. After beginning with the simple concept of independent-cavity linacs, we describe the more common multicell linac structures including the Wideröe linac, the Alvarez drift-tube linac (DTL), and the coupled-cavity linacs (CCLs). To understand the structure physics of CCLs, we use the coupled-circuit model and first apply it to the case of a three-cell coupled-cavity system. This allows us to illustrate the remarkable properties of the $\pi/2$ normal mode of a periodic coupled-cavity array, which leads us to the coupling approach known as *resonant coupling*, illustrated in Chapter 3. We see how the field-stability advantages of the $\pi/2$ mode can be combined with the shunt-impedance advantage of the π mode, through introduction of a biperiodic coupled-cavity structure such as the side-coupled linac.

4.1
Independent-Cavity Linacs

Linacs have been built using arrays of identical independent cavities, each containing only one or two accelerating gaps. Such linacs have been used mainly for acceleration of heavy ions for nuclear physics research, but may also become important as the most attractive design approach for superconducting proton linacs. Well-known examples of heavy-ion linacs include the single-gap cavities in the high-energy section of the UNILAC linac [1] at GSI, Darmstadt, Germany, and the ATLAS [2] superconducting linac at Argonne (see Fig. 4.1).

Each cavity is driven by a separate RF generator and the cavity phases and amplitudes can be set independently. The transit-time factor of a cavity with just a few cells has a broad velocity acceptance, and if each cavity is excited

RF Linear Accelerators. 2nd, completely revised and enlarged edition.
Thomas P. Wangler
Copyright © 2008 Wiley-VCH Verlag GmbH & Co. KGaA, Weinheim
ISBN: 978-3-527-40680-7

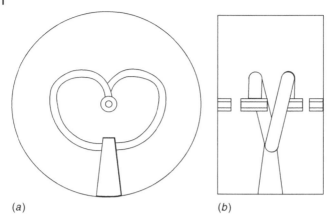

Figure 4.1 Split-ring resonator used at the ATLAS superconducting linac at Argonne.

by its own RF generator, each cavity phase can be adjusted independently to maximize the acceleration for the injected beam. These properties allow for flexible linac operation, especially important for a linac that accelerates ions with different charge-to-mass ratios. The cavities used for very low velocities in heavy-ion linacs are often transverse electromagnetic (TEM) coaxial-type structures, such as quarter-wave and half-wave resonators, loaded at the end by a drift tube. Examples are shown in Figs. 4.1 and 4.2. Typically, a single resonator element is contained within a cylindrical cavity, and it gives two accelerating gaps with opposite polarity (π mode). The spacing between the gap centers is designed to equal $\beta_s \lambda/2$, where β_s is the velocity of a reference particle that travels between the two gap centers in half an RF period, and λ is the RF wavelength.

For a simple analysis of the acceleration in the two-gap cavities, we choose the origin at the center of the cavity, as shown in Fig. 4.3, and we assume that the particle has a constant velocity through the cavity. It is convenient to express the time dependence of the field using the sine rather than the cosine, as

$$E_z(r, z, t) = E(r, z) \sin(\omega t + \phi) \tag{4.1}$$

For an arbitrary particle of velocity β, we write $\omega t = 2\pi z/\beta\lambda$, where $t = 0$ is the time at which the particle is at the center of the cavity at $z = 0$. This is consistent with the phase convention used before, because the reference particle has its maximum acceleration from both gaps at $\phi = 0$.

The energy gain is

$$\Delta W = q \int_{-L/2}^{L/2} E(0, z) \sin(2\pi z/\beta\lambda + \phi) \, dz \tag{4.2}$$

4.1 Independent-Cavity Linacs

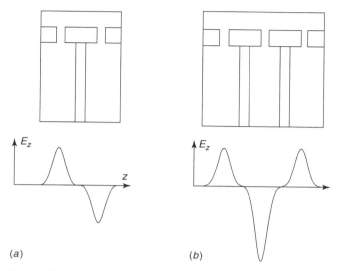

Figure 4.2 Schematic drawing of two- and three-gap cavities.

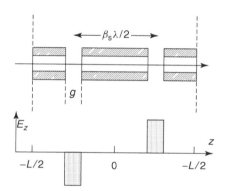

Figure 4.3 Two-gap π-mode cavity with uniform fields in the gap.

If we use a trigonometric identity and assume $E(0, z)$ is an odd function about the origin, then

$$\Delta W = q \cos \phi \int_{-L/2}^{L/2} E(0, z) \sin(2\pi z/\beta\lambda) \, dz \tag{4.3}$$

It is convenient to define a multicell transit-time factor as we did for a single gap in Section 2.2. However, if $E(0, z)$ is an odd function, the denominator of the transit-time factor, as defined in Eq. (2.14), is zero. Suppose that for the multicell case, we define an average field as

$$E_0 = \frac{1}{L} \int_{-L/2}^{L/2} |E(0, z)| \, dz \tag{4.4}$$

Then, the energy gain is

$$\Delta W = qE_0 T \cos\phi L$$

where

$$T = \frac{\int_{-L/2}^{L/2} E(0, z) \sin(2\pi z/\beta\lambda)\, dz}{\int_{-L/2}^{L/2} |E(0, z)|\, dz} \tag{4.5}$$

Assuming that the electric field is uniform over the gap, as shown in Fig. 4.3, we obtain the axial transit-time factor

$$T = \frac{\sin\pi g/\beta\lambda}{\pi g/\beta\lambda} \sin\frac{\pi\beta_s}{2\beta} \tag{4.6}$$

The first factor in T is the usual gap factor, which depends on the gap g and the particle velocity β. The second factor is associated with the degree of synchronism of the particle with respect to the phase of the field in the two gaps. When $\beta = \beta_s$ this factor is unity, indicating perfect synchronism. The synchronism factor strongly affects the range of β over which T is large; it becomes a narrower function of β as the number of cells per cavity increases. In general for a cavity with N identical cells, and an electric field that is uniform over the gap, the axial transit-time factor is

$$T = T_g S(N, \beta_s/\beta) \tag{4.7}$$

where $T_g = \sin(\pi g/\beta\lambda)/(\pi g/\beta\lambda)$ and is the synchronism factor, which can be written as

$$S(N, \beta/\beta_s) = \begin{cases} \dfrac{1}{N}\left[1 + \displaystyle\sum_{m=1}^{(N-1)/2} (-1)^m 2\cos(m\pi\beta_s/\beta)\right], & N \text{ odd} \\ \dfrac{2}{N}\left[\displaystyle\sum_{m=0}^{N/2-1} (-1)^m \sin(\{m+1/2\}\pi\beta_s/\beta)\right], & N \text{ even} \end{cases} \tag{4.8}$$

which has the property that when $\beta = \beta_s$, $S(N, 1) = 1$. If the cavities have large apertures, and if the N-cavity array is excited in a π mode, which is the usual mode for superconducting cavities, it may be preferable to use a different expression for the gap factor. The simple form $T_g = \sin(\pi g/\beta\lambda)/(\pi g/\beta\lambda)$ can be replaced by the result of Eq. (2.52), for which the result on axis can be written as

$$T_g = \frac{\pi}{4}\left[\frac{\sin[(\beta_s/\beta - 1)\pi/2]}{(\beta_s/\beta - 1)\pi/2} + \frac{\sin[(\beta_s/\beta + 1)\pi/2]}{(\beta_s/\beta + 1)\pi/2}\right] \tag{4.9}$$

Section 6.9 presents more details of the longitudinal beam dynamics for independent-cavity linacs.

Independent-cavity linacs can be designed to provide relatively efficient acceleration over a broad velocity range, using an array of identical cavities designed for a single velocity β. Consider a design procedure for the case of superconducting cavities, where the RF power per cavity, most of which is delivered to the beam, is often held constant. One may choose the design β to obtain equal values of T at the two ends of the velocity range, where T is minimum. The minimum T can be chosen equal to some fraction of the peak transit-time factor, which is obtained near (slightly larger than) the design β. Each cavity in the section can be operated at the value of the average axial accelerating field, E_0, required to maintain the fixed beam-power value for each cavity. Thus, E_0 is maximum for the cavities at the ends of the section and is minimum where T is maximum. If the design β is increased, the cavity becomes longer, and since at fixed beam power the voltage per cavity seen by the beam is constant, the accelerating gradient and peak field decrease. However, at the low-velocity end of the section, the accelerating gradient and peak field may increase, because T falls off rapidly as the difference between the beam velocity and design velocity increases. On average there may be more cavities with lower gradient, and a few cavities at the low-velocity end of the section that have higher gradient. If the objective is to minimize the maximum peak field in the section, the optimum solution corresponds to choosing the design β so the peak fields are equal at the two ends of the velocity range. If instead the objective is to minimize the average gradient, one would choose the design β to be somewhat larger. The optimum solution might be obtained by choosing the design β such that the maximum number of cavities meet the accelerating-gradient specifications. This would require a matching of the gradient distribution required by the design with the estimated experimental gradient distribution.

If more cells are added to the cavity, there are two competing effects that affect the accelerating gradient. At fixed input power, the voltage gain seen by the beam is unchanged, and if T was constant, the gradient and peak field would be smaller. However, the velocity acceptance of T is reduced as the number of cells is increased. At beam velocities away from the velocity that produces maximum T, whether the gradient and peak field are higher or lower depends on which effect is larger. The penalties for adding more cells are increased cavity and cryostat lengths, and possibly an increase in the number of different design-β values to cover the same velocity range.

4.2
Wideröe Linac

For acceleration of a single particle species from one fixed energy to another, a simpler and more economical linac may be obtained by using a longer multicell cavity, where the relative phasing between cells is determined by the structure geometry and the cavity mode. A multicell structure requires

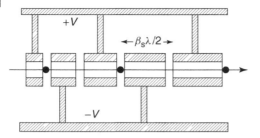

Figure 4.4 Wideröe or interdigital structure.

fewer RF drive lines and this simplifies the operation. RF power sources with higher output power capacity can be used to drive the larger cavities, and these larger power sources generally result in lower cost RF systems. We discuss multicell structures in more detail, beginning with the Wideröe linac [3, 4] shown schematically in Fig. 4.4 in the form of an *interdigital structure*.

As discussed in Chapter 1, the Wideröe linac was the first successful linear accelerator, and it is still used as a low-frequency structure for acceleration of low-velocity heavy ions. The beam moves within an array of hollow cylindrical electrodes to which an RF voltage is applied. Acceleration takes place in the gaps between the electrodes. The characteristic property of the Wideröe linac is that the voltages applied to successive electrodes alternate in sign. The electric field reverses in successive gaps, giving the Wideröe linac the appearance of a π-mode standing-wave structure. The linac is designed so that the synchronous particle with velocity β_s travels from the center of one gap to the center of the next in half an RF period, making the cell lengths equal to $\beta_s \lambda/2$.

The *Wideröe structure*, also called the *Sloan–Lawrence structure*, is used for heavy ions with velocity below about $\beta = 0.03$, for which a relatively low-frequency structure, usually <100 MHz, is needed to keep the gap spacings practical, and to maintain a large aperture without reducing the transit-time factor. An example is the 27-MHz Wideröe at the UNILAC linac [5] at GSI, Darmstadt. The main advantage of the Wideröe is that it can be operated at relatively low frequencies, but the cavity can be fabricated with limited transverse dimensions, whereas a DTL, based on the TM_{010} cylindrical cavity mode, would have a transverse radius >3 m at 27 MHz.

In principle, the voltages can be supplied to the electrodes by alternately connecting them to two conductors parallel to the beam line and driven by a high-frequency oscillator. At these high frequencies, however, where the accelerator length is comparable to or larger than the RF wavelength, the voltage will vary along the conductors. In this case quarter-wave resonators can be used to charge the electrodes. Because the voltage on the quarter-wave structure varies sinusoidally as a function of position, higher average accelerating fields can be obtained by bending the low-voltage end of the quarter-wave resonators away from the beam axis, and attaching the electrodes only to the high-voltage end. This results in an array of bent quarter-wave

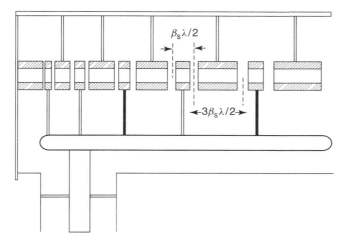

Figure 4.5 Wideröe or Sloan–Lawrence coaxial-line structure in a $\pi-3\pi$ configuration.

resonator sections, which can be joined end to end and loaded with drift tubes. An example of this structure is shown in Fig. 4.5.

Focusing can be provided by installing quadrupole magnets within the electrodes, but often the space available is limited. One common arrangement to create the necessary space is to use an alternating $\pi-3\pi$ configuration, in which long electrodes containing quadrupoles are alternated with short electrodes with no quadrupoles; the spacing between gaps alternates from $\beta_s\lambda/2$ to $3\beta_s\lambda/2$. Another approach to focusing, recently adopted at GSI [6] is to install a sequence of about a dozen thin drift tubes without quadrupoles, followed by a quadrupole triplet and a few gaps to provide longitudinal matching to the next section. As the particle velocity increases, the electrode spacing increases, and if the voltage between electrodes is constant, the average accelerating field will decrease with particle velocity. Eventually, at higher velocity a higher-frequency Alvarez DTL will provide better efficiency.

4.3
H-Mode Structures

H-mode accelerating structures, which could also be called *transverse-electric (TE)-mode structures*, are structures with a predominant RF longitudinal magnetic field especially in the outer regions of the cavity. The RF electric field is concentrated in the cavity inner regions, and would be transverse to the cavity axis in a simple pillbox cavity. But, an effect of the drift-tube loading is to produce a longitudinal electric-field component near the beam axis, as necessary for acceleration of the beam. Historically, H-mode structures such as the interdigital or *interdigital H-mode (IH) structure*, shown in Fig. 4.4, have also been called *Wideroe-linac structures*. The nominal gap-to-gap spacing along the beam axis is $\beta\lambda/2$, and the gap-to-gap phase difference is π radians. Since the beam propagates along the beam axis, the beam sees the structure as operating in an effective π mode. Two different kinds of H-mode structures have been developed in recent years, the *IH structure* (Fig. 4.6a), and the *crossbar H-mode (CH) structure*. The IH mode may be considered as similar to a TE_{110}-like mode of a pillbox cavity, and the CH mode is similar to a TE_{210} pillbox cavity mode.

The advantage of both H-mode structures is very high shunt impedance resulting in high RF power efficiency, compared with the Alvarez DTL that is discussed in detail in Section 4.4. Two contributing factors to the high shunt impedance are the use of very compact drift tubes resulting in very small capacitance, as well as the enhanced transit-time factor from π-mode operation, discussed in Section 2.4. The shunt impedance can also be increased by increasing the cross section of the stems. Either structure could be used immediately after an initial radiofrequency–quadrupole (RFQ) linac structure, which is discussed in detail in Chapter 8. On the basis of shunt impedance, the IH structure is better for velocities below about $\beta = 0.1$,

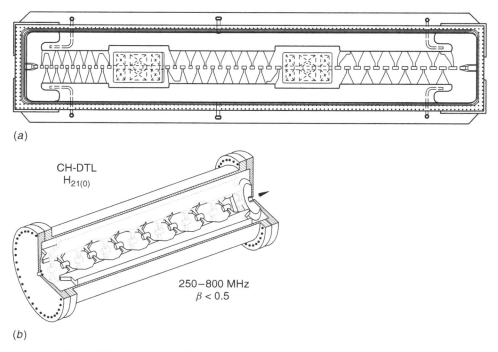

Figure 4.6 (a) Interdigital H-mode (IH) structure showing regions with a long sequence of electrodes for acceleration with no transverse focusing lenses separated by triplet quadrupoles to provide transverse focusing (courtesy of U. Ratzinger). (b) Crossbar H-Mode or CH structure (courtesy of U. Ratzinger).

whereas the CH structure is better for velocities in the range of about $\beta = 0.1$ to $\beta = 0.5$. Transverse focusing can be supplied from triplet quadrupole lenses (Fig. 4.6a) or magnetic solenoids. Longitudinal beam dynamics is discussed in Section 6.12. Transverse focusing can also be provided at the expense of reduced shunt impedance by use of compact permanent magnet quadrupoles installed in some of the drift tubes. Both IH structures exceed the mechanical rigidity of the Alvarez DTL. In principle they can be used for light as well as heavy ions, and for a normal-conducting copper linac or for a superconducting linac. For the latter case the transverse focusing lenses would be installed between tanks to avoid trapping magnetic flux in the superconducting cavities.

4.4
Alvarez Drift-Tube Linac

Suppose we want to accelerate the beam to high energies in a long TM_{010} pillbox cavity. If the cavity is made longer than the distance a particle travels in half an RF period ($\beta\lambda/2$), the beam will experience deceleration as well as acceleration. The solution proposed by Alvarez et al. [7] was to install hollow conducting drift tubes along the axis, as shown in Fig. 4.7, into which the RF electric field decays to zero, as in a waveguide below the cutoff frequency. This creates field-free regions that shield the particles, when the polarity of the axial electric field is opposite to the beam direction. The drift tubes allow us to divide the cavity into cells of nominal length $\beta\lambda$, extending from the center of one drift tube to the center of the next. As the particle velocities increase, the cell lengths must also increase. The fields in all cells have the same phase so that the multicell structure can be said to operate in a zero mode. The resulting accelerating structure is known as the *DTL*.

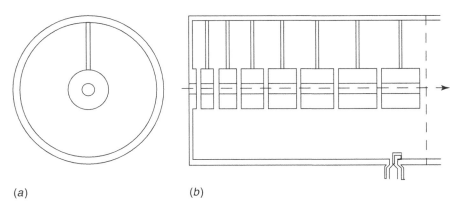

(a) (b)

Figure 4.7 Alvarez DTL cavity.

Applying the integral form of Faraday's law within a single cell over a rectangular path that includes the beam axis and the outer wall, one finds that

$$E_0 = -\frac{j\omega\Phi}{\beta\lambda} \tag{4.10}$$

where Φ is the magnetic flux per unit length that circulates azimuthally in the cell. If the cavity is tuned so that the magnetic flux per unit length is the same for all cells, then E_0 is the same for all cells in the structure. Magnetic fields can be measured from standard perturbation techniques (see Section 5.13) and it is possible to tune the cells to obtain the same B_θ near the wall. If the cavity radius is constant along the length, and the cavity has the same radial distribution of B_θ for all cells, then the magnetic flux per unit length and E_0 will be the same for all cells. In practice these conditions can be approximately obtained and the DTL typically is tuned for a constant E_0 in each cell rather than a constant voltage across each cell, as for a Wideröe linac. The voltage across each cell varies with the cell length $\beta\lambda$, because $V_0 = E_0\beta\lambda$. The behavior is similar to that of a voltage divider, where there is a fixed voltage across the whole structure, given by $E_0 L$, where L is the length of the structure, and the cell voltages are determined by the fraction of the length occupied by each cell.

The drift tubes are supported mechanically by stems attached to the outer wall. The currents flow longitudinally on the outer walls and on the drift tubes, and are everywhere in phase, as shown in Fig. 4.8. There is no net charge on the drift tubes. There is no net current on the stems at any time, but the time-varying magnetic fields at the surface of the stems induce eddy currents that produce power dissipation. Although E_0 is the same for all cells, the electric field in the gaps and the peak surface electric field on the drift tubes will vary, depending on the gap length and other details of the cell geometry. If desired, the structure can also be tuned so that E_0 is not uniform.

Adding the drift tubes makes the cavity look somewhat like a coaxial resonator especially when the gaps are small. This suggests that the coaxial line may also be used as a model to understand the DTL structure. Conduction current flows through each drift tube and becomes displacement current in the gaps. The magnetic field outside the drift-tube radius falls off with increasing

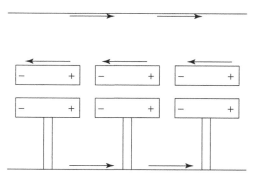

Figure 4.8 Charges and currents in a DTL.

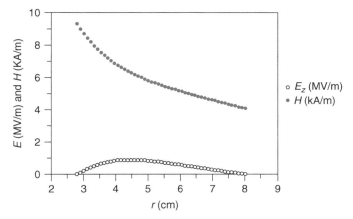

Figure 4.9 Typical magnetic field (upper curve) and electric field (lower curve) versus radius in a DTL in the transverse plane passing through the center of a drift tube [8]. The beam axis is at $r = 0$, the outer radius of the drift tube is at 2.75 cm, and the outer wall is at $r = 8.0$ cm. The average axial electric field in the cell is 2.2 MV/m.

radius, as in a coaxial line. Typical radial profiles of magnetic and electric fields are shown in Fig. 4.9.

It is useful to consider the DTL as an array of cavity cells. In fact, if a conducting plane, perpendicular to the cavity axis is inserted at the electrical center of each drift tube, where by symmetry the electric field has only a longitudinal z component, the field pattern is undisturbed and the cell resonant frequencies are unchanged. Conducting end walls are undesirable because they dissipate power. However, they are useful conceptually for visualizing the cells as individual cavities. Each effective cavity can be thought of as a lumped circuit consisting of the drift tubes as capacitors, and inductance associated with a current loop from one drift tube to the cylindrical outer wall and back to the other drift tube. This lumped-circuit model is not good enough for an accurate quantitative analysis of the structure properties, which requires use of a numerical electromagnetic-field-solver code such as SUPERFISH [9]. The lumped-circuit picture often remains useful for qualitative understanding. It would predict an approximate capacitance from the parallel plate formula of $C_0 = \varepsilon_0 \pi d^2/4g$ and an inductance (ratio of magnetic flux to current on drift tube) $L_0 = \mu_0 \beta \lambda \ln(D/d)/2\pi$. The geometry parameters are the gap g, cell length $\beta \lambda$, drift-tube diameter d, and tank diameter D. The resonant frequency of the cell is given in this model by $\omega_0^2 = 1/L_0 C_0$.

Figure 4.10 shows the electric-field pattern in DTL cells with the same frequency, cavity diameter, drift-tube diameter, and different β. The electric-field strength is not exactly proportional to the density of lines shown in the figure. However, the electric field is largest near the drift tubes, as is suggested in the figure. At low β the field lines concentrate on the front face of the drift tube. As β increases, the cell length increases, and the lumped-circuit

94 | *4 Standard Linac Structures*

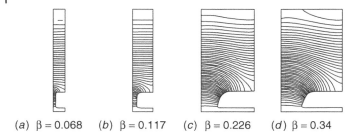

(a) β = 0.068   (b) β = 0.117   (c) β = 0.226   (d) β = 0.34

Figure 4.10 Electric-field lines shown in one quarter of the projections of DTL cells as calculated by the program SUPERFISH (courtesy of J. H. Billen).

model predicts that the inductance should increase. To keep the resonant frequency the same, the capacitance must decrease. If the gap increases with the cell length, the model predicts an unchanged frequency. But, at higher β the pictures show that more field lines terminate on the outer diameter of the drift tube than in the gap. It becomes harder to reduce the capacitance of the drift tubes by increasing the gap, and larger gaps are required to maintain a constant frequency. As energy increases, the larger gaps per unit cell length cause the transit-time factor and the effective shunt impedance to become smaller, and the DTL becomes less efficient. Table 4.1 lists parameters for the LANSCE [10] 201.25-MHz DTL at Los Alamos. The LANSCE DTL uses four tanks to accelerate the proton beam from 0.75 to 100 MeV.

Transverse focusing (discussed further in Chapter 7) is provided by installing magnetic quadrupole lenses in the drift tubes. At low velocities it becomes more difficult to provide focusing because the drift tubes are smaller, and less space is available for installing quadrupole magnets. If the focusing needs to be increased, one solution is the $2\beta\lambda$ DTL, which is built to provide longer drift tubes by using twice the cell period or one accelerating gap per $2\beta\lambda$. We have already seen that the average axial electric field is determined by the magnetic flux per unit length, and this is still true for the $2\beta\lambda$ linac. We would then expect that the ratio of surface magnetic field on the outer wall to E_0 would be unchanged if we exchange our $1\beta\lambda$ drift tubes for the longer $2\beta\lambda$ variety. However, for the same resonant frequency and the same diameters for the tank and drift tube, the longer cells will have twice the inductance and therefore will require half the capacitance, compared with the $1\beta\lambda$ case. This means a larger gap and smaller transit-time factor and effective shunt impedance, which reduce the efficiency.

Another proposal for providing increased focusing at lower velocities for heavy-ion acceleration is known as the *quasi-Alvarez DTL* [11]. The quasi-Alvarez approach is to install quadrupoles at only every third or fourth drift tube. The drift tube with a quadrupole has a larger diameter and is longer than normal, occupying the space of two cells instead of only one. The remaining small-diameter drift tubes contain no quadrupoles and occupy only one cell. At

Table 4.1 DTL parameters for the LANSCE proton accelerator.

	Tank 1	Tank 2		Tank 3	Tank 4
Cell number	1 to 31	32 to 59	60 to 97	98 to 135	136 to 165
Energy in (MeV)	0.75	5.39		41.33	72.72
Energy out (MeV)	5.39	41.33		72.72	100.00
Energy gain (MeV)	4.64	35.94		31.39	27.28
Tank length (cm)	326.0	1968.8		1875.0	1792.0
Tank diameter (cm)	94.0	90.0		88.0	88.0
Drift-tube diameter (cm)	18.0	16.0		16.0	16.0
Drift-tube corner radius (cm)	2.0	4.0		4.0	4.0
Bore radius (cm)	0.75	1.0	1.5	1.5	1.5
Bore corner radius (cm)	0.5	1.0		1.0	1.0
g/L	0.21–0.27	0.16–0.32		0.30–0.37	0.37–0.41
Number of cells	31	66		38	30
Number of quads	32	29	38	20	16
Quad gradient (kG/cm)	8.34–2.46	2.44–1.89	1.01–0.87	0.90–0.84	0.84–0.83
Quad length (cm)	2.62–7.88	7.88	16.29	16.29	16.29
E_0 (MV/m)	1.60–2.30	2.40		2.40	2.50
ϕ_s (°)	−26	−26		−26	−26
Power (MW)	0.305	2.697		2.745	2.674
Intertank space (cm)	15.90	85.62		110.95	–
Transit-time factor, T	0.72–0.84	0.87–0.80		0.82–0.74	0.74–0.68
Mean ZT^2 (MV/m)	26.8	30.1		23.7	19.2

Total length including intertank spaces = 61.7 m

200 MHz, a quasi-Alvarez structure can be designed for operation at a velocity as low as $\beta = 0.025$.

The modern DTL also uses posts of electrical length $\lambda/4$ to provide resonant coupling between the cells, as shown in Fig. 4.11. This is desirable to make the field distribution less sensitive to errors and to eliminate power-dependent phase shifts between cells in long DTLs. Later in this chapter, we say more about resonant coupling.

After the machining and assembly of a DTL tank, a resonant frequency of several hundred megahertz is typically within about a few megahertz of the design value. For $Q = 5 \times 10^4$ and a resonant frequency of 200 MHz, the bandwidth is 4 kHz. The frequency must be brought within the bandwidth by tuning procedures. Generally, tuning is done in two stages. The first stage of coarse tuning is required to bring the frequency to within a few tens of kilohertz of the correct value. This may be done by machining of demountable tuning bars, which are installed in the cavity near the outer walls. Removing material from the tuning bars increases the volume in the outer, magnetic-field region of the cavity, increasing the effective inductance and lowering the resonant frequency. Frequency errors in the range of a couple of megahertz

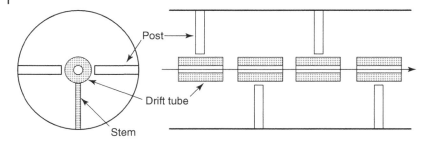

Figure 4.11 Drift-tube linac with post couplers for field stabilization.

can be corrected with the tuning bars. A smaller coarse-tuning range can be obtained by using a finite number of fixed slugs instead of removable tuning bars, which are also machined to lower the resonant frequency. The tuning range of the slugs is usually about 100 kHz. The machining that is done during the coarse-tuning stage also can be used to correct field tilts in the tank. Finer tuning can be easily obtained by controlling the temperature of the cooling water. The temperature coefficient is typically about a few kilohertz per degree Fahrenheit, and temperature control to about a degree is straightforward. The fine tuning allows for slow dynamic corrections, and results in resonant-frequency control to an accuracy of nearly 1 kHz, and within the bandwidth.

The DTL is used exclusively for accelerating protons and heavier ions in the velocity region near $0.05 < \beta < 0.4$. It is not needed for electron linacs because the injected beam velocity from a dc electron gun is usually higher than this velocity range. The advantages of the DTL include (1) an open structure without cell end walls, generally resulting in high ZT^2, and (2) focusing quadrupoles within the drift tubes, providing strong focusing and permitting high beam-current limits. Its main disadvantages are that (1) T and ZT^2 decrease significantly at high β values as gap lengths increase; (2) T and ZT^2 decrease at low β for fixed aperture as fields penetrate into drift tubes; and (3) focusing vanishes at low β where drift tubes are too short to hold the quadrupoles of adequate length.

4.5
Design of Drift-Tube Linacs

For a DTL with constant synchronous phase, the synchronous particle travels from the center of a drift tube to the center of the gap in half an RF period, and continues to the center of the next drift tube in the next half period. Owing to the acceleration, the length of each cell must increase to maintain the synchronism. The required cell-length profile depends on the synchronous velocity β_s, which increases because of the energy gain in each cell. The energy gain in each cell depends on the electric field, and on the length of the cell. Because of the interdependency of the cell length and energy gain, the cell design is usually done by a method of successive approximations. First, the cell geometry at each β_s must be chosen, generally based on the criterion of maximum effective shunt impedance, consistent with (1) obtaining the correct resonant frequency, (2) allowing room within the drift tubes for quadrupole focusing lenses, and (3) keeping the peak surface electric and magnetic fields within the technological limits, determined by electric breakdown and drift-tube cooling requirements. The fields, power, transit-time factor for the synchronous particle, and the shunt-impedance calculations are usually done using electromagnetic-field-solver codes like SUPERFISH. This procedure results in an optimum cell geometry in which the gap length, drift-tube shape, and tank diameter are determined. Because of the energy gain in the accelerating gap, the gap is always displaced toward the low-energy end of the cell. Fortunately, the field calculations relative to the gap are nearly independent of the precise location of the accelerating gap within the cell. Therefore, the SUPERFISH calculation can be done for an equivalent symmetric cell, which has the same length and geometry as the real cell, but has the gap in the geometric center of the cell.

After the electrical properties of the cells as a function of β_s have been calculated, the positions of the gap centers must be determined. This requires calculations that allow for the change of β_s in the cell, which depends on the magnitude of the average accelerating field E_0 and on the synchronous phase ϕ_s. The choice of E_0 is important because it affects the accelerator length, the power dissipation, the probability of electric breakdown, and the longitudinal focusing. The choice of synchronous phase affects both the accelerator length and the longitudinal focusing. The positions of electrical centers of successive gaps and the successive cell lengths still must be determined by numerical integration of the synchronous particle trajectory. A simple algorithm for integrating through the cell that gets fairly accurate results is called the *drift–kick–drift method*. In this approximation the particle is assumed to drift at constant velocity β_{s1} from the center of the drift tube to the center of the gap, where it receives the full acceleration as an impulse at the gap center, after which it drifts at the new constant velocity β_{s2} to the center of the next drift tube. The drift distances are $x_1 = \beta_{s1}\lambda/2$ and $x_2 = \beta_{s2}\lambda/2$, and the total cell length is $L = x_1 + x_2$. The energy gain at the center of the gap is $\Delta W_s = qE_0T(\beta_s)\cos\phi_s L$. This calculation may be continued from cell to cell until the final energy is reached. Similar methods are used in the DTL design code PARMILA [12] to lay out the gap locations and cell lengths. The DTL is usually configured as a sequence of independent multicell tanks, each with its own RF drive. The tank lengths are determined mainly by the power available from the RF generator, by mechanical constraints, and by the ability to control the field distribution along the tank to a sufficient precision. The actual field distribution is affected by fabrication errors and other perturbations. Field tilts and other distortions can occur, and these effects increase as the total cavity length increases. This subject can be understood from the application of perturbation theory, and will be discussed in more detail in connection with the RFQ structure in Chapter 8.

4.6
Coupled-Cavity Linacs

The coupled-cavity linac or CCL consists of a linear array of resonant cavities, coupled together to form a multicavity accelerating structure. In Fig. 4.12 the side-coupled linac structure is shown as an example. The CCL is used for acceleration of higher velocity beams of electrons and protons in the typical velocity range $0.4 < \beta < 1.0$. The individual cavities are sometimes called *cells*, and each cell usually operates in a TM_{010}-like standing-wave mode. CCL structures provide two accelerating gaps per $\beta\lambda$. Most of the properties of the CCL can be understood from a model of $N+1$ coupled electrical oscillators [13]. There will be $N+1$ normal modes of the system, each with a characteristic resonant frequency and a characteristic pattern of the relative amplitudes and phases for the different oscillators. The properties of these $N+1$ normal modes can be determined by solving an eigenvalue problem. This is done in the following way. Kirchoff's law is applied to the $N+1$ circuits, and the sum of the voltages around each loop is set to zero. The resulting $N+1$ simultaneous equations are solved for the eigenfrequencies and the corresponding eigenvectors. The eigenvector components give the currents in the individual oscillators for each normal mode. Many general details of the behavior of such a system can be deduced from this coupled-circuit

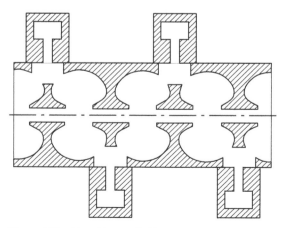

Figure 4.12 The side-coupled linac structure as an example of a coupled-cavity linac structure. The cavities on the beam axis are the accelerating cavities. The cavities on the side are nominally unexcited and stabilize the accelerating-cavity fields against perturbations from fabrication errors and beam loading.

model, including the relative phases and amplitudes of the cells for each normal mode, the nature of the dispersion relation, the effects of errors in the resonant frequencies of the cells, and the effects of power losses. Because we anticipate the applicability of this model to a chain of coupled-cavity resonators, we often refer to the eigenvectors as *fields* instead of currents.

4.7
Three Coupled Oscillators

To get a better feeling for the physics before discussing the general problem of $N+1$ oscillators, we present the results for the chain of three coupled oscillators, shown in Fig. 4.13.

The three-cell array consists of one central oscillator with mutual inductances on each side for coupling to each of the end oscillators. The end oscillators have only half the inductance of the middle oscillator, but have twice the capacitance to give equal resonant frequencies in the limit of zero coupling strength. The end oscillators are called *half-cells* [14]. The circuit equations, obtained by summing the voltages around each circuit, can be written as

$$x_0 \left[1 - \frac{\omega_0^2}{\Omega^2}\right] + x_0 k = 0, \quad \text{oscillator } n = 0 \tag{4.11}$$

$$x_1 \left[1 - \frac{\omega_0^2}{\Omega^2}\right] + (x_0 + x_2)\frac{k}{2} = 0, \quad \text{oscillator } n = 1 \tag{4.12}$$

$$x_2 \left[1 - \frac{\omega_0^2}{\Omega^2}\right] + x_1 k = 0, \quad \text{oscillator } n = 2 \tag{4.13}$$

The x_n are normalized currents, defined in terms of the real currents i_n by $x_n = i_n \sqrt{2L_0}$, ($n = 0, 1, 2$), $k = M/L_0$ is the coupling constant, $\omega_0 = 1/\sqrt{2L_0 C_0}$ is the resonant frequency of the individual oscillators if they were uncoupled, and Ω is the frequency of a normal mode. The circuit equations can be put into matrix form as

$$LX_q = \frac{1}{\Omega_q^2} X_q \tag{4.14}$$

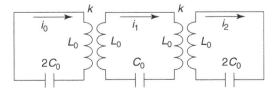

Figure 4.13 Three coupled oscillators.

where L is the matrix operator, a function of the cavity frequency and the coupling constant k; X_q is the eigenvector whose elements are the currents (fields) x_n, and $1/\Omega_q^2$ is the eigenvalue. One finds that

$$L = \begin{bmatrix} 1/\omega_0^2 & k/\omega_0^2 & 0 \\ k/2\omega_0^2 & 1/\omega_0^2 & k/2\omega_0^2 \\ 0 & k/\omega_0^2 & 1/\omega_0^2 \end{bmatrix} \quad (4.15)$$

and

$$X_q = \begin{pmatrix} x_1 \\ x_2 \\ x_3 \end{pmatrix} \quad (4.16)$$

The three normal-mode eigenfrequencies Ω_q and the normalized eigenvectors X_q in order of increasing mode frequency are given for the three modes as follows:

(a) Mode $q = 0$, which is called the *zero mode* because all oscillators have zero relative phase difference, is specified by

$$\Omega_0 = \frac{\omega_0}{\sqrt{1+k}}, \quad X_0 = \begin{bmatrix} 1 \\ 1 \\ 1 \end{bmatrix} \quad (4.17)$$

where the elements of the normalized eigenvector give the relative fields in each oscillator.

(b) Mode $q = 1$, the $\pi/2$ mode, is specified by

$$\Omega_1 = \omega_0, \quad X_1 = \begin{bmatrix} 1 \\ 0 \\ -1 \end{bmatrix} \quad (4.18)$$

(c) Mode $q = 2$, called the π *mode*, and has the highest frequency, is specified by

$$\Omega_2 = \frac{\omega_0}{\sqrt{1-k}}, \quad X_2 = \begin{bmatrix} 1 \\ -1 \\ 1 \end{bmatrix} \quad (4.19)$$

The bandwidth, defined as the frequency difference between the highest and lowest normal-mode frequencies, for $k \ll 1$ is $\delta\omega \approx \omega_0 k$. The $\pi/2$ mode is different from the other modes, because it has an unexcited oscillator. Indeed, we see later that the $\pi/2$ mode has unusual properties that would make it attractive for the operating mode of an accelerating structure.

4.8
Perturbation Theory and Effects of Resonant-Frequency Errors

Perturbation theory can be used to study the effects of frequency errors of the individual oscillators [15]. The errors are contained in a perturbation matrix P, defined so that when P is added to the unperturbed L matrix, the corrected L matrix is obtained. The first-order eigenvector corrections are calculated from the series

$$\Delta X_q = \sum_{r \neq q} a_{qr} X_r \qquad (4.20)$$

where the X_r are the unperturbed eigenvectors, and the amplitudes are

$$a_{qr} = \frac{[X_q P X_r]}{\dfrac{1}{\Omega_r^2} - \dfrac{1}{\Omega_q^2}} \qquad (4.21)$$

This shows that the effect of the errors is to create a new eigenvector, which is equal to the unperturbed eigenvector, with corrections that come from adding or mixing contributions from all the other unperturbed eigenvectors. The amount by which the other eigenvectors are mixed depends on the appropriate element of a matrix $X_q P X_r$, where P is the perturbation matrix, given in terms of the coupling constant k, the resonant frequency of the uncoupled oscillators ω_0, and the error parameters $\delta\omega_0$ and $\delta\omega_1$, as

$$P = \begin{bmatrix} -(2\delta\omega_0/\omega_0^3) & -k(2\delta\omega_0/\omega_0^3) & 0 \\ -k(2\delta\omega_1/\omega_0^3)/2 & -(2\delta\omega_1/\omega_0^3) & -k(2\delta\omega_1/\omega_0^3)/2 \\ 0 & k(2\delta\omega_0/\omega_0^3) & (2\delta\omega_0/\omega_0^3) \end{bmatrix} \qquad (4.22)$$

Next, we apply the perturbation-theory results to the problem of three coupled oscillators treated in Section 4.6. If all three frequencies are different, we are free to choose the unperturbed frequency in a convenient way. We choose the unperturbed frequency as the average over the end-cell frequencies, which we call ω_0. We define $\pm\delta\omega_0$ as the frequency error of the end cells relative to ω_0, and define $\delta\omega_1$ as the frequency error of the middle cell. We present the results through first order in the fractional frequency errors for both the zero and the π modes. For the $\pi/2$ mode, the first-order theory gives a nonzero correction only for the field in the middle oscillator. Therefore, for the $\pi/2$ mode we have carried out the calculations through second order. The results for the three modes, perturbed by oscillator frequency errors are as follows:

(a) *The zero mode, valid through first order:*

$$\Omega_0 = \frac{\omega_0}{\sqrt{1+k}}\sqrt{1+\frac{\delta\omega_1}{\omega_0}} \qquad (4.23)$$

$$X_0 = \begin{bmatrix} 1 + \frac{1+k}{2k}\left\{\frac{\delta\omega_1}{\omega_0} - 4\frac{\delta\omega_0}{\omega_0}\right\} \\ 1 - \frac{1+k}{2k}\frac{\delta\omega_1}{\omega_0} \\ 1 + \frac{1+k}{2k}\left\{\frac{\delta\omega_1}{\omega_0} + 4\frac{\delta\omega_0}{\omega_0}\right\} \end{bmatrix} \qquad (4.24)$$

(b) *The $\pi/2$ mode, valid through second order* [16]:

$$\Omega_1 = \frac{\omega_0}{\sqrt{1 - 4(\delta\omega_0/\omega_0)^2}} \qquad (4.25)$$

$$X_1 = \begin{bmatrix} 1 + \frac{4}{k^2}\frac{\delta\omega_1}{\omega_0}\frac{\delta\omega_0}{\omega_0} - \frac{2}{k^2}\left(\frac{\delta\omega_0}{\omega_0}\right)^2 \\ -\frac{2}{k}\frac{\delta\omega_0}{\omega_0} \\ -1 + \frac{4}{k^2}\frac{\delta\omega_1}{\omega_0}\frac{\delta\omega_0}{\omega_0} + \frac{2}{k^2}\left(\frac{\delta\omega_0}{\omega_0}\right)^2 \end{bmatrix} \qquad (4.26)$$

(c) *The π mode, valid through first order:*

$$\Omega_2 = \frac{\omega_0}{\sqrt{1-k}}\sqrt{1+\frac{\delta\omega_1}{\omega_0}} \qquad (4.27)$$

$$X_2 = \begin{bmatrix} 1 - \frac{1-k}{2k}\left\{\frac{\delta\omega_1}{\omega_0} - 4\frac{\delta\omega_0}{\omega_0}\right\} \\ -1 - \frac{1-k}{2k}\frac{\delta\omega_1}{\omega_0} \\ 1 - \frac{1-k}{2k}\left\{\frac{\delta\omega_1}{\omega_0} + 4\frac{\delta\omega_0}{\omega_0}\right\} \end{bmatrix} \qquad (4.28)$$

We express the perturbed eigenvectors as

$$X_{q,n} = Ae^{j\phi}\cos\frac{\pi q n}{N}e^{j\Omega_q t} \qquad (4.29)$$

where the $Ae^{j\phi}$ factor corresponds to the ratio of the elements of the perturbed to the unperturbed eigenvectors. The above results show that the perturbed amplitudes A are generally no longer unity, but since the expressions are still all pure real numbers, the phases are still $\phi = 0$.

The presence of first-order corrections to the elements of the eigenvectors for the zero and π modes shows that the field distributions for these modes can be sensitive to resonant-frequency errors of the individual oscillators. The expressions show that the errors result in nonuniformity in the magnitude of the field excitations in the individual oscillators, which increase with the fractional error in the frequencies and decrease with increasing coupling strength k. By contrast, the effect of such frequency errors for the $\pi/2$ mode appears as a first-order correction only to the field of the middle oscillator, which is nominally unexcited, and the errors affect the nominally excited cells only in second order. Therefore the fields in the nominally excited cells in the $\pi/2$ mode are very insensitive to any small frequency errors. We describe later how this insensitivity can be used beneficially in accelerating structures.

4.9
Effects from Ohmic Power Dissipation

Next we present the steady-state solution with resistive power losses included, and ignore the oscillator frequency errors discussed in Section 4.8. When energy is dissipated, a generator is needed to produce a steady-state solution. The system model is modified by adding generators to the three individual circuits in Fig. 4.13, and applying Kirchoff's laws. One obtains Eqs. (4.11) to (4.13) modified with the generator voltages on the right side of the equations. For simplicity, we assume the quality factor Q is the same for all three oscillators. It is useful to discuss the case where the only generator is in the end oscillator, $n = 0$. If the driving frequency is a normal-mode frequency, that normal mode will be excited and the relative fields in individual oscillators will be proportional to the corresponding eigenvector elements. We obtain the following results in lowest order for the fields, after normalizing the results to the field in the driven oscillator.

(a) *Zero mode:*

$$X_0 = \begin{bmatrix} 1 \\ \exp\left\{-j\dfrac{3\sqrt{1+k}}{kQ}\right\} \\ \exp\left\{-j\dfrac{4\sqrt{1+k}}{kQ}\right\} \end{bmatrix} \quad (4.30)$$

(b) *$\pi/2$ mode, with the second-order term kept for third oscillator:*

$$X_1 = \begin{bmatrix} 1 \\ \dfrac{1}{kQ}\exp\left\{j\dfrac{\pi}{2}\right\} \\ -1 + \dfrac{2}{(kQ)^2} \end{bmatrix} \quad (4.31)$$

(c) π mode:

$$X_2 = \begin{bmatrix} 1 \\ -\exp\left\{j\dfrac{3\sqrt{1-k}}{kQ}\right\} \\ \exp\left\{j\dfrac{4\sqrt{1-k}}{kQ}\right\} \end{bmatrix} \quad (4.32)$$

We have expressed the field elements as an amplitude times the phase factor. As before, we express the perturbed normalized eigenvectors as

$$X_{q,n} = Ae^{j\phi}\cos\frac{\pi q n}{N} e^{j\Omega_q t} \quad (4.33)$$

The above results show that, when losses are included, the amplitudes of the zero and π mode are unchanged to the lowest order. Losses produce a relative phase shift ϕ between adjacent oscillators in both the zero and π modes, which increases as the Q decreases, and decreases with increasing coupling strength k. This is called the *power-flow phase shift*, because it is associated with the flow of power, directed away from the generator through the oscillator chain. In the $\pi/2$ mode, power losses do not produce this phase shift. Instead, they produce an excitation of the middle, nominally unexcited, oscillator that is 90° out phase with the fields in the excited oscillators. In the $\pi/2$ mode, there is only a second-order amplitude decrease in the third oscillator, called the *power-flow droop*.

We have treated the effect of the frequency errors separately from that of the power losses. In the general case both effects will occur together. The resulting normalized eigenvector will have the form

$$X_{q,n} = A_{q,n}\cos\frac{\pi q n}{N} e^{j(\Omega_q t + \phi_{q,n})} \quad (4.34)$$

where, to the lowest order the deviations from unity of the amplitude $A_{q,n}$ increase with the oscillator resonant-frequency errors. The power-flow phase shifts $\phi_{q,n}$ increase with decreasing Q of the resonators. Both effects decrease with increasing coupling strength k. The three-coupled-oscillator problem illustrates the sensitivity of the zero and π modes to oscillator frequency errors. Such errors produce distortions of the nominally uniform field amplitudes of the zero and π modes. Power flow along the oscillator chain in the lossy case causes relative phase shifts between adjacent oscillators. We have also observed a very attractive feature of the $\pi/2$ mode, an insensitivity of the field amplitudes and phases for those oscillators that are nominally excited in the unperturbed case.

4.10
General Problem of $N + 1$ Coupled Oscillators

Now that we have looked in some detail at the results for three coupled oscillators, we summarize the general results for $N + 1$ coupled oscillators. In the case we are considering, there are $N - 1$ identical internal oscillators and an oscillator called a *half-cell* on each end, for a total of $N + 1$ coupled oscillators, as shown in Fig. 4.14.

By carrying out the same calculation as for three coupled oscillators, it is found that the eigenvector components for each mode number q can be summarized in a compact form. For $N + 1$ coupled oscillators, it is found that, including the time dependence, the eigenvector elements in the qth normal mode are

$$X_{q,n} = \cos\frac{\pi q n}{N} e^{j\Omega_q t}, \quad n = 0, 1, 2 \text{(oscillator)},$$
$$\text{and } q = 0, 1, 2 \text{(mode)} \quad (4.35)$$

The form of Eq. (4.35) describes a standing wave, but it can also be expressed in terms of forward and backward traveling waves as

$$X_{q,n} = \cos\frac{\pi q n}{N} e^{j\Omega_q t} = \frac{e^{j(\Omega_q t - \pi q n/N)} + e^{j(\Omega_q t + \pi q n/N)}}{2} \quad (4.36)$$

The quantity $\pi q/N$ is the phase advance of the traveling waves per oscillator. The eigenfrequencies can be expressed by the dispersion relation

$$\Omega_q = \frac{\omega_0}{\sqrt{1 + k\cos(\pi q/N)}}, \quad q = 0, 1, 2 \text{(mode)} \quad (4.37)$$

The coupling constant k represents the coupling between adjacent oscillators, and the resonant frequency of the individual cavities is ω_0. The bandwidth, defined as the frequency difference between the highest and lowest normal-mode frequencies for $k \ll 1$ is approximately $\delta\omega \cong \omega_0 k$, independent of N. The dispersion relation of Eq. (4.37) is illustrated in Fig. 4.15 for the case $N = 6$. There are $N + 1$ normal modes centered about ω_0. As the number of normal modes increases, the bandwidth remains fixed at $k\omega_0$, and the modes become

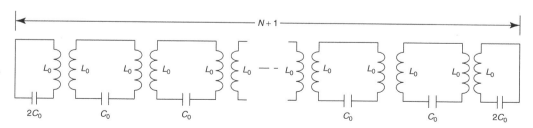

Figure 4.14 $N + 1$ coupled oscillators.

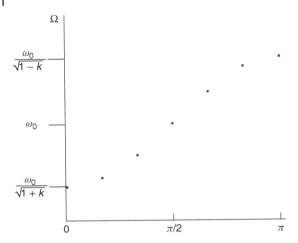

Figure 4.15 Normal-mode spectrum of coupled oscillator system with seven oscillators.

more dense and closely spaced, which can lead to two problems. First, when power losses are included, each normal mode has a natural width which is $\Delta\omega = \omega/Q$. If the natural width of the normal mode is larger than the spacing to the nearest normal mode, the normal modes will overlap, and the generator will excite both modes rather than only the desired one. Second, as described earlier, perturbations resulting from frequency errors produce corrections to each eigenvector. The corrections come in perturbation theory by adding or mixing specific fractions of each eigenvector of the unperturbed state to the original unperturbed eigenvector. The contribution from each mode of the unperturbed system depends not only on the specific perturbation, but also on the relative frequencies of these modes. For a given set of oscillator errors, modes that are closer together in frequency contribute larger perturbations to the fields than modes that are farther apart. When N is large, the frequency spacing between the π mode and the closest mode is small because the π mode is at a point on the dispersion curve where the slope is zero. When $k \ll 1$, this spacing is approximately

$$\frac{\delta\Omega}{\Omega} = k\left[\frac{\pi}{2N}\right]^2 \tag{4.38}$$

For the $\pi/2$ mode, the closest-mode spacing is larger for $N > 2$. Assuming $k \ll 1$, the spacing to the mode closest to the $\pi/2$ mode is given by

$$\frac{\delta\Omega}{\Omega} = k\left[\frac{\pi}{2N}\right] \tag{4.39}$$

Thus, Eq. (4.39) illustrates another advantage of the $\pi/2$ mode, the closest-mode spacing is greater than the spacing for the other normal modes. The advantages described in the previous section for three coupled oscillators have nothing to do with the improved normal-mode spacing, because there are not enough modes on the dispersion curve for this to matter when $N=2$. The advantage for $N=2$ results from a different effect, which is the symmetric central position of the $\pi/2$ mode in the mode spectrum. For the $\pi/2$ mode the perturbation-theory contributions from the nearest frequency modes cancel, leaving only second-order terms.

The practical implementation of the $\pi/2$ mode concept to a linac structure will be discussed in the following section, where we discuss the biperiodic structure. For a periodic structure, the most commonly used coupled-cavity accelerating mode is the π mode, which is used for cases where the structure consists of only a few cells. General perturbation-theory results have been derived for the π mode, [17] which we now present. The main effect of power losses in the π mode is to introduce the cell-to-cell power-flow phase shift, which adds to the π phase shift per cell for the ideal lossless solution. The cell-to-cell phase difference relative to the ideal π phase shift is

$$\phi_m - \phi_{m-1} = \frac{2\sqrt{1-k}(N-m+0.5)}{kQ_L} \quad (4.40)$$

where the drive cell is labeled 1, the cell index m varies from 2 to N, k is the intercell coupling constant, and Q_L is the loaded quality factor. Equation (4.40) is called a *power-flow phase shift* because the quantity in parenthesis divided by Q_L is proportional to the power flow across the center of the cell. The total phase shift from the drive cell to cell N is

$$\Delta\phi = \phi_1 - \phi_N = \frac{(N-1)^2\sqrt{1-k}}{kQ_L} \quad (4.41)$$

The intercell coupling strength k can be chosen to limit the total power-flow phase shift to any specific value. Note that if we require $\Delta\phi \ll 1$, and if $k \ll 1$, then

$$kQ_L \gg (N-1)^2 \quad (4.42)$$

From this, it follows that to control the power-flow phase shift, the coupling must be large enough that the width of the normal-mode passband must be much larger than the square of the number of cells times the resonance width $\delta\omega_\pi$. The required value of the coupling constant increases with the square of the number of cells after the drive point, and is inversely proportional to Q_L.

Individual cell resonant-frequency perturbations cause first-order deviations from the nominally flat field amplitude. A simple general formula can be obtained for a symmetric perturbation of the end cells, where one end-cell frequency is increased by $\Delta\omega$, and the other is decreased by the same amount. For $N+1$ cells in a π mode, this produces an end-to-end tilt of the field amplitude, which is given in the coupled-circuit model [18] by

$$\frac{\Delta A}{A} = \frac{2N}{k}\frac{\Delta\omega}{\omega_\pi} \quad (4.43)$$

Note that if we require $\Delta A/A \ll 1$, we obtain

$$k\frac{\omega_\pi}{\Delta\omega} \gg 2N \tag{4.44}$$

Thus, to limit a field tilt, the coupling must be large enough that the width of the passband must be much larger than the number of cells times the fractional frequency error of the end cells.

A more accurate description of the mode spectrum than that given by Eq. (4.37) is obtained by including the coupling between next nearest neighbors. This type of coupling typically modifies the dispersion relation by eliminating the exact symmetry of the frequency spacings about the $\pi/2$ mode.

4.11
Biperiodic Structures for Linacs

We have seen that the $\pi/2$ normal mode of a chain of coupled oscillators has unique properties that would be especially important when the number of cells is large. It would be attractive to use this mode for a linac if we could devise a suitable geometry satisfying a synchronous condition for the particles, and resulting in high shunt impedance. To ensure synchronism in the $\pi/2$ normal mode in a periodic array of cavities, one can choose the cavity lengths so that the spacing between sequential excited cavities is $\beta\lambda/2$ corresponding to half an RF period. This gives the configuration shown in Fig. 4.16a, which contains equal-length cavities on axis. There are two excited accelerating cells per $\beta\lambda$ as for a π-mode structure, but these are separated by unexcited cells. This configuration does not lead to a high shunt impedance, because with the fields concentrated in only half of the available space, the net power dissipation is larger than for the π mode, where all of the cells can contribute acceleration to produce the same energy gain.

A better solution for retaining the advantages of the $\pi/2$ mode, while maintaining high shunt impedance, is to form a biperiodic chain in which the geometry of the excited accelerating cavities is optimized for best shunt impedance, and the unexcited cavities, also called *coupling cavities*, are chosen to occupy less axial space and are tuned to the same resonant frequency as the excited cavities. The two most popular geometries that have been invented (see Figs. 4.16b and 4.16c) are the on-axis coupled structure, [19] where the coupling cavities occupy a smaller axial space than the accelerating cavities, and the side-coupled cavity, [20] where the coupling cavities are removed from the beam line, leaving the beam axis completely available to the excited accelerating cavities. For both geometries the coupling cavities are coupled magnetically to the accelerating cavities, through slots cut in the outer walls. Both these solutions retain the unique properties of the $\pi/2$ mode of the periodic cavity chain, and provide the flexibility for obtaining the high shunt impedance that is characteristic of the π mode. The biperiodic cavity chain can also be analyzed using the coupled-circuit model. This model shows that to recover the $\pi/2$-mode properties, another step is required. To explain this, we present the results of the coupled-circuit model for a biperiodic array for $2N$ (coupling) cavities with frequency ω_c, which alternate with $2N + 1$ accelerating cavities with frequency ω_a; the total number of cavities is $4N + 1$. The dispersion relation from the model is

$$k^2 \cos^2 \frac{\pi q}{2N} = \left[1 - \frac{\omega_a^2}{\Omega_q^2}\right]\left[1 - \frac{\omega_c^2}{\Omega_q^2}\right], \quad q = 0, 1, \ldots, 2N \quad (4.45)$$

where k is the coupling constant between the accelerating and coupling cells, called *nearest neighbors*, and Ω_q is the mode frequency for normal mode q. The next-nearest-neighbor coupling between adjacent accelerating cells or adjacent coupling cells is ignored in our treatment. The quantity $\pi q/2N$ is an effective the phase advance per cavity of a traveling wave. The $\pi/2$ normal mode corresponds to $q = N$, which makes the right side of Eq. (4.45) equal to zero. Then, the dispersion relation for the $\pi/2$ mode has two solutions, $\Omega_q = \omega_a$ and $\Omega_q = \omega_c$. This situation is more clearly seen in the dispersion relation, shown in Fig. 4.17, where at $\pi/2$ there is a discontinuity in the dispersion curve, unless $\omega_a = \omega_c$. In general there are two branches, which are called the *lower* and *upper passbands*. Between them is the stopband, within which there are no normal-mode solutions. The discontinuity can be removed by tuning all cavities to have the same frequency so that $\omega_a = \omega_c$.

After tuning, the two passbands are joined at the $\pi/2$ mode, using methods discussed in the next section. It is generally found that the dispersion curve at this point, after tuning for equal resonant frequencies, changes to produce a nonzero slope, looking just as it would for a periodic structure. This effect was illustrated in Section 3.8, when we studied a periodic array of cavities with a resonant-coupling element.

To present the general results, we number the even or excited cavities $2n$, and the odd or unexcited cavities $2n + 1$, where $n = 0, 1, 2, \ldots$. The quality factor for the excited (accelerating) cavities is Q_a and for the unexcited (coupling) cavities it is Q_c. When beam is present we interpret Q_a as the loaded quality factor. The total number of cavities is odd, and the end boundary conditions are defined by half-cell accelerating cavities. We assume that the drive is located in an accelerating cavity labeled $2m$ and express the results for coupling cells with $n < m$ and for accelerating cells with $n \leq m$. Thus the numbering begins at an end cell where $n = 0$ and advances toward the drive cell located at $2n = 2m$. The frequency difference between the coupling and accelerating cavities, $\delta\omega = \omega_c - \omega_a$, is also the width of the stopband in this approximation. The field X_{2n} in accelerating cavity $2n$ is given to second-order relative to the field X_{2m} in the drive cavity as

$$X_{2n} \cong (-1)^{n-m} X_{2m} \left[1 - \frac{2(m^2 - n^2)}{k^2 Q_a Q_c} + j\frac{4(m^2 - n^2)}{k^2 Q_a}\frac{\delta\omega}{\omega_a}\right] \quad (4.46)$$

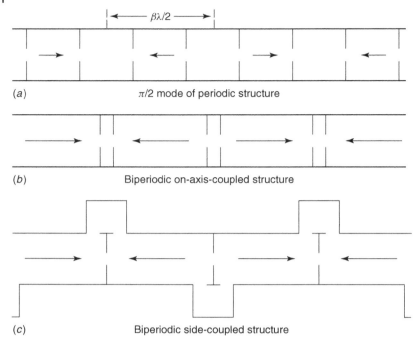

Figure 4.16 $\pi/2$-like-mode operation of a cavity resonator chain. (a) A periodic structure in $\pi/2$ mode, (b) a biperiodic on-axis coupled-cavity structure in $\pi/2$ mode, and (c) a biperiodic side-coupled cavity in $\pi/2$ mode.

The second and third terms in the bracket are power-flow correction terms and are both of second order. The second is the power-flow droop and the third is the power-flow phase shift. Both vary quadratically with n, as n decreases from the drive point. The result may also be written as

$$X_{2n} \cong (-1)^{n-m} X_{2m} \left[1 - \frac{2(m^2 - n^2)}{k^2 Q_a Q_c} \right] \exp\left[j \frac{4(m^2 - n^2)}{k^2 Q_a} \frac{\delta\omega}{\omega_a} \right] \quad (4.47)$$

When the structure is tuned to close the stopband, so $\delta\omega = 0$, the power flow-phase shift vanishes. The amplitude of the accelerating cells is largest at the drive cell, where $n = m$, and is smallest at the end cell $n = 0$.

The field in coupling cavity $2n + 1$ has a first-order term which is

$$X_{2n+1} \cong (-1)^{n-m} X_{2m} \left[\frac{2n+1}{k Q_a} \right] \exp(j\pi/2) \quad (4.48)$$

The amplitude of the coupling cells is largest at the coupling cell adjacent to the drive cell, where $n = m - 1$ and is smallest at the end cell, where $n = 0$. Note that up to first order Q_c does not affect the coupling-cell amplitude.

The application of the formulas we have presented is sometimes confusing because of the cavity numbering, for which we have retained the original nomenclature. We present an example that is hoped to clarify the notation. Consider a nine-cell structure with five accelerating cells and four coupling cells. The total number of cavities is $4N + 1 = 9$, and $N = 2$. The middle cell is numbered $2N = 4$, and it is typically chosen for the drive cell. Thus, with $2m = 4$ we have the drive cell index $m = 2$. The general result for the drive cell in the middle is $m = N$, and the drive cell number is $2m = 2N$. The results of Eqs. (4.46) through (4.48) are valid for cell index values $n = 0$ through $n = 2$. The end cell is an accelerating cell, which is numbered $2n = 0$, and the other accelerating cells are numbered $2n = 2$ and $2n = 4$. The coupling cells are numbered $2n + 1 = 1$ and $2n + 1 = 3$. The fields are symmetric with respect to the drive cell in this case, but in general, to get the results for cells with $n > m$, one can renumber the cells to count from the other end towards the drive point, and then apply the same formulas for these cells, using the new numbering.

4.12
Design of Coupled-Cavity Linacs

Four different types of biperiodic CCL structures have received the most study for linac applications, the side-coupled structure, the on-axis coupled structure, the annular coupled structure, [21] and the disk and washer structure [22]. These are shown in Fig. 4.18. Of these the side-coupled structure is the most widely used and its application in the LANSCE linac at Los Alamos is the most well-known example. The basic accelerating unit of the CCL is a multicell tank, ideally consisting of identical accelerating cells and identical coupling cells. The simplification of using identical length cells in a given tank is possible because the velocity change in a high-beam-velocity CCL tank is much smaller than in a DTL, which generally accelerates lower-velocity particles. The end cavities are actually not built as half-cells, even though in our treatment it was convenient to model them this way. To build a cavity as a half-cell that provides a proper termination to make the cavity look like an infinite periodic structure, it would be necessary to place a conducting plane at the center of each end cavity to provide perfect reflecting symmetry. But a real accelerating cavity needs a beam aperture, which would destroy the perfect symmetry anyway. In practice, full cells are used at each end, which are different than the interior cells because they couple to only one coupling cavity, and have a beam pipe connected to the other end. The departure from an exact periodic structure can cause changes in the dispersion curves, some of which may be related to ambiguities in the definition of an equivalent phase advance per period. Procedures have been developed for describing the modes for these conditions [23].

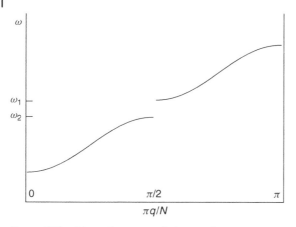

Figure 4.17 Dispersion curve of a biperiodic structure.

The cell lengths are equal to $\beta_d \lambda/2$, where β_d represents an average design velocity for the beam particle in that tank. The cell design is then somewhat different from that for a DTL (beam dynamics is discussed in more detail in Section 6.9). The choice of length for the CCL tanks results from a compromise among several factors. For protons the requirement for focusing usually sets a limit on the tank length, because a focusing element must be placed at the end of a tank and the spacing of focusing elements affects the overall focusing strength. Two tanks can still be electromagnetically coupled by means of a special cavity called a *bridge coupler*, a nominally excited cavity which is displaced from the beam line to make room for the quadrupole [24]. The bridge coupler is magnetically coupled through coupling cells to the end cells of the two adjacent tanks. A combination of separate tanks coupled together by bridge couplers and driven by one RF generator is usually called an *accelerating module*, and the overall length of the sequence of coupled tanks may be limited by the capacity of the RF power source. The other main consideration in setting the length of an accelerating module is the desire to limit the power-flow droop in the field distribution.

To get a rough idea of the order of magnitude of the typical frequency error Δf in a CCL cavity with resonant frequency f, we assume that $\Delta f/f \approx \Delta \ell / \lambda$, where $\Delta \ell$ is the machining error in the tank diameter, and λ is the RF wavelength. If $\Delta \ell = 0.03$ mm, $f = 1$ GHz, and $\lambda = 0.3$ m, we obtain $\Delta f = 100$ kHz. With $Q \approx 20,000$, the bandwidth of the resonator is about 50 kHz, and frequency tuning will be required to make the resonant-frequency error small compared with the bandwidth. Furthermore, after the initial construction of a multicell cavity, such as the side-coupled linac structure, systematic errors cause the average accelerating-cell frequency to be different from the average coupling-cell frequency. A systematic difference between the resonant frequencies of the accelerating and coupling cells results in a finite stopband between the upper and lower branches of the dispersion relation, which must be at least approximately closed by a suitable tuning procedure to realize the desired field stability. An added concern is the possibility of thermal instability that can occur with a nonzero stopband, when the accelerating mode lies in the upper branch above the stopband. This case is sometimes called a *negative stopband*, and if a small field tilt is present, the end-cell temperature at the high-field end will increase, which will produce local thermal expansion and a local resonant-frequency decrease. For the negative stopband case, a resonant-frequency decrease produces a local amplitude increase, and the field tilt will grow. If the accelerating mode lies in the lower branch below the stopband (*positive stopband*), a resonant-frequency decrease produces a field decrease, which gives thermal stability. Although a safe procedure would be to tune for a small positive stopband, in practice, thermal stability can often be achieved even in the presence of a small negative stopband. A more detailed description of these topics has been discussed by Dome [25] and summarized by Lapostolle [26]. Typically, the accelerating cells in the side-coupled linac are initially machined slightly too large, resulting in a low accelerating-cell frequency. The accelerating cells can be tuned up in frequency by producing small deformations on the outer walls at selected points where the walls have been made thin; the resulting deformations reduce the effective inductances. Likewise, the coupling cells are machined with a gap that is slightly too large. They can be tuned down in frequency by deforming the walls behind the gaps to reduce the average gap spacings, thus raising the capacitances of the coupling cells. Typically a 10-kHz residual stopband may remain after the tuning process is completed. The fine tuning of the frequency is accomplished by feedback control of the temperature of the cooling water, which produces dimensional changes in the cavity. The temperature coefficient is several kilohertz per degree Fahrenheit, and temperature control to about 1 °F is straightforward. The feedback loop uses the phase difference between the drive signal and an RF pickup signal as a measure of the frequency difference between the drive and the resonant frequencies. Resonant-frequency variations, as determined from a nonzero phase-error signal can be corrected by opening or closing a valve to control the temperature of the cooling water. The thermal time constant of this system is typically a few seconds, so that slow frequency drifts can be corrected.

The choice of transition energy from a DTL to a CCL is generally based on power efficiency. A CCL cell is only half as long as a DTL cell at the same velocity, and therefore for the same diameter has about half the inductance. The capacitance must be twice as large, which means the gap spacing is reduced. The smaller CCL gap is beneficial because it raises the transit-time factor. This must be weighed against another property. Unlike DTL cells, CCL cells have end walls on which additional power is dissipated. The effective shunt impedance increases as a function of β as the cell length increases. For low velocities the wall power loss is a disadvantage for the CCL, which favors the more open DTL structure, whereas at higher velocities the transit-time

4.13
Intercell Coupling Constant

It is known that the perturbation to the electromagnetic fields in a cavity, caused by an aperture on a cavity wall, may be represented by electric and magnetic dipoles [27]. For a circular aperture centered on the axis of the end walls of a pillbox cavity, excited in the TM_{010} mode, the field perturbations are represented by electric dipoles, and based on this theory, the intercell coupling constant for an array of electrically coupled cavities that form the iris-loaded waveguide, was presented in Section 3.11. Similarly, an aperture on the cavity outer wall, where the RF magnetic fields are dominant, can be represented by magnetic dipoles, and from this theory the intercell coupling constant for magnetically coupled cavities can be obtained. Consider an elliptical aperture or slot, which is oriented such that the major axis of the slot is aligned along the unperturbed magnetic field, or equivalently the slot is oriented to provide the maximum impediment to the flow of the unperturbed surface current. Consider N identical slots per cavity with identical geometry, as shown in Fig. 4.19, where slots in an azimuthally symmetric pillbox cavity, excited in a TM_{010} mode, are shown at different azimuthal angles, but at the same radius. The dimensionless intercell coupling constant can be shown [28] to equal

$$k = \frac{\pi N e_0^2}{6[K(e_0) - E(e_0)]} \frac{\mu_0 H_1 H_2 \ell^3}{U} e^{-\alpha d} \qquad (4.49)$$

where $e_0 = (1 - s^2/\ell^2)^{1/2}$, and s and ℓ are the minor and major axes of the slot. The quantity U is the stored energy of a cavity, d is the wall thickness, and α is the attenuation factor of a TE_{11}-mode wave that is transmitted through the slot from one cavity to the next. The attenuation factor is given in the approximation of a circular slot by

$$\alpha = \frac{2\pi}{\lambda} \left[\left(\frac{\lambda}{\lambda_c}\right)^2 - 1 \right]^{1/2} \qquad (4.50)$$

where $\lambda_c = 2\pi a/1.841 = 3.41a$ is the cutoff wavelength for a circular radial aperture $a = \ell = s$. We are picturing the slot as a uniform waveguide, whose length equals the thickness d. The TE_{11} mode is used, because it has the largest cutoff wavelength of all the uniform waveguide modes, and with $\lambda > \lambda_c$, the TE_{11} mode will have the smallest attenuation through any given wall thickness [29]. The magnetic fields H_1 and H_2 in amperes per meter are the unperturbed fields of the two cavities at the center of the slot, and $K(e_0)$

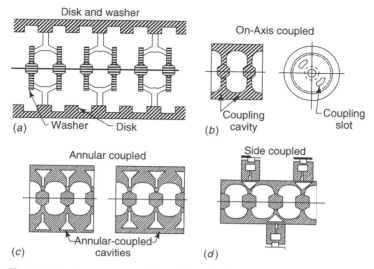

Figure 4.18 Four examples of coupled-cavity linacs are shown as labeled.

and $E(e_0)$ are the complete elliptic integrals of the first and second kinds. These functions are defined as

$$K(e_0) = \frac{\pi}{2}\left(1 + \left(\frac{1}{2}\right)^2 e_0^2 + \left(\frac{1 \cdot 3}{2 \cdot 4}\right)^2 e_0^4 + \left(\frac{1 \cdot 3 \cdot 5}{2 \cdot 4 \cdot 6}\right)^2 e_0^6 + \cdots\right) \quad (4.51)$$

and

$$E(e_0) = \frac{\pi}{2}\left(1 - \left(\frac{1}{2}\right)^2 e_0^2 - \left(\frac{1 \cdot 3}{2 \cdot 4}\right)^2 \frac{e_0^4}{3} - \left(\frac{1 \cdot 3 \cdot 5}{2 \cdot 4 \cdot 6}\right)^2 \frac{e_0^6}{5} - \cdots\right) \quad (4.52)$$

The numerical value of the coupling constant in Eq. (4.49) is typically a few percent, and is independent of frequency, if all the cavity dimensions, including the slot dimensions, scale with the RF wavelength. For the case where the two coupled cavities are geometrically different, such as for the side-coupled linac, Eq. (4.49) can be applied approximately by replacing the factor $H_1 H_2/U$ in Eq. (4.49) by $H_1 H_2/\sqrt{U_1 U_2}$, where U_1 and U_2 are the unperturbed stored energies in the two cavities. The coupling constant from Eq. (4.49) enters in the expression for the dispersion relation for magnetically coupled cavities,

$$\omega^2 = \omega_{\pi/2}^2 (1 + k \cos \psi) \quad (4.53)$$

where $\omega_{\pi/2}$ is 2π times the frequency of the $\pi/2$ mode, and ω is 2π times the frequency of the mode with phase advance per cell ψ, which has its usual range from zero to π. Note that the zero-mode frequency with $\psi = 0$ is higher than for the π mode in magnetically coupled structures, unlike the case with

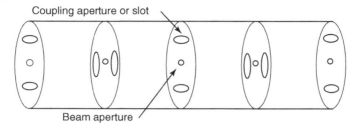

Figure 4.19 Coupled TM_{010}-mode pillbox cavities that are magnetically coupled through elliptical slots oriented with major axis parallel to the unperturbed surface magnetic field. In this example the parameter $N = 4$.

electrically coupled structures. Equation (4.49) shows that k is sensitive to the slot major axis ℓ, which is assumed to be oriented parallel to the unperturbed surface magnetic field, and k is weakly dependent on the minor axis of the slot. Generally, it is found that increasing the slot size increases the coupling constant at the expense of lowering the shunt impedance. A rule of thumb is that the shunt impedance decreases by about 3% for every percent increase of the coupling constant. Thus, if $k = 0.05$, the effective shunt impedance can be expected to decrease by about 15% from the value calculated for a cavity without coupling slots.

4.14
Decoupling of Cavities Connected by a Beam Pipe

Consider two identical cavities connected by a beam pipe each of which are excited in the same mode by different RF sources. To guarantee that these cavities are electrically independent, it is necessary to decouple the cavities electromagnetically by making the pipe long enough that electric coupling through the beam pipe will be negligible. The decoupling is important for superconducting cavities installed close together in the same cryomodule, whose phases and amplitudes must be independently controlled. Physically, one expects that the effects of the cavity–cavity coupling will be small, if the separation of the two modes of the two coupled cavities is small compared with the width of the cavity mode. This implies that the decoupling condition can be expressed as $kQ_L \ll 1$, where k is the coupling constant for the two cavities, and Q_L is the loaded quality factor of the mode of the individual cavities. Using the methods of Section 3.11, we find that decoupling the two pillbox cavities requires that

$$kQ_L = \frac{a^3 \varepsilon_0 E_0^2 e^{-\alpha h}}{3U} Q_L \ll 1 \qquad (4.54)$$

where a is the pipe radius, h is the pipe length, E_0 is the unperturbed electric field at the wall, and U is the cavity stored energy. Assuming TM_{01} mode in the pipe, the attenuation factor α is given by Eq. (4.50) with $\lambda_c = 2\pi a/2.405$. If the cavity parameters, and the beam pipe radius are fixed, Eq. (4.54) puts a lower bound on the choice of pipe length h that is required for decoupling. A more quantitative analysis of this problem was carried out by Spalek, [30] who concluded that for the amplitude difference of two identical cavities driven by equal-strength generators to be less than 1%, one must have $kQ_L < 0.005$. Assuming two identical N-cell cavities with cell-to-cell coupling k_0, Spalek concluded $k = k_0 e^{-\alpha h}/2N$, and the two cavities act independently, if the decoupling length is

$$h > -\ln(0.01 N/k_0 Q_L)/\alpha \qquad (4.55)$$

4.15
Resonant Coupling

The general approach of using resonant oscillators as coupling elements is called *resonant coupling*. We have already discussed the case of resonant coupling in a $\pi/2$-like mode of a biperiodic cavity chain. In general the dispersion relation is described by two passbands, separated by a stopband, as shown in Fig. 4.20. For an infinite structure, the stopband is eliminated or closed, when the frequency of the highest mode of the lower passband coincides with the frequency of the lowest mode of the upper passband. The modes that are joined may be either the zero or the π modes, shown in Fig. 4.21. The joining of the two passbands in this way is called *confluence*. The method is sometimes called *structure compensation* or *structure stabilization*, because when confluence is obtained, the slopes of the two dispersion curves where the passbands join are increased from zero to a maximum value.

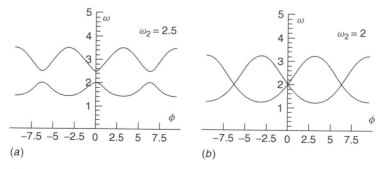

Figure 4.20 Dispersion curves of a structure with resonant coupling. (a) The curves when not at confluence, and (b) the curves at confluence.

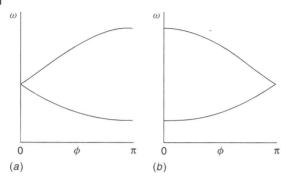

Figure 4.21 (a) Confluence at zero mode and (b) confluence at π mode.

The steepening effect of the slope at the point of confluence reflects the increase in the group velocity associated with the improved power flow along the tuned structure.

At first it may seem counterintuitive that we can improve the accelerating-mode field stability by tuning the structure for confluence, which brings modes closer in frequency. What happens as a result of confluence is that the tuning process improves the power flow, which is directly related to the local mode spacing. The creation of a finite slope where the passbands join, produces the desired increased mode spacing at the operating point, and usually this procedure produces a symmetrical or nearly symmetrical spacing between the nearest modes. The symmetry produces a first-order cancellation of the error-induced contributions from the nearest modes according to perturbation theory.

Another important example of resonant coupling is the stabilization of the fields in the zero-mode DTL cell array, where the accelerating mode of the cells is a TM_{010}-like mode, and where instead of external coupling cavities, the resonant-coupling elements are internal posts called *post couplers*, of electrical length $\lambda/4$. The post couplers, shown in Fig. 4.22, mount on the cylindrical wall across from the center of each drift tube, with typically one post coupler per cell. The posts extend radially inward, leaving a small gap near the body of the drift tube.

Each post may be thought of as being capacitively coupled through the small gap to each end of the opposing drift tube. An unbalanced excitation of the two adjacent cells results in a net excitation of the post. The posts do provide additional coupling between adjacent cells, but unlike the example of the side-coupled linac, the accelerating cells are already strongly coupled without the posts. Nevertheless, the posts function effectively as resonant-coupling elements. A distinctive feature of the post-coupled structure is that the post fields share the same structure volume as the fields of the accelerating cell. This sharing of the same cavity volume results in a more complicated interaction between all the different elements than one finds in

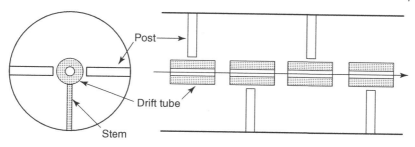

Figure 4.22 Post-coupled drift-tube linac structure.

a coupled-cavity structure. It is found experimentally that if the azimuthal positions of sequential posts are alternated by 180° from one side of the drift tubes to the other, the post-mode passband is widened, resulting in a larger mode spacing near the accelerating mode at confluence, and a better-stabilized structure. The drift tubes are each supported by a stem, which is oriented at 90° to the post to reduce the coupling between the posts and stems. Because of their location at the symmetry points of the structure, the posts have little effect on the $^0TM_{010}$ accelerating mode, [31] but they capacitively load the $^0TE_{110}$-like mode of each accelerating cell. At confluence the upper frequency of the post-mode passband is tuned to the accelerating-mode frequency, which is the lowest frequency of the TM_{010} passband. To maintain a high degree of field stabilization it is found that one should avoid letting the $^0TE_{110}$-mode frequency from becoming larger than the accelerating-mode frequency [32].

We note that the resonant-coupling elements for the DTL and side-coupled linac function in different ways. The coupling of the DTL accelerating cells is very large, as is evidenced by a large bandwidth when no posts are present. But this large coupling does not promote electromagnetic energy flow for the $^0TM_{010}$ accelerating mode. For this mode, the Poynting vector is directed transversely, which means that the electromagnetic-field energy cycles between the inside and the outsides of the cavity volume. The flow of energy longitudinally is zero, which means that the axial group velocity is also zero. To produce a large mode spacing and a nonzero axial group velocity in the vicinity of the accelerating mode, the accelerating mode must be perturbed to create a longitudinal component of Poynting vector. This is accomplished very effectively with the post couplers after bringing the TE post mode into confluence with the accelerating mode. The nonzero Poynting-vector longitudinal component, resulting from the electric field of the post mode and the magnetic field of the accelerating mode, produces a finite axial group velocity and a dispersion relation with a nonzero slope. The posts provide the mechanism that generates the electromagnetic energy flow, needed to produce a longitudinal component of the Poynting vector. For the CCL, the direct coupling of adjacent accelerating cavities through the beam holes is usually very small. The main coupling between adjacent accelerating cavities is produced through the coupling slots and through the coupling cavities. For the side-coupled linac, the longitudinal energy flow is a consequence of providing a bypass for the electromagnetic energy flow through the resonant side cavities.

If the direct coupling between the individual coupling elements is very strong, it is more convenient to describe the coupling resonators using the normal modes of an array of coupling elements. The fields of the normal modes are then distributed throughout the structure. An example of this case is the disk and washer structure. This is a structure with disks that connect to the outer wall, and washers, supported by longitudinal stems from the disks, that are concentric with the beam axis and are mounted halfway between the disks. The accelerating mode is the πTM_{020}-like mode, where the electric field is concentrated in the accelerating gaps that are centered in the same plane as the disks, and with a node in the plane of the washers. The coupling mode is a πTM_{010} mode that has most of the electric field concentrated between the disks at large radius. There is a considerable fringe field near the washers at large radii, with a node in the plane of the disks. The electric-field patterns of these modes are shown in Fig. 4.23.

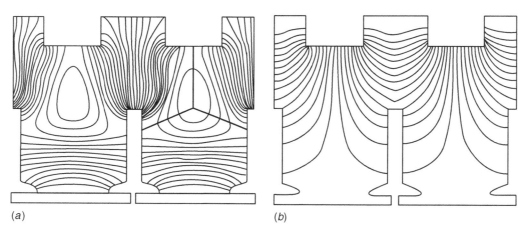

(a) (b)

Figure 4.23 Electric-field pattern of disk and washer modes that are degenerate at confluence. (a) The πTM_{020}-like accelerating mode and (b) the πTM_{010}-like coupling mode. For a structure with finite length, the boundary conditions prevent the excitation of the πTM_{010} mode in the error-free case. It is not easy to look at the detailed field distribution in this structure and to see a $\pi/2$-like mode pattern, since the entire physical space of the structure contains the superposition of the fields from both modes, if both were excited. The coupling modes must have low fields near the beam axis to avoid undesirable perturbations on the beam. This is generally not considered to be a problem for the post-coupled DTL, or the disk and washer structure. The concept of resonant coupling is an important development in the field of RF linacs that has enabled the construction of long standing-wave structures, which is an economical choice for many accelerator applications.

4.16 Accelerating Structures for Superconducting Linacs

Another basic application of resonant coupling is to electromagnetically couple together two cavities that nominally have the same frequency, by introduction of a resonant-coupling element that is coupled to both cavities. As seen by the RF system this effectively creates a single very stable electromagnetic structure, which can be excited by a single RF drive. The basic principle to make this work is to provide a resonant-coupling element, such as a post in a DTL or a coupling cell in a CCL, that is unexcited when driven by the two cavities to be coupled together. Generally, it is necessary to tune the three-resonator assembly to operate in a $\pi/2$ mode. With perturbations present an unbalanced net drive to the resonant-coupling element will produce excitation of that element and promotes power flow between the two cavities that compensates for the imbalance.

4.16
Accelerating Structures for Superconducting Linacs

Accelerating structures based on a TEM mode are commonly used for acceleration of low- and medium-velocity beams. We now discuss these TEM mode structures, making use of the excellent review of medium-β superconducting accelerating structures that has been presented by Delayen [33, 34]. There are two general types of TEM resonant structures, those based on electrical length $\lambda/4$, and those based on electrical length $\lambda/2$.

$\lambda/4$ Superconducting Structures

Delayen [33] provides examples of $\lambda/4$ resonators. Figure 4.24 shows an example of a superconducting $\lambda/4$ resonator with a single center conductor, more generally called a *loading element*. Delayen classifies the split-ring resonator shown in Fig. 4.25, as an example of a quarter-wave resonator containing two loading elements. The $\lambda//4$ resonators have been typically designed for use over a velocity range of $\beta = 0.05$ to about $\beta = 0.15$.

$\lambda/2$ Superconducting Structures

Half-wave resonators include both the coaxial geometry, as shown in Fig. 4.26, and the spoke geometry as shown in Figs. 4.27 and 4.28. Spoke resonators with multiple spokes that are sequentially rotated by 90° as shown in Fig. 4.28 are also called *crossbar H-type resonators*. Half-wave resonators have been designed for use from near $\beta = 0.1$ to near to $\beta = 0.5$.

122 4 Standard Linac Structures

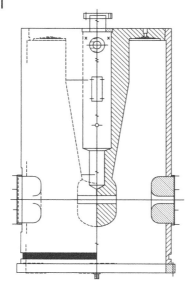

Figure 4.24 University of Washington, 150-MHz $\beta = 0.2$ coaxial quarter-wave resonator (courtesy of J. R. Delayen, Ref. 33).

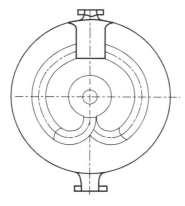

Figure 4.25 Example of a split-ring resonator concept for a 115-MHz $\beta = 0.13$ resonator with two loading elements for a rare-isotope accelerator (courtesy of J. R. Delayen, Ref. 33).

TM Superconducting Structures

The Alvarez DTL and the CCL may be considered as examples of normal-conducting coupled multicell TM_{010} cavities. Examples of coupled superconducting multicell TM_{010} cavities are shown in Fig. 4.29. These are also called *elliptical cavities*. Cavities of this type are designed for use over a velocity range near $\beta = 0.5$ to $\beta = 1.0$.

Some important differences between the transverse magnetic (TM) and TEM resonators include (1) a significant size difference, (2) a difference in Lorentz detuning and microphonics, and (3) a difference in multipacting. The inner diameter of TM resonators is about 0.9λ, whereas $\lambda/2$ structures have

Figure 4.26 350-MHz $\beta = 0.12$ coaxial half-wave resonator with a single loading element (courtesy of J. R. Delayen, Ref. 33).

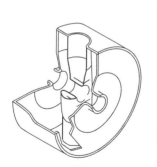

Figure 4.27 850-MHz, $\beta = 0.28$ spoke resonator (courtesy of J. R. Delayen, Ref. 33).

an inner diameter of nearly a factor of 2 less. The $\lambda/2$ resonator is smaller and lighter. For the same transverse size it would have half the frequency, and for the same length would have fewer cells, and would have a larger velocity acceptance. The $\lambda/2$ resonators are less sensitive to microphonics and Lorentz force detuning than TM resonators of the same frequency and same β. Multipacting has limited the performance of TM resonators, but is nearly eliminated in the TM elliptical resonators. Furthermore simulation programs

124 | 4 Standard Linac Structures

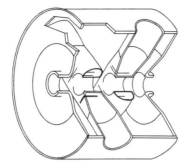

(a)

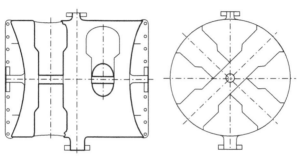

(b)

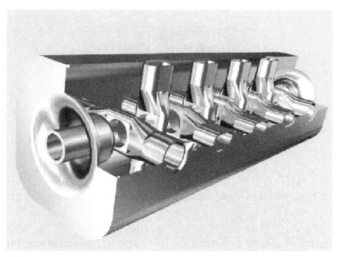

(c)

Figure 4.28 Spoke cavities with multiple loading elements. (a) An 850-MHz, $\beta = 0.28$ double spoke concept. (b) A 345-MHz, $\beta = 0.4$ double spoke concept. (c) A 700-MHz, $\beta = 0.2$ eight-spoke concept (courtesy of J. R. Delayen. Ref. 33).

Figure 4.29 TM$_{010}$ cavities for $\beta < 1$. (a) 805-MHz, $\beta = 0.82$, six-cell cavity, (b) 805-MHz, $\beta = 0.62$, six-cell cavity (courtesy of J. R. Delayen, Ref. 33).

have been developed that allow the designer to avoid multipacting. But $\lambda/2$ resonators almost always exhibit multipacting, and no such simulation codes are available that are reliable enough to guarantee no multipacting. Nevertheless, multipacting almost always occurs in TEM resonators, but no cases are known for which multipacting prevented operation of the cavity.

RF Properties and Scaling Laws for TM and $\lambda/2$ Superconducting Structures

Delayen [33] has compiled data and calculations for TM and $\lambda/2$ superconducting resonators, shown in Figs. 4.30 to 4.35. Figure 4.30 shows the peak surface electric field to accelerating-gradient ratio versus the design velocity β. The peak surface electric field is important because it can be a source of field emission that absorbs RF power. This electron loading is an important practical limitation for achieving high gradients. In both TM and TEM structures the peak surface electric field occurs near the iris. For TM structures the ratio is about 2 at $\beta = 1$ and increases as β decreases. For TEM structures the geometry of the loading element near the beam aperture can be optimized to obtain a nearly constant peak surface field value around its circumference that results in a ratio of about 3.3 independent of β. This optimization can be made with minimal effect on the peak surface magnetic field.

The peak surface magnetic field to accelerating-gradient ratio versus design velocity β is shown in Fig. 4.31. The peak surface magnetic field is important

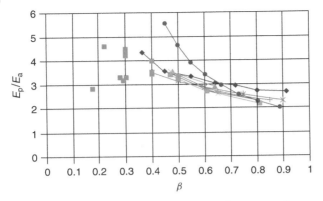

Figure 4.30 Peak surface electric field to accelerating-gradient ratio versus design velocity β compiled by Delayen. Points joined by lines are for TM resonators, while the squares not connected by lines are for $\lambda/2$ resonators (courtesy of J. R. Delayen, Ref. 33).

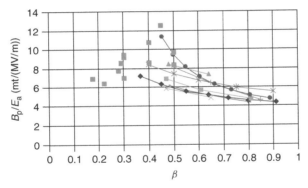

Figure 4.31 Peak surface magnetic field to accelerating-gradient ratio versus design velocity β, compiled by Delayen. Points joined by lines are for TM resonators, while the squares not connected by lines are for $\lambda/2$ resonators (courtesy of J. R. Delayen, Ref. 33).

because too large a value will ultimately result in a quench that destroys the superconductivity. For TM structures the peak surface magnetic field occurs at the equator. The ratio typically increases with decreasing β from about 4 to 4.5 mT at 1 MV/m at $\beta = 1$ to 6 to 8 mT at 1 MV/m for $\beta = 0.5$. For TEM structures the maximum occurs where the loading element intersects the outer wall, and can be optimized to achieve nearly constant peak surface magnetic-field values around the circumference of the loading element. Delayen quotes achievable values of 7 to 8 mT, independent of β.

The quality factor $Q = \omega U/P$ is proportional to the ratio of stored energy to power dissipation per cycle. The geometry factor, defined as $G = QR_s$, where

R_s is the RF surface resistance, is a very useful parameter for comparing different cavity shapes. This is because QR_s depends on the cavity's shape, but not its size or wall material. The quality factor Q does depend on the cavity size because it depends on R_s which depends on the resonant frequency, and the cavity size also depends on frequency. For TM structures the stored energy is approximately proportional to the cavity volume, which means that since the length depends on β, the stored energy will be approximately proportional to β. The power dissipation is approximately proportional to surface area, and for a typical TM structure like a low-β elliptical structure, the surface area is dominated by the end-wall surface area and is more weakly dependent on β. Neglecting this weak β dependence, QR_s should be approximately linearly dependent on β for a TM structure, and Delayen gives

$$QR_s \approx 275\beta (\Omega) \tag{4.56}$$

independent of the number of cells. In Ref. 34, Delayen uses a transmission line model to obtain a scaling formula for $\lambda/2$ structures,

$$QR_s \approx 270\beta (\Omega) \tag{4.57}$$

Figure 4.32 shows the results for QR_s as a function of β.

Shunt Impedance for TM and $\lambda/2$ Superconducting Structures

Following Delayen [33] we use the definition of shunt impedance $R_{sh} = (V_0 T)^2 / P$ where $V_0 T$ is the maximum voltage gained by a particle as

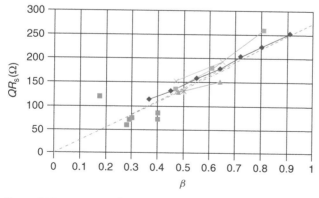

Figure 4.32 Geometry factor QR_s versus design velocity β, compiled by Delayen. Points joined by lines are for TM resonators, while the squares not connected by lines are for $\lambda/2$ resonators. The dashed line shows the scaling formula for TM resonators, and the dashed-dotted line shows the scaling formula for $\lambda/2$ resonators (courtesy of J. R. Delayen, Ref. 33).

a function of input velocity at the crest of the RF phase. The shunt impedance is a measure of the voltage gained by a particle of optimum velocity to the power dissipation averaged over an RF cycle. It depends on the wall material. There are two shunt-impedance-related quantities that are independent of wall material, R_{sh}/Q and $R_{sh}R_s$. These parameters also depend only on the shape and not on size or wall material. As we shall see, for TM structures these quantities are proportional to the number of cells and are quoted per cell. For TEM structures they are proportional to the number of loading elements and are quoted per loading element.

Suppose we begin with an infinitely long periodic structure operating in a π mode, and we wish to obtain from this a finite length cavity by inserting two conducting end walls that are perpendicular to the beam axis. These end walls would properly terminate the finite length structure if the fields are unchanged as a result of inserting the plates. If the end walls contained no beam aperture, the way to terminate the cavities without distorting the fields would be to end with half-cells that intersect the outer wall at the equator. However, perturbations such as those caused by the beam aperture can complicate matters, depending on the magnitude of the perturbation. Delayen argues that the importance of the perturbation to the fields from the beam aperture on the end walls depends on the strength of the cell-to-cell coupling. For the case of the TEM structure where the outer cell-to-cell openings and the corresponding magnetic intercell coupling are very large, typically about 20% or more, the field profile is very insensitive to perturbations such as the end-wall beam aperture. Since the fields are approximately unchanged by such a perturbation, the proper end-wall placement for the TEM structures is still nearly that which corresponds to half-cell terminations. However, for the TM structures that use electric cell-to-cell coupling through the beam holes, the coupling is small, typically about 2%. In this case the end-cell perturbation caused by the beam aperture has a large effect on the fields. One finds that the resulting termination for the TM structures is no longer a half-cell, but in this case the end cells are approximately the same as the interior cells, which are approximately full cells.

Figure 4.33 shows R_{sh}/Q per cell or loading element versus design velocity β. Delayen obtains a scaling formula for R_{sh}/Q using the following arguments. For a single cell of a TM structure,

$$\frac{R_{sh}}{Q} = \frac{(V_0 T)^2}{\omega U} = \frac{(E_0 T L)^2}{\omega U} \quad (4.58)$$

The stored energy is approximately proportional to volume and therefore proportional to β. The cell length L is also proportional to β. The simplest argument is that R_{sh}/Q is proportional to β. But, Delayen finds that this argument is not correct. One needs to take into account a decrease in the on-axis field E_0 with decreasing cell length, relative to the electric field at an

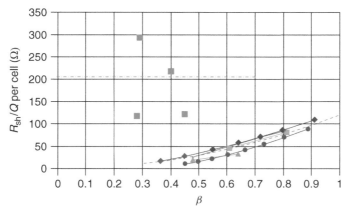

Figure 4.33 R_{sh}/Q per cell or loading element versus design velocity β, compiled by Delayen. Points joined by lines are for TM resonators, while the squares not connected by lines are for $\lambda/2$ resonators. The dashed line shows the scaling formula for TM resonators, and the dashed-dotted line shows the scaling formula for $\lambda/2$ resonators (courtesy of J. R. Delayen, Ref. 33).

approximately fixed iris radius. R_{sh}/Q decreases more rapidly than linearly, and the TM data are best fit by a quadratic dependence,

$$\frac{R_{sh}}{Q} \approx 120\beta^2 (\Omega) \tag{4.59}$$

per cell. For $\lambda/2$ structures Delayen's transmission line model gives

$$\frac{R_{sh}}{Q} \approx 205 (\Omega) \tag{4.60}$$

per loading element.

Figure 4.34 shows the product of shunt impedance R_{sh} times RF surface resistance R_s per cell or loading element versus design velocity β. By multiplying the scaling formulas already obtained for R_{sh}/Q and QR_s, we have per cell for TM or per loading element for $\lambda/2$

$$\begin{aligned} R_{sh}R_s &\approx 33 \times 10^3 \beta^3 (\Omega)^3, \quad TM \\ R_{sh}R_s &\approx 55 \times 10^3 \beta (\Omega)^3, \quad \lambda/2 \end{aligned} \tag{4.61}$$

Stored Energy for TM and $\lambda/2$ Superconducting Structures

Figure 4.35 shows the electromagnetic stored energy per cell or per loading element at $E_{acc} = 1\,\text{MV/m}$ and at 500 MHz. Delayen also obtains scaling

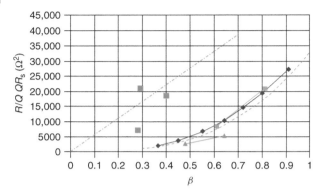

Figure 4.34 Product of shunt impedance R_{sh} times RF surface resistance R_s per cell or loading element versus design velocity β, compiled by Delayen. Points joined by lines are for TM resonators, while the squares not connected by lines are for $\lambda/2$ resonators. The dashed lines show the scaling formulas for the two types of resonators (courtesy of J. R. Delayen, Ref. 33).

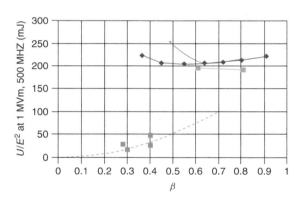

Figure 4.35 Stored energy per cell or loading element versus design velocity β at 1 MV/m for 500-MHz structures, compiled by Delayen [33]. Points joined by lines are for TM resonators, while the squares not connected by lines are for $\lambda/2$ resonators. The dashed line is the scaling law developed by Delayen for $\lambda/2$ resonators (courtesy of J. R. Delayen, Ref. 33).

formulas for these which can be written as

$$U(J) \approx 0.2 E_{acc}^2 (\text{MV/m}) \left[\frac{500}{f(\text{MHz})}\right]^3, \quad TM$$

$$U(J) \approx 0.2 \beta^2 E_{acc}^2 (\text{MV/m}) \left[\frac{500}{f(\text{MHz})}\right]^3, \quad \lambda/2 \tag{4.62}$$

Scaling Formulas for $\lambda/4$ Superconducting Structures

Delayen [35] obtains scaling formulas for $\lambda/4$ resonators from his scaling formulas for $\lambda/2$ resonators given above by using a simple model where a $\lambda/4$ resonator is assumed to be a $\lambda/2$ resonator cut in two equal parts. His resulting formulas are

$$QR_s(\Omega) \approx 270\beta \qquad (4.63)$$

independent of the number of loading elements.

$$R_{sh}/Q \approx 410(\Omega) \qquad (4.64)$$

per loading element.

$$R_{sh}R_s \approx 111\beta(\Omega^2) \qquad (4.65)$$

per loading element.

Problems

4.1. Plot the transit-time factor for a π-mode two-gap cavity versus β/β_s in the range $0.5 < \beta/\beta_s < 3$. Assume a square-wave field distribution in the gaps, and assume $g/\beta_s\lambda = 1/4$. Is the transit-time factor maximum when $\beta = \beta_s$?

4.2. Suppose you want to accelerate deuterons synchronously in a DTL that has been designed and operated for protons in a $1\beta\lambda$ mode. Assume the deuteron and proton masses are related by $m_D = 2m_P$, and assume the synchronous phase ϕ is unchanged. Calculate the required value of E_0T for the following two cases. (a) Use a $1\beta\lambda$ mode for the deuteron operation. If E_0 was already as high as you could reliably operate, is this solution practical? (b) Use a $2\beta\lambda$ mode for the deuteron operation. (c) Suppose the ratio of gap to cell length is $1/4$. Calculate the transit-time factor for the $1\beta\lambda$ and $2\beta\lambda$ cases using the simple formula $T = (\sin x)/x$, where $x = \pi g/\beta\lambda$. What is the ratio of E_0 for the $2\beta\lambda$ deuteron case to that for proton operation?

4.3. Plot the dispersion relation (mode frequency $\Omega_q/2\pi$ versus phase advance per cavity $\pi q/N$) for $N+1 = 5$ coupled cavities with uncoupled cavity frequency $\omega_0/2\pi = 400$ MHz and coupling constant $k = 0.05$.

4.4. Consider $N+1$ coupled cavities with an uncoupled cavity frequency ω_0 and coupling constant $k \ll 1$. (a) Write the expressions for the frequencies of the π mode and the mode closest to the π mode. (b) Calculate the fractional difference between the two mode frequencies. Obtain an approximation for this result when $N \gg 1$. How does this depend on k and N? (c) If $k = 0.02$ and $N+1 = 30$, calculate the fractional difference between the frequencies of the two modes. Compare this with the width of the π mode assuming the Q value for the π mode is 20,000. Are the two modes adequately separated?

4.5. Repeat parts (a), (b), and (c) of Problem 4.4. for the $\pi/2$ mode of $N+1$ coupled oscillators.

4.6. A CCL module that operates in a $\pi/2$ mode has 135 excited (accelerating) cells and 134 nominally unexcited coupling cells, and is driven in the central excited cell by a single klystron. The parameters are $Q_a = 20{,}000$ for the accelerating cells, $Q_c = 2000$ for the coupling cells, the frequency is 1280 MHz, and the coupling constant is $k = 0.05$. Assume the beam power is negligible. (a) Calculate the power-flow droop in the field excitation of the end cavities and sketch the field excitation along the module relative to the drive cell (for the accelerating cells only). (b) Calculate the power-flow phase shift in the end cavities and sketch the power-flow phase shift along the module (for the accelerating cells only), assuming the stopband is open by $\delta\omega/2\pi = 100$ kHz. (c) Calculate the field excitation of the two coupling cells adjacent to the drive cell relative to the drive cell amplitude. Sketch the excitation of the coupling cells along the module relative to the drive cell excitation.

4.7. Resonant-frequency errors in a coupled-cavity system that are small compared with the spacing of the normal modes might not seem to be serious, because the modes still remain well separated. But even small cavity-frequency errors can cause nonuniformity of the field distribution that may be unacceptable for accelerators. Consider the weakly coupled three-cavity system operating in π mode, with an end-cell frequency error $\delta\omega_0/\omega_0 = 10^{-3}$ ($2\delta\omega_0$ is the frequency difference of the two end cells), a middle-cell frequency error $\delta\omega_1 = 0$ (the middle cell has same frequency as the average of the end cells: ω_0), and coupling constant $k = 0.01$. (a) Ignoring the frequency error, calculate the frequency spacing of the three normal modes relative to $\omega_0/2\pi$. Is the cavity-frequency error small compared with the frequency spacing of the modes? (b) Calculate the perturbed π-mode field vector for this case. Describing the perturbed distribution as a linear field tilt, what is the fractional end-to-end field tilt? (c) Calculate the perturbed $\pi/2$-mode field vector. If only the end cells are used for acceleration, would the $\pi/2$ mode give a field distribution that is more uniform in the presence of the same cavity-frequency errors?

4.8. We consider the effects of power losses in the coupled three-cavity system. Assume there are no frequency errors ($\delta\omega_0 = \delta\omega_1 = 0$), a coupling constant $k = 0.01$, and $Q = 10^4$ for each mode. Assume the RF drive is located in one end cell. (a) Calculate the field vector for the zero, $\pi/2$, and π modes. (b) Calculate the power-flow phase shift in the end cell without the drive for the zero, $\pi/2$, and π modes. Calculate the $\pi/2$-mode power-flow droop in the end cell without the drive. (c) Assume the loaded Q is reduced to $Q = 10^3$, and repeat part (b).

References

1. Bohne, D., Proc. 1976 Linear Accel. Conf., September 14–17, (1976), Chalk River, Ontario, Canada, AECL-5677, p. 2.
2. For a review of superconducting heavy-ion linacs, see Bollinger, L.M., *Ann. Rev. Nucl. Part. Sci.* **36**, 375 (1986); or Bollinger, L.M. Proc. 1992 Linear Accel. Conf., August 24–28, 1992, Ottawa, Ontario, Canada, AECL-10728, p. 13.
3. Wideröe, R., *Arch. Electrotech.* **21**, 387 (1928).
4. Sloan, D.H. and Lawrence, E.O., *Phys. Rev.* **38**, 2021–2032 (1931).
5. Kaspar, K., Proc. 1976 Linear Accel. Conf., September 14–17, 1976, Chalk River, Ontario, Canada, AECL-5677, p. 73.
6. Ratzinger, U., Proc. 1988 Linear Accel. Conf., 1988, CEBAF~Rep. 89-001, p. 185.
7. Alvarez, L.W., *Phys. Rev.* **70**, 799 (1946); Alvarez, L.W. et al., *Rev. Sci. Instrum.* **26**, 111–133 (1955).
8. Calculated by Billen, J. from the electromagnetic-field-solver program SUPERFISH.
9. Halbach, K. and Holsinger, R.F., *Part. Accel.* **7**, (No. (4)), 213–222 (1976).
10. Prior to 1995, the linac at Los Alamos was called LAMPF for Los Alamos Meson Physics Facility. The new name is LANSCE, which means Los Alamos Neutron Science Center.
11. Warner, D.J., Proc. 1988 Linear Accel. Conf., 1988, CEBAF Rep. 89-001, p. 109.
12. Swenson, D.A. and Stovall, J., Los Alamos Internal Report, MP-3-19, January, 1958; Boicourt, G.P. AIP Conf. Proc. 177: *Linear Accelerator and Beam Optics Codes*, ed. Eminhizer, C.R., American Institute of Physics, New York, 1988, pp. 1–21.
13. Nagel, D.E. Knapp, E.A. and Knapp, B.C., *Rev. Sci. Instrum.* **38**, 1583–1587 (1967).
14. Strictly speaking, in an accelerator the end oscillator would correspond to half-length cells with conducting planes inserted at the center plane of symmetry and with no beam holes.
15. Mathews, Jon. and Walker, R.L., *Mathematical Methods of Physics*, 2nd ed., W. A. Benjamin, Inc., 1970, p. 286; Wangler, T.P. *Calculations for Three Coupled Oscillators-Including Eigenvalue Problem, Perturbation Theory and Losses*, Los Alamos Group AT-1 Report AT-1 : 84–299, August 29, 1984.
16. Note that the second-order terms appear in the eigenfrequency and in the first and third elements of the eigenvector.
17. Nagel, D.A., unpublished Los Alamos technical memo P-11/DEN-2, August 21, 1963.
18. Knapp, E.A., Proc. 1964 Linac Conf., MURA, Stoughton, Wis. 1964, p. 35.
19. Schriber, S.O., Heighway, E.A. and Funk, L.W., Proc. 1972 Linear Accel. Conf., Los Alamos Laboratory Report LA-5115, p. 140; Schriber, S.O., Proc. 1976 Linear Accel. Conf., September 14–17, 1976, Chalk River, Ontario, Canada, AECL-5677, p. 405.
20. Knapp, E.A., Knapp, B.C. and Potter, J.M., *Rev. Sci. Instrum.* **39**, 979–991 (1968).
21. The annular coupled structure is discussed in *Radio-Frequency Structure Development for the Los Alamos/NBS Racetrack Microtron*, compiled and edited by Jameson, R.A., Stokes, R.H., Stovall, J.E. and Taylor, L.S., Los Alamos National Laboratory Report LA-UR-83-95, January 21, 1983.
22. Andreev, V.G. et al., Proc. 1972 Linear Accel. Conf., Los Alamos, N.M., Los Alamos Scientific Laboratory Report LA-5115, p. 114.
23. Schriber, S.O., *Fitting of an Ordered Set of Mode Frequencies*, Chalk River Nuclear Laboratories Report AECL-3669, Chalk River, Ontario, October, 1970.
24. Knapp, E.A., High energy structures, in *Linear Accelerators*, Lapostolle, P.M. and Septier, A.L., North-Holland Publishing Company, Amsterdam, and John Wiley & Sons, New York, 1970, p. 615.

25 Dome, G. Review and survey of accelerating structures, in *Linear Accelerators*, eds. Lapostolle, P.M. and Septier, A.L., North-Holland Publishing Company, Amsterdam, and John Wiley & Sons, New York, 1970, p. 615.

26 Lapostolle, P.M., *Proton Linear Accelerators: A Theoretical and Historical Introduction*, Los Alamos Report, LA-11601-MS, July, 1989, p. 64.

27 Bethe, H.A., *Phys. Rev.* **66**, 163 (1944).

28 See Eqn. (44) from Gao, J., *Nucl. Instrum. Methods* **A311**, 437–443 (1992).

29 In the case of TM_{010} cavities that are electrically coupled through a hole on the axis, the symmetry of the unperturbed field on the axis, prohibits the excitation of the TE_{11} waveguide mode in the hole. For the case treated in Section 3.11, we assumed that the waveguide mode with the smallest attenuation factor was the TM_{01} mode.

30 Spalek, G., Los Alamos Group AT-1 memorandum, AT-1 : 91–124, March 26, 1991.

31 We adopt a nomenclature for the operating mode, denoted by $^0TM_{010}$, where the superscript gives the specific mode of the passband corresponding to normal mode spectrum of the TM_{010} cavity mode.

32 Lapostolle, P.M., *Proton Linear Accelerators: A Theoretical and Historical Introduction*, Los Alamos Report, LA-11601-MS, July 1989, pp. 64–65.

33 Delayen, J.R., *Medium-β Superconducting Accelerating Structures*, Proc. of the 10th Superconducting RF Workshop, Tsukuba, Japan, 2001.

34 Delayen, J.R., Design of low velocity superconducting accelerating structures using quarter-wavelength resonant lines, *Nucl. Instrum. Methods* **A259**, 341–357 (1987).

35 Delayen, J.R. private communication.

5
Microwave Topics for Linacs

In this chapter, we present some general topics from the field of microwaves and microwave systems that are important for the linac application. We will begin with descriptions of the resonant cavity that constitutes the most fundamental element in an RF linac. This is followed by discussions of coupling of waveguides to cavities, including a waveguide model of an iris-coupled cavity that illustrates the wave-interference effects in the drive line, and the build up of the cavity fields as a result of reflections of waves inside the cavity. This is followed by elementary discussions of the klystron, multipacting, field emission, and electric breakdown. Finally, we present some special accelerator-physics topics including the Boltzmann–Ehrenfest theorem of adiabatic invariants of an oscillator, the Slater perturbation theorem used as the basis of cavity-field measurements, the quasistatic approximation for solving Maxwell's equations, and the Panofsky–Wenzel theorem describing transverse deflections of particles from RF fields.

5.1
Shunt Resonant Circuit Model

A parallel resonant circuit driven by a current generator is the simplest model for describing a single mode of an accelerating cavity. This is because a cavity behaves like a parallel resonant circuit, when viewed from the plane of a detuned short circuit in the input waveguide, as is discussed in Section 5.4. We consider a circuit, shown in Fig. 5.1, with an inductance L, a capacitance C, and circuit shunt resistance R, driven by a simple current generator that supplies a current $I(t)$ at frequency ω. In Section 5.4, we will improve the model to include the input waveguide and the waveguide-to-cavity coupling mechanism. The driving current from the generator produces a voltage $V(t)$ across the circuit that can be identified with the axial voltage in the cavity. The total current is

$$I(t) = C\dot{V} + \frac{\int V\, dt}{L} + \frac{V}{R} \tag{5.1}$$

RF Linear Accelerators. 2nd, completely revised and enlarged edition.
Thomas P. Wangler
Copyright © 2008 Wiley-VCH Verlag GmbH & Co. KGaA, Weinheim
ISBN: 978-3-527-40680-7

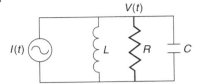

Figure 5.1 Circuit model of cavity resonator.

and by differentiation

$$\frac{\dot{I}}{C} = \ddot{V} + \frac{\omega_0 \dot{V}}{Q} + \omega_0^2 V \qquad (5.2)$$

where $\omega_0 = 1/\sqrt{LC}$ is the resonant frequency. The stored energy $U = CV_0^2/2$, the dissipated power $P = V_0^2/2R$, where V_0 is the peak voltage, and the quality factor $Q = \omega_0 U/P = \omega_0 RC$. Equation (5.2) is the equation for a damped, driven oscillator.

To derive the steady-state solution, we express $I(t) = I_0 e^{j\omega t}$ and $V(t) = V_0 e^{j(\omega t + \phi)}$, where I_0 and V_0 are real amplitudes, and ϕ is the phase of the resonator voltage relative to the driving current. Substituting these expressions into Eq. (5.2), we obtain the steady-state solution,

$$V(t) = \frac{RI_0 e^{j(\omega t + \phi)}}{\sqrt{1 + y^2}} \qquad (5.3)$$

where $\phi = -\tan^{-1} y$, and y is called the *detuning factor*, defined by

$$y \equiv Q\left(\frac{\omega}{\omega_0} - \frac{\omega_0}{\omega}\right) \qquad (5.4)$$

We define $\delta\omega = \omega - \omega_0$ and a useful approximation when $Q \gg 1$ is $y \cong 2Q\delta\omega/\omega_0$. From Eq. (5.3) the circuit shunt impedance [1] is identified as

$$Z = \frac{V}{I} = \frac{Re^{j\phi}}{\sqrt{1 + y^2}} \qquad (5.5)$$

On resonance, $y = \phi = 0$, and from Eqs. (5.3) and (5.5) we can see that the cavity voltage and impedance increase with the circuit shunt resistance R. Off resonance, if $\omega < \omega_0$, ϕ is positive, the resonator voltage leads the driving current. By convention, the bandwidth of an oscillator is defined as the frequency difference $\Delta\omega$ between the two points on each side of the peak of the resonance curve, where the voltage is lower than the peak value by $\sqrt{2}$, or where $y \cong 2Q\delta\omega/\omega_0 = 1$. These two points are called the *half-power points*, and the resonator bandwidth is defined as

$$\Delta\omega \equiv 2\delta\omega = \frac{\omega_0}{Q} \qquad (5.6)$$

In a real cavity, this result provides a useful way to measure the Q; measurement of the resonance curve allows the determination of both ω_0 and $\Delta\omega$, and $Q = \omega_0/\Delta\omega$.

The general solution of Eq. (5.2) can be written as

$$V(t) = e^{-\omega_0 t/2Q}[c_1 e^{j\omega_1 t} + c_2 e^{-j\omega_1 t}] + \frac{RI_0 e^{j(\omega t+\phi)}}{\sqrt{1+\gamma^2}} \quad (5.7)$$

where c_1 and c_2 are constants that depend on the initial conditions, and $\omega_1 \equiv \omega_0\sqrt{1 - 1/(2Q)^2}$. The resonator time constant is defined as $\tau \equiv 2Q/\omega_0$. If we choose $c_1 = c_2 = -RI_0/2$, assume $Q \gg 1$, and take the real part of Eq. (5.7), we find

$$V(t) = RI_0(1 - e^{-t/\tau})\cos(\omega_0 t) \quad (5.8)$$

which shows the time dependence of the resonator voltage, when the driving current is turned on at $t = 0$ with an amplitude I_0. After an interval of several time constants, the voltage is nearly the steady-state value of $V(t) = RI_0 \cos(\omega_0 t)$. Returning to Eq. (5.7), suppose that at $t = 0$, the driving current is suddenly turned off with $V(0) = RI_0 \cos(\delta)$, where δ is an arbitrary phase angle. In Eq. (5.7) the last term is zero for $t > 0$, and the initial condition is satisfied by choosing $c_1 = RI_0 e^{j\delta}/2$ and $c_2 = RI_0 e^{-j\delta}/2$. The solution is

$$V(t) = RI_0 e^{-t/\tau} \cos(\omega_0 t + \delta) \quad (5.9)$$

that describes the decay of the voltage, when the generator is turned off at $t = 0$.

5.2
Theory of Resonant Cavities

Another description of an accelerating cavity is based directly on Maxwell's equations, following the approach of Condon [2]. We consider an ideal cavity with zero power losses, and no charge sources in the volume. As described in Section 1.8, the fields can be derived from a vector potential $\mathbf{A}(\mathbf{r}, t)$ that satisfies the wave equation

$$\nabla^2 \mathbf{A}(\mathbf{r}, t) - \frac{1}{c^2}\frac{\partial^2}{\partial t^2}\mathbf{A}(\mathbf{r}, t) = -\mu_0 \mathbf{J}(\mathbf{r}, t) \quad (5.10)$$

The ideal loss-free cavity will have an infinite number of eigenmodes that satisfy Eq. (5.10) with $\mathbf{J}(\mathbf{r}, t) = 0$, so that each mode n has a harmonic time dependence at a frequency ω_n, and a characteristic vector potential $\mathbf{A}_n(\mathbf{r}, t) = \mathbf{A}_n(\mathbf{r})e^{j\omega_n t}$ that satisfies the Helmholtz equation $\nabla^2 \mathbf{A}_n(\mathbf{r}) + (\omega_n/c)^2 = 0$. An arbitrary field within the cavity that satisfies the boundary conditions can be expanded in terms of the complete set of eigenmodes,

$$\mathbf{A}(\mathbf{r}, t) = \sum_n q_n(t)\mathbf{A}_n(\mathbf{r}) \qquad (5.11)$$

The $\mathbf{A}_n(\mathbf{r})$ satisfy $\int \mathbf{A}_m \cdot \mathbf{A}_n \, dv = 2\delta_{mn}\sqrt{U_m U_n}/\varepsilon_0 \, \omega_m \omega_n$, where δ_{mn} is the Kronecker δ, and $U_n = (\varepsilon_0/2)\int \mathbf{E}_n^2(\mathbf{r})\, dv = (\varepsilon_0 \omega_n^2/2)\int \mathbf{A}_n^2(\mathbf{r})\, dv$ is the stored energy for mode n. Suppose that an external generator excites the cavity through some coupling mechanism, [3] that establishes a current-density distribution $\mathbf{J}(\mathbf{r}, t)$ inside the cavity. To determine the effect of the driving current on amplitude q_n, we substitute Eq. (5.11) into Eq. (5.10), multiply by $\mathbf{A}(\mathbf{r})$, and integrate to obtain

$$\ddot{q}_n + \omega_n^2 q_n = \frac{1}{\varepsilon_0} \frac{\int \mathbf{J}(\mathbf{r}, t) \cdot \mathbf{A}_n(\mathbf{r})\, dv}{\int \mathbf{A}_n^2 \, dv} \qquad (5.12)$$

The effect of ohmic losses in the cavity walls introduces a damping term that depends on the quality factor for the mode, $Q_n = \omega_n U_n / P_n$, where P_n is the power loss averaged over a cycle [4]. Then,

$$\ddot{q}_n + \frac{\omega_n}{Q_n}\dot{q}_n + \omega_n^2 q_n = \frac{1}{\varepsilon_0} \frac{\int \mathbf{J}(\mathbf{r}, t) \cdot \mathbf{A}_n(\mathbf{r})\, dv}{\int \mathbf{A}_n^2 \, dv} \qquad (5.13)$$

which is the equation for a damped, driven oscillator. The integral on the right side of Eq. (5.13) determines the effectiveness of the current distribution for exciting the nth mode. Physically, Eq. (5.13) shows that the driving term on the right is proportional to the overlap of the current density produced by the coupling device with the vector potential for the mode. The differential equation, Eq. (5.13), describing the dynamical performance of the cavity, has the same form as the differential equation Eq. (5.2) that describes the shunt resonant circuit model of the cavity. Thus, a physical interpretation of Eq. (5.13) can be made, where the right side of Eq. (5.13) is interpreted as a quantity proportional to the effective current that drives the mode, and the coefficient q_n is proportional to the effective voltage for the mode.

5.3
Coupling to Cavities

The most common methods for coupling electromagnetic energy into or out of a cavity are shown in Fig. 5.2 as (1) a magnetic-coupling loop at the end of a coaxial transmission line, (2) a hole or iris in a cavity wall to which a waveguide is connected, (3) an electric-coupling probe or antenna, using an open-ended center conductor of a coaxial transmission line. The loop-coupling method can be analyzed in terms of Eq. (5.13). Suppose we assume that the driving current $I(t)$ is carried by a loop coupler with a single-turn wire loop. Then, $J(\mathbf{r}, t)\, dv = I(t) d\ell$, where $d\ell$ is an element of length along the wire. The right side of Eq. (5.13) contains an integral over the wire, which is

$$\int \mathbf{J}(\mathbf{r}, t) \cdot \mathbf{A}_n(\mathbf{r})\, dv = I(t) \oint \mathbf{A}_n(\mathbf{r}) \cdot d\boldsymbol{\ell} \tag{5.14}$$

where $\mathbf{A}_n(\mathbf{r})$ is the vector potential of the unperturbed mode in the wire. Using Stoke's theorem

$$\oint \mathbf{A}_n \cdot d\boldsymbol{\ell} = \int \nabla \times \mathbf{A}_n \cdot d\mathbf{S} = \Phi_n \tag{5.15}$$

where $d\mathbf{S}$ is the area element in the plane of the loop and Φ_n is the magnetic flux for the mode that threads the loop. Substituting the results of Eqs. (5.14) and (5.15) into Eq. 5.13, we find a simple expression for the steady-state amplitude,

$$|q_n| = \frac{Q_n I \Phi_n}{2 U_n} \tag{5.16}$$

The quantities Q_n, Φ_n, and U_n are all constants that are properties of the mode and the coupler geometry.

Another common method of coupling is the electric probe or antenna. The probe may consist of an open center conductor of a transmission line. The general treatment, using the methods of Section 5.2, is complicated by the need to include a scalar potential to describe the fields produced by the charges at the end of the probe, as discussed by Condon. Physically, this coupling method may be pictured as a conduction current I on the center conductor that flows as displacement current from the end of the conductor. Based on this physical picture, one might expect that the steady-state result for an antenna coupler, corresponding to the result of Eq. (5.16) for a coupling loop, should have a form such as $q_n \propto Q_n I \Phi_{E,n}/cU_n$, where $\Phi_{E,n} = \int \mathbf{E}_n \cdot d\mathbf{S}$ is the electric flux from mode n that is intercepted on the surface of the probe.

5.4
Equivalent Circuit for a Resonant-Cavity System

While the direct application of Maxwell's equations is useful for obtaining detailed information about the resonant frequencies and field distributions of real cavities, the method is cumbersome for describing the complete system consisting of an RF generator, a transmission line or waveguide to transport the electromagnetic wave from the generator to the cavity, and a coupling mechanism that couples electromagnetic energy between the guide and the cavity. Instead, we use an equivalent ac-circuit model similar to that of Section 5.1 to describe the steady-state behavior of the system. As before, we represent a cavity by a shunt resonant circuit with components L, C, and R. The accelerator shunt impedance, defined in Section 2.5, is related to R by $r_s = 2R$. For the external circuit we will introduce a matched current generator, a coupling mechanism (see Fig. 5.2), a waveguide with characteristic impedance Z_0, and a circulator, as shown in Fig. 5.3. The coupling mechanism and the waveguide are represented by a transformer with a turns ratio of $1:n$. The external circuit with a circulator or equivalent isolator transmits forward waves going into the cavity but absorbs all backward waves in the matched load Z_0. In general, the cavity does not present an exactly matched load to the input waveguide; the input waveguide between the circulator and the cavity contains a standing wave with voltage minima (and maxima) separated by $\lambda/2$, and whose locations vary as the drive frequency is varied. The end of the input waveguide in this model is defined to correspond to the plane of a detuned short circuit in the waveguide, from which the cavity appears as a shunt resonant circuit. Figure 5.3a shows generator, circulator, matched load, power coupler, and cavity.

We denote quantities in the external circuit with a primed symbol, and in the resonator circuit with no primes. The impedance Z_c of the shunt resonant circuit is given by Eq. (5.5). The resonator voltage V is given by $V = Z_c i$, where i is the driving current, transformed into the resonator circuit. From the transformer coupling, we write

$$V = nV', \quad i = i'/n \tag{5.17}$$

and

$$Z_c = \frac{V}{i} = n^2 \frac{V'}{i'} = n^2 Z'_c \tag{5.18}$$

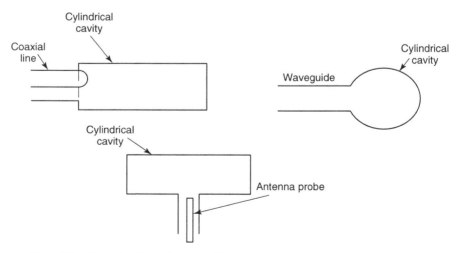

Figure 5.2 Methods of coupling to cavities.

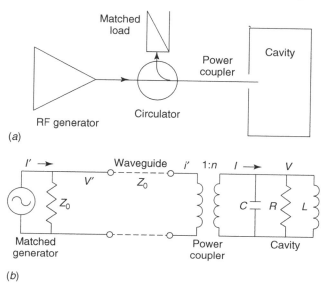

Figure 5.3 (a) Block diagram of RF system components and (b) the equivalent circuit.

where Z'_c is the resonator impedance transformed into the generator circuit. Then,

$$Z'_c = \frac{Z_c}{n^2} = \frac{1}{j\omega C n^2 + n^2/j\omega L + n^2/R} \tag{5.19}$$

From Eq. (5.19) we identify the transformed components as $C' = n^2 C$, $L' = L/n^2$, and $R' = R/n^2$, and the circuit as transformed into the primary circuit can be represented as shown in Fig. 5.4.

The resonator stored energy is $U = n^2 C V'^2/2$, the average power dissipated in the resonator is $P_c = n^2 V'^2/2R$, and the power dissipated in the external load is $P_{ex} = V'^2/2Z_0$ [5]. The quality factor, unloaded by the external resistance, is called the *unloaded Q*, and is written

$$Q_0 = \frac{\omega_0 U}{P_c} = \omega_0 RC \tag{5.20}$$

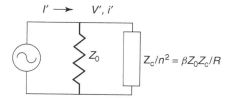

Figure 5.4 Equivalent circuit transformed into the generator circuit.

The quality factor associated with the external load is called the *external Q*, and is written

$$Q_{ex} = \frac{\omega_0 U}{P_{ex}} = \omega_0 n^2 Z_0 C \qquad (5.21)$$

The average power dissipated is $P = P_c + P_{ex}$ and the *loaded Q* is defined as

$$Q_L = \frac{\omega_0 U}{P} \qquad (5.22)$$

From Eqs. (5.20) to (5.22), it can be shown that the Q values are related by

$$\frac{1}{Q_L} = \frac{1}{Q_0} + \frac{1}{Q_{ex}} \qquad (5.23)$$

In the circuit problem, the coupling between the generator and resonator circuits is determined by the transformer–turns ratio n that enters into the expression for Q_{ex} in Eq. (5.21). For a cavity coupled to a waveguide, it is conventional to define a general parameter that is a measure of the waveguide-to-cavity coupling strength, known as the parameter β that is defined by [6]

$$\beta \equiv \frac{P_{ex}}{P_c} = \frac{Q_0}{Q_{ex}} \qquad (5.24)$$

When $\beta < 1$ the waveguide and cavity are said to be *undercoupled*. When $\beta > 1$, the waveguide and cavity are said to be *overcoupled*. When $\beta = 1$, the waveguide and cavity are said to be *critically coupled*, and we have $Q_{ex} = Q_0$ and $Q_L = Q_0/2$. Substituting Eq. (5.24) into Eq. (5.23), we find

$$Q_L = \frac{Q_0}{1+\beta} \qquad (5.25)$$

Substituting Eqs. (5.20) and (5.21) into Eq. (5.24), we find that for the equivalent circuit, the relationship between n and β is $\beta = R/n^2 Z_0$. Thus, larger n implies smaller β or smaller coupling strength, because for larger n, the external load resistance transforms into the resonator as a larger shunt resistance, compared with the resonator shunt resistance, and produces a smaller loading of the Q_L. It is convenient to express the circuit model results in terms of the waveguide-to-cavity coupling parameter β, instead of the transformer–turns ratio n. We find $R' = \beta Z_0$, $C' = RC/\beta Z_0$, and $L' = \beta Z_0 L/R$.

It is also convenient to associate the forward power from the generator with the available power to a matched load in the equivalent circuit. The available power to a matched load corresponds to setting $\beta = 1$ and driving on resonance, so that $Z_c = R$. Then, referring to Fig. 5.4, the matched load impedance is Z_0, the peak current flowing through the matched load impedance is $I'/2$, and the peak voltage across the load is $I'Z_0/2$. The average available power to the load

5.4 Equivalent Circuit for a Resonant-Cavity System

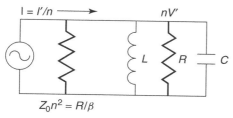

Figure 5.5 Equivalent circuit transformed into the resonator circuit.

is $P_+ = Z_0 I'^2/8$. It is also useful to be able to transform the equivalent circuit of Fig. 5.3 into the resonator circuit. The result is shown in Fig. 5.5.

Using the result that $I = I'/n = I'\sqrt{\beta Z_0/R}$, where in general I depends on the coupling strength β, the available power from the generator can be expressed in terms of quantities transformed to the resonator circuit as $P_+ = I^2 R/8\beta$. When the generator drives the resonator at the resonant frequency, the steady-state resonator voltage is

$$V = \frac{IR}{1+\beta} = \sqrt{P_+ r_s} \frac{2\beta^{1/2}}{1+\beta} \tag{5.26}$$

The equivalent circuit has made it possible to discuss the characteristics of the wave in the input guide, because the model has given us an expression for the cavity load impedance as seen in the generator circuit. For simplicity we restrict our treatment to consider a cavity driven on resonance, where the load impedance for the guide is $Z_L = R' = Z_0\beta$. If we consider a wave that is emitted from the generator into the guide, the reflection coefficient produced by the cavity load impedance, becomes

$$\Gamma = \frac{Z_L - Z_0}{Z_L + Z_0} = \frac{\beta - 1}{\beta + 1} \tag{5.27}$$

The standing wave ratio is $S = (1+|\Gamma|)/(1-|\Gamma|)$. For the undercoupled case with $\beta < 1$, $|\Gamma| = (1-\beta)/(1+\beta)$ and $S = 1/\beta$. For the overcoupled case with $\beta > 1$, $|\Gamma| = (\beta-1)/(1+\beta)$ and $S = \beta$. For the critically coupled case with $\beta = 1$, $|\Gamma| = 0$, and $S = 1$, $\Gamma = 0$ is defined as the condition for the matched state. Thus, critical coupling results in a matched cavity load as seen from the waveguide. As $\beta \to 0$, $\Gamma \to -1$, which corresponds to a short-circuit load, and as $\beta \to \infty$, $\Gamma \to +1$, which is an open-circuit load.

If P_+ is the incident power from the generator, the power propagating back from the input coupler toward the generator is [7]

$$P_- = P_+ \Gamma^2 = P_+ \left(\frac{\beta-1}{\beta+1}\right)^2 \tag{5.28}$$

From energy conservation, the power delivered into the cavity through the input power coupler is

$$P_c = P_+(1 - \Gamma^2) = P_+ \frac{4\beta}{(1+\beta)^2} \quad (5.29)$$

It is easy to show that for a given P_+, P_c is maximum, when $\beta = 1$. The voltage response of the complete circuit to the driving current I' that is supplied by the generator is $V' = Z'_c i'$, where Z'_c can be expressed as a function of β by returning to Eq. (5.19), and substituting $\beta = R/n^2 Z_0$. We can write a result that is analogous to Eq. (5.5),

$$Z'_c = \frac{R'_L e^{j\phi}}{\sqrt{1 + y_L^2}} \quad (5.30)$$

where $R'_L = R/(1+\beta)$, $y_L = Q_L(\omega/\omega_0 - \omega_0/\omega) \cong 2Q_L(\omega - \omega_0)/\omega_0$, and $\phi = -\tan^{-1} y_L$. We note that Q_L, rather than Q_0, determines the voltage response of the total circuit. It can also be shown that the resonator time constant becomes $\tau = 2Q_L/\omega_0$.

5.5
Equivalent Circuit for a Cavity Coupled to two Waveguides

A more common configuration consists of a cavity coupled to external two transmission lines or waveguides. One external circuit usually contains the RF generator, and the other may serve as an RF pickup to monitor the fields in the cavity. The method of Section 5.4 can be used to treat this problem, using the equivalent ac circuit of Fig. 5.6. Both external guides are assumed to be terminated with a matched load resistance Z_0 as shown in the figure. The input guide and generator are represented in the circuit on the left, and the output guide is contained in the circuit on the right. We can simplify the equivalent circuit by reflecting the external circuit on the right of Fig. 5.6 into the resonator circuit, and then reflecting all of the components into the circuit on the left that contains the generator. The final equivalent circuit is shown in Fig. 5.7.

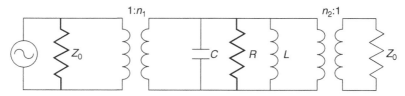

Figure 5.6 Equivalent circuit of a cavity coupled to two external circuits, which are transformers coupled to the cavity resonator in the center.

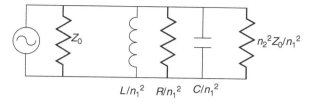

Figure 5.7 Equivalent circuit with all components reflected into the external circuit containing the generator.

In this section, we will drop the prime symbols that were used in the previous section to denote quantities seen in the primary circuit. Referring to Fig. 5.7, the resonator power dissipated is $P_c = n_1^2 V^2/2R$. The power dissipation in the external circuits is $P_{ex1} = V^2/2Z_0$, and $P_{ex2} = n_1^2 V^2/2n_2^2 Z_0$, and the stored energy is $U = n_1^2 CV^2/2$. The Q values are $Q_0 = \omega_0 U/P_c = \omega_0 RC$, $Q_{ex1} = \omega_0 U/P_{ex1} = \omega_0 n_1^2 Z_0 C$, and $Q_{ex2} = \omega_0 U/P_{ex2} = \omega_0 n_2^2 Z_0 C$. The total power dissipated is $P = P_c + P_{ex1} + P_{ex2}$, and the loaded Q is $Q_L = \omega_0 U/P$. The Q values are related by

$$\frac{1}{Q_L} = \frac{1}{Q_0} + \frac{1}{Q_{ex1}} + \frac{1}{Q_{ex2}} \tag{5.31}$$

Now, we introduce the resonator-to-guide coupling factors, defined as

$$\beta_1 = \frac{P_{ex1}}{P_c} = \frac{Q_0}{Q_{ex1}}, \beta_2 = \frac{P_{ex2}}{P_c} = \frac{Q_0}{Q_{ex2}} \tag{5.32}$$

From the equivalent circuit model, we find that $\beta_1 = R/n_1^2 Z_0$, $\beta_2 = R/n_2^2 Z_0$, $Q_{ex1} = Q_0/\beta_1$, $Q_{ex2} = Q_0/\beta_2$, and Eq. (5.31) can be expressed as

$$Q_L = \frac{Q_0}{1 + \beta_1 + \beta_2} \tag{5.33}$$

Following the derivations of the previous section, we find that on resonance the resonator presents a load impedance to the input guide, $Z_L = Z_0 \beta_1/(1 + \beta_2)$. Now we take this result for the load impedance from the circuit model, and use it to calculate the reflection coefficient for the input guide that is

$$\Gamma_1 = \frac{\beta_1 - \beta_2 - 1}{1 + \beta_1 + \beta_2} \tag{5.34}$$

If the input guide is matched to the resonator, we must have $\Gamma_1 = 0$, and the matched solution is

$$\beta_1 = 1 + \beta_2 \tag{5.35}$$

Because $\beta_2 \geq 0$, the matched condition of the input guide implies that $\beta_1 \geq 0$. Only in the weakly coupled limit for guide 2, where $\beta_2 \cong 0$, does $\beta_1 \cong 1$ for the matched case.

5.6
Transient Behavior of a Resonant-Cavity System

The steady-state solution of a resonant-cavity system was treated with an equivalent ac-circuit model in Section 5.4, and in this section we discuss the transient behavior. The average power in the incident wave is $P_+ = V_+^2/2Z_0, t \geq 0$. Equation (5.28) expresses the result that for an incident wave from the generator of amplitude V_+, the effective reflected wave [8] from the effective load impedance at the coupling mechanism has a steady-state amplitude

$$V_- = \frac{\beta - 1}{\beta + 1} V_+ \tag{5.36}$$

and the total steady-state voltage across the effective resonator load impedance at the coupler is

$$V = n(V_+ + V_-) = \frac{2\beta n}{\beta + 1} V_+ \tag{5.37}$$

This represents the effective voltage on resonance across the resonator impedance. Now, suppose that at time $t = 0$, the generator of Fig. 5.3 is turned on to launch an incident wave with amplitude V_+. We assume that the wave will be partially reflected at the coupler with amplitude V_- and partially transmitted to the resonator. The time dependence of the resonator fields during the turn-on transient was shown in Section 5.1 to contain the factor $1 - e^{-t/\tau}$, where τ is the cavity time constant, $\tau = 2Q_L/\omega_0$. Combining the known time dependence with the steady-state result of Eq. (5.37), the time-dependent resonator voltage is

$$V_0(t) = V_+(1 - e^{-t/\tau}) \frac{2\beta n}{1 + \beta} \tag{5.38}$$

We have assumed that V_+ is constant after $t = 0$, so the reflected wave must have a time-dependent amplitude given by

$$V_-(t) = \frac{V_0(t)}{n} - V_+ = V_+ \left[(1 - e^{-t/\tau}) \frac{2\beta}{1 + \beta} - 1 \right] \tag{5.39}$$

which corresponds to a time-dependent reflection coefficient

$$\Gamma(t) = \frac{V_-(t)}{V_+} = (1 - e^{-t/\tau}) \frac{2\beta}{1 + \beta} - 1 \tag{5.40}$$

These results can also be expressed in terms of traveling-wave power. The average reflected power is

$$P_-(t) = P_+ \Gamma^2(t) = P_+ \left[(1 - e^{-t/\tau}) \frac{2\beta}{1+\beta} - 1 \right]^2 \qquad (5.41)$$

This power will be dissipated in the load resistor Z_0, shown in Fig. 5.3. The time-dependent power delivered to the resonator is

$$P_c(t) = P_+ (1 - \Gamma^2(t)) = P_+ \left[1 - \left\{ (1 - e^{-t/\tau}) \frac{2\beta}{1+\beta} - 1 \right\}^2 \right] \qquad (5.42)$$

We consider the limiting cases predicted by the equations of this section. At $t = 0$, when the incident wavefront reaches the coupler, $V_0(0) = 0$, $V_-(0) = -V_+$, $\Gamma(0) = -1$, $P_-(0) = P_+$, and $P_c(0) = 0$. These results tell us that at $t = 0$ the resonator presents a short-circuit load, and all of the incident power is reflected. Similarly, as $t \to \infty$ the steady-state results are recovered. The transient results depend on β and are illustrated in Fig. 5.8. For $\beta \leq 1$, the backward power $P_-(t)$ decreases monotonically as a function of time from its initial value P_+ to the steady-state value. For the overcoupled case with $\beta > 1$, $P_-(t)$ decreases as a function of time until it reaches zero; thereafter it increases and approaches the steady-state value. From Eq. (5.42) we find that when $t \ll \tau$, the power delivered to the resonator, as well as the resonator stored energy is proportional to $4\beta t P_+ / (\beta + 1)\tau$. This means that it is more difficult to excite a resonator with a pulse that is short compared to the resonator time constant.

The results of Eqs. (5.39) to (5.42), which show a time-dependent reflected wave and reflection coefficient need further explanation. This effect is caused

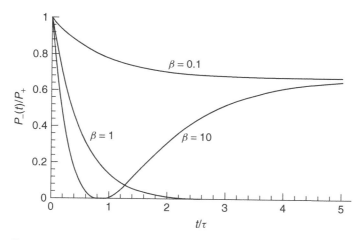

Figure 5.8 Backward power versus time for different values of the coupling factor β.

by the time dependence of the resonator. We are interested in understanding this effect at the high frequencies of interest for accelerators, where the dimensions of the circuit are comparable to or greater than the wavelength, and the fields vary spatially as well as temporally. For this case, an explicit traveling-wave description that is presented in Section 5.7, is needed to provide an adequate explanation of the physics. There we will learn that the cavity builds up stored energy as a result of multiple internal reflections of the injected traveling wave, reinforced at each pass by a continuous flow of energy from the generator through the coupler. The cavity not only receives an input wave from the generator, but radiates a wave back through the coupler. As the cavity stored energy increases, the amplitude of the radiated wave also increases. The effective reflected voltage and power in the input waveguide is really the sum of two traveling waves; one is the direct-reflected wave from the coupler, and the other is the radiated wave from the cavity. These two waves are out of phase and whether the coupling regime is undercoupled, critically coupled, or overcoupled depends on the relative strengths of the two waves. The time dependence of the effective reflected wave is associated with the radiated wave.

For a simple model, we describe the total wave for times $t > 0$ as the sum of three monochromatic waves that propagate in the input waveguide. The incident wave from the generator is

$$V_{in} = V_+ e^{j(\omega t - kz)} \tag{5.43}$$

The wave reflected from the coupler is approximately [9]

$$V_{ref} \cong \pm V_+ e^{j(\omega t + kz)} \tag{5.44}$$

The wave radiated from the cavity is

$$V_{rad} = \mp V_+ \frac{2\beta}{1+\beta}(1 - e^{-t/\tau})e^{j(\omega t + kz + \phi)} \tag{5.45}$$

where $\tau \gg 2\pi/\omega$, and where in general $\phi = -\tan^{-1}(2Q_L \delta\omega/\omega)$ is the phase shift of the cavity field relative to its value when driven on resonance. The model, described by Eqs. (5.43) to (5.45), ignores the details of the wavefront propagation near $t = 0$. The signs in Eqs. (5.44) and (5.45) are always opposite, and the choice relative to Eq. (5.43) depends on the type of coupler. The backward wave given in Eq. (5.39) is really $V_-(t) = V_{ref} + V_{rad}(t)$.

5.7
Wave Description of a Waveguide-to-Cavity Coupling

In this section, we reexamine the cavity system from a more physical point of view without reference to the ac-circuit model. In the circuit model the true wave behavior of the cavity system is concealed. In this section we

5.7 Wave Description of a Waveguide-to-Cavity Coupling

will be describing the waves, their propagation, transmission, reflection, and interference, both in the input waveguide and in the cavity. We begin the discussion by introducing an important tool for the analysis, called the *scattering matrix* [10]. Consider a waveguide that contains an obstacle or discontinuity. To be specific, we will describe the voltage across the waveguide resulting from a transverse electric field, although the general description is not restricted to this case. An incident voltage wave coming from the left can be expressed as $v_+ e^{j(\omega t - kz)}$, and a reflected wave from the obstacle is $\Gamma v_+ e^{j(\omega t - kz)}$, where Γ is the reflection coefficient. Some microwave devices have more than two incident channels, and some are more complex elements than simple obstacles or discontinuities. For each channel there is generally an incident wave and an outgoing wave. The total outgoing wave in any channel is the sum of the reflected wave in that channel plus the contributions of the transmitted waves from all the other channels. The general description has been formulated in terms of a scattering matrix [11]. To define the scattering matrix for an n-port device, we begin by defining normalized incident waves, such that the incoming traveling waves are represented as a_i, $i = 1, 2, \ldots, n$, and the outgoing waves are represented by b_i, $i = 1, 2, \ldots, n$. The ith normalized incident voltage wave is defined as $a_i = v_+^i e^{j(\omega t - kz)}/\sqrt{Z_0}$, and the outgoing wave is $b_i = v_-^i e^{j(\omega t + kz)}/\sqrt{Z_0}$. The powers in the ith incident and outgoing waves are $P_+^i = (v_+^i)^2/2Z_0 = |a_i|^2/2$, and $P_-^i = (v_-^i)^2/2Z_0 = |b_i|^2/2$. The scattering matrix S is defined as the transformation of the column vector **a** of the incident waves into the column vector **b** of the outgoing waves. Thus,

$$\begin{pmatrix} b_1 \\ b_2 \\ \cdots \\ b_n \end{pmatrix} = \begin{bmatrix} s_{11} & s_{12} & \cdots & s_{1n} \\ s_{21} & s_{22} & \cdots & s_{2n} \\ \cdots & \cdots & \cdots & \cdots \\ s_{n1} & s_{n2} & \cdots & s_{nn} \end{bmatrix} \begin{pmatrix} a_1 \\ a_2 \\ \cdots \\ a_n \end{pmatrix} \qquad (5.46)$$

Each diagonal element s_{ii} is the reflection coefficient Γ_i in the ith channel. Each off-diagonal element s_{ij} is the transmission coefficient from channel i into channel j. The number of channels is also called the *number of ports*. Thus, an n-port element has n input channels, and is described by an $n \times n$ scattering matrix. Reciprocal n-port elements are symmetric and satisfy $s_{ij} = s_{ji}$. It can be shown that S is a unitary matrix when the element is free of power losses.

Next, we describe a simple cavity. We know that two conducting plates can be inserted into a waveguide to create a microwave cavity in the enclosed volume. To excite such a cavity there must be some coupling mechanism to provide a way of delivering the electromagnetic energy from a microwave generator into the cavity. One simple way of doing this is to place a hole in one of the plates, through which at least some fraction of the incident wave in the transmission line can be transmitted into the enclosed cavity. The scattering matrix can be used to describe the effects of a conducting plate with a small hole or iris that is inserted into a waveguide perpendicular to the axis, as shown in Fig. 5.9. We consider the iris to be a two-port device that has two input channels, one from the left and one from the right. We describe the iris by a 2×2 scattering

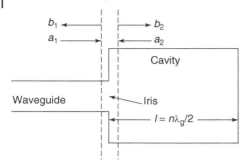

Figure 5.9 Waveguide and cavity section separated by a conducting plate with an iris.

matrix,

$$\begin{pmatrix} b_1 \\ b_2 \end{pmatrix} = \begin{bmatrix} s_{11} & s_{12} \\ s_{21} & s_{22} \end{bmatrix} \begin{pmatrix} a_1 \\ a_2 \end{pmatrix} \tag{5.47}$$

We can determine the properties of the scattering matrix for the iris, using a simple treatment based on two assumptions. The iris is assumed to be left–right symmetric, so that if the iris orientation is reversed, its behavior is unchanged. This property is known as *reciprocity*, and implies that $s_{11} = s_{22}$, and $s_{21} = s_{12}$. We have already noted that the diagonal elements are equal to the reflection coefficient Γ. As the hole size increases until the obstacle recedes into the waveguide wall, the magnitude of Γ will be expected to decrease to zero. If the hole size vanishes, we know from transmission-line theory that the reflection coefficient for the voltage wave from a perfectly conducting short circuit is $\Gamma = -1$. In general we write $s_{11} = s_{22} = \Gamma$, where Γ is a negative number. The iris itself is assumed to be free of energy dissipation. Thus, energy conservation implies that the total wave power coming into the iris equals the total wave power going out, or

$$|a_1|^2 + |a_2|^2 = |b_1|^2 + |b_2|^2 \tag{5.48}$$

To see what energy conservation implies about the scattering matrix, we form $b_1 = s_{11}a_1 + s_{12}a_2 = \Gamma a_1 + s_{12}a_2$. The complex conjugate is $b_1^* = \Gamma a_1^* + s_{12}^* a_2^*$. Then we obtain

$$b_1 b_1^* = \Gamma^2 |a_1|^2 + s_{12}s_{12}^*|a_2|^2 + \Gamma s_{12}^* a_1 a_2^* + \Gamma s_{12} a_2 a_1^* \tag{5.49}$$

and interchanging subscripts 1 and 2 we obtain the corresponding result for $b_2 b_2^*$. Then,

$$b_1 b_1^* + b_2 b_2^* = (\Gamma^2 + s_{12}s_{12}^*)|a_1|^2 + (\Gamma^2 + s_{12}s_{12}^*)|a_2|^2$$
$$+ \Gamma(s_{12} + s_{12}^*)a_1 a_2^* + \Gamma(s_{12} + s_{12}^*)a_2 a_1^* \tag{5.50}$$

To satisfy energy conservation as in Eq. (5.48), we require

$$\Gamma^2 + |s_{12}|^2 = 1 \text{ and } s_{12} + s_{12}^* = 0 \tag{5.51}$$

Equation (5.51) are satisfied if $\text{Re}(s_{12}) = 0$, and $|s_{12}|^2 = 1 - \Gamma^2$. A solution is $s_{12} = j\sqrt{1 - \Gamma^2}$, and the scattering matrix for the iris becomes

$$S = \begin{bmatrix} \Gamma & j\sqrt{1 - |\Gamma|^2} \\ j\sqrt{1 - |\Gamma|^2} & \Gamma \end{bmatrix} \quad (5.52)$$

The properties of symmetry or reciprocity, and conservation of energy in the waves, imply that S is determined in terms of a single parameter, the reflection coefficient.

Suppose that on the left side of the iris is an input waveguide through which an RF generator delivers a traveling electromagnetic wave to drive the cavity. When the wave from the generator is incident upon the iris there will be some fraction that is reflected back into the waveguide, and the remaining fraction that is transmitted into the cavity. The transmitted wave will propagate in the cavity to the plane of the short circuit, where it will be reflected back to the iris. At the iris the wave is again partially reflected back into the cavity, and partially transmitted back into the incident waveguide. The input waveguide will have both incident and outgoing traveling waves. If the generator frequency is varied, there will exist discrete resonant frequencies at which we will observe the build up of a high-field standing wave within the cavity region as a result of the constructive interference of the multiply reflected waves inside the cavity. Eventually, a steady state will be reached, where the energy lost from ohmic power dissipation in the cavity walls, and wave energy propagating out of the iris, becomes equal to the energy entering the cavity from the generator through the iris. From Eqs. (5.47) and (5.52) we write

$$b_1 = \Gamma a_1 + j(1 - \Gamma^2)^{1/2} a_2$$
$$b_2 = j(1 - \Gamma^2)^{1/2} a_1 + \Gamma a_2 \quad (5.53)$$

The expressions for b_1 and b_2 have two terms, a reflected wave and a transmitted wave through the iris.

To examine the build up of the field, suppose that at time $t = 0$, the wave from the generator appears at the iris with amplitude a_1, and we will also assume that $a_2 = 0$ initially. Shortly afterwards, at time $t = 0^+$, the outgoing amplitudes, obtained from the scattering matrix are $b_1 = \Gamma a_1$ and $b_2 = j(1 - \Gamma^2)^{1/2} a_1$. One cycle later at time $T = 2\ell/v_G$, the wave b_2 will have propagated to the short, where it is reflected with a phase reversal, and propagated back to the iris, where it becomes an incoming wave from the right with amplitude a_2. A small fraction of the a_2 wave is transmitted through the iris to contribute to the outgoing wave b_1, and the largest fraction is reflected back with a phase reversal and produces a step in the wave b_2. If we assume an ohmic attenuation for the round trip is given by $e^{-2\alpha\ell}$, where α is the attenuation constant per unit length, we have $a_2 = -b_2 e^{-(2\alpha\ell + j2\phi)}$, where 2ϕ is the round-trip phase shift of the wave. The phase shift $\phi = 2\pi\ell/\lambda$ is determined by ℓ/λ, the electrical length of the cavity, where λ is the free-space wavelength. Subsequent round-trip passes of the internal wave will reinforce each other and produce resonance when $\phi = 2\pi n$, where n is a positive integer. At resonance we have

$$a_2 = -b_2 e^{-2\alpha\ell} \tag{5.54}$$

At time $t = 0^+ + T$, the updated amplitudes become $a_2 = -e^{-2\alpha\ell}$ $b_2 = -je^{-2\alpha\ell}\sqrt{1-\Gamma^2}a_1$, $b_1 = (\Gamma + (1-\Gamma^2)e^{-2\alpha\ell})a_1$, and $b_2 = j\sqrt{1-\Gamma^2}$ $(1-\Gamma e^{-2\alpha\ell})a_1$. Following the buildup of the waves, after N cycles we find

$$a_2(N) = -j\sqrt{1-\Gamma^2}a_1 e^{-2\alpha\ell} \sum_{n=0}^{N-1} x^n = -j\sqrt{1-\Gamma^2}a_1 e^{-2\alpha\ell} \left[\frac{1-x^N}{1-x}\right] \tag{5.55}$$

$$b_1(N) = a_1 \left[\Gamma + (1-\Gamma^2)e^{-2\alpha\ell} \sum_{n=0}^{N-1} x^n\right]$$

$$= a_1 \left[\Gamma + (1-\Gamma^2)e^{-2\alpha\ell} \left(\frac{1-x^N}{1-x}\right)\right] \tag{5.56}$$

and

$$b_2(N) = j\sqrt{1-\Gamma^2}a_1 \sum_{n=0}^{N} x^n = j\sqrt{1-\Gamma^2}a_1 \left(\frac{1-x^{N+1}}{1-x}\right) \tag{5.57}$$

where $x \equiv -\Gamma e^{-2\alpha\ell}$ is a positive number. The steady-state solution corresponds to $N \to \infty$, and since $|x| < 1$, $x^N \to 0$ and the steady-state solution is

$$a_2(\infty) = \frac{-j\sqrt{1-\Gamma^2}a_1 e^{-2\alpha\ell}}{1-x} \tag{5.58}$$

$$b_1(\infty) = a_1 \left[\Gamma + \frac{(1-\Gamma^2)e^{-2\alpha\ell}}{1-x}\right] \tag{5.59}$$

and

$$b_2(\infty) = \frac{j\sqrt{1-\Gamma^2}a_1}{1-x} \tag{5.60}$$

Thus, we have an expression for the growth of the traveling-wave amplitude in the cavity as a stairstep profile given by $b_2(N)/b_2(\infty) = 1 - x^{N+1}$, and as shown in Fig. 5.10. Assuming $\Gamma \approx -1$, we can approximate $(-\Gamma)^N \cong e^{-N(1+\Gamma)}$ or $x^N = (-\Gamma e^{-2\alpha\ell})^N \cong e^{-(1+\Gamma+2\alpha\ell)N}$. We approximate this result as a continuous function of time using a smooth approximation $t = NT$, where $T = 2\ell/v_G$ is the round-trip transit time of the wave or twice the fill time, and v_G is the group velocity. The cavity time constant τ_c is defined as the time required for the exponent to increase from zero to unity; thus $x^N \cong e^{-t/\tau_c}$, where

$$\tau_c = \frac{T}{1+\Gamma+2\alpha\ell} \tag{5.61}$$

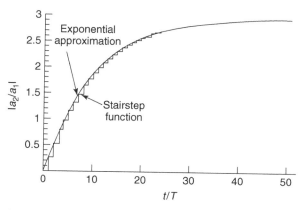

Figure 5.10 Transient build up of the cavity field seen at the input iris, and the exponential approximation for the buildup. The parameters $\Gamma = -0.95$ and $2\alpha l = 0.05$ were chosen to facilitate an illustration of the stairstep buildup. The cavity time constant is $\tau = 10T$.

We can also write $x = e^{-T/\tau_c}$, so that

$$\frac{b_2(N)}{b_2(\infty)} = 1 - x^{t/T+1} = 1 - e^{-(t+T)/\tau_c} \cong 1 - e^{-t/\tau_c} \tag{5.62}$$

where the last step is valid when $t \gg T$. Therefore, we find that the stairstep growth of the traveling-wave amplitude of the cavity is approximated by an exponential growth. It will be shown that the time constant is $\tau_c = 2Q_L/\omega_0$, the same result that we obtained from the equivalent circuit model in Section 5.1.

Critical coupling corresponds to the value of Γ that makes the outgoing-wave amplitude $b_1 = 0$. For this condition all of the incident power is absorbed in the cavity. At critical coupling, $b_1 = 0$ is the result of the reflected wave in the input waveguide being exactly canceled by destructive interference with the wave emitted from the cavity. Any other condition results in an outgoing wave in the input waveguide, and less power absorbed in the cavity. The steady-state value of b_1 from Eq. (5.59) can be re-expressed as

$$b_1 = \left[\frac{\Gamma + e^{-2\alpha \ell}}{1 + \Gamma e^{-2\alpha \ell}}\right] a_1 \tag{5.63}$$

Thus, $b_1 = 0$ for critical coupling occurs when

$$\Gamma = -e^{-2\alpha \ell} \tag{5.64}$$

For critical coupling the reflection coefficient equals the round-trip field attenuation factor for the traveling wave in the cavity. Suppose $2\alpha \ell \approx 10^{-4}$, and we approximate $\Gamma \cong -1 + 2\alpha \ell$. Equations (5.58) and (5.60) give $|a_2| \cong |b_2| \cong |a_1|/\sqrt{4\alpha \ell} \cong 10^2 a_1$. Then the iris is nearly a short circuit that almost completely reflects the incident wave. The outgoing wave b_1 is the sum of a reflected wave plus an emitted wave from the cavity. The emitted wave is a small fraction of the large internal wave a_2 that has built up within the cavity. Because Γ is negative, the reflected and emitted waves are of opposite phase. The phase difference can be understood by adding the phase shifts that occur in a round-trip through the cavity, including the 180° shift at the reflections, and the 90° phase shift that occurs both at injection into the cavity and at emission from the cavity. When $\Gamma = -e^{-2\alpha\ell}$, the reflected and emitted waves cancel. When $|\Gamma| > e^{-2\alpha\ell}$, the reflected wave dominates, which is the undercoupled case. When $|\Gamma| < e^{-2\alpha\ell}$, the emitted wave from the cavity dominates, which is the overcoupled case.

We now relate these results to conventional cavity parameters for comparison with the results from the equivalent circuit model. The standing wave in the cavity is composed of the two waves, traveling in opposite directions. For the wave propagating in the $+z$ direction, the traveling-wave power is $P_+ = U_+ v_G/\ell$, where U_+ is the stored energy in the $+z$ wave. Similarly, for the wave propagating in the $-z$ direction, $P_- = U_- v_G/\ell$ is the power in the wave propagating in the $-z$ direction. We have $P_+ \cong P_-$, so the total stored energy in the cavity is

$$U = U_+ + U_- = \frac{2P_+ \ell}{v_G} \tag{5.65}$$

The fractional decrease in the field amplitude of the traveling wave in one round-trip transit of the cavity is $(V_+ - \Delta V_+)/V_+ = e^{-2\alpha\ell}$, and the fractional decrease in the traveling-wave power is $(P_+ - \Delta P_+)/P_+ = e^{-4\alpha\ell}$. The wave energy crossing any plane decreases by $\Delta U = \Delta P_+ T \cong 4\alpha\ell P_+ T$ in a round-trip transit time $T = 2\ell/v_G$. The power dissipation is $P = \Delta U/T = 4\alpha\ell P_+$. Then, the unloaded Q, the quality factor associated only with cavity power dissipation is

$$Q_0 = \frac{\omega_0 U}{P} = \frac{\omega_0}{2\alpha v_G} \tag{5.66}$$

We note that Q_0 is independent of ℓ and inversely proportional to v_G, because smaller v_G implies more stored energy per unit wave power.

Another important quality factor is related to the strength of the waveguide-to-cavity coupling through the iris, and this is defined by the external quality factor or Q_{ex}, defined at resonance as the ratio of $\omega_0 U$ to the energy radiated from the cavity through the iris, when *the generator is turned off from P_+ to 0*. From Eq. (5.53), when $a_1 = 0$, the emitted wave is

$$b_1 = j(1 - \Gamma^2)^{1/2} a_2 \tag{5.67}$$

Then the emitted or externally dissipated power is

$$P_{ex} = \frac{|b_1|^2}{2} = (1 - \Gamma^2)\frac{|a_2|^2}{2} = (1 - \Gamma^2) P_+ \tag{5.68}$$

5.7 Wave Description of a Waveguide-to-Cavity Coupling

The external Q is

$$Q_{ex} = \frac{\omega_0 U}{P_{ex}} = \frac{2\omega_0 \ell}{v_G(1-\Gamma^2)} \qquad (5.69)$$

Note the important requirement that the generator be turned off. If this condition is forgotten, one might mistakenly conclude that for critical coupling Q_{ex} is infinite, since for critical coupling, when the generator is on, there is no power radiated into the external circuit. The absence of radiated power is because of the complete destructive interference of the radiated wave with the reflected wave from the iris, when the generator is on. The external Q would seem to be an artificial concept for describing the performance of the cavity, when the generator drives the cavity exactly on resonance. Nevertheless, we will see that Q_{ex} contributes to the loaded Q that determines the real cavity time constant, when the cavity is loaded by the external waveguide [12]. Whereas Q_0 depends on the attenuation constant α, Q_{ex} depends on the power transmission through the iris, $1 - \Gamma^2$. The waveguide-to-cavity coupling parameter is

$$\beta \equiv \frac{Q_0}{Q_{ex}} = \frac{1-\Gamma^2}{1-e^{-4\alpha\ell}} \cong \frac{1-\Gamma^2}{4\alpha\ell} \qquad (5.70)$$

Physically, β corresponds to the ratio of power that would be radiated out through the iris, if the incident wave from the generator suddenly collapsed to the power dissipated in the walls of the cavity. Recall that the condition for critical coupling is $-\Gamma = e^{-2\alpha\ell}$, so given the assumption that $\alpha\ell \ll 1$, we have $1-\Gamma^2 \cong 4\alpha\ell$, and from Eq. (5.70) this corresponds to $\beta = 1$, and $Q_{ex} = Q_0$. Similarly, undercoupling corresponds to $\beta < 1$, and overcoupling corresponds to $\beta > 1$.

Finally the loaded Q or Q_L, is the ratio at resonance of cavity stored energy to energy dissipated in the cavity per radian plus the energy radiated from the cavity per radian through the iris, when *the generator is turned off*. The loaded Q is given by

$$Q_L = \frac{\omega_0 U}{P + P_{ex}} \cong \frac{\omega_0 \ell}{v_G} \frac{1}{\left[2\alpha\ell + \frac{(1-\Gamma^2)}{2}\right]} \qquad (5.71)$$

Using $1 - \Gamma^2 \cong (1-\Gamma)(1+\Gamma) \cong 2(1+\Gamma)$, we find that

$$Q_L \cong \frac{\omega_0 \ell}{v_G} \frac{1}{[1+\Gamma+2\alpha\ell]} \qquad (5.72)$$

It is easy to show that $Q_L = Q_0/(1+\beta)$, and for critical coupling where $\beta = 1$, we have $Q_L = Q_0/2$. Comparing Eqs. (5.61) and (5.72), and using the definition $T = 2\ell/v_G$, shows that the cavity time constant is

$$\tau_c = \frac{2Q_L}{\omega_0} \qquad (5.73)$$

which is the same result already deduced from the equivalent circuit model. Thus, the equivalent circuit model of Section 5.1 is seen to be a valid representation of the average or smoothed time dependence of the cavity fields. Analysis of the cavity decay process also shows a stairstep solution for the decay that is approximated by a smooth exponential decay. The same argument that was used above for the growth of the fields, shows that the decay time constant is again $\tau_c = 2Q_L/\omega_0$, the same as for the growth.

5.8
Microwave Power Systems for Linacs

The material of this section will provide a short introduction to the subject of microwave power systems. We begin with a discussion of microwave power sources. The interested reader is referred to the other books that deal with this subject, such as Smith and Phillips (1995), and Gandhi (1987). RF power must be supplied to the linac cavities to establish and maintain the electromagnetic fields [13]. The choice of an RF power source for an accelerator cavity depends on many factors, including frequency, peak and average power, efficiency, reliability, and cost, and generally, the RF power system is a major contributor to the cost of a linac. High-power vacuum-tube amplifiers, designed for operation in the frequency range of importance for linacs (typically from a few tens of megahertz to a few tens of gigahertz) include triodes, tetrodes, and klystrons. The gridded tubes, including triodes and tetrodes, are used as amplifiers at frequencies below about 300 MHz, and the application of the gridded tube has been almost exclusively for pulsed operation. Magnetrons have been used for accelerators with a single section, but their inherent lack of phase stability has prevented their application to multisection linacs. The most commonly used linac RF power source is the klystron amplifier, which is normally used at frequencies above about 300 MHz, where its size is practicable. Klystrons have been operated at pulse lengths ranging from about 1 μs to a continuous wave or CW.

In the basic two-cavity klystron, the beam is injected from an electron gun, accelerated by a dc potential V, and focused along the length of the klystron by an axial magnetic field. The input cavity is excited at the resonant frequency by a low-voltage generator. As the dc electron beam enters the input cavity, the electrons are either accelerated or decelerated by the RF axial electric field of the input cavity, depending on the phase of the accelerating field when the electrons enter the gap. As the velocity-modulated electron beam travels along the drift tube following the input cavity, the accelerated electrons from one RF cycle tend to catch up with decelerated electrons from the previous cycle. This results in the formation of bunches of electrons at the frequency of the input cavity fields. If the output cavity is tuned to the same frequency, it will be excited by the bunched beam, and if the output cavity is coupled to an output waveguide, RF output power will be radiated into the guide. The output power level can be much larger than the input power, resulting in amplified RF power. Maximum gains of up to 50 to 60 dB are typical. Klystrons have been designed to operate successfully from a few hundred megahertz up to 100 GHz. For room-temperature accelerators operating at accelerating field levels of a few megavolts per meter, and over lengths of a few meters, the RF power required is about 1 MW. At frequencies of a few hundred megahertz, CW klystrons near 1 MW are commercially available. For pulsed accelerators in the 3-GHz frequency range, klystrons with peak power of 60 MW are used; the SLAC linac uses a total of 240 such klystrons. Klystron efficiencies, defined as the ratio of RF output power to dc beam power, near about 40% are typical for pulsed operation, and efficiencies near 60% have been achieved for cw operation. The output circuit must be designed to limit the reflected power from the cavity load to prevent deterioration of the performance of the device, and to avoid arcing damage from the enhanced fields. It is common in high-power systems to install an arc detector, and a reflected power monitor in the output line, and to switch off the dc power within microseconds of an abnormal condition. Such a condition can develop when the accelerator cavity that represents the load impedance for the klystron, experiences an arc-down condition. Another method of protection is to install a waveguide device called a *circulator* to isolate the klystron from power reflected from the load impedance. The circulator, installed between the amplifier and the cavity, is a special microwave device that transmits RF power traveling from the RF amplifier to the cavity. Any power coming back from the cavity to the circulator is delivered to a matched load rather than back to the RF amplifier, where it could cause damage or problems with the performance. Because of the circulator, the drive system appears like a matched RF generator. Klystrons are reliable devices with a mean time before failure that is greater than about 30,000 h.

A low-level RF control system is used to maintain the phase and amplitude of linac-cavity fields, given the thermal and mechanical perturbations that cause resonant frequency variations, and beam-current variations that directly perturb the cavity fields [14]. Because an accelerator cavity has a high Q and can only accept power over a narrow frequency band, the low-level control system must also provide a resonance tracking and adjustment system that retunes the cavity, allowing it to accept RF power at the design frequency. Figure 5.11 shows a schematic drawing of an RF system for an accelerator cavity. Conceptually, the cavity RF pickup signal is fed back to the field-control electronics, where it is subtracted from a stable RF reference signal to produce an error signal. The field-control electronics may augment the error signal with a feedforward signal that allows disturbances in the beam current that are detected upstream of the cavity to be anticipated and compensated. This approach is based on the use of past error information to predict a suitable feedforward correction function. Feedforward regulation works for nonrelativistic beams, where the signal can propagate from the point of detection to the point of application, faster than the beam. The resulting RF error signal has an amplitude that

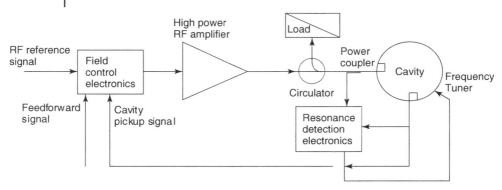

Figure 5.11 Schematic drawing of an RF system for an accelerator cavity.

is adjusted in proportion to the cavity-amplitude error, and a frequency that allows the cavity phase to catch up to the reference phase. This error signal is applied to the input of the high-gain, high-power RF amplifier that delivers RF power to the cavity. The details by which the error signal is produced may be realized with some standard microwave components such as a doubly balanced mixer, used to compare two RF signals and generate a voltage level proportional to the phase difference, and a voltage-controlled oscillator that produces an RF output wave with a frequency that depends on the input phase-error level. The RF pickup signal is also used by the resonance-detection electronics, which compares the relative phases of the pickup signal and the drive signal, and produces an error signal proportional to the phase difference. The difference can be used to implement a slow retuning correction to the cavity. For a normal conducting cavity, the resonant frequency is usually maintained by regulating the temperature of the cooling water. For a superconducting cavity, the resonant frequency is usually maintained by applying a mechanical force, for example, to push or pull on the ends of the cavity.

Most perturbations that affect the cavity fields occur at frequencies below about 1 MHz. The typical 1-MHz upper limit occurs because noise frequencies corresponding to periods that are short compared with the cavity time constant will be ineffective at producing significant field variations. In feedback control, phase shifts in the loop, especially at high frequencies, which cause a roll off in the loop gain, establish a characteristic time constant and an associated bandwidth limit above which the feedback is positive. Consequently, feedback is effective only for regulating low-frequency field perturbations; in practice feedback works well for frequencies below about 10 kHz. Feedforward is especially helpful for control of field variations induced by beam noise at frequencies above about 10 kHz, where feedback control is only partially effective. Using feedback and feedforward, the amplitude and phase of the cavity field may typically be controlled to less than 1% and 1°.

5.9
Multipacting

Electrons can be driven from the cavity walls by bombardment of charged particles or X rays, or produced by ionizing processes in the residual gas of the imperfect vacuum. These electrons can absorb energy from their interaction with the electromagnetic fields. Such effects are often referred to as *electron loading*, a process that can limit the fields, and result in X-ray emission. Three effects can be identified as contributing to electron loading: (1) multipacting, (2) field emission, and (3) RF electric breakdown. The word *multipacting* [15] is a contraction of the phrase *multiple impact*. To understand this phenomenon, consider an ac electric field perpendicular to two plane-parallel metallic surfaces with a field strength such that an electron originating from one surface at $x = 0$ takes exactly one-half period to travel across the gap between the surfaces, as shown in Fig. 5.12. Suppose the conditions are such that this electron knocks out more than one secondary electron from the second surface at $x = x_0$. The field will reverse, and these electrons will return to the first surface, where they may each knock out more electrons. Once this process has begun, it will continue until the avalanche is limited by the space charge of the electrons. Multipacting requires two conditions: (1) a kinematic condition for resonant buildup must be satisfied (i.e., the electrons must travel from surface to surface in an integer number of half periods of the ac voltage and (2) a physical condition must be satisfied, i.e., the secondary electron coefficient must exceed unity).

The simple theory of two-conductor or *two-point multipacting* is based on nonrelativistic electron dynamics in a pure ac electric field between two plane-parallel conducting plates. Because electrons can be born at different phases of the RF voltage, the kinematic conditions for multipacting can occur at a range of voltages $V = E_0 x_0$ between

$$V_{\max} = \frac{m\omega^2 x_0^2}{2e} \tag{5.74}$$

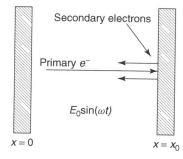

Figure 5.12 Initial stage of two-point multipacting between two plane surfaces. Electrons traverse gap in half an rf period.

and

$$V_{\min} = \frac{m\omega^2 x_0^2}{e} \frac{1}{\sqrt{4 + (2n+1)^2\pi^2}} \qquad (5.75)$$

where e and m are the electron charge and mass. The index n is the order of the multipacting level, and is related to the transit time between the two surfaces $t = (n + 1/2)T$, where T is the RF period, and $n = 0, 1, 2, \ldots$. There is a maximum electric field, above which multipacting cannot occur, given by

$$E_{\max} = \frac{V_{\max}}{x_0} = \frac{m\omega^2 x_0}{2e} \qquad (5.76)$$

For larger fields the electron transit times are too short to satisfy the resonant condition. There is no minimum field below which the kinematic conditions for multipacting cannot be satisfied, because the order n can be increased indefinitely. It may seem surprising that the maximum voltage is independent of n. This is because multipacting at the maximum voltage corresponds to the electron leaving a surface, when the electric field is maximum, and arriving at the opposite surface, when the field is maximum in the reversed direction. This electron is accelerated and decelerated by the same amount, and the integrated force is zero, independent of n. Electrons born at other phases require less peak voltage, because they are given a net velocity increase by the field. Regions where kinematic conditions for multipacting between two parallel plates are satisfied are shown in Fig. 5.13, where Eq. (5.74) is plotted. For values of V/f^2 larger than that given by Eq. (5.74), multipacting is not kinematically possible. For values less than the Eq. (5.74) result, multipacting can always exist for some order n.

It is useful to summarize some characteristics of multipacting.
  1. Multipacting often occurs at much lower electric field levels than those typically used for acceleration of the beam. It is often encountered when raising the cavity field to the operating level.

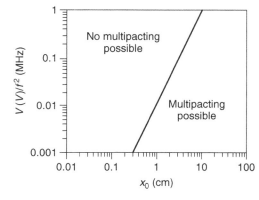

Figure 5.13 Regions where kinematic conditions between two parallel plates are or are not satisfied. The quantity $f = 1/T$ is the RF frequency in megahertz, V is in volts, and x_0 is in centimeters. The curve that separates the two regions is Eq. (5.74).

2. The simple theory does not account for the effects of the RF magnetic fields on the electrons.
3. There is evidence that dc magnetic fields perpendicular to the surfaces can enhance multipacting, whereas magnetic fields parallel to the surface can suppress the effect.
4. Overcoupling to reduce the time constant for the buildup of the cavity fields can suppress the effect by bringing the field past the multipacting region, before the resonance has time to build up.
5. Keeping the surfaces thoroughly clean and oil-free by using good vacuum practice such as use of oil-free vacuum pumps, can suppress multipacting by keeping the secondary electron coefficient small.
6. RF conditioning, which means operating at the multipacting resonance with relatively high-power levels, may help by letting the multipactor discharge burn until the secondary electron coefficient reduces to a sufficiently small value.
7. The most serious multipacting condition arises when multipacting occurs at a level near the operating level of the cavity. A problem has occurred in the low-field coupling cavities of coupled-cavity linacs, sometimes exacerbated by the presence of solenoidal dc magnetic fields used to focus the beam. The solution for this case has been to detune the coupling cavities, which raises the fields above the resonant values.
8. Surface coatings of titanium and titanium nitride can reduce the secondary electron coefficient.
9. A plot of the secondary electron coefficient versus electron kinetic energy shows that the coefficient typically exceeds unity over a range of impact kinetic energies extending from a few tens of electron volts to several kilovolts.
10. Symmetry in both the cavity geometry and the cavity fields can result in strong multipacting, because electrons from a large surface area can be involved.

Another form of multipacting, known as *single-point multipacting*, was discovered in superconducting cavities, and occurs at the outer wall of cavities operating in the TM_{010}-like mode, where the RF magnetic fields are large. Characteristics of single-point multipacting include the following. (1) The electrons are emitted from, and return to the same surface, where they knock out secondary electrons and build up a discharge. (2) The time between surface hits is an integer multiple of an RF period, rather than an odd number of half periods, as for the two-point multipacting [16]. Multipacting can occur only if the electrons impact the surface with sufficient energy that the secondary electron coefficient exceeds unity. The single-point multipacting theory yields an expression for the impact energy

$$W_{imp}(eV) = 7.6 \times 10^3 \left[\frac{E_\perp(MV/m)}{f(GHz)}\right]^2 \tag{5.77}$$

Single-point multipacting has limited the performance of pillbox-like superconducting cavities operating in the TM_{010} mode. The method that has been most successful in suppressing single-point multipacting for the superconducting case is to use a spherical cavity geometry at the outer wall. This cavity geometry effectively suppresses single-point multipacting by creating a configuration such that electrons, emitted from the walls, drift toward the equator, where $E_\perp \approx 0$.

5.10
Electron Field Emission

Electron field emission is an effect that limits the high-field performance of superconducting cavities. Electrons are emitted from material surfaces in the presence of a strong surface electric field. These electrons are further accelerated by the cavity fields, and when they strike the cavity walls their kinetic energy is converted to heat and X rays. Electron loading absorbs RF energy and lowers the Q of superconducting cavities at high fields. Classical field emission follows the *Fowler–Nordheim* law for the emitted current density

$$j \propto \frac{E^2}{\Phi} \exp\left(-\frac{a\Phi^{3/2}}{E}\right) \tag{5.78}$$

where E is the electric field and Φ is the work function. For niobium $\Phi = 4.3$ eV. An approximate formula for the field-emitted current density for niobium is

$$j(10^{12} A/m^2) \cong (6E)^2 \exp\left(-\frac{6}{E}\right) \tag{5.79}$$

where E is in units of 10^{10} V/m. In practice E is not the ideal surface field E_s, but is a surface-field-enhanced value $E = \beta E_s$, where β is a field enhancement factor that can be as large as $\beta \approx 250$. It is believed that the onset of field emission at lower than expected fields is often associated with low-work-function dust particles as well as needles or other protrusions. The most important rule for suppressing field emission in superconducting cavities is cleanliness. This may include rinsing of cavity surfaces in ultrapure water, and assembly in laminar dust-free airflow systems as used in the semiconductor industry. Statistically, the probability of field emission is expected to increase with the surface area that is exposed to high fields. Thus, smaller cavities tend to achieve higher fields. An approximate scaling formula for peak surface electric field in superconducting elliptical cavities is $E_{\text{peak}} \propto (\text{surface area})^{-1/4}$.

5.11
RF Electric Breakdown: Kilpatrick Criterion

At sufficiently high fields, room-temperature copper cavities will suffer electric breakdown or sparking. The detailed mechanism of this breakdown is not well understood, but it may be initiated by electron field emission and it has been suggested that protons, originating on the surfaces or perhaps from hydrogen in the residual gas, are involved in the discharge. In the 1950s, W. D. Kilpatrick [17] analyzed the data on RF breakdown, and defined the conditions that would result in breakdown-free operation. The Kilpatrick results were expressed in a convenient formula by T. J. Boyd [18] given as

$$f(MHz) = 1.64 E_K^2 e^{-8.5/E_K} \tag{5.80}$$

where f is the frequency, and E_K in megavolts per meter is known as the *Kilpatrick limit*. This is plotted in Fig. 5.14. Note that for a given frequency, the equation must be solved iteratively for E_K. Also, notice the similarity of the Kilpatrick-limit formula and the Fowler–Nordheim field-emission formula. In the Kilpatrick formula, E_K increases with increasing frequency.

The Kilpatrick criterion is based on experimental results that were obtained in an era before clean vacuum systems were common. Therefore, the Kilpatrick criterion is considered conservative by today's standards. Nevertheless, the same expression is commonly used for choosing the design field level for

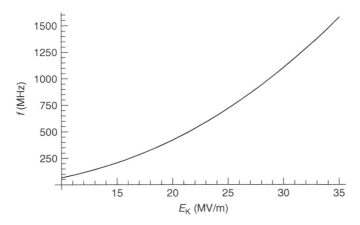

Figure 5.14 The Kilpatrick formula from Eq. (5.80).

accelerating cavities, except that the actual peak surface field E_s is expressed as

$$E_s = bE_K \qquad (5.81)$$

where b is known as the *bravery factor*. Typical values chosen for b, by accelerator designers, range from 1.0 to 2.0. Larger values of b are often chosen for pulsed accelerators with pulse lengths less than about 1 ms. For very short pulses, valid near 1 μs or less, an empirical scaling has been observed, given by [19]

$$E_s \propto f^{1/2}/t^{1/4} \qquad (5.82)$$

5.12
Adiabatic Invariant of an Oscillator

If a perturbation is introduced in the cavity geometry, either from the displacement of the cavity walls, from the introduction of a hole or aperture, or from the introduction of some object, the resonant frequency will generally change, and we will be interested in calculating this change. First we look at a general theorem concerning the effect of adiabatic changes on the frequency, and then we will present the Slater perturbation theorem, which is the basis of microwave field measurements.

Consider an oscillator in which a parameter varies as a result of some external action at a rate that is slow compared to the period of the oscillation. Such changes are said to be adiabatic [20]. It can be shown [21] that for a frictionless or Hamiltonian system,

$$I \equiv \frac{\oint p\,dq}{2\pi} = \frac{\iint dq\,dp}{2\pi} \qquad (5.83)$$

is invariant when any parameter is varied adiabatically. To illustrate this consider a one-dimensional simple harmonic oscillator with displacement q, momentum p, mass m, and resonant frequency ω_0. The Hamiltonian is

$$H(q,p) = \frac{p^2}{2m} + \frac{m\omega_0^2 q^2}{2} \qquad (5.84)$$

and from conservation of energy $H(q,p) = U$, where U is the constant total energy. The path of a trajectory in phase space is an ellipse, given by

$$\frac{p^2}{2mU} + \frac{m\omega_0^2 q^2}{2U} = 1 \qquad (5.85)$$

corresponding to semiaxes $p_0 = \sqrt{2mU}$, and $q_0 = \sqrt{2U/m\omega_0^2}$. If there are adiabatic changes of the parameters ω_0 or m, the energy is no longer constant, and it can be shown that an adiabatic invariant is

$$I = \frac{\pi p_0 q_0}{2\pi} = \frac{U}{\omega_0} \qquad (5.86)$$

or

$$\frac{\Delta\omega_0}{\omega_0} = \frac{\Delta U}{U} \qquad (5.87)$$

Thus, when the parameters of the oscillator vary slowly, the frequency is proportional to the energy of the oscillator. This result, also known as the *Boltzmann–Ehrenfest theorem*, [22] applies to resonant cavities as well as mechanical oscillators, and describes the effects of frequency tuners used in microwave cavities. It is easy to see from the result, that when the resonant frequency changes, so also do the amplitudes q_0 and p_0, where we have $p_0 \propto U$ and $q_0 \propto U^{-1}$. An adiabatic process can be thought of as proceeding approximately through a sequence of equilibrium steps. The maximum kinetic and potential energies remain equal to each other and equal to the total energy U.

5.13
Slater Perturbation Theorem

From the discussion of the previous section, we have learned that perturbations of a simple oscillator resulting in a change in the stored energy will generally result in a resonant frequency shift. For a cavity on resonance, the electric and magnetic stored energies are equal. If a small perturbation is made on the cavity wall, this will generally produce an unbalance of the electric and magnetic energies, and the resonant frequency will shift to restore the balance. The *Slater perturbation theorem* [23] describes the shift of the resonant frequency, when a small volume ΔV is removed from the cavity of volume V. The general result is

$$\frac{\Delta\omega_0}{\omega_0} = \frac{\int_{\Delta V}(\mu_0 H^2 - \varepsilon_0 E^2)\,dV}{\int_V(\mu_0 H^2 + \varepsilon_0 E^2)\,dV} = \frac{\Delta U_m - \Delta U_e}{U} \qquad (5.88)$$

where U is the total stored energy, given in terms of the unperturbed field amplitudes E and H by

$$U = \frac{1}{4}\int_V(\mu_0 H^2 + \varepsilon_0 E^2)\,dV \qquad (5.89)$$

$\Delta U_m = \int_{\Delta V}\mu_0 H^2\,dV/4$ is the time average of the stored magnetic energy *removed*, and $\Delta U_e = \int_{\Delta V}\varepsilon_0 E^2\,dV/4$ is the time average of the stored electric energy *removed* as a result of the reduced volume [24]. The frequency increases if the magnetic field is large where the walls are pushed in, and it decreases if the electric field is large there. This result is easier to remember if one identifies a decrease in the effective inductance where the magnetic field is large and an increase in the effective capacitance where the electric field is large. There can be cases where the electric and magnetic effects cancel. For example, this would occur if the end walls of a pillbox cavity, excited in a TM_{010} mode, are uniformly pushed in. The contribution from the electric fields near the axis is exactly canceled by the dominant magnetic field effect at larger radius. This is why the frequency is independent of the cavity length for this case.

The Slater theorem provides the basis for field measurements in cavities [25]. If a small bead is inserted into the cavity, the perturbation shifts the resonant frequency. For a spherical bead of volume ΔV, the shift is given as a function of the unperturbed field amplitudes E and H, which are assumed to be constant over the bead, by

$$\frac{\Delta \omega_0}{\omega_0} = -\frac{3\Delta V}{4U}\left[\frac{\varepsilon_r - 1}{\varepsilon_r + 2}\varepsilon_0 E^2 + \frac{\mu_r - 1}{\mu_r + 2}\mu_0 H^2\right] \qquad (5.90)$$

where ε_r and μ_r are the dielectric constants, and magnetic permeability of the material relative to vacuum. For a spherical dielectric bead, we choose $\mu_r = 1$, and

$$\frac{\Delta \omega_0}{\omega_0} = -\frac{3\Delta V}{4U}\frac{\varepsilon_r - 1}{\varepsilon_r + 2}\varepsilon_0 E^2 \qquad (5.91)$$

Thus, if the shift in the resonant frequency is measured, and the stored energy is known from measurement of the Q and the power dissipation, the magnitude of the electric field can be calculated. Sapphire is a commonly used dielectric that has a relative dielectric constant $\varepsilon_r \approx 9$. The results for a spherical perfectly conducting bead are obtained by letting $\varepsilon_r \to -j\infty$ (Inside a conductor the relative dielectric constant is $\varepsilon_r = 1 - j\sigma/\varepsilon_0\omega$.), and $\mu_r \to 0$ for a diamagnetic metal, and we find

$$\frac{\Delta \omega_0}{\omega_0} = -\frac{3\Delta V}{4U}\left[\varepsilon_0 E^2 - \frac{\mu_0 H^2}{2}\right] \qquad (5.92)$$

Because this result depends on both E and H, unless it is already known which field is dominant, it would be necessary to do measurements using both dielectric and metallic beads to determine both field values. Similar equations have been derived for beads of other geometries. For example, a metallic needle allows one to obtain some directional information about the fields.

In the standard bead-perturbation technique for measuring the RF electric and/or magnetic fields versus position, a small bead is moved through the cavity using a typical setup as shown in Fig. 5.15. Typically, the beam is attached to a thin nylon line that is driven through the cavity by a motor. The perturbation at any beam position causes a frequency shift that is proportional to the square of the local field. The frequency shift can easily be measured, by driving the cavity using a phase-locked loop that locks the driving frequency to the resonant frequency of the cavity. The measurement determines the ratio of squared field to the stored energy. The stored energy can be determined from

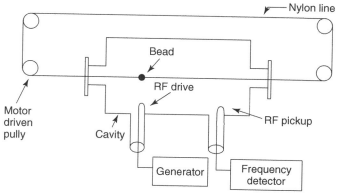

Figure 5.15 A schematic diagram of a field-measuring apparatus based on the Slater perturbation method.

the relation $U = QP/\omega_0$, and from separate measurements of Q and the cavity power dissipation P, using standard microwave measurement techniques.

5.14
Quasistatic Approximation

For some applications involving time-dependent electromagnetic fields, the spatial distribution of the fields is nearly the same as for the static problem, even though the frequency may be in the microwave region. In these cases the electric-field components can be derived from a time-dependent scalar potential that satisfies Laplace's equation. This method is called the *quasistatic approximation*, and is usually valid when the geometrical variations are small compared to the free-space wavelength. This result follows from consideration of the wave equation in charge-free space, which for the case of harmonic time dependence becomes the Helmholtz equation

$$\nabla^2 \mathbf{E} + k^2 \mathbf{E} = 0 \tag{5.93}$$

where the wave number is $k = \omega/c = 2\pi/\lambda$. Expanding the Helmholtz equation into its three components gives

$$\frac{\partial^2 \mathbf{E}}{\partial x^2} + \frac{\partial^2 \mathbf{E}}{\partial y^2} + \frac{\partial^2 \mathbf{E}}{\partial z^2} + \left(\frac{2\pi}{\lambda}\right)^2 \mathbf{E} = 0 \tag{5.94}$$

When field variations develop on a scale determined by structural elements, whose spacings are small compared to the wavelength, the values of derivative terms near those elements will dominate over the last term, and the equation then reduces approximately to Laplace's equation

$$\nabla^2 \mathbf{E} \cong 0 \tag{5.95}$$

The quasistatic approximation consists of neglecting the last term of the Helmholtz equation, and assuming that the electric-field components satisfy Laplace's equation. The electric field must also satisfy the basic vector relationship $\nabla \times (\nabla \times \mathbf{E}) = \nabla(\nabla \cdot \mathbf{E}) - \nabla^2 \mathbf{E}$. Since the last term is approximately zero, and in charge-free space $\nabla \cdot \mathbf{E} = 0$, Eq. (5.95) implies that $\nabla \times (\nabla \times \mathbf{E}) = 0$. The latter result is satisfied if $\mathbf{E} = -\nabla U$, where U is a scalar potential, because of the vector identity $\nabla \times \nabla U = 0$. Therefore, in the quasistatic approximation, the time-dependent electric-field components are derivable from a scalar potential, and since

$$\nabla \cdot \mathbf{E} = -\nabla^2 U = 0 \tag{5.96}$$

the potential must also satisfy Laplace's equation. Notice that the quasistatic approximation decouples the magnetic and electric fields in Maxwell's equations, and this implies that the contribution to the electric fields from Faraday's law is negligible. For the radiofrequency quadrapole or RFQ the quasistatic approximation applies very accurately to the region near the four poles, including the region of the beam, where the spacings of the poles in that region are typically only a few thousandths of the wavelength.

5.15
Panofsky–Wenzel Theorem

The Panofsky–Wenzel theorem [26] is a general theorem pertaining to the transverse deflection of particles by an RF field. A transverse deflection in a linac is undesirable if it leads to the beam-breakup instability, or it may be desirable if the cavity is to be used as a beam chopper to pulse the beam. Consider a particle with velocity v and charge q, traveling through an RF cavity near and parallel to the axis, and assume that the particle is moving fast enough that the transverse forces, which produce a transverse momentum impulse, have a negligible effect on the trajectory in the cavity. The transverse momentum change over a length d, is approximately

$$\Delta \mathbf{p}_\perp = \frac{q \int_{z=0}^{d} [\mathbf{E}_\perp + (\mathbf{v} \times \mathbf{B})_\perp] dz}{v} \tag{5.97}$$

where the fields are those seen by the particle. In a cavity where the electric and magnetic fields are time-dependent, generally not in phase, and where cancellation of the electric and magnetic deflections is possible, it is not obvious what kind of cavity mode gives the largest deflection. Following the original derivation, we introduce the vector potential. In the cavity volume with no free charges, we may write the electric and magnetic fields as

$$\mathbf{E} = -\frac{\partial \mathbf{A}}{\partial t}, \quad \mathbf{B} = \nabla \times \mathbf{A} \tag{5.98}$$

5.15 Panofsky–Wenzel Theorem

where **A** is the vector potential. We can write

$$(\mathbf{v} \times \mathbf{B})_\perp = [\nabla(\mathbf{v} \cdot \mathbf{A}) - (\mathbf{v} \cdot \nabla)\mathbf{A}]_\perp = \nabla_\perp(\mathbf{v} \cdot \mathbf{A}) - (\mathbf{v} \cdot \nabla)\mathbf{A}_\perp \quad (5.99)$$

We substitute Eqs. (5.98) and (5.99) into Eq. (5.97), and using the fact that v is a constant, and directed along the z-axis, we have

$$\Delta \mathbf{p}_\perp = q \int_0^d \left[-\left(\frac{1}{v}\frac{\partial \mathbf{A}_\perp}{\partial t} + \frac{\partial \mathbf{A}_\perp}{\partial z}\right) + \nabla_\perp A_z \right] dz \quad (5.100)$$

The total derivative of $\mathbf{A}_\perp$ along the path of the particle is

$$d\mathbf{A}_\perp = \left(\frac{1}{v}\frac{\partial \mathbf{A}_\perp}{\partial t} + \frac{\partial \mathbf{A}_\perp}{\partial z}\right) dz \quad (5.101)$$

Substituting Eq. (5.101) into Eq. (5.100), we find

$$\Delta \mathbf{p}_\perp = -q \int_{\mathbf{A}_\perp(z=0)}^{\mathbf{A}_\perp(z=d)} d\mathbf{A}_\perp + q \int_0^d \nabla_\perp A_z \, dz \quad (5.102)$$

If we assume an $e^{j\omega t}$ harmonic time dependence for **E**, Eq. (5.98) requires that $\mathbf{A} = j\mathbf{E}/\omega$, and Eq. (5.102) becomes [27]

$$\Delta \mathbf{p}_\perp = -q\frac{jq}{\omega} \int_{\mathbf{A}_\perp(z=0)}^{\mathbf{A}_\perp(z=d)} d\mathbf{E}_\perp + \frac{jq}{\omega} \int_0^d \nabla_\perp E_z \, dz \quad (5.103)$$

The first term of Eq. (5.103) vanishes for any case with a conducting wall at $z = 0$ and d, such as for a pillbox cavity without beam holes. It also vanishes for any cavity with a beam pipe at each end, if $z = 0$ and d are chosen far enough beyond the cavity end walls that the fields have attenuated, so that approximately $\mathbf{E}_\perp = 0$. For these cases

$$\Delta \mathbf{p}_\perp = \frac{jq}{\omega} \int_0^d \nabla_\perp E_z \, dz \quad (5.104)$$

Equation (5.104) is the Panofsky–Wenzel theorem that says that the total momentum impulse is proportional to the transverse gradient of the longitudinal electric field. Because E_z is the field seen by the particle, one should substitute $t = z/v$. The quantities **A**, **B**, and $\Delta \mathbf{p}_\perp$ are in phase, and are 90° ahead of **E**. The right side of Eq. (5.104) is zero for a pure TE mode because E_z and $\nabla_\perp E_z = 0$ are then zero everywhere in the cavity. Physically, the pure TE mode of a pillbox cavity gives zero deflection, because the electric and magnetic forces cancel. A pure TM mode can give a nonzero $\nabla_\perp E_z$, and can give a nonzero transverse deflection. For example, the TM_{010} mode produces a nonzero $\nabla_\perp E_z$ for an off-axis particle, and the TM_{110} mode produces a large nonzero $\nabla_\perp E_z$ on axis. For a cavity with beam holes, there will generally be neither pure TE nor pure TM modes [28]. Note that the derivation did not assume that the particle had to be traveling at the extreme relativistic limit;

only that it traveled fast enough that the particle trajectory was approximately unperturbed by the deflecting field.

Problems

5.1. A circular cylindrical cavity with radius R and length l operating in a TM_{010} mode is excited by a coupling loop inserted through a hole in the cylindrical wall. Assume that the loop is located at the wall and that the effective loop area A is small enough so that the magnetic field is uniform over the loop. (a) Use the expression for the power dissipated in the TM_{010} mode and show that the transformed load resistance presented by the cavity at resonance to the line is $R_L = [\omega\mu_0 A]^2/[2\pi R(\ell + R)R_s]$. (b) Consider a room-temperature copper cavity operated in the TM_{010} mode with a radius of 0.287 m, and a length of 0.345 m. Calculate the effective loop area required so that at resonance the cavity presents a matched load to a coaxial line with a 50 Ω characteristic impedance. (c) Suppose we consider the cavity and loop as an equivalent transformer that transforms the axial cavity voltage $V_0 = E_0 \ell$ to the loop voltage $V = \omega B A$. Show that the effective transformer–turns ratio $n = V_0/V = \ell c/(J_1(2.405)\omega A)$. Using the result $J_1(2.405) = 0.5191$ and the other numerical values of this problem, calculate the numerical value of n.

5.2. A spherical conducting bead of diameter 3/16 in. is placed near the outer cylindrical wall of a cylindrical cavity of length $l = 0.5$ m, operating in a TM_{010} mode at a resonant frequency of 400 MHz. (a) Assuming we can ignore any effect from the electric field near the outer wall, and evaluating the magnetic field at the outer wall, show that the fractional frequency shift is $\Delta\omega_0/\omega_0 = (3/4)\Delta V/(\pi R^2 \ell)$ where ΔV is the volume of the bead, and R is the cavity radius. (b) Calculate the numerical value of the shift in the resonant frequency $\Delta f_0 = \Delta\omega_0/2\pi$. (c) Suppose the bead is moved to a new radius where the unperturbed magnetic field is larger by 1% for the same excitation level of the cavity. Calculate the change in the resonant frequency shift. (d) If the cavity length was increased by a factor of 4 and the cavity-field level remains the same, how would the frequency shifts be affected? Would you expect the relative field changes obtained from the measurements of frequency shifts to be as easy to measure for a longer cavity?

5.3. An accelerator cavity operates in the TM_{010} mode at a resonant frequency of 400 MHz, and has a measured (unloaded) Q_0 of 40,000. The cavity is matched to a transmission line with a characteristic $Z_0 = 50$ Ω, through which the generator supplies an average power $P = 12$ kW to excite the cavity to a peak axial voltage of $V_0 = 0.5$ MV. (a) Use the axial voltage and the power to calculate the cavity shunt resistance R, and the value of R/Q_0. (b) The input power coupler behaves like a transformer with turns ratio n, transforming the low voltage in the input transmission line to the high voltage in the cavity. Use the shunt resistance R and the cavity load resistance at the transmission line to calculate the effective transformer–turns ratio n. (c) Use the value of n to calculate the peak voltage of the incident wave in the input transmission line. What is the corresponding peak forward current needed to supply the cavity power?

5.4. A cavity, operating at a resonant frequency of 350 MHz with an unloaded quality factor $Q_0 = 20,000$, is excited on resonance by an RF source through a transmission line and an input power coupler. The input coupler is not adjustable, and to measure the degree of coupling the RF source is turned off and the time constant of the cavity decay is determined from measurement to be $\tau = 12.0$ µs. What are the loaded and external quality factors, Q_L and Q_{ex}, and the coupling factor β. Assuming all other ports connected to the cavity are weakly coupled, is the cavity undercoupled, matched, or overcoupled to the input transmission line?

5.5. A cavity is characterized by $Q_0 = 20,000$, an R/Q parameter (ratio of shunt resistance of R to Q_0) = 250 Ω, and the input transmission line is matched at its characteristic impedance value of $Z_0 = 50\ \Omega$, when looking back from the power coupler. The cavity coupling to the transmission line is characterized by the coupling factor $\beta = 2$. (a) Draw the equivalent circuit as seen by the generator circuit (primary) and show the numerical values of the resistances. (b) Draw the equivalent circuit as seen by the cavity (secondary) and show the numerical values of the resistances.

5.6. Consider a cavity with $R/Q = 240\ \Omega$ and $Q_0 = 20,000$ and assume that the driving generator emits a forward wave P_+ that supplies the cavity power P_c with reflected power P_- at the power coupler. (a) Using the coupling factor $\beta = 2$, calculate the reflection coefficient of the input transmission line. (b) Assuming the cavity voltage is $V = 1$ MV, calculate the power dissipated in the cavity. (c) Calculate the forward and reflected power in the transmission line. (d) Calculate the voltages in the transmission line, including the forward and reverse waves and the voltage at the load (equal to the transformed cavity voltage).

5.7. A cavity operating on resonance is coupled to an input transmission line with a coupling factor β_1, and to an output transmission line with a coupling factor β_2. The cavity resistance transformed into the generator circuit can be shown to equal $Z_0\beta_1$, and the transformed output resistance is $Z_0\beta_1/\beta_2$. These two resistances combined in parallel constitute the load resistance for the input line. (a) Express this input load resistance as a function of Z_0, β_1, and β_2, and show that the reflection coefficient is given by $\Gamma = (\beta_1 - (1 + \beta_2))/(\beta_1 + \beta_2 + 1)$. (b) If $\beta_2 = 0.01$, calculate the value of β_1 that makes the reflection coefficient equal to zero. (c) Suppose the input and output coupling factors are adjusted so that $\beta_1 = \beta_2 = 1$. Calculate the reflection coefficient in the input transmission line. Is the cavity matched to the input line? If β_1 could be adjusted for a match, what value should it be?

5.8. A room-temperature pillbox cavity is designed to operate in a TM_{010} mode at 400 MHz at the highest surface electric field E_s allowed by the Kilpatrick criterion, modified to allow an additional bravery factor $b = 2.0$, such that $E_s = b E_K$. (a) Calculate the Kilpatrick field E_K, and the peak surface electric field E_s. Where is the electric field maximum for this mode? (b) Calculate the corresponding maximum value for the maximum surface magnetic field. Where on the cavity wall does this occur?

5.9. Suppose the gap length of the first gap of the LANSCE 201.25-MHz drift-tube linac is $g = 1.05$ cm. Assume a plane-parallel geometry is a good approximation for the opposing drift tubes. (a) Calculate the minimum and maximum gap voltage and electric field for which the lowest $n = 0$ multipacting mode can occur. (b) What is the maximum field value, above which no multipacting of any order can occur in the gap?

5.10. (a) Derive the nonzero field components in complex exponential form for the TM_{010} mode of a cylindrical cavity resonator using the vector potential $A_x = 0$, $A_y = 0$, and $A_z = E_0 J_0(k_r r) j e^{j\omega t}/\omega$, where $\omega = k_r c$, and show that they are $E_z = E_0 J_0(k_r r) e^{j\omega t}$, and $B_\theta = E_0 J_1(k_r r) j e^{j\omega t}/c$. (b) Consider a cavity with length $d \ll \beta \lambda$, and a particle crossing the cavity parallel to the axis at fixed radius r. Calculate the transverse momentum delivered to the particle by the cavity RF fields, directly from the Lorentz force, Eq. (5.97), and from the Panofsky–Wenzel theorem, Eq. (5.104), and show that these are equal. Although the Panofsky–Wenzel theorem expresses the transverse deflection as a function of an electric-field gradient, is the transverse deflection for this problem caused by the electric or the magnetic force?

5.11. Suppose a generator launches an RF wave in a coaxial transmission line to drive an initially unexcited cavity. Assume the line is coupled to the cavity through a coaxial antenna-type coupler, which, in the extreme undercoupled limit, is assumed to be an open circuit at its electrical end ($z = 0$). (a) As was discussed in Section 5.6, ignoring the details of the wavefront propagation, write the expression for the total wave in the line for times $t > 0$ as a sum of three traveling waves, an incident wave, a direct-reflected wave, and a time-dependent radiated wave from the cavity. Assume that the peak of the total voltage wave at $t = 0$ occurs at $z = 0$. (b) Express the total wave in the undercoupled limit, when $\beta = 0$, and show that it is a pure standing wave. (c) Express the total wave at critical coupling, when $\beta = 1$. What is the nature of this wave at both $t = 0$ and ∞? (d) Express the total wave in the overcoupled limit, $\beta \to \infty$. What is the nature of this wave at both $t = 0$ and ∞? Show that momentarily at time $t = \tau \ln 2$, there is zero effective reflected wave (only an incident wave).

5.12. An RF window is used to separate the beam-line high vacuum from atmospheric-pressure air in the RF waveguide that electromagnetically connects the generator to the cavity. The windows are made from an insulating material such as ceramic, and their location in the waveguide should be chosen to minimize high electric fields, which can produce damage from thermal stress caused by dielectric heating. Consider the RF system and cavity in problem 5.11, and describe the locations of the voltage minima in the coaxial line for the different coupling regimes: **(a)** undercoupled limit, **(b)** critical coupling, and **(c)** overcoupled limit. For any of these cases do the voltage minima change with time t?

References

1. To conform to common terminology, the accelerator shunt impedance is defined as twice the circuit shunt impedance.
2. Condon, E.U., *J. Appl. Phys.* **11**, 502 (1940); Condon, E.U., *J. Appl. Phys.* **12**, 129 (1941).
3. A common example of a suitable coupling mechanism is a loop at the end of a coaxial transmission line that is inserted through a small hole in the cavity wall.
4. Energy conservation can be used to justify the damping term.
5. We will learn later that for a waveguide-cavity system, this is a virtual power that is dissipated when the generator is suddenly turned off.
6. For a waveguide-cavity system, this definition needs to be qualified. The power P_{ex} that enters in the definition of β is defined as the external power radiated from the cavity, when the generator is turned off.
7. P_- is generally not equal to P_{ex}, unless the generator is off and the cavity fields are decaying.
8. Strictly speaking, we will find that this is not simply a reflected wave, but is really the superposition of a reflection from the input coupler, plus the emitted wave from the cavity.
9. This amplitude is slightly less than V_+ because a small fraction of incident energy is transmitted to the cavity.
10. The approach of this section closely follows that given in Altman, J.L., *Microwave Circuits*, D. Van Nostrand, Princeton, N.J., 1964.
11. Ramo, S., Whinnery, J.R. and Van Duzer, T., *Fields and Waves in Communication Electronics*, 2nd ed., John Wiley & Sons, New York, 1984.
12. The cancellation of reflected and radiated waves that produces the critical-coupling condition occurs only for the resonant frequency. For the transient behavior, while the cavity fields are building or decaying, there is power dissipated in the external load, because other frequencies besides the resonant frequency will contribute.
13. A summary of accelerator issues for RF power systems is given by Taylor, C.S., in *Linear Accelerators*, Lapostolle, P.M. and Septier, A.L., North Holland Publishing, 1970, pp. 905–933.
14. Ziomek, C.D., Regan, A.H., Lynch, M.T. and Bowling, P.S., Low-level RF control system issues for an ADTT accelerator, *Int. Conf. Accelerator Driven Transmutation Technol. Appl.*, Las Vegas, NV, July, 1994, ed., Arthur, E.D., Rodruiguez, A. and Schriber, S.O., *AIP Conf. Proc.* **346**, p. 453.
15. Hatch, A.J. and Williams, H.B., *Phys. Rev.* **112**, 68 (1958). It is common usage to refer to this phenomenon as either multipacting or multipactoring.
16. Klein, U. and Proch, D., *Proc. Conf. on Future Possibilities for Electron Accel.*, eds. McCarthy, J.S. and Whitney, R.R., Charlottesville, VA, 1979.
17. Kilpatrick, W.D., *Rev. Sci. Instr.* **28**, 824 (1957).

18 Boyd, T.J. Jr., *Kilpatrick's Criterion*, Los Alamos Group AT-1 report AT-1:82-28, February 12, 1982.
19 Palmer, R.B., Prospects for high energy e^+ e^- linear colliders, *Ann. Rev. Nucl. Part. Sci.* **40**, 568 (1990).
20 Adiabatic in this context should not be confused with its common use in thermodynamics, where it is associated with rapid processes. However, the two seemingly contradictory uses of this term may not really be incompatible. In physics textbooks, adiabatic processes are defined as being sufficiently rapid that the body is thermally isolated. However, the condition of slowness compared to the internal time constants is also tacitly assumed for any reversible process that takes the system through quasi-equilibrium states. The reader is referred to the discussion by Landau, L.D. and Lifshitz, E.M., *Statistical Physics*, 2nd ed., Pergamon Press, 1970, p. 38.
21 Landau, L.D. and Lifshitz, E.M., *Mechanics*, 3rd ed., Pergamon Press, 1989, p. 154.
22 See Boltzmann, L., *Vorlesungen uber Mechanik II*, J. A. Barth, Leipzig, 1903, p. 48; Ehrenfest, P., *Ann. Phys.* **36**, 91 (1911); Ehrenfest, P., *Proc. Amsterdam Acad.* **16**, 591 (1914); Maclean, W.R., *Q. Appl. Math.* **2**, 329 (1945).
23 Slater, J.C., *Microwave Electronics*, D. Van Nostrand Company, New York, 1950, pp. 80–81.
24 Slater's derivation does not assume the process is adiabatic. However, a presentation of the theorem in the context of an adiabatic theorem is given by Klein, H., *CERN Accelerator School on RF Engineering for Particle Accelerators*, CERN 92-03, Vol. 1, 1992, p. 112–118.
25 Maier, L.C., Jr. and Slater, J.C., *J. Appl. Phys.* **23**(No. 1), 68–83 (1952); Klein, H., *CERN Accelerator School on RF Engineering for Particle Accelerators*, CERN 92-03, Vol. 1, 1992, pp. 112–118.
26 Panofsky, W.K.H. and Wenzel, W.A., *Rev. Sci. Instrum.* **27**, 967 (1956). The treatment presented here follows that of M. Jean Browman, Los Alamos, 1991, private communication.
27 The phase of **A** is obtained simply by using the identity $j = \exp[j\pi/2]$ so that **A** is proportional to $\exp[j(\omega t + \pi/2)]$. Thus **A** is $\pi/2$ ahead of **E**.
28 The original paper discusses the theorem only in the context of pure TE and TM modes, which at that time may have represented the only available solutions, and those solutions only applied to simple pillbox cavities without beam holes. In general, the introduction of axially symmetric beam holes results in solutions with mixed TE and TM terms, except for axially symmetric modes such as the TM_{010} mode.

6
Longitudinal Particle Dynamics

An ion linac is designed for acceleration of a single particle, which remains in synchronism with the accelerating fields and is called the *synchronous particle*. For acceptable output beam intensity, restoring forces must be present so that those particles near the synchronous particle will have stable trajectories. Longitudinal restoring forces are produced when the beam is accelerated by an electric field that is increasing with time, and these forces produce phase and energy oscillations about the synchronous particle. The final energy of an ion that undergoes phase oscillations is approximately determined, not by the field, but by the geometry of the structure that is tailored to produce a specific final synchronous energy. An exception is an ion linac built from an array of short independent cavities; each one is capable of operating over a wide velocity range. In this case, the final energy depends on the field and the phasing of the cavities. The longitudinal dynamics is different for relativistic electron linacs. After beam injection into electron linacs, the velocities approach the speed of light so rapidly that hardly any phase oscillations take place. The electrons initially slip relative to the wave and rapidly approach a final phase that is maintained all the way to high energy. The final energy of each electron with a fixed phase depends on the accelerating field and on the value of the phase. In this chapter we will develop the general equations of longitudinal dynamics in a linac, including the Liouvillian gap transformations used in computer simulation codes. This will be followed by special treatments of the dynamics in coupled-cavity ion linacs, independent-cavity ion linacs, and low-energy electron beams injected into conventional electron-linac structures.

6.1
Longitudinal Focusing

Longitudinal focusing is provided by an appropriate choice of the phase of the synchronous particle relative to the crest of the accelerating wave. A longitudinal restoring force exists when the synchronous phase is chosen corresponding to a field that is rising in time, as shown in Fig. 6.1. For this case, the early particles experience a smaller field and the late particles a larger field

RF Linear Accelerators. 2nd, completely revised and enlarged edition.
Thomas P. Wangler
Copyright © 2008 Wiley-VCH Verlag GmbH & Co. KGaA, Weinheim
ISBN: 978-3-527-40680-7

than the synchronous particle. The accelerated particles are formed in stable bunches that are near the synchronous particle. Those particles outside the stable region slip behind in phase and do not experience any net acceleration.

To develop these ideas, we consider an array of accelerating cells (Fig. 6.2), containing drift tubes and accelerating gaps, designed at the nth cell for a particle with synchronous phase ϕ_{sn}, synchronous energy W_{sn}, and synchronous velocity β_{sn}. We express the phase, energy, and velocity of an arbitrary particle in the nth cell as ϕ_n, W_n, and β_n. The particle phase in the nth cell is defined as the phase of the field when the particle is at the center of the nth gap, and the particle energy for the nth cell is the value at the end of the nth cell at the center of the drift tube. We now investigate the motion of particles with phases and energies that deviate from the synchronous values. We assume that the synchronous particle always arrives at each succeeding gap at the correct phase, and we consider particles with velocities that are close enough to the synchronous velocity that all particles have about the same transit-time factor.

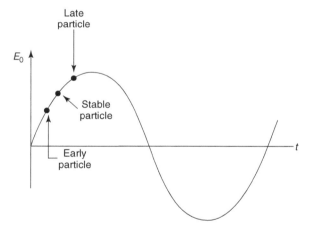

Figure 6.1 Stable phase.

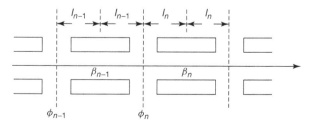

Figure 6.2 Accelerating cells for describing the longitudinal motion.

6.2
Difference Equations of Longitudinal Motion for Standing-Wave Linacs

We consider the particle motion through each cell to consist of a sequence of drift spaces where the particles see no forces, follow thin gaps where the forces are applied as impulses. The thin gaps are assumed to be located at the geometrical center of each cell. From gap $n-1$ to gap n, the particle has a constant velocity β_{n-1}, as indicated in Fig. 6.2. The RF phase changes as the particle advances from one gap to the next according to

$$\phi_n = \phi_{n-1} + \omega \frac{2\ell_{n-1}}{\beta_{n-1} c} + \begin{cases} \pi & \text{for } \pi \text{ mode} \\ 0 & \text{for 0 mode} \end{cases} \quad (6.1)$$

where the half-cell length is

$$\ell_{n-1} = N \beta_{s,n-1} \lambda / 2, \quad \text{where } N = \begin{cases} \frac{1}{2} & \text{for } \pi \text{ mode} \\ 1 & \text{for 0 mode} \end{cases} \quad (6.2)$$

By definition of the synchronous velocity, the cell length, measured from the center of one drift tube to the center of the next, must be $L_n = N(\beta_{s,n-1} + \beta_{s,n})\lambda/2$. The phase change during the time an arbitrary particle travels from gap $n-1$ to gap n relative to that of the synchronous particle is

$$\Delta(\phi - \phi_s)_n = \Delta \phi_n - \Delta \phi_{s,n} = 2\pi N \beta_{s,n-1} \left[\frac{1}{\beta_{n-1}} - \frac{1}{\beta_{s,n-1}} \right] \quad (6.3)$$

Using the Taylor expansion to write

$$\frac{1}{\beta} - \frac{1}{\beta_s} = \frac{1}{\beta_s + \delta\beta} - \frac{1}{\beta_s} \cong -\frac{\delta\beta}{\beta_s^2}, \quad \text{for } \delta\beta \ll 1 \quad (6.4)$$

and using $\delta\beta = \delta W / mc^2 \gamma_s^3 \beta_s$, we find the difference equation relating to the change in the relative phase,

$$\Delta(\phi - \phi_s)_n = -2\pi N \frac{(W_{n-1} - W_{s,n-1})}{mc^2 \gamma_{s,n-1}^3 \beta_{s,n-1}^2} \quad (6.5)$$

Next, we write the difference equation for the energy change of a particle relative to that of the synchronous particle as

$$\Delta(W - W_s)_n = \Delta W_n - \Delta W_{s,n} = qE_0 TL_n (\cos \varphi_n - \cos \varphi_{s,n}) \quad (6.6)$$

Eqs. (6.5) and (6.6) form two coupled difference equations for relative phase and energy change that can be solved numerically for the motion of any particle.

6.3
Differential Equations of Longitudinal Motion

To study the stability of the motion, it is convenient to convert the difference equations to differential equations. In this approximation, we replace the discrete action of the standing-wave fields by a continuous field. We let

$$\Delta(\phi - \phi_s) \to \frac{d(\phi - \phi_s)}{dn}, \quad \text{and} \quad \Delta(W - W_s) \to \frac{d(W - W_s)}{dn} \quad (6.7)$$

where n is now treated as a continuous variable. We can change variables from n to the axial distance s, using $n = s/(N\beta_s\lambda)$. The coupled difference equations, Eqs. (6.5) and (6.6) become the coupled differential equations

$$\gamma_s^3 \beta_s^3 \frac{d(\phi - \phi_s)}{ds} = -2\pi \frac{(W - W_s)}{mc^2\lambda} \quad (6.8)$$

and

$$\frac{d(W - W_s)}{ds} = qE_0 T(\cos\phi - \cos\phi_s) \quad (6.9)$$

6.4
Longitudinal Motion when Acceleration Rate is Small

We can differentiate Eq. (6.8) and substitute Eq. (6.9), to obtain a second-order differential equation for longitudinal motion:

$$\frac{d}{ds}\left[\gamma_s^3 \beta_s^3 \frac{d(\phi - \phi_s)}{ds}\right] = -2\pi \frac{qE_0 T}{mc^2\lambda}(\cos\phi - \cos\phi_s) \quad (6.10)$$

that can also be written as

$$\gamma_s^3 \beta_s^3 \frac{d^2(\phi - \phi_s)}{ds^2} + 3\gamma_s^2 \beta_s^2 \left[\frac{d}{ds}(\gamma_s\beta_s)\right]\left[\frac{d(\phi - \phi_s)}{ds}\right]$$

$$+ 2\pi \frac{qE_0 T}{mc^2\lambda}(\cos\phi - \cos\phi_s) = 0 \quad (6.11)$$

This is a nonlinear second-order differential equation for the phase motion. Before we investigate the properties of the solution, we comment that, because phase is proportional to time, the more negative the phase, the earlier the particle arrival time relative to the crest of the wave. Also, the phase difference $\phi - \phi_s$ between a particle and the synchronous particle is proportional to a spatial separation

$$z - z_s = -\frac{\beta_s\lambda}{2\pi}(\phi - \phi_s) \quad (6.12)$$

6.4 Longitudinal Motion when Acceleration Rate is Small

We assume that the acceleration rate is small, and that $E_0 T$, ϕ_s, and $\beta_s \gamma_s$ are constant. It is convenient to introduce the following notation:

$$w \equiv \delta\gamma = \frac{W - W_s}{mc^2}, \quad A \equiv \frac{2\pi}{\beta_s^3 \gamma_s^3 \lambda}, \quad \text{and} \quad B \equiv \frac{qE_0 T}{mc^2} \tag{6.13}$$

Using this notation, Eqs. (6.8), (6.9), and (6.10) are

$$w' \equiv dw/ds = B(\cos\phi - \cos\phi_s) \tag{6.14}$$

$$\phi' \equiv d\phi/ds = -Aw \tag{6.15}$$

and

$$\phi'' \equiv d^2\phi/ds^2 = -AB(\cos\phi - \cos\phi_s) \tag{6.16}$$

where $AB = 2\pi q E_0 T/(mc^2 \beta_s^3 \gamma_s^3 \lambda)$. Integrating Eq. (6.16) gives

$$d\phi' = -AB(\cos\phi - \cos\phi_s) ds \tag{6.17}$$

Using $ds = d\phi/\phi'$, multiplying Eq. (6.17) by ϕ', and integrating, we obtain, after substituting Eq. (6.15),

$$\frac{Aw^2}{2} + B(\sin\phi - \phi\cos\phi_s) = H_\phi \tag{6.18}$$

where the quantity H_ϕ is a constant of integration that we identify as the Hamiltonian. The first term of Eq. (6.18) is a kinetic energy term, and the second is the potential energy. The potential energy V_ϕ is

$$V_\phi = B(\sin\phi - \phi\cos\phi_s) \tag{6.19}$$

and it can be verified that there is a potential well when $-\pi < \phi_s < 0$. Recalling that there is acceleration for $-\pi/2 \leq \phi_s \leq \pi/2$, one obtains simultaneous acceleration and a potential well when $-\pi/2 \leq \phi_s \leq 0$. The stable region for the phase motion extends from $\phi_2 < \phi < -\phi_s$, where the lower phase limit ϕ_2 can be obtained numerically by solving for ϕ_2 using $H_\phi(\phi_2) = H_\phi(-\phi_s)$.

Figure 6.3 shows longitudinal phase space and the longitudinal potential well. At the potential maximum, where $\phi = -\phi_s$, we have $\phi' = 0$, and therefore from Eq. (6.15), $w = 0$. Then at $\phi = -\phi_s$

$$B(\sin(-\phi_s) - (-\phi_s \cos\phi_s)) = H_\phi \tag{6.20}$$

Eq. (6.20) can be used to determine the constant H_ϕ, and we find that points on the separatrix must satisfy

$$\frac{Aw^2}{2} + B(\sin\phi - \phi\cos\phi_s) = -B(\sin\phi_s - \phi_s\cos\phi_s) \tag{6.21}$$

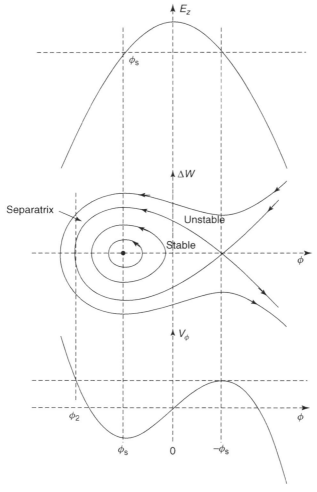

Figure 6.3 At the top, the accelerating field is shown as a cosine function of the phase; the synchronous phase ϕ_s is shown as a negative number that lies earlier than the crest where the field is rising in time. The middle plot shows some longitudinal phase-space trajectories, including the separatrix, the limiting stable trajectory that passes through the unstable fixed point at $\Delta W = 0$ and $\phi = -\phi_s$. The stable fixed point lies at $\Delta W = 0$ and $\phi = \phi_s$, where the longitudinal potential well has its minimum as shown in the bottom plot.

The separatrix defines the area within which the trajectories are stable, and it can be plotted if the constants A and B are given. In common accelerator jargon, the separatrix is also called the *fish*, and the stable area within is called the *bucket*. There are two separatrix solutions for $w = 0$, which determine the maximum phase width of the separatrix. One solution is $\phi_1 = -\phi_s$, which is a positive number for stable motion because ϕ_s is negative, and this point gives the maximum phase for stable motion. The point at $\phi = \phi_2$ and $w = 0$ is the other solution that gives the minimum phase for stable motion. The equation for the separatrix for this case becomes

$$\sin \phi_2 - \phi_2 \cos \varphi_s = \phi_s \cos \varphi_s - \sin \phi_s \tag{6.22}$$

and this can be solved numerically for $\phi_2(\phi_s)$. The total phase width of the separatrix is

$$\Psi = |\phi_s| + |\phi_2| = -\phi_s - \phi_2 \tag{6.23}$$

Then

$$\sin \phi_2 = -\sin(\phi_s + \Psi) = -[\sin \phi_s \cos \Psi + \sin \Psi \cos \phi_s] \tag{6.24}$$

Substituting Eq. (6.24) into Eq. (6.22), we find an equation for $\tan \phi_s$:

$$\tan \phi_s = \frac{\sin \Psi - \Psi}{1 - \cos \Psi} \tag{6.25}$$

When $\psi \ll 1$ and $\phi_s \ll 1$, $\sin \psi = \psi - \psi^3/6 + \ldots$ and $\cos \psi = 1 - \psi^2/2 + \ldots$, and we obtain

$$\tan \phi_s = \frac{\sin \Psi - \Psi}{1 - \cos \Psi} \cong -\frac{\Psi}{3} \tag{6.26}$$

which turns out to be a good approximation even upto $|\phi_s| \approx 1$.

Then with $\tan \phi_s \cong \phi_s$, we have $\Psi \cong 3|\phi_s|$ and $\phi_2 \cong 2\phi_s$. At $\phi_s = -90°$ the phase acceptance is maximum, extending over the full 360° (Fig. 6.4).

A linac is designed to provide a specific synchronous-velocity profile. For a linac that is designed to accelerate relativistic electrons, the synchronous-velocity profile is easily maintained, regardless of field or energy errors, because all relativistic particles have a velocity approximately equal to c.

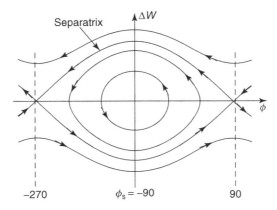

Figure 6.4 Separatrix for $\phi_s = -90°$ (no acceleration).

For this case field errors cause a shift in the final energy. For a linac that accelerates nonrelativistic ions, a field error that changes the particle-velocity gain causes a shift to a new synchronous phase. To see this, consider a design-energy profile given approximately by $W_s(z) = q(E_0 T)_{\text{design}} \cos(\phi_s) z$. If the actual accelerating field $E_0 T$ is different from the design value, a different particle phase can still satisfy the synchronous energy profile, and the new synchronous phase is given by $\cos\phi = \cos\phi_s (E_0 T)_{\text{design}}/E_0 T$. For this case the final energy is unchanged. At $\phi = 0$ the phase acceptance vanishes, and the corresponding accelerating field is the threshold field for forming a stable bucket for synchronous acceleration. Thus

$$(E_0 T)_{\text{threshold}} = (E_0 T)_{\text{design}} \cos\phi_s \tag{6.27}$$

Finally, we find the energy half-width of the separatrix, $w = w_{\max}$ that occurs for $\phi = \phi_s$. Solving Eq. (6.21) for $w = w_{\max}$,

$$w_{\max} = \frac{\Delta W_{\max}}{mc^2} = \sqrt{\frac{2q E_0 T \beta_s^3 \gamma_s^3 \lambda}{\pi mc^2}(\phi_s \cos\phi_s - \sin\phi_s)} \tag{6.28}$$

6.5
Hamiltonian and Liouville's Theorem

The equations of longitudinal motion can also be expressed in Hamiltonian form. We identify the Hamiltonian as the energy invariant given in Eq. (6.18). If the canonically conjugate variables are identified as ϕ and $p_\phi = -w$, the second-order differential equation of motion can be derived from the Hamilton Equations

$$p'_\phi = -\frac{\partial H_\phi}{\partial \phi} \quad \text{and} \quad \phi' = \frac{\partial H_\phi}{\partial p_\phi} \tag{6.29}$$

Liouville's theorem states that the density in phase space of noninteracting particles in a conservative or Hamiltonian system, measured along the trajectory of a particle is invariant [1]. It follows that the phase-space volume enclosed by a surface of any given density is conserved. The shape may change, but not the volume. The Hamiltonian description applies to the motion ignoring acceleration. When acceleration is introduced as an adiabatic effect, the parameters β_s and γ_s change slowly, and the phase-space area is an adiabatic invariant as was discussed in Section 5.12. The separatrix, including acceleration, looks like a golf club rather than a fish, as shown in Fig. 6.5. We will use phase-space area conservation later to calculate the change of the phase width of the beam during the adiabatic acceleration process.

If we consider any transformation of the system from one time to another, it can be shown that the area preserving property is equivalent to the Jacobian determinant, which has a geometrical interpretation of a phase-space area

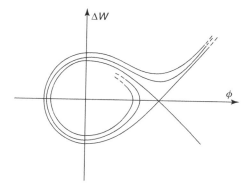

Figure 6.5 Longitudinal phase-space trajectories when acceleration is taken into account. The separatrix shape is called the *golf club*.

magnification factor, being equal to unity [2]. If the initial state of a Hamiltonian system with phase-space coordinates (q, p) is transformed to coordinates (Q, P), the Jacobian determinant is defined as

$$\frac{\partial(Q, P)}{\partial(q, p)} \equiv \begin{vmatrix} \dfrac{\partial Q}{\partial q} & \dfrac{\partial P}{\partial q} \\ \dfrac{\partial Q}{\partial p} & \dfrac{\partial P}{\partial p} \end{vmatrix} = 1 \tag{6.30}$$

The solution for any particle trajectory can always be obtained in principle by direct numerical integration through the known fields. More rapid computer calculations can be made of multiparticle dynamics through a linac, if the particles are transformed through a sequence of equivalent thin gaps, where the momentum impulses are applied, and field-free drift spaces, where the momentum components produce position changes. The problem of calculating accurate transformations is constrained significantly by Liouville's theorem. Consider the transformation of a nonrelativistic particle through an accelerating gap of length L, which is represented by two equal-length drift spaces, separated by an infinitesimally thin accelerating gap. Suppose we consider the transformation equations for energy and phase of a particle through the thin gap alone, written as

$$W_f = W_i + qV_0 T(k) \cos \phi_i \tag{6.31}$$

and

$$\phi_f = \phi_i \tag{6.32}$$

where the subscripts i and f refer to the initial and final coordinates, and $T(k)$ is the transit-time factor,

$$T(k) = \frac{\int_{-L/2}^{L/2} E_z(z) \cos(kz)\, dz}{\int_{-L/2}^{L/2} E_z(z)\, dz} \tag{6.33}$$

which depends on the particle energy W_i, through the relations $k = 2\pi/\beta\lambda$ and $\beta c = \sqrt{2W_i/m}$. However, Liouville's theorem must be satisfied, and if the transformation is correct, the Jacobian determinant must have equal unity. Making use of the nonrelativistic relations $W_i = mc^2\beta^2/2$ and $dk = -2\pi\, d\beta/\beta^2\lambda = -k\,dW/2W_i$, the Jacobian of this transformation is

$$\frac{\partial(W_f, \phi_f)}{\partial(W_i, \phi_i)} = \begin{vmatrix} \dfrac{\partial W_f}{\partial W_i} & \dfrac{\partial \phi_f}{\partial W_i} \\ \dfrac{\partial W_f}{\partial \phi_i} & \dfrac{\partial \phi_f}{\partial \phi_i} \end{vmatrix} = 1 - \frac{qV_0}{2\,W_1} kT' \cos\phi_i \qquad (6.34)$$

where

$$T'(k) = \frac{dT(k)}{dk} = -\frac{\displaystyle\int_{-L/2}^{L/2} zE_z(z)\sin(kz)\,dz}{\displaystyle\int_{-L/2}^{L/2} E_z(z)\,dz} \qquad (6.35)$$

For typical cases, the integrals in Eq. (6.35) are positive, the sign of T' is negative, and the Jacobian is greater than unity, which implies that the transformation increases the phase-space area. Because the Jacobian is not unity, we know that the proposed transformation is not exactly correct. The Jacobian can artificially be made equal to unity, if we replace the velocity-dependent transit-time factor with a constant transit-time factor, equal to that of the synchronous particle, $T(k_s)$, where $k_s = 2\pi/\beta_s\lambda$. Such an approximation may be acceptable if the velocity spread of the beam is sufficiently small. It can be considered as a first approximation for computer simulations of linac particle dynamics, and it preserves Liouville's theorem, which we know is satisfied for the real system. The modified transformation becomes

$$W_f = W_i + qV_0 T(k_s)\cos\phi_i \qquad (6.36)$$

and

$$\phi_f = \phi_i \qquad (6.37)$$

This approach is simple but not very satisfying, because we know that the correct calculation for the energy gain is one that allows us to use the correct transit-time factor, a function of the particle velocity. This problem was studied in the 1960s and it was found that to satisfy Liouville's theorem, the equivalent thin-gap transformation for the phase must include a discontinuous jump [3–7]. The jumps required in both energy and phase are illustrated in Fig. 6.6, where the solid curves show the actual trajectories and the separations of the dashed lines show the required jumps. To illustrate this result, we express the approximate nonrelativistic thin-gap transformation for particles on the axis, using the lowest-order terms obtained from this earlier work. We replace the transformation of Eqs. (6.31) and (6.32) by

$$W_f = W_i + qV_0 T(k)\cos\phi_i \qquad (6.38)$$

and include a phase jump given by

$$\phi_f = \phi_i + \frac{qV_0}{2W_1} kT'(k) \sin \phi_i \qquad (6.39)$$

The Jacobian of this transformation is

$$\frac{\partial(W_f, \phi_f)}{\partial(W_i, \phi_i)} = \begin{vmatrix} \frac{\partial W_f}{\partial W_i} & \frac{\partial \phi_f}{\partial W_i} \\ \frac{\partial W_f}{\partial \phi_i} & \frac{\partial \phi_f}{\partial \phi_i} \end{vmatrix} = 1 + \left\{ \text{terms of order } \left(\frac{qV_0}{W_i}\right)^2 \right\} \qquad (6.40)$$

Thus the addition of the phase jump results in a transformation that preserves Liouville's theorem up to terms of second order in the ratio of the voltage gain to the particle energy.

The geometrical picture of the effects of the two transformations on particle coordinates in longitudinal phase space is illustrated simply in Fig. 6.7a and b. The figures show an area enclosed by four particles distributed around the synchronous or reference particle. Particles A and B have the same energy as the reference particle, but have different phases, and particles C and D have the same phase, but have different energies. In the transformation of Eqs. (6.31) and (6.32), shown in Fig. 6.7a, an effect of the gap is to displace A and B by equal and opposite amounts in energy, which produces little change in the area. Particle C, which has the larger energy, has a larger transit-time factor, and gains more energy than the reference particle, while D, which has the smaller energy, gains less energy. Figure 6.7a shows that, because of particles C and D, the transformation primarily stretches the area along the energy axis, and the enclosed area increases. Figure 6.7b, representing the transformation

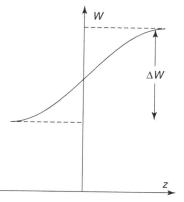
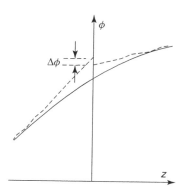

Figure 6.6 The solid curves show the evolution of energy W and phase ϕ across a gap centered at $z = 0$. The dashed curves show the progression of the approximate drift-kick-drift treatment, where the separation of the dashed lines at $z = 0$ shows the energy and phase jumps required at the gap.

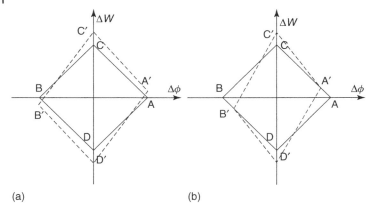

Figure 6.7 (a) The transformation of Eqs. (6.31) and (6.32) does not satisfy Liouville's theorem and the dashed parallelogram has a larger area than the solid one. (b) In the transformation of Eqs. (6.38) and (6.39), the areas are equal and Liouville's theorem is satisfied.

of Eqs. (6.38) and (6.39), shows that particles A and B move inward in phase to compensate for the outward movement of particles C and D, and the total phase-space area is conserved.

6.6
Small Amplitude Oscillations

We return to the equation of motion, Eq. (6.16) to consider the motion of particles with coordinates close to the synchronous particle. For a phase difference that is small relative to the synchronous phase

$$\cos\phi \cong \cos\phi_s - (\phi - \phi_s)\sin\phi_s - \frac{(\phi - \phi_s)^2}{2}\cos\phi_s \quad (6.41)$$

and we obtain an equation of motion for small longitudinal oscillations

$$\phi'' + k_{\ell 0}^2 \left[(\phi - \phi_s) - \frac{(\phi - \phi_s)^2}{2\tan(-\phi_s)} \right] = 0 \quad (6.42)$$

where

$$k_{\ell 0}^2 = \frac{2\pi q E_0 T \sin(-\phi_s)}{mc^2 \beta_s^3 \gamma_s^3 \lambda} \quad (6.43)$$

The corresponding angular frequency of small longitudinal oscillations is $\omega_{\ell 0} = k_{\ell 0} \beta_s c$, or

$$\left[\frac{\omega_{\ell 0}}{\omega} \right]^2 = \frac{q E_0 T \lambda \sin(-\phi_s)}{2\pi mc^2 \gamma_s^3 \beta_s} \quad (6.44)$$

where $\omega = 2\pi c/\lambda$ is the RF angular frequency. The longitudinal oscillation frequency is usually small compared with the RF frequency. As the beam becomes relativistic, the longitudinal oscillation frequency approaches zero. The quadratic term in Eq. (6.42) is the lowest-order nonlinear term. Because it is quadratic, it produces an asymmetric potential well, and the minus sign means that the nonlinear part of the restoring force weakens the focusing.

Now we examine a trajectory of small oscillations in longitudinal phase space. The general trajectory equation is Eq. (6.18). In the approximation $|\phi - \phi_s| \ll 1$, we can write

$$\sin\phi - \phi\cos\phi_s \cong \sin\phi_s - \phi_s\cos\phi_s - \frac{(\phi - \phi_s)^2}{2}\sin\phi_s \qquad (6.45)$$

Substituting Eq. (6.45) into Eq. (6.18), we obtain

$$Aw^2 + B\sin(-\phi_s)(\phi - \phi_s)^2 = 2(H_\phi + \phi_s\cos\phi_s - \sin\phi_s) \qquad (6.46)$$

Since ϕ_s is negative for stable motion, Eq. (6.46) is the equation for an ellipse in longitudinal phase space, centered at $w = 0$, and $\phi = \phi_s$. We define the maximum phase half-width as $\Delta\phi_0 = \phi_0 - \phi_s$, where ϕ_0 is the maximum phase. The right side of Eq. (6.46) is constant for fixed synchronous phase. Because the point of maximum phase corresponds to a point on the ellipse with $w = 0$, we can then use Eq. (6.46) to determine the constant H_ϕ. Then the equation for the ellipse can be written as

$$\frac{w^2}{w_0^2} + \frac{(\phi - \phi_s)^2}{\Delta\phi_0^2} = 1 \qquad (6.47)$$

where

$$w_0 = \frac{\Delta W_0}{mc^2} = \sqrt{qE_0 T \beta_s^3 \gamma_s^3 \lambda \sin(-\phi_s)\Delta\varphi_0^2/2\pi mc^2} \qquad (6.48)$$

which is the normalized energy on the ellipse that corresponds to $\phi - \phi_s = 0$.

6.7
Adiabatic Phase Damping

If the acceleration rate is small, the parameters of the phase-space ellipses for small oscillations vary slowly. As discussed in Section 5.12, the area of the ellipse describing the small amplitude oscillations is an adiabatic invariant during the acceleration process. We can use Eq. (6.48) to express the constant area of the ellipse for any particle as

$$\text{Area} = \pi \Delta\phi_0 \Delta W_0 = \pi \Delta\phi_0^2 \sqrt{qE_0 Tmc^2 \beta_s^3 \gamma_s^3 \lambda \sin(-\phi_s)/2\pi} \qquad (6.49)$$

or

$$\Delta\phi_0 = \frac{\text{Constant}}{[qE_0 T mc^2 \beta_s^3 \gamma_s^3 \lambda \sin(-\phi_s)/2\pi]^{1/4}} \quad (6.50)$$

If the accelerating field and the synchronous phase are fixed, we have

$$\Delta\phi_0 = \frac{\text{Constant}}{(\beta_s\gamma_s)^{3/4}} \quad (6.51)$$

and since the phase-space area is an adiabatic invariant, from Eqs. (6.49) and (6.51) we conclude that

$$\Delta W_0 = \text{Constant} \times (\beta_s\gamma_s)^{3/4} \quad (6.52)$$

These results describe a decrease of the phase amplitude and an increase of the energy amplitude of phase oscillations during acceleration in a linac, an effect called *phase damping*, as illustrated in Fig. 6.8. If an initial amplitude is $\Delta\phi_{0,i}$ when the synchronous velocity is β_{si}, the amplitude $\Delta\phi_0$ at some later point in the accelerator, where the synchronous velocity is β_s, is

$$\frac{\Delta\phi_0}{\Delta\phi_{0,i}} = \left[\frac{(\beta_s\gamma_s)_i}{\beta_s\gamma_s}\right]^{3/4} \quad (6.53)$$

Similarly, the energy half-width scales as

$$\frac{\Delta W_0}{\Delta W_{0,i}} = \left[\frac{\beta_s\gamma_s}{(\beta_s\gamma_s)_i}\right]^{3/4} \quad (6.54)$$

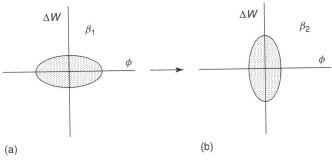

Figure 6.8 Phase damping of a longitudinal beam ellipse caused by acceleration. The phase width of the beam decreases and the energy width increases, while the total area remains constant.

6.8
Longitudinal Dynamics of Ion Beams in Coupled-Cavity Linacs

In the high-energy section of ion linacs, where the fractional velocity increase in a multicell accelerating structure is small, one can simplify the fabrication by designing each structure with a constant cell length throughout the structure. The length of each cell is $\beta_s \lambda/2$, where β_s is the design velocity. For each structure the β_s can be chosen to equal the velocity of a specific reference particle at the center of the structure. Because all cell lengths are equal, the true synchronous phase is $\phi_s = -90°$ that would correspond to no acceleration in the structure. However, the accelerated reference particle is not a synchronous particle, and the reference-particle phase in each cell varies along the structure. We can express the total energy gain of the particle in each structure as

$$\Delta W_r = qE_0 T \cos \phi_r N_c \beta_s \lambda/2 \qquad (6.55)$$

where N_c is the number of cells in the structure, and ϕ_r is an average reference-particle phase, defined so that $\cos \phi_r = \overline{\cos \phi}$, where ϕ is the cell-dependent phase of the reference particle, and the average is taken over all the cells. The reference-particle velocity is less than the design velocity in the first half of the structure, and its phase gradually increases from cell to cell. The phase decreases in the second half of the structure, where the reference-particle velocity exceeds the design velocity. The injection phase can be chosen so that the reference particle enters and exits the structure with the same phase.

The maximum phase excursion of the reference particle can be approximately calculated in the following way. We assume that the reference particle moves on the constant energy trajectory given by Eq. (6.18). The trajectory is illustrated in Fig. 6.9, where the separatrix is also shown that corresponds to the true synchronous phase of $\phi_s = -90°$. The structure design and the entrance phase are chosen so that the entrance coordinates at the center of the first cell are $\phi = \phi_1$ and $w = w_1 = -\Delta W_r/2mc^2$, where ΔWr is the total energy gain of the reference particle in the structure. The reference-particle trajectory equation satisfies

$$\frac{Aw^2}{2} + B \sin \phi = \frac{Aw_1^2}{2} + B \sin \phi_1 \qquad (6.56)$$

As is shown in Fig. 6.9, the maximum phase excursion $\phi = \phi_0$ occurs at $w = \Delta W/mc^2 = 0$. Substituting this point into Eq. (6.56), gives

$$\sin \phi_0 - \sin \phi_1 = \frac{A}{2B} w_1^2 \qquad (6.57)$$

For small phase variations through the structure, we can expand the two sine expressions about ϕ_r, and we obtain

$$\sin \phi_0 - \sin \phi_1 \cong \cos \phi_r (\phi_0 - \phi_1) \qquad (6.58)$$

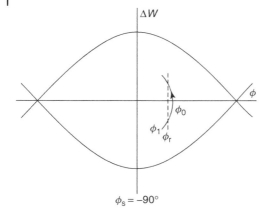

Figure 6.9 Phase-space trajectory in a coupled-cavity linac structure with constant length cells.

Equating Eqs. (6.57) and (6.58), we find the approximate phase excursion, or phase slip, through the structure, which is convenient to express as

$$\phi_0 - \phi_1 = \frac{\pi N_c}{8} \frac{\Delta W_r}{\gamma_s^3 \beta_s^2 mc^2} \tag{6.59}$$

We can approximate

$$\phi_r \cong \frac{\phi_0 + \phi_1}{2} \tag{6.60}$$

for calculation of the energy gain ΔW_r from Eq. (6.55). The reference particle moves from ϕ_1 to ϕ_0 in the first half of the structure and back to ϕ_1 in the second half. The phase excursion is usually no more than a few degrees.

6.9
Longitudinal Dynamics in Independent-Cavity Ion Linacs

An independent-cavity ion linac, introduced in Section 4.1, is comprised of an array of short cavities, each with a broad velocity profile for the transit-time factor, and with independent settings for the phases and amplitudes. Such a linac is capable of flexible operation, including energy variability, which is not available in long multicell ion linacs, whose transit-time-factor velocity profiles are very narrow, requiring operation only for a unique velocity profile. The flexibility is useful for heavy-ion linacs where beams with different charge-to-mass ratios must be accelerated, and can be exploited to advantage in superconducting ion linacs, where for several reasons the cavities are usually built using a small number of cells. The longitudinal dynamics differs from those of other cases we have examined in this chapter, because the synchronous particle must be replaced by a reference particle, whose velocity β_r may differ by a large amount from the design or geometric velocity β_G.

6.9 Longitudinal Dynamics in Independent-Cavity Ion Linacs

To describe the longitudinal dynamics in terms of difference equations as we did earlier in the chapter, we represent the particle motion through each cell of an N-cell cavity by a sequence of drift spaces where the particles see no forces, follow thin gaps at the electrical centers of the cells, where the energy gains are delivered as impulses. From gap n to gap $n+1$, the reference particle has a constant velocity $\beta_{r,n}$. The RF phase changes as the reference particle advances from cell n to $n+1$ according to $\phi_{r,n+1} - \phi_{r,n} = (z_{n+1} - z_n) 2\pi/\beta_{r,n}\lambda - \pi$, where z_n is the coordinate of the electrical center of cell n, and a constant phase π is subtracted to compensate for the cell-to-cell phase difference associated with operation in the π mode. The energy gain, received at the electrical center of cell n is $\Delta W_n = q E_{0n} T_n(\beta_{r,n-1}) \cos(\phi_{r,n}) \beta_G \lambda/2$, where the average axial field over the cell is $E_{0n} = (2/\beta_G \lambda) \int_{cell} E_z(z) dz$, the cell transit-time factor is $T_n(\beta_r) = \int_{cell} E_z(z) \cos(2\pi z/\beta_r \lambda)/\int_{cell} E_z(z) dz$, and $\phi_{r,n}$ is the phase of the field in cell n, when the reference particle is at the electrical center. The quantities E_{0n} and T_n may be calculated for any cell by a standard electromagnetic field solver code such as SUPERFISH. The main cause of cell-to-cell parameter differences is the geometrical difference between the end cells and the inner cells of the cavity. As the beam becomes more relativistic so that the velocity change over the n-cell cavity is small, the method can be simplified by applying a single kick at the center of the cavity instead of a kick for every cell.

The description of the longitudinal dynamics of any other particle in the bunch with velocity β_n near the reference velocity β_r is straightforward. The equation for the relative phase change during the time the particles move from one electrical center to the next is

$$\phi_{n+1} - \phi_{s,n+1} = \frac{2\pi(z_{n+1} - z_n)}{\lambda} \left[\frac{1}{\beta_n} - \frac{1}{\beta_{s,n}} \right] + (\phi_n - \phi_{s,n}) \quad (6.61)$$

and the relative energy gain at the electrical center of cell n is approximately

$$\Delta W_n - \Delta W_{s,n} = q E_{0n} T(\beta_{s,n-1})[\cos(\phi_n) - \cos(\phi_{s,n})]\beta_G \lambda/2 \quad (6.62)$$

Equations (6.61) and (6.62) describe longitudinal oscillations of particles about the reference particle.

Another computational approach for independent-cavity linacs, which may also be convenient for coupled-cavity linacs, is used in the DYNAC beam-dynamics code [8]. If the axial field E_z is known in a complex structure with several accelerating gaps, one can compute $T(k)$ and its first and second derivatives for any k. One can introduce an equivalent wave field of uniform amplitude, for which $T(k)$ and its first and second derivatives are the same, near the design value of k, as for the real field distribution. The equation of motion of the design particle is integrated using the field of the simple equivalent wave, and individual particle trajectories are derived with high accuracy from expansions around the design particle.

6.10
Longitudinal Dynamics of Low-Energy Beams Injected into a $v = c$ Linac

Low-energy electrons (50 keV) are typically injected into a linac with a velocity nearly half the speed of light. It is attractive to use the same $v = c$ accelerating structure that is used for acceleration of extremely relativistic electrons for the lowest energy electrons, and in this case the particle phases will slip on the wave. In this section we discuss the physics of the capture, bunching, and acceleration of an initial $v < c$ electron beam in a $v = c$ traveling-wave structure. The equation of motion of a particle with position z at time t, accelerated by a traveling wave with a longitudinal electric-field amplitude E_0 is

$$\frac{d}{dt} mc\beta\gamma = qE_0 \cos\phi(z, t) \tag{6.63}$$

where the phase of the traveling wave with velocity c, as a function of z and t is

$$\phi(z, t) = \omega t - 2\pi z/\lambda \tag{6.64}$$

The phase motion is described by the equation

$$\frac{d\phi}{dt} = \frac{2\pi c}{\lambda}(1 - \beta) \tag{6.65}$$

Note that since $\beta < 1$, ϕ increases with time, which means the particle falls further behind the initial phase on the wave. It is convenient to change the independent variable in Eq. (6.63) from time to phase, and using the result that $d(\gamma\beta) = \gamma^3 d\beta$, and we obtain

$$mc\gamma^3 \frac{d\beta}{d\phi} \frac{d\phi}{dt} = qE_0 \cos\phi \tag{6.66}$$

Substituting Eq. (6.65) into Eq. (6.66), and expressing γ as a function of β, we find

$$\frac{1}{(1+\beta)\sqrt{1-\beta^2}} \frac{d\beta}{d\phi} = \frac{qE_0\lambda}{2\pi mc^2} \cos\phi \tag{6.67}$$

We can integrate both sides of Eq. 6.67 to find the dependence of phase on velocity during the acceleration process. To carry out the integration we will change to a new variable α defined by $\beta = \cos\alpha$. We also use the integral

$$\int \frac{d\alpha}{1 + \cos\alpha} = \tan\frac{\alpha}{2} \tag{6.68}$$

and the identity

$$\tan\frac{\alpha}{2} = \sqrt{\frac{1-\cos\alpha}{1+\cos\alpha}} = \sqrt{\frac{1-\beta}{1+\beta}} \tag{6.69}$$

6.10 Longitudinal Dynamics of Low-Energy Beams Injected into a $v = c$ Linac

to obtain the final result

$$\sin\phi = \sin\phi_i + \frac{2\pi mc^2}{qE_0\lambda}\left[\sqrt{\frac{1-\beta_i}{1+\beta_i}} - \sqrt{\frac{1-\beta}{1+\beta}}\right] \quad (6.70)$$

In Eq. (6.70) we have the interesting result that there is no oscillatory solution for the phase motion. Because of the acceleration, $\beta > \beta_i$, the second term on the right is positive and increases with increasing β. Therefore, $\sin\phi > \sin\phi_i$, and the phase becomes more positive as β increases. If we want the particle to approach the crest where $\phi = 0$, the particle must be injected at a negative phase. This process is illustrated in Fig. 6.10. We see also that as β approaches unity, the phase ϕ approaches a constant asymptotic value, which is

$$\sin\phi_\infty = \sin\phi_i + \frac{2\pi mc^2}{qE_0\lambda}\sqrt{\frac{1-\beta_i}{1+\beta_i}} \quad (6.71)$$

It is convenient to define the second term as

$$F \equiv \frac{2\pi mc^2}{qE_0\lambda}\sqrt{\frac{1-\beta_i}{1+\beta_i}} \quad (6.72)$$

From Eq. (6.71) we see that the minimum field gives the maximum phase slip. Suppose we choose $\phi_i \approx -90°$, or slightly more positive to ensure acceleration. Choosing $F = 1$ corresponds to placing the asymptotic phase at the crest, which implies that

$$E_0 = \frac{2\pi mc^2}{q\lambda}\sqrt{\frac{1-\beta_i}{1+\beta_i}} \quad (6.73)$$

and shows that the smaller the initial velocity, the larger the accelerating field must be.

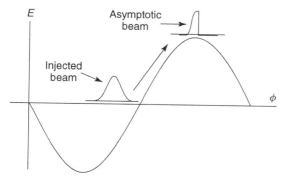

Figure 6.10 Longitudinal dynamics of a low-energy electron beam after injection into a $v = c$ linac structure.

During the capture process, as the injected beam moves up to the crest, the beam is also bunched, which is caused by velocity modulation, as the early particles experience a lower accelerating field than later particles. To estimate the magnitude of the bunching effect, we again assume that $\phi_\infty \approx 0$, and suppose that the initial phase width is $2\delta\phi_i$, with the initial centroid at $-\pi/2 + \delta\phi_i$. We find that $F = 1 - \delta\phi_i^2/2$, and the particle with the earliest initial phase will have an asymptotic phase $\phi_\infty = -\delta\phi_i^2/2$, which is just earlier than the crest. The particle with the latest initial phase will be asymptotically at $\phi_\infty = 3\delta\phi_i^2/2$. The final half-width of the beam is $\delta\phi_\infty = \delta\phi_i^2$. If $\delta\phi_i = 15°$ is the initial half-width, the asymptotic half-width is about 3.9°. Summarizing, when low-velocity electrons are injected into a structure whose phase velocity equals the velocity of light, they slip in phase, but if the accelerating field is chosen correctly, the particles will asymptotically approach the crest, where they can be efficiently accelerated to high energies. The capture process can also result in significant bunching of the captured beam.

6.11
Rf Bunching

Prior to acceleration, the RF linac input beam is formed into bunches. Bunching is most commonly accomplished by introducing an RF cavity that produces a velocity modulation of the beam at the same frequency as the linac; the beam particles are either accelerated or decelerated in the buncher cavity depending on the phase. In the drift distance that follows the faster particles catch up with the slower particles, so that the velocity modulation at the buncher cavity is converted into a bunch compression. The beam arrives at a phase focus near the entrance of the linac where the bunch compression is maximum. Sometimes more than one cavity is used to increase the bunch compression. Debunchers, comprised of a drift space followed by an RF cavity are also used, typically at the output of the linac to reduce the energy spread of the beam. In practice, the phase compression of a buncher or the energy compression of a debuncher are limited by the nonlinearity of the RF waveform, or by the space-charge force.

A simple model illustrates how an RF cavity and drift distance can be used to compress the phase width of the beam. We approximate the initial beam, which may in reality have a uniform continuous distribution, as a sequence of bunches each with upright ellipses in longitudinal phase space, whose total phase width for the ellipse that we will follow may be chosen equal to 360°. This initial bunch arrives at the buncher cavity, whose peak effective voltage is V_0. For simplicity we replace the real cavity with a thin cavity located at the gap center. Assume that the semiaxes of the phase-energy ellipse are φ_1 and E_1, as shown in Fig. 6.11. The reference particle is at the origin $\varphi = 0$ and $\Delta W = 0$. For a cavity to perform as a buncher, the reference particle arrives at the gap when the voltage is rising in time and is zero, so there will be essentially zero

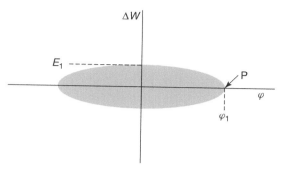

Figure 6.11 Longitudinal phase-space ellipse of the incident bunch at the buncher cavity, prior to the kick. The reference particle is at the origin. Point P is shown with coordinates φ_1 and energy $\Delta W = 0$ relative to the reference particle.

average energy gain. We will follow the point P that has initial phase-space coordinates ϕ_1 and $\Delta W = 0$.

The buncher cavity delivers a phase-dependent kick which changes the upright ellipse to a tilted ellipse as shown in Fig. 6.12, after which the beam propagates in the drift space that follows and arrives at a phase focus where the ellipse is upright again with a compressed phase width, as shown in Fig. 6.13. The phase-space pictures show what happens to particle P.

The phase-space area occupied by the beam is conserved, so $E_1 \varphi_1 = E_2 \varphi_2$. Before the kick the phase-space coordinates of particle P are $(\varphi_1, 0)$. After the kick its coordinates are (φ_1, E_2) where $E_2 = qV_0 \sin(\omega t_1) \cong qV_0 \varphi_1$ and $\varphi_1 = \omega t_1$. After the drift to the phase focus, the coordinates of particle P have rotated in phase space by $90°$ and are $(0, E_2)$. Assuming the initial semiaxes ϕ_1 and E_1 are given, the phase width of the bunch at the phase focus is given

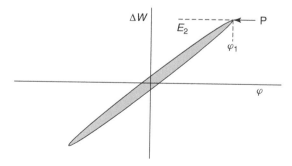

Figure 6.12 Longitudinal phase-space ellipse of the bunch at the buncher cavity after kick. The reference particle is at the origin. Point P is shown with coordinates φ_1 and energy E_2 relative to the reference particle.

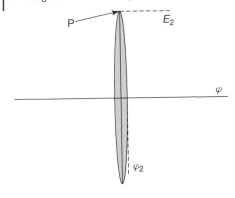

Figure 6.13 Longitudinal phase-space ellipse of the bunch at the phase focus. The reference particle is at the origin. Point P is shown with coordinates $\varphi = 0$ and energy E_2 relative to the reference particle.

by phase-space area conservation

$$\varphi_2 = \frac{E_1 \varphi_1}{E_2} \cong \frac{E_1}{qV_0} \tag{6.74}$$

The phase change of particle P relative to the reference particle at the origin determines the location L of the phase focus relative to the cavity. The time difference for two particles with velocity difference $\delta\beta$ to travel a distance L between the cavity and the phase focus is

$$\delta t = -\frac{L}{c}\frac{\delta\beta}{\beta^2} = \frac{L\delta W}{mc^3\beta^3\gamma^3} \tag{6.75}$$

We write

$$\delta\varphi = \varphi_1 = -\frac{2\pi L}{\lambda}\frac{E_2}{mc^2\beta^3\gamma^3} = -\frac{2\pi L}{\lambda}\frac{qV_0\varphi_1}{mc^2\beta^3\gamma^3} \tag{6.76}$$

Solving for L,

$$L = \frac{\lambda mc^2 \beta^3 \gamma^3}{2\pi qV_0} \tag{6.77}$$

The greater the kick V_0, the shorter the drift distance.

6.12
Longitudinal Beam Dynamics in H-Mode Linac Structures

In conventional longitudinal dynamics, the reference particle and the synchronous particle are the same. Longitudinal focusing is obtained by operating at a negative synchronous phase, which means that the reference particle arrives at the center of each gap earlier than the crest of the accelerating waveform. Then, early particles with more negative phase than the reference particle will see a lower accelerating field than the reference particle and will move back toward the reference particle. A method called *KONUS* for "Kombinierte Null Grad Struktur" or "Combined Zero-Degree Structure" was developed for application to the efficient operation of H-mode structures that were discussed in Section 4.3 [9]. The KONUS approach is intended to provide transverse and longitudinal beam focusing for relatively long H-mode linac sections, where the defocusing effects of the transverse RF fields (see Chapter 7) and space charge must be compensated in a way that does not require quadrupole focusing lenses in the drift tubes. Not installing quadrupole lenses in the drift tubes allows the drift-tube diameters and the drift-tube capacitances to be reduced, which greatly increases the efficiency of the accelerating structure. In the KONUS longitudinal-focusing method, the linac is comprised of a sequence of multigap accelerating sections, with section lengths corresponding to approximately one quarter of a longitudinal-focusing period. In this scheme the reference particle and the synchronous particle are not the same. The gap-to-gap spacings are adjusted so that the synchronous phase for each section is zero degrees, which means that a synchronous particle would arrive at the center of each gap when the field is maximum. This would correspond to zero transverse RF focusing, but also zero longitudinal focusing, if the design or reference particle was identical with the synchronous particle.

To explain the KONUS concept, suppose that the beam is injected into an accelerating section so that the beam energy of the reference particle is larger than the synchronous energy, and the injected reference-particle phase is near the zero-degree synchronous phase where there is no longitudinal focusing. Then, consider the motion in longitudinal phase space, where the origin is centered on the zero-degree synchronous phase and on the synchronous energy. The higher energy reference particle is initially above the origin and on or near the boundary between the first and second quadrant. As the particles advance from gap to gap, the reference particle will move counterclockwise in the second quadrant of longitudinal phase space. The reference-particle phase becomes more negative indicating that it arrives earlier at the next gap than would a synchronous particle. Also, the reference-particle energy decreases from its initial value; thus the reference-particle energy approaches the synchronous energy. With phases that are becoming more negative from gap to gap, the particles near the reference particle not only are accelerated but experience longitudinal focusing, since early particles near the reference particle see a lower accelerating field than that of the reference particle. When the reference particle advances to the third quadrant, early particles would experience a larger energy gain than the reference particle, which implies longitudinal defocusing. Thus, before the reference particle arrives at the third quadrant, the multigap section is ended so that the longitudinal motion remains in the second quadrant as is required to avoid longitudinal defocusing. In the next multigap section the gap spacings are reset so that synchronous phase and energy are suitable for obtaining the proper initial conditions for a new section.

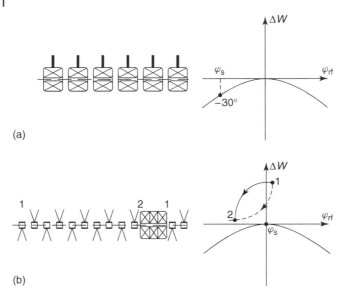

Figure 6.14 Drift tubes and longitudinal phase space with the origin at the position of the synchronous particle. The figure is described in the text. (Courtesy of U. Ratzinger.).

An IH structure can begin with an accelerating section with radially compact low-capacitance drift tubes containing no transverse-focusing elements. This can be followed by a quadrupole triplet lens that provides transverse focusing outside the drift-tube array; the triplet is followed by the next drift-tube array for further acceleration. The triplet focusing between drift-tube arrays maintains a circular beam as desired for keeping small aperture of the drift tubes.

Figure 6.14a shows the conventional arrangement for a synchronous particle at $-30°$ for an array of six drift tubes that contain quadrupole lenses. Figure 6.14b shows a basic KONUS scheme showing a ten drift-tube IH drift-tube array with no lenses in the drift tubes, followed by a triplet lens, followed by the first two drift tubes of the next array. Also shown is the phase advance of the reference particle from location 1 to 2. The cycle begins at location 1 and the gap lengths are adjusted to produce the new operating point to begin a new cycle.

Problems

6.1. The parameters of a 400-MHz proton drift-tube linac at the injection energy of 2.5 MeV are $E_0 T = 2$ MV/m and $\phi_s = -30°$. What is the ratio of the energy acceptance (full height of the initial separatrix) to the injection energy?

6.2. For the drift-tube linac of Problem 6.1, what is $(E_0 T)_{\text{threshold}}$ for forming an initial bucket for synchronous acceleration?

6.3. The drift-tube linac of Problem 6.1, with constant $E_0 T = 2$ MV/m and $\phi_s = -30°$, accelerates a proton beam from 2.5 to 25 MeV. (a) Calculate the longitudinal phase advance per unit length $k_{\ell 0}$ in radians per meter at the injection energy of 2.5 MeV. Neglecting the increase of cell lengths along the linac, how many RF periods are in one longitudinal oscillation at injection? (b) Repeat part (a) at the output energy of 25 MeV.

6.4. A 1284-MHz coupled-cavity linac accelerates a proton beam from 70 to 600 MeV. Ignore the effects of space charge. (a) If the maximum phase amplitude $\Delta\phi_0$ of the longitudinal oscillations is 10° at injection, what is its value at 600 MeV? (b) If the energy amplitude ΔW_0 of the longitudinal oscillations is 0.2 MeV at injection, what is its value at 600 MeV? Does the fractional energy spread increase or decrease? (c) Calculate the phase width ψ and the energy half-width $\Delta W_{\max}$ of the separatrix at 70 and 600 MeV, assuming $E_0 T = 7$ MV/m and $\phi_s = -25°$. How do the changes in the maximum dimensions of the separatrix between 70 and 600 MeV compare with the changes in the phase and energy widths of the beam?

6.5. Suppose a coupled-cavity linac is composed of 1284-MHz tanks with 16 cells each, and has the parameters $E_0 T = 7$ MV/m and an effective synchronous phase of $\phi_s = -25°$ (this is really the mean phase of the reference particle in each structure). At 100 MeV what is the phase excursion of the reference particle as it travels through the structure, and what is an approximate value of the injection phase?

6.6. A 0.25-MeV electron beam is injected into a 3-GHz accelerating structure that supports a traveling wave with phase velocity equal to the velocity of light. (a) What accelerating field is required to capture a beam injected at a phase near $\phi_i = -90°$ and to reach an asymptotic phase at the crest. (b) What is the asymptotic phase width of the beam if the initial phase width is 30°?

References

1 Lawson J.D., *The Physics of Charged-Particle Beams*, 2nd ed., Clarendon Press, Oxford, 1988, p. 154.

2 Percival I. and Richards D., *Introduction to Dynamics*, Cambridge University Press, 1987, pp. 84–85.

3 Lapostolle P., CERN/AR/int. SG/5-11, 1965.

4 Carne A., Lapostolle P., and Prome M., Proc. 5th Int. Conf. on High Energy Accel. Frascati, 1965, pp. 656–662.

5 Carne A., Lapostolle P., Prome M., and Schnizer B., in *Linear Accelerators*, Lapostolle P.M. and Septier A. North Holland, Amsterdam, 1970, pp. 747–783.

6 Prome M., thesis presented at Universite-Paris-Sud, Center d'Orsay, 1971, English translation in Los Alamos report LA-R-79-33 (1979).

7 Lapostolle P.M., *Proton Linear Accelerators: A Theoretical and Historical Introduction*, Los Alamos Report LA-11601-MS, July, 1989.

8 Lapostolle P., Tanke E., and Valero S., A new method in beam dynamics computations for electrons and ions in complex accelerating elements, *Part. Accel.* **44**, 215–255 (1994).

9 Ratzinger U. and Tiede R., Status of the HIIF RF linac study based on H-mode cavities, *Nucl. Instrum. Methods Phys. Res.* **A415**, 229–235 (1998).

7
Transverse Particle Dynamics

It is necessary that off-axis particles neither drift away, nor be subjected to defocusing forces that take them further from the axis. First, we analyze the effect of the radio-frequency (RF) electric and magnetic fields on the transverse dynamics. We find that the requirement for longitudinal stability, which is especially important for ion linacs, results in transverse RF electric fields that defocus the beam, a result known as the *incompatibility theorem*. There are two general methods for addressing this problem. The first is to exploit loopholes in the incompatibility theorem, which will allow some form of overall stable motion, using only the RF fields. The second approach is to provide external focusing elements. Historically, the first approach was used in the form of conducting foils or grids, installed at the exit end of the gaps. This method eliminated the transverse electric field component at the gap exit, which eliminated the defocusing. It was found that this method was not practical, because the foils or grids were unable to simultaneously withstand the beam-induced damage, and efficiently transmit the beam. Two other approaches for using the RF fields to circumvent the focusing difficulties have been proposed. One is alternating-phase focusing [1], where the bunch phase alternates on both sides of the crest, resulting in alternating focusing and defocusing in each plane (Section 12.1). The alternating-phase focusing approach is an application of the alternating-gradient focusing principle. The other method uses the radio-frequency quadrupole, described in detail in Chapter 8. The most common method for providing external focusing is to use a periodic lattice of quadrupole lenses. Our discussion of quadrupole focusing in this chapter is followed by elementary treatments of random quadrupole misalignments, ellipse transformations and beam matching, solenoid focusing, a comparison of lens and lattice properties, and a discussion of radial motion for relativistic beams.

7.1
Transverse RF Focusing and Defocusing

As can be seen from the electric-field lines in Fig. 7.1, when off-axis particles enter a gap and are accelerated by a longitudinal RF electric field, they also

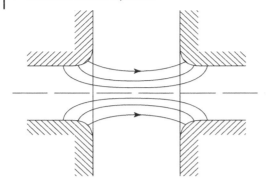

Figure 7.1 Electric-field lines in an RF gap.

experience radial RF electric and magnetic forces. At first it might be thought that the oppositely directed radial electric forces in the two halves of the gap will produce a cancellation in the total radial momentum impulse. On closer examination, it is found that generally there will be a net radial impulse, which occurs as a result of three possible mechanisms: (1) the fields vary in time as the particle cross the gap; (2) the fields also depend on the radial particle displacement, which varies across the gap; and (3) the particle velocity increases, while the particle crosses the gap, so that the particle does not spend equal times in each half of the gap. For longitudinal stability we have seen in Chapter 6 that ϕ_s must be negative, which means the field is rising when the synchronous particle is injected. This means that most particles experience a field in the second half of the gap that is higher than the field in the first half, resulting in a net defocusing force. This corresponds to mechanism (1), and for ion linacs this is the dominant effect, known as the *RF-defocusing force*. Mechanisms (2) and (3) are relatively more important in electron linacs than in most proton linacs.

It has been shown [2] that the condition for longitudinal focusing, which requires that the RF fields are increasing in time, is generally incompatible with local radial focusing from those fields. This is a consequence of *Earnshaw's theorem*, a more general principle that follows from Laplace's equation, which states that the electrostatic potential in free space cannot have a maximum or a minimum [3,4]. Suppose a Lorentz transformation is made to the rest frame of the particle. In the rest frame the magnetic forces are zero, and only the electrostatic forces can act. From Laplace's equation

$$\frac{\partial^2 V}{\partial^2 x} + \frac{\partial^2 V}{\partial^2 y} + \frac{\partial^2 V}{\partial^2 z} = 0 \qquad (7.1)$$

For a potential minimum all three partial derivatives $\partial^2 V/\partial x^2$, $\partial^2 V/\partial y^2$, and $\partial^2 V/\partial z^2$ must be positive, and for a potential maximum they must all be negative. Both requirements are incompatible with Eq. (7.1). Thus, if the longitudinal forces provide focusing at a given point, the two transverse-force components cannot both be focusing at the same point.

7.2
Radial Impulse from a Synchronous Traveling Wave

We can write an expression for the transverse momentum impulse delivered to a particle from the RF fields of a synchronous traveling wave. The nonzero field components of the synchronous space harmonic experienced by a particle of phase ϕ were obtained in Section 3.2 as

$$E_z = E_0 T I_0(Kr) \cos\phi \tag{7.2}$$

$$E_r = -\gamma_s E_0 T I_1(Kr) \sin\phi \tag{7.3}$$

and

$$B_\theta = -\frac{\gamma_s \beta_s}{c} E_0 T I_1(Kr) \sin\phi \tag{7.4}$$

where $K = 2\pi/\gamma_s \beta_s \lambda$, and $E_0 T$ is interpreted as the axial electric field of the traveling wave. The radial Lorentz force component is

$$\frac{dp_r}{dt} = q(E_r - \beta c B_\theta) = -q\gamma_s(1 - \beta\beta_s) I_1(Kr) E_0 T \sin\phi \tag{7.5}$$

The radial momentum can be written as $p_r = mc\gamma\beta r'$ where $r' = dr/dz$, and the radial momentum impulse delivered to a particle over a length L is

$$\Delta(\gamma\beta r') = \frac{1}{mc} \int_0^L -q\gamma_s(1-\beta\beta_s) E_0 T I_1(Kr) \sin\phi \frac{dz}{\beta c}$$

$$= -q\frac{\gamma_s(1-\beta\beta_s)}{mc^2 \beta} E_0 T L I_1(Kr) \sin\phi \tag{7.6}$$

If we assume that $\beta \approx \beta_s$ then

$$\Delta(\gamma\beta r') = -\frac{q E_0 T L I_1(Kr) \sin\phi}{mc^2 \gamma_s \beta_s} \tag{7.7}$$

For $Kr \ll 1$ we have $I_1(Kr) \approx Kr/2$, and

$$\Delta(\gamma\beta r') = -\frac{\pi q E_0 T L \sin\phi}{mc^2 \gamma_s^2 \beta_s^2 \lambda} r \tag{7.8}$$

For longitudinal stability the sign of ϕ_s is negative, and on average all particles executing stable oscillations about the synchronous particle will have the same negative phase. Therefore the radial impulse is positive, which means an outward or defocusing impulse. We see that the RF-defocusing impulse is largest at low velocities, and vanishes in the extreme relativistic limit, where the electric force is canceled by the magnetic force. Equation (7.6) may be used as the basis for a thin-lens approximation for the transformation of the transverse coordinates through an accelerating gap. The distance L may be chosen equal to the cell length of a multicell structure. A simple computer algorithm, known as the *drift–kick–drift*, consists of first advancing the particles through a drift distance $L/2$, then applying a thin-lens kick, and finally applying a drift through the remaining distance $L/2$. The momentum kick changes the transverse momentum component, so that we may write

$$(\gamma\beta r')_\text{f} = (\gamma\beta r')_\text{i} - \frac{q\gamma_\text{s}(1-\bar{\beta}\beta_\text{s})E_0 T L I_1(Kr)\sin\phi}{mc^2\bar{\beta}} \quad (7.9)$$

or

$$r'_\text{f} = \frac{(\gamma\beta r')_\text{i} - q\gamma_\text{s}(1-\bar{\beta}\beta_\text{s})E_0 T L I_1(Kr)\sin\phi/mc^2\bar{\beta}}{\gamma_\text{f}\beta_\text{f}} \quad (7.10)$$

where i and f refer to the initial and final values, r is the radius at the center of the gap, and $\bar{\beta}$ is the mean velocity in the gap. We note that the effect of the particle energy gain in the gap has been included in the transformation.

Equation (7.5) may also be used for a smooth approximation, in which we replace the impulses in the gaps by an average effective force that acts continuously on the beam. It is convenient to change variables to replace time with the axial position s by substituting $ds = \beta c dt$. We approximate $\beta \approx \beta_\text{s}$, and expand $I_1(Kr)$ keeping only the leading term, to obtain a smoothed differential equation of motion

$$\frac{1}{\gamma_\text{s}\beta_\text{s}}\frac{d}{ds}\gamma_\text{s}\beta_\text{s} r' - \frac{k_{\ell 0}^2}{2}r = 0 \quad (7.11)$$

where

$$k_{\ell 0}^2 = \frac{2\pi q E_0 T \sin(-\phi)}{mc^2 \lambda (\gamma\beta)^3} \quad (7.12)$$

is the longitudinal wave number, previously defined in Section 6.6.

7.3
Radial Impulse near the Axis in an Accelerating Gap

For ion linacs, the previous simple treatment using the synchronous traveling wave is usually a good approximation to describe the RF-defocus effect. But for cases where the energy gain in each gap is a large fraction of the particle energy, a more accurate calculation may be needed, which includes electrostatic-focusing terms. To begin we obtain some important relationships for the field components near the axis, assuming that for the mode of interest E_z, E_r, and B_θ are the only nonzero field components. The nonzero components of Maxwell's equations are

$$\frac{1}{r}\frac{\partial(rE_r)}{\partial r} + \frac{\partial E_z}{\partial z} = 0 \quad \text{from } (\boldsymbol{\nabla}\cdot\mathbf{E}=0) \quad (7.13)$$

7.3 Radial Impulse near the Axis in an Accelerating Gap

$$\frac{\partial E_r}{\partial z} - \frac{\partial E_z}{\partial r} = -\frac{\partial B_\theta}{\partial t} \quad \text{from } (\nabla \times \mathbf{E})_\theta = -\frac{\partial B_\theta}{\partial t} \tag{7.14}$$

$$-\frac{\partial B_\theta}{\partial z} = \frac{1}{c^2}\frac{\partial E_r}{\partial t} \quad \text{from } (\nabla \times \mathbf{B})_r = \frac{1}{c^2}\frac{\partial E_r}{\partial t} \tag{7.15}$$

$$\frac{1}{r}\frac{\partial (rB_\theta)}{\partial r} = \frac{1}{c^2}\frac{\partial E_z}{\partial t} \quad \text{from } (\nabla \times \mathbf{B})_z = \frac{1}{c^2}\frac{\partial E_z}{\partial t} \tag{7.16}$$

We assume that near the axis, E_z is approximately independent of r. Then from Eq. (7.13)

$$rE_r = -\frac{\partial E_z}{\partial z}\int_0^r r\, dr \tag{7.17}$$

or

$$E_r = -\frac{\partial E_z}{\partial z}\frac{r}{2} \tag{7.18}$$

and differentiating with respect to r gives

$$\frac{\partial E_r}{\partial r} = -\frac{1}{2}\frac{\partial E_z}{\partial z} \tag{7.19}$$

From Eq. (7.15)

$$\frac{\partial B_\theta}{\partial z} = \frac{r}{2c^2}\frac{\partial}{\partial z}\frac{\partial E_z}{\partial t} \tag{7.20}$$

Integrating with respect to z gives

$$B_\theta = \frac{r}{2c^2}\frac{\partial E_z}{\partial t} \tag{7.21}$$

or

$$\frac{\partial B_\theta}{\partial r} = \frac{1}{2c^2}\frac{\partial E_z}{\partial t} \tag{7.22}$$

Using these results, we derive expressions that are valid near the axis. We assume that the standing-wave electric-field solution for E_z near the axis looks like

$$E_z(r, z, t) = E_a(z)\cos(\omega t + \phi) \tag{7.23}$$

If we look at the shape of the amplitude $E_a(z)$ in a typical accelerating gap, shown in Fig. 7.2, we see that when E_a is increasing with respect to z at the beginning of the gap, E_r is negative, which implies radial focusing. When E_a is decreasing with respect to z at the end of the gap, E_r is positive, which implies radial defocusing. These results are in accordance with Eq. (7.18).

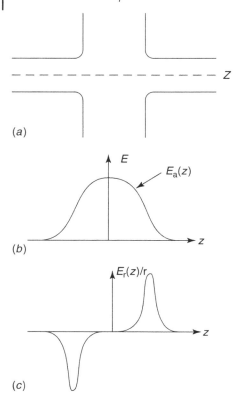

Figure 7.2 Longitudinal and radial electric fields in an RF gap.

If the field is rising in time, as is necessary for longitudinal focusing, the defocusing field experienced by a particle at the exit will be larger than the focusing field at the entrance. But, especially for very low velocities when there are large velocity changes in the gap, the particle spends less time in the exit half of the gap than in the entrance half. This can give rise to a net focusing impulse, which is called *electrostatic focusing*, because it is the same focusing mechanism that is the basis for electrostatic lenses. The momentum impulse near the axis is

$$\Delta p_r = q \int_{-L/2}^{L/2} (E_r - \beta c B_\theta) \frac{dz}{\beta c} = -\frac{q}{2} \int_{-L/2}^{L/2} r \left[\frac{\partial E_z}{\partial z} + \frac{\beta}{c} \frac{\partial E_z}{\partial t} \right] \frac{dz}{\beta c} \quad (7.24)$$

First, we examine the case where the velocity and position changes over the gap are small. To simplify the integral in Eq. (7.24), it is convenient to use

$$\frac{dE_z}{dz} = \frac{\partial E_z}{\partial z} + \frac{1}{\beta c} \frac{\partial E_z}{\partial t} \quad (7.25)$$

which expresses the fact that the total rate of change of the field experienced by a moving particle is the sum of the contributions from the position change of the particle plus the time change. The radial momentum impulse is

$$\Delta p_r = -\frac{qr}{2\beta c}\int_{-L/2}^{L/2}\left[\frac{dE_z}{dz} - \left(\frac{1}{\beta c} - \frac{\beta}{c}\right)\frac{\partial E_z}{\partial t}\right]dz \quad (7.26)$$

The integral over the total derivative vanishes if the interval L extends to zero field at both ends, or if the field is periodic with period L. Then

$$\Delta p_r = \frac{-qr\omega}{2\gamma^2\beta^2c^2}\int_{-L/2}^{L/2} E_z(z)\sin(\omega t + \phi)\,dz \quad (7.27)$$

Consider a particle at the origin at time $t = 0$, so that $\omega t = kz$, where $k = 2\pi/\beta\lambda$. Using a trigonometric identity, and taking the origin at the electrical center of the gap, we obtain

$$\Delta p_r = -\frac{qr\omega}{2\gamma^2\beta^2c^2}\sin\phi\int_{-L/2}^{L/2} E_z(z)\cos kz\,dz \quad (7.28)$$

Finally, introducing the definition of E_0 and T from Chapter 2 and substituting $p_r = mc\gamma\beta r'$, we obtain

$$\Delta(\gamma\beta r') = -\frac{\pi qE_0 T\sin\phi L}{mc^2\gamma^2\beta^2\lambda}r \quad (7.29)$$

When $\beta = \beta_s$, the result of Eq. (7.8) for the deflection of a particle near the axis from a synchronous wave, is the same as that of Eq. (7.29).

7.4
Including Electrostatic Focusing in the Gap

We now include the effects of position and velocity changes in the gap, which produces *electrostatic focusing* [5]. First we consider a nonrelativistic treatment, neglecting the magnetic force. The radial momentum impulse is

$$\Delta p_r = q\int_{-L/2}^{L/2} E_r\frac{dz}{\beta c} = -\frac{q}{2}\int_{-L/2}^{L/2}\frac{r}{\beta c}\frac{\partial E_z}{\partial z}dz \quad (7.30)$$

For a specific example we consider the square-wave field defined by

$$E_z = \begin{cases} E_g\cos(\omega t + \phi), & |z| \leq g/2 \\ 0, & |z| \geq g/2 \end{cases} \quad (7.31)$$

Then

$$\frac{\partial E_z}{\partial z} = E_g\left[\delta\left(z + \frac{g}{2}\right) - \delta\left(z - \frac{g}{2}\right)\right]\cos(\omega t + \phi) \quad (7.32)$$

where δ is the Dirac delta function. Next, we calculate the momentum change of a particle with $z = \beta ct$, and obtain

$$\Delta p_r = -\frac{q}{2}E_g \int \frac{r}{\beta c}\left[\delta\left(z+\frac{g}{2}\right) - \delta\left(z-\frac{g}{2}\right)\right]\cos\left(\frac{2\pi z}{\beta\lambda} + \phi\right)dz \quad (7.33)$$

or

$$\Delta(\gamma\beta r') = -\frac{q}{2mc^2}E_g\left[\frac{r_1}{\beta_1}\cos\left(\frac{\pi g}{\beta\lambda} - \phi\right) - \frac{r_2}{\beta_2}\cos\left(\frac{\pi g}{\beta\lambda} + \phi\right)\right] \quad (7.34)$$

where 1 and 2 refer to the entrance and exit planes of the gap, respectively. Because of the acceleration, $\beta_1 < \beta_2$. For ϕ near zero (near the crest), the overall sign will be negative, if the radius $r_1 > r_2$. The latter condition is expected on average, because the entrance field functions as a thin focusing lens and the exit lens as a thin defocusing lens. Three competing effects, contained in Eq. (7.34), determine the sign of the focusing: (1) for negative ϕ, needed for longitudinal focusing, the field is rising and is larger at the exit; (2) the beam has a smaller radius at the exit than at the entrance because of the focusing at the entrance; and (3) more time is spent at the entrance or focusing end than at the exit or defocusing end.

A general formulation of the transverse dynamics in the gap, derived by Lapostolle, [6] includes the RF and electrostatic terms, and explicitly takes into account the acceleration or momentum increase, which reduces the divergence angles. Instead of displacement r, Lapostolle uses a "reduced coordinate" R, given by $R = r\sqrt{\beta\gamma}$. The general transverse equation of motion becomes

$$\frac{d^2 R}{dz^2} = \left[\frac{q}{2mc^2(\beta\gamma)^3}\frac{\partial E_z}{\partial t} - \frac{(\gamma^2+2)}{(\beta\gamma)^4}\left(\frac{q}{2mc^2}\right)^2 E_z^2\right]R \quad (7.35)$$

The first term on the right side is the RF-defocusing term, and for ions this term is usually dominant. The second term, which is quadratic in the accelerating field is the electrostatic term, which is often found to be larger for electron linacs, although the RF-defocusing term is not always negligible for that case.

7.5
Coordinate Transformation through an Accelerating Gap

The transformation of particle coordinates through an accelerating gap was introduced in Section 6.5 for the longitudinal case, and in Section 7.2 for the transverse case. Such transformations form the basis for linac beam-dynamics codes. It has been shown [7–9] that the particle trajectory in a gap may be replaced by an equivalent trajectory in which the particle drifts at the constant initial velocity to the center of the gap, receives an impulse at the center, and drifts at a constant final velocity to the end of the gap. The relativistically valid forms for the transformation at the center of the gap are

$$\Delta W \cong qV\cos(\phi)\left\{T_s\left[1+\left(\frac{k_s r}{2\gamma_s}\right)^2\right]\right.$$
$$\left.+\left[k_s T_s'\left(1+\left(\frac{k_s r}{2\gamma_s}\right)^2\right)+\frac{T_s}{2}\left(\frac{k_s r}{\gamma_s}\right)^2\right]\frac{\beta_s-\beta}{\beta}\right\} \qquad (7.36)$$

$$\Delta\phi \cong \frac{qV\cos(\phi)}{mc^2\gamma_s^3\beta_s^2}\left[k_s T_s'\left(1+\left(\frac{k_s r}{2\gamma_s}\right)^2\right)+\frac{T_s}{2}\left(\frac{k_s r}{\gamma_s}\right)^2\right] \qquad (7.37)$$

$$\Delta r \cong \frac{-qVr\cos\phi}{2mc^2\gamma_s^3\beta_s^2}\left[\gamma_s^2 T_s + k_s T_s'\right] \qquad (7.38)$$

and

$$\Delta r' \cong -\frac{qV}{2mc^2\gamma_s^3\beta_s^2}[k_s T_s r \sin(\phi) + (\gamma_s^2 T_s - k_s T_s')r'\cos(\phi)] \qquad (7.39)$$

where the subscript s refers to the synchronous particle.

In these equations $k_s = 2\pi/\beta_s\lambda$, $V = E_0 L$, and $T_s' \equiv dT_s/dk_s$, where

$$T_s = \frac{2}{V}\int_0^{L/2} E_z(z)\cos(k_s z)\,dz \qquad (7.40)$$

and

$$T_s' = -\frac{4\pi}{VL}\int_0^{L/2} zE_z(z)\sin(k_s z)dz \qquad (7.41)$$

The phase and position impulses for $\Delta\phi$ and Δr, given by Eqs. (7.37) and (7.38) would be zero in an ideal thin lens, but are necessary to satisfy Liouville's theorem. We note that these terms decrease rapidly as the particle energy increases. For maximum accuracy the values of β_s, γ_s, and ϕ that appear on the right sides of the equations are the values at the gap center rather than those at the gap entrance, which makes the equations difficult to use. To address this problem equations have also been derived, which transform the parameters to the center of the gap and provide the required values so that Eqs. (7.36) through (7.39) can be evaluated in the most accurate way. Equations (7.38) and (7.39) can be converted to Cartesian coordinates by using $x = r(x/r)$, and $x' = r'(x/r)$, and similarly for the y coordinates.

7.6
Quadrupole Focusing in a Linac

The most common method of compensating for the transverse RF-defocusing effects in linacs has been the use of magnetic lenses, of which the magnetic-quadrupole lens is the most common [10–13]. In the next six sections we review the dynamics of quadrupole focusing systems, and apply the results to linacs. A cross section of a quadrupole magnet showing the magnetic-field pattern is shown in Fig. 7.3. The quadrupoles are installed between the tanks in coupled-cavity linacs, and within the drift tubes in conventional drift-tube linacs (DTLs), with the necessary connections made through the drift-tube support stems.

In an ideal quadruple field the pole tips have hyperbolic profiles, and produce a constant transverse quadrupole gradient

$$G = \frac{\partial B_x}{\partial y} = \frac{\partial B_y}{\partial x} \tag{7.42}$$

For a particle moving along the z direction with velocity v and with transverse coordinates (x, y), the Lorentz force components are

$$F_x = -qvGx, \quad F_y = qvGy \tag{7.43}$$

If qG is positive, the lens focuses in x and defocuses in y. For a pole-tip with radius a_0 and pole-tip field B_0, the gradient is $G = B_0/a_0$. Although individual quadrupole lenses focus in only one plane, they can be combined with both polarities to give overall strong focusing in both transverse planes. For a particle with charge q moving parallel to the beam axis with velocity β and transverse coordinates x and y, the equations of motion for a perfect quadrupole lens with the axial position s as the independent variable are

$$\frac{d^2x}{ds^2} + \kappa^2(s)x = 0, \quad \text{and} \quad \frac{d^2y}{ds^2} - \kappa(s)y = 0 \tag{7.44}$$

where

$$\kappa^2(s) = \frac{|qG(s)|}{mc\gamma\beta} \tag{7.45}$$

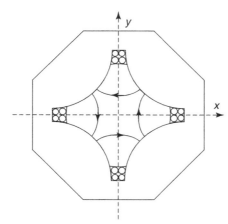

Figure 7.3 Quadrupole magnet cross section showing the four poles, the coils, and the magnetic-field pattern. The beam axis is at the center.

Equations (7.44) and (7.45) show that the quadrupole effects are greater at lower momentum. This is because the magnetic rigidity increases with momentum, and over a fixed distance, high-momentum particles are more difficult to bend.

The effect of the RF defocusing in the gap may be included in a continuous-force approximation from the equivalent traveling wave, using the results of Eqs. (7.11) and (7.12). Including the quadrupole term we obtain

$$\frac{d^2x}{ds^2} + \kappa^2(s)x - \frac{k_{\ell 0}^2}{2}x = 0, \quad \text{and} \quad \frac{d^2y}{ds^2} - \kappa^2(s)y - \frac{k_{\ell 0}^2}{2}y = 0 \quad (7.46)$$

The quadrupoles are typically arranged in a regular lattice, which is either periodic or quasiperiodic to allow for the increasing cell lengths as the beam is accelerated. The most common lattice configuration is the FODO array, shown in Fig. 7.4. Other common groupings are FOFODODO, or FDO, which uses doublets.

Equation (7.46) is based on a restoring force that is a linear function of the displacement from the equilibrium trajectory, and is an example of Hill's equation. For an ideal focusing lattice in a linac, it can be expressed in a normalized form as

$$\frac{d^2x}{d\tau^2} + (\theta_0^2 F(\tau) + \Delta)x = 0 \quad (7.47)$$

where $\theta_0^2 = q\beta G\lambda^2/\gamma mc$, is a dimensionless measure of the quadrupole focusing strength, $\Delta = \pi q E_0 T \lambda \sin\phi / \gamma^3 mc^2 \beta$ is a dimensionless measure of the RF-defocusing force, $\tau = s/\beta\lambda$ is the normalized axial variable, and the periodic function $F(\tau) = 1, 0,$ or -1. The equation of motion for y can be obtained from Eq. (7.47) by replacing $F(\tau)$ by $-F(\tau)$. Hill's equation has regions of stable and unstable motion. To study the properties of Hill's equation, it is convenient to introduce a matrix solution.

7.7
Transfer-Matrix Solution of Hill's Equation

We describe a quadrupole-transport channel by the equation of motion

$$x'' + K(s)x = 0 \quad (7.48)$$

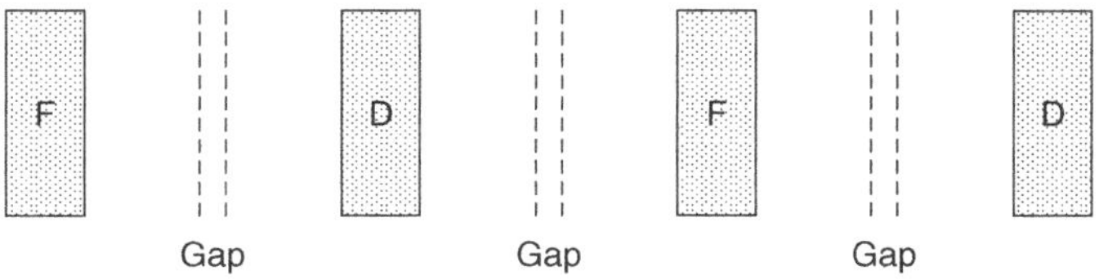

Figure 7.4 FODO quadrupole lattice with accelerating gaps.

where s is the axial direction, x is the displacement and $x'' = d^2x/ds^2$, and we introduce the divergence angle $x' = dx/ds$. Because Eq. (7.48) is a linear second-order differential equation, its solution can be written in matrix form, which we write as

$$\begin{bmatrix} x \\ x' \end{bmatrix} = \begin{bmatrix} a & b \\ c & d \end{bmatrix} \begin{bmatrix} x_0 \\ x'_0 \end{bmatrix} \tag{7.49}$$

where x_0 and x'_0 are the initial displacement and divergence and x and x' are the final values. The 2 × 2 matrix is called a *transfer matrix*, and, because Hill's equation does not contain any first-derivative term, its determinant can be shown to equal unity. We are interested in field-free drift spaces with $K = 0$, and in quadrupole magnets that give transverse focusing when $K > 0$ and defocusing when $K < 0$. Denoting the transfer matrix by R, the following results can be derived.

(a) *Drift space* ($K = 0$):

$$R = \begin{bmatrix} 1 & \ell \\ 0 & 1 \end{bmatrix} \tag{7.50}$$

where ℓ is the drift length.

(b) *Focusing quadrupole* $\left(K = \dfrac{|qG|}{mc\beta\gamma} > 0 \right)$:

$$R = \begin{bmatrix} \cos\sqrt{K}\ell & \dfrac{\sin\sqrt{K}\ell}{\sqrt{K}} \\ -\sqrt{K}\sin\sqrt{K}\ell & \cos\sqrt{K}\ell \end{bmatrix} \tag{7.51}$$

where ℓ is the length, and G is the quadrupole gradient.

(c) *Defocusing quadrupole* $\left(K = \dfrac{qG}{mc\beta\gamma} < 0 \right)$:

$$R = \begin{bmatrix} \cosh\sqrt{|K|}\ell & \dfrac{\sinh\sqrt{|K|}\ell}{\sqrt{|K|}} \\ \sqrt{|K|}\sinh\sqrt{|K|}\ell & \cosh\sqrt{|K|}\ell \end{bmatrix} \tag{7.52}$$

(d) *Thin lens*:

$$R = \begin{bmatrix} 1 & 0 \\ \pm\frac{1}{f} & 1 \end{bmatrix} \tag{7.53}$$

where f is the focal length. When the sign is negative this is a focusing lens, and when it is positive it is defocusing. The quadrupole lens approaches the thin-lens form when $\sqrt{|K|}\ell \to 0$ while $|K|\ell$ remains finite. Then for a thin-lens quadrupole,

$$\frac{1}{f} = |K|\ell = \frac{qG\ell}{mc\beta\gamma} \tag{7.54}$$

The total transfer matrix through a sequence of piecewise constant elements is obtained by forming the product of the individual R matrices, taken in the correct order. If the beam is transported through the elements 1, 2, 3, ..., n, the total R matrix is $R = R_n, \ldots, R_3, R_2, R_1$.

7.8
Phase-Amplitude Form of Solution to Hill's Equation

When $K(s)$ is a periodic function, we are interested in a solution to Hill's equation that has a form similar to that of a harmonic oscillator. The general solution is sometimes called the *phase-amplitude form* of the solution and is written as

$$x(s) = \sqrt{\varepsilon_1 \tilde{\beta}(s)} \cos(\phi(s) + \phi_1) \tag{7.55}$$

where $\tilde{\beta}(s)$ and $\phi(s)$ are called the *amplitude* and *phase functions*, and ε_1 and ϕ_1 are constants determined by the initial conditions. In this section and the next, we summarize the properties of the solution. The functions $\tilde{\beta}(s)$ and $\phi(s)$ are related by

$$\phi(s) = \int \frac{ds}{\tilde{\beta}(s)} \tag{7.56}$$

It is customary to define two other functions of $\tilde{\beta}(s)$ as

$$\tilde{\alpha}(s) = -\frac{1}{2}\frac{d\tilde{\beta}(s)}{ds}, \quad \text{and } \tilde{\gamma}(s) = \frac{1 + \tilde{\alpha}(s)^2}{\tilde{\beta}(s)} \tag{7.57}$$

The quantities $\tilde{\alpha}(s)$, $\tilde{\beta}(s)$, and $\tilde{\gamma}(s)$ are called either *Twiss* or *Courant–Snyder parameters*, and can all be shown to be periodic functions with the same period as $K(s)$. The functions $\tilde{\beta}(s)$ and $\tilde{\gamma}(s)$ are always positive. The coordinates x and x' satisfy the equation

$$\tilde{\gamma}(s)x^2 + 2\tilde{\alpha}(s)xx' + \tilde{\beta}(s)x'^2 = \varepsilon_1 \tag{7.58}$$

Equation (7.58) is the general equation of an ellipse centered at the origin of $x - x'$ phase space, whose area is $A = \pi \varepsilon_1$. The general ellipse is shown in Fig. 7.5. When $\tilde{\alpha}(s)$ is nonzero, the ellipse is tilted. Because $\tilde{\alpha}$, $\tilde{\beta}$ and $\tilde{\gamma}$ depend on s, the ellipses vary with s. But, because these parameters are periodic in s, if the coordinates lie on a particular ellipse at location s then, the coordinates at a location one full period away lie on an identical ellipse. It can also be shown that the area of the ellipse is invariant with respect to s.

7 Transverse Particle Dynamics

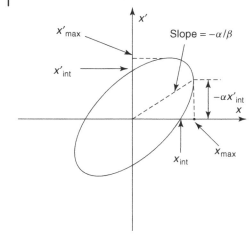

Figure 7.5 The general trajectory ellipse and its parameters. The maximum projections are $x_{max} = \sqrt{\varepsilon\beta}$, and $x'_{max} = \sqrt{\varepsilon\gamma}$, and the intercepts are $x_{int} = \sqrt{\varepsilon/\gamma}$ and $x'_{int} = \sqrt{\varepsilon/\beta}$.

7.9
Transfer Matrix through One Period

The transfer matrix M through one full period has the form [14]

$$P \equiv R(s \to s + P) = \begin{bmatrix} \cos\sigma + \tilde{\alpha}\sin\sigma & \tilde{\beta}\sin\sigma \\ -\tilde{\gamma}\sin\sigma & \cos\sigma - \tilde{\alpha}\sin\sigma \end{bmatrix} \quad (7.59)$$

where $\sigma = \Delta\phi$ is the phase advance per period of every particle, each lying on its own trajectory ellipse. If the focusing is too strong the solution is unstable. The requirement for stability is $|\text{Tr}\,P| < 2$ where Tr P is the trace of the P matrix. From Eq. (7.59) we find that the stability requirement corresponds to $|\cos\sigma| < 1$, or $0 < \sigma < \pi$. The phase advance σ per period L is related to the $\tilde{\beta}$ function by

$$\sigma = \int^L \frac{ds}{\tilde{\beta}(s)} \quad (7.60)$$

The average $\tilde{\beta}$ value over the period is approximately $\langle\tilde{\beta}\rangle = L/\sigma$, where L is the period. Typically, if σ is equal to about 1 rad, then $\langle\tilde{\beta}\rangle = L$. Equation (7.59) for the P matrix is useful because it provides a simple method for obtaining the $\tilde{\alpha}(s)$, $\tilde{\beta}(s)$, $\tilde{\gamma}(s)$ functions, and the phase advance per period σ. By transporting two orthogonal trajectories, beginning at any location s over one full period, either by numerical integration through the specified fields, or by matrix multiplication for piecewise constant elements, all four transfer-matrix elements per period can be calculated. These can then be compared with the elements of the P matrix in Eq. (7.59), obtained from numerical integration or matrix multiplication, and in this way one can determine the functions $\tilde{\alpha}(s)$, $\tilde{\beta}(s)$, $\tilde{\gamma}(s)$, and the constant σ.

7.10
Thin-Lens FODO Periodic Lattice

The FODO-lattice structure is the most common focusing structure used in accelerators. Each period contains a single focusing lens, and a single defocusing lens of equal strengths, as shown in Fig. 7.6. For a simple application of the methods of Section 7.9, we assume thin lenses separated by equal drift distances of length L, so that the resulting period is $2L$, and neglect the effects of any accelerating gaps between the lenses. We calculate the Courant–Snyder parameters at four different places in the lattice, by starting at four different locations and calculating the transfer matrix through a period. We summarize the results below.

(a) Beginning at the center of the focusing lens:

$$P = \begin{bmatrix} 1 & 0 \\ -(2f)^{-1} & 1 \end{bmatrix} \begin{bmatrix} 1 & L \\ 0 & 1 \end{bmatrix} \begin{bmatrix} 1 & 0 \\ f^{-1} & 1 \end{bmatrix} \begin{bmatrix} 1 & L \\ 0 & 1 \end{bmatrix}$$
$$\times \begin{bmatrix} 1 & 0 \\ -(2f)^{-1} & 1 \end{bmatrix} \quad (7.61)$$

After multiplying the matrices above and equating the result to the transfer matrix P in Eq. (7.59) through one period, we obtain

$$\cos\sigma = 1 - \frac{L^2}{2f^2}, \quad \sin\frac{\sigma}{2} = \frac{L}{2f},$$
$$\tilde{\beta} = 2L\frac{1+\sin\sigma/2}{\sin\sigma}, \quad \tilde{\alpha} = 0 \quad (7.62)$$

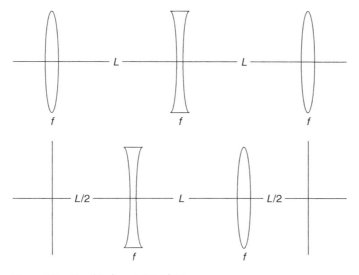

Figure 7.6 The thin-lens FODO lattice.

(b) Beginning at the center of the defocusing lens:

$$P = \begin{bmatrix} 1 & 0 \\ (2f)^{-1} & 1 \end{bmatrix} \begin{bmatrix} 1 & L \\ 0 & 1 \end{bmatrix} \begin{bmatrix} 1 & 0 \\ -f^{-1} & 1 \end{bmatrix} \begin{bmatrix} 1 & L \\ 0 & 1 \end{bmatrix}$$
$$\times \begin{bmatrix} 1 & 0 \\ (2f)^{-1} & 1 \end{bmatrix} \tag{7.63}$$

$$\cos\sigma = 1 - \frac{L^2}{2f^2}, \quad \sin\frac{\sigma}{2} = \frac{L}{2f},$$
$$\tilde{\beta} = 2L\frac{1-\sin\sigma/2}{\sin\sigma}, \quad \tilde{\alpha} = 0 \tag{7.64}$$

(c) Beginning at the center of drift after the focusing lens:

$$P = \begin{bmatrix} 1 & \frac{L}{2} \\ 0 & 1 \end{bmatrix} \begin{bmatrix} 1 & 0 \\ -f^{-1} & 1 \end{bmatrix} \begin{bmatrix} 1 & L \\ 0 & 1 \end{bmatrix} \begin{bmatrix} 1 & 0 \\ +f^{-1} & 1 \end{bmatrix} \begin{bmatrix} 1 & \frac{L}{2} \\ 0 & 1 \end{bmatrix} \tag{7.65}$$

$$\cos\sigma = 1 - \frac{L^2}{2f^2}, \quad \sin\frac{\sigma}{2} = \frac{L}{2f} \tag{7.66}$$

$$\tilde{\beta} = \frac{L}{\sin\sigma}[2 - \sin^2(\sigma/2)], \quad \tilde{\alpha} = \frac{2\sin(\sigma/2)}{\sin\sigma} \tag{7.67}$$

(d) Beginning at the center of drift after the defocusing lens:

$$P = \begin{bmatrix} 1 & \frac{L}{2} \\ 0 & 1 \end{bmatrix} \begin{bmatrix} 1 & 0 \\ +f^{-1} & 1 \end{bmatrix} \begin{bmatrix} 1 & L \\ 0 & 1 \end{bmatrix} \begin{bmatrix} 1 & 0 \\ -f^{-1} & 1 \end{bmatrix} \begin{bmatrix} 1 & \frac{L}{2} \\ 0 & 1 \end{bmatrix} \tag{7.68}$$

$$\cos\sigma = 1 - \frac{L^2}{2f^2}, \quad \sin\frac{\sigma}{2} = \frac{L}{2f} \tag{7.69}$$

$$\tilde{\beta} = \frac{L}{\sin\sigma}[2 - \sin^2(\sigma/2)], \quad \tilde{\alpha} = \frac{-2\sin(\sigma/2)}{\sin\sigma} \tag{7.70}$$

The β function is maximum in the focusing lens and minimum in the defocusing lens. The ratio of the maximum and minimum β functions is

$$\frac{\tilde{\beta}_{\max}}{\tilde{\beta}_{\min}} = \frac{1 + \sin(\sigma/2)}{1 - \sin(\sigma/2)} \tag{7.71}$$

If the quadrupole lens is focusing in the x plane, it is defocusing in the y plane. Therefore the same ratio applies in the two different planes. Comparing cases (c) and (d), we obtain the result at the center of the drift spaces that

$\tilde{\beta}_x = \tilde{\beta}_y$ and $\tilde{\alpha}_x = -\tilde{\alpha}_y$. The solution is unstable for $\sigma \geq \pi$, which means that for stable motion the focal length must satisfy $f \geq L/2$. At $\sigma = \pi/3$, the focal length is equal to the period.

7.11
Transverse Stability Plot in a Linac

The transverse equation of motion in a linac satisfies Hill's equation, which we expressed in dimensionless form in Eq. (7.47) in terms of a quadrupole strength parameter, and an RF-defocusing parameter $\theta_0^2 = q\beta G\lambda^2/\gamma m$, and $\Delta = \pi q E_0 T\lambda \sin\phi/\gamma^3 mc^2\beta$. To study the stability properties of Hill's equation, it is convenient to introduce a matrix solution, where each element is represented by a transfer matrix. The matrices can be multiplied together in the proper order to obtain the total transfer matrix over a period, and the stability of the beam is determined.

In this section we extend the results to include the RF fields in the gap. The transfer matrix for an equivalent thin lens to represent the transverse impulse from the accelerating gap can be obtained by integrating Eq. (7.11). If the change of velocity in the gap is small, the effective focal length for a cell of length $\beta\lambda$ is

$$\frac{1}{f_g} = \frac{\Delta x'}{x} = \frac{\pi q E_0 T \sin(-\phi)}{\gamma^3 \beta^2 mc^2} \qquad (7.72)$$

The transfer matrix through a period beginning at the center of a focusing lens is

$$P = F_{1/2} \cdot L \cdot G \cdot L \cdot D \cdot L \cdot G \cdot L \cdot F_{1/2} \qquad (7.73)$$

where

$$F_{1/2} = \begin{bmatrix} \cos(\sqrt{K}\ell/2) & \frac{\sin(\sqrt{K}\ell/2)}{\sqrt{K}} \\ -\sqrt{K}\sin(\sqrt{K}\ell/2) & \cos(\sqrt{K}\ell/2) \end{bmatrix} \qquad (7.74)$$

$$D = \begin{bmatrix} \cosh\sqrt{|K|}\ell & \frac{\sinh\sqrt{|K|}\ell}{\sqrt{|K|}} \\ \sqrt{|K|}\sinh\sqrt{|K|}\ell & \cosh\sqrt{|K|}\ell \end{bmatrix} \qquad (7.75)$$

$$G = \begin{bmatrix} 1 & 0 \\ \frac{1}{f_g} & 0 \end{bmatrix} \qquad (7.76)$$

and

$$L = \begin{bmatrix} 1 & d \\ 0 & 0 \end{bmatrix} \qquad (7.77)$$

where $K = qG/\gamma mv$, and d is the drift space between the end of a quadrupole and the accelerating gap. A stability plot can be produced, plotting θ_0^2 as the ordinate, and Δ as the abscissa, and identifying the curves, which correspond to the stability limit, where $\text{Tr}|P| = 2$. The regions with $\text{Tr}|P| \leq 2$ are stable and those with $\text{Tr}|P| > 2$ correspond to overfocusing and are unstable. It is customary, following the definitions introduced by Smith and Gluckstern, but changing the notation to correspond to modern usage, to introduce the quantities $\tilde{\beta}_{\max}$, the value of $\tilde{\beta}$ at the center of the focusing lens, and $\tilde{\beta}_{\min}$, the value of $\tilde{\beta}$ at the center of the defocusing lens. We also define $\psi = \sqrt{\tilde{\beta}_{\max}/\tilde{\beta}_{\min}}$, which is called the *flutter factor*, $\Lambda = \ell/(\ell + 2d)$ the quadrupole filling factor, and N, the polarity reversal index, where N corresponds to the number of quadrupoles between polarity alternations. For example, $N = 1$ for FODO, $N = 2$ for FOFODODO, and so on.

The Smith and Gluckstern stability charts for a DTL with $\Lambda = 0.5$ for the FODO, or $N = 1$ case is shown in Fig. 7.7a, and for $N = 2$ is shown in Fig. 7.7b. In the notation of Smith and Gluckstern, the symbol γ_N is used, where $\gamma_N = \tilde{\beta}_{\max} c/\nu\lambda$ and μ_N is used instead of σ. We are only concerned with the lowest passband, because it alone has sufficient width for a practical focusing system. The parameters can be chosen by using the stability chart to pick an operating line as a function of particle phase, so that all particles in the bunch lie within the stable region. The use of larger N tends to reduce the required quadrupole gradients, but the height and width of the stable region also decrease with increasing N. If one uses the criterion that the strongest focusing is obtained when the phase advance per unit length is maximum, it is found that larger N is preferable. However, including the additional constraint that for stability, the phase advance is limited to $\sigma < \pi$, it is found that N will be limited. The most typical choices are $N = 1$ or 2. Finally, there is a resonance between the radial and phase motion, when the longitudinal oscillation frequency is twice the transverse oscillation frequency. In general it is not difficult to avoid this resonance [15].

7.12
Effects of Random Quadrupole Misalignment Errors

During the commissioning of a newly constructed accelerator, most large errors are discovered and corrected. Nevertheless, there are always practical or economic limits to an attainable precision, and within such limits, the parameters may exhibit random deviations from their design values. To achieve the desired performance, the effects on the beam, caused by such random errors, must be understood. Some very useful results, showing the effects of random errors, have been obtained by Crandall for periodic or quasiperiodic accelerator systems [16]. In this section we discuss random quadrupole misalignments.

7.12 Effects of Random Quadrupole Misalignment Errors

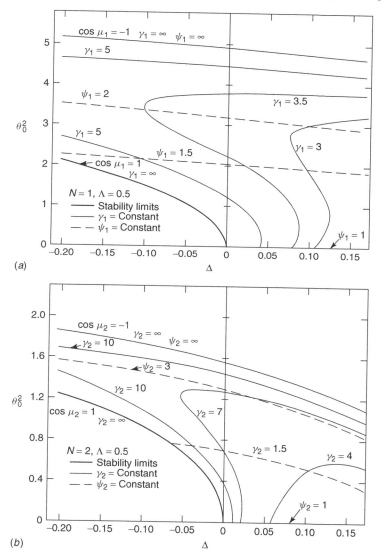

Figure 7.7 Stability chart for transverse motion from Smith and Gluckstern for (a) $N = 1$ and (b) $N = 2$. In the notation of Smith and Gluckstern, $\gamma_N = \tilde{\beta}_{max} c / \nu \lambda$ and μ_N is used instead of σ. The other notation is explained in the text. [Reprinted with permission from L. W. Smith and R. L. Gluckstern, *Rev. Sci. Instrum.* **26**, 220 (1955). Copyright 1955 American Institute of Physics.]

First, we consider the effect of random transverse misalignments of thin-lens quadrupoles in a beam transport line, when no acceleration is present. If there are no misalignments, but the injected beam centroid is displaced from the axis, the centroid trajectory will behave as a single particle. If the beam line comprises a sequence of randomly misaligned quadrupoles, the beam centroid will receive deflections in every quadrupole, and the evolution of the centroid may be viewed as a random walk. It is found that on average the area of the trajectory ellipse on which the centroid moves will grow as the beam passes through more quadrupoles. The process is a statistical one, and a mean ellipse area may be defined at any point along the beam. If the amplitude A of the centroid motion in the x plane is defined as the maximum displacement in a period, we can write $A^2 = \varepsilon \tilde{\beta}_{\max}$, where $\tilde{\beta}_{\max}$ is the maximum value of the Courant–Snyder ellipse parameter, and ε is the ellipse area divided by π. If the mean value of ε is $\langle \varepsilon \rangle$, the mean square value of the amplitude at any location along the beam line is

$$\langle A^2 \rangle = \tilde{\beta}_{\max} \langle \varepsilon \rangle \tag{7.78}$$

For a system of identical thin-lens quadrupoles, after N periods one obtains

$$\langle \varepsilon_N \rangle = \langle \varepsilon_0 \rangle + N \langle (\delta x)^2 \rangle \sum_{i=1}^{n} \frac{\tilde{\beta}_i}{f_i^2} \tag{7.79}$$

where f_i is the focal length of the ith lens in a period, given by $1/f_i = qG\ell/mc\beta\gamma$, $\tilde{\beta}_i$ is the Courant–Snyder ellipse parameter at the ith lens, ε_0 is the initial ellipse area, which is zero if the beam is aligned on the axis initially, δx is the quadrupole displacement error, and n is the number of lenses in a single period. Equation (7.79) shows that misalignment of the focusing lenses is more serious than that of the defocusing lenses, because $\tilde{\beta}_i$ is larger in the focusing lenses. Furthermore, misalignments are more serious as the strength of the lenses, as measured by $1/f_i$, increases. Substituting Eq. (7.79) in Eq. (7.78), the root-mean-square (rms) value of the amplitude after N periods is

$$A_{N,\mathrm{rms}} = \sqrt{A_N^2} = \sqrt{\langle A_0^2 \rangle + N \tilde{\beta}_{\max} \langle (\delta x)^2 \rangle \sum_{i=1}^{n} \frac{\tilde{\beta}_i}{f_i^2}} \tag{7.80}$$

For identical thin-lens quadrupoles in a FODO lattice, Eq. (7.79) simplifies to

$$\langle \varepsilon_N \rangle = \langle \varepsilon_0 \rangle + \frac{N}{f^2} (\tilde{\beta}_{\max} + \tilde{\beta}_{\min}) \langle (\delta x)^2 \rangle \tag{7.81}$$

where $\tilde{\beta}_{\max}$ and $\tilde{\beta}_{\min}$ are evaluated at the focusing and defocusing lenses, respectively. The rms value of the amplitude after N periods is

$$A_{N,\mathrm{rms}} = \sqrt{A_N^2} = \sqrt{\langle A_0^2 \rangle + \frac{N \tilde{\beta}_{\max}}{f^2} \langle (\delta x)^2 \rangle (\tilde{\beta}_{\max} + \tilde{\beta}_{\min})} \tag{7.82}$$

These equations can be modified to include the effects of quasiperiodicity, and of acceleration. For quasiperiodic systems, where the parameters are slowly changing with energy, the second term of Eq. (7.80) is modified. Also, when acceleration is included, the ellipse area is no longer constant in the absence of errors, but decreases, since it is the ellipse area in the phase space of position and momentum that is constant. The general result, including all these effects, is

$$A_{N,\mathrm{rms}} = \sqrt{\langle A_N^2 \rangle} = \sqrt{\frac{\tilde{\beta}_N(\beta_0\gamma_0)}{\tilde{\beta}_0(\beta_N\gamma_N)}\langle A_0^2\rangle + \frac{\tilde{\beta}_{\max}\langle(\delta x)^2\rangle}{(\beta_N\gamma_N)}\left(\frac{qG\ell}{mc}\right)^2 \sum_{i=1}^{n}\frac{\tilde{\beta}_i}{(\beta_i\gamma_i)}}$$

(7.83)

Crandall has studied numerically the distribution of rms amplitudes of the centroid, associated with a randomly displaced FODO sequence of thin lenses. The results are only weakly dependent on the number of lattice periods. It was found that between 60 and 65% of the cases have amplitudes less than $A_{N,\mathrm{rms}}$, about 90% have amplitudes less than $1.5 A_{N,\mathrm{rms}}$, and about 97.5% have amplitudes less than $2 A_{N,\mathrm{rms}}$. Although the amplitude of the centroid oscillation produced by misaligned quadrupoles is generally larger for stronger quadrupoles, weaker quadrupoles result in an increase in the width of the beam, which suggests that when errors are taken into account, there is an optimum focusing strength.

7.13
Ellipse Transformations

We have discussed the transfer-matrix formalism to transport particle coordinates between two points in an accelerator. The formalism is valid when the forces acting on the beam are linear. The matrix formalism can also be extended to transport the beam ellipses between two locations, as shown in Fig. 7.8. The general equation of an ellipse in $x - x'$ phase space is

$$\tilde{\gamma}x^2 + 2\tilde{\alpha}xx' + \tilde{\beta}x'^2 = \varepsilon \qquad (7.84)$$

where $\tilde{\gamma}\tilde{\beta} - \tilde{\alpha}^2 = 1$. It is convenient to express this equation in matrix form as

$$X^T \sigma^{-1} X = \varepsilon \qquad (7.85)$$

where $X = \begin{bmatrix} x \\ x' \end{bmatrix}$, $X^T = [xx']$ is the transpose matrix, and we define

$$\sigma^{-1} = \begin{bmatrix} \tilde{\gamma} & \tilde{\alpha} \\ \tilde{\alpha} & \tilde{\beta} \end{bmatrix} \qquad (7.86)$$

The inverse matrix of σ^{-1} is

$$\sigma = \begin{bmatrix} \tilde{\beta} & -\tilde{\alpha} \\ -\tilde{\alpha} & \tilde{\gamma} \end{bmatrix} \qquad (7.87)$$

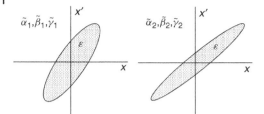

Figure 7.8 Phase-space ellipse transformation between two locations.

We know that for any particle the coordinate transformation from position 1 to 2 can be expressed in the form of a 2 × 2 transfer matrix R, such that

$$X_2 = RX_1 \tag{7.88}$$

The equation for the ellipse at location 1 is $X_1^T \sigma_1^{-1} X_1 = \varepsilon$, and that for the ellipse at location 2 is $X_2^T \sigma_2^{-1} X_2 = \varepsilon$, where, because of the linear transformation, the ellipse emittance is assumed to be unchanged. It can be shown that σ_2 is related to σ_1, as

$$\sigma_2 = R\sigma_1 R^T \tag{7.89}$$

An equivalent expression, obtained by carrying out the above matrix multiplication, is

$$\begin{bmatrix} \tilde{\beta}_2 \\ \tilde{\alpha}_2 \\ \tilde{\gamma}_2 \end{bmatrix} = \begin{bmatrix} R_{11}^2 & -2R_{11}R_{12} & R_{12}^2 \\ -R_{11}R_{21} & 1+2R_{12}R_{21} & -R_{12}R_{22} \\ R_{21}^2 & -2R_{21}R_{22} & R_{22}^2 \end{bmatrix} \begin{bmatrix} \tilde{\beta}_1 \\ \tilde{\alpha}_1 \\ \tilde{\gamma}_1 \end{bmatrix} \tag{7.90}$$

Equation (7.89) or (7.90) is useful for designing the matching optics to inject a matched beam ellipse into an accelerator with periodic focusing. They are valid even in the presence of space-charge forces, as long as those forces are linear.

7.14
Beam Matching

If the injected beam ellipse is not matched to the focusing system, there will be additional oscillations of the rms beam projections, which produce a larger beam at some locations and a smaller beam at other locations. Generally, beam matching means that the beam-density contours coincide with the ellipses corresponding to particle trajectories. For this discussion, we assume that the beam is described by a well-defined phase-space ellipse. In Chapter 9, this is refined in terms of an rms definition for the beam ellipse. At every point in the focusing lattice, there exists a matched beam condition, defined by the Courant–Snyder ellipse parameters $\tilde{\alpha}_m$, $\tilde{\beta}_m$, and $\tilde{\gamma}_m$. In general, the beam will have ellipse parameters that may differ from the matched values.

7.14 Beam Matching

It is convenient to define a *mismatch factor*, [17] which is a measure of the increase in the maximum beam size, resulting from a mismatch. Suppose that the matched beam ellipse is defined by

$$\tilde{\gamma}_m x^2 + 2\tilde{\alpha}_m xx' + \tilde{\beta}_m x'^2 = \varepsilon \tag{7.91}$$

and a mismatched beam ellipse with the same area is defined by

$$\tilde{\gamma} x^2 + 2\tilde{\alpha} xx' + \tilde{\beta} x'^2 = \varepsilon \tag{7.92}$$

The mismatch factor is defined by

$$M = \left[1 + \frac{\Delta + \sqrt{\Delta(\Delta + 4)}}{2} \right]^{1/2} - 1 \tag{7.93}$$

where

$$\Delta = (\Delta\tilde{\alpha})^2 - \Delta\tilde{\beta}\Delta\tilde{\gamma} \tag{7.94}$$

and $\Delta\tilde{\alpha} = \tilde{\alpha} - \tilde{\alpha}_m$, $\Delta\tilde{\beta} = \tilde{\beta} - \tilde{\beta}_m = \Delta\tilde{\gamma} = \tilde{\gamma} - \tilde{\gamma}_m$. It is useful to express Δ as

$$\Delta = -\det \Delta\sigma \tag{7.95}$$

where

$$\Delta\sigma = \begin{bmatrix} \Delta\tilde{\beta} & -\Delta\tilde{\alpha} \\ -\Delta\tilde{\alpha} & \Delta\tilde{\gamma} \end{bmatrix} \tag{7.96}$$

is the difference between the actual beam matrix σ, and the matched beam matrix σ_m, given by

$$\sigma_m = \begin{bmatrix} \tilde{\beta}_m & -\tilde{\alpha}_m \\ -\tilde{\alpha}_m & \tilde{\gamma}_m \end{bmatrix} \tag{7.97}$$

Although all physics effects associated with beam mismatch cannot necessarily be described in terms of the mismatch factor alone, the mismatch factor is nevertheless a very useful parameter. It can be shown that the mismatch factor has the property that the maximum beam size for the mismatched beam is larger than that for a matched beam by a factor $1 + M$. Thus, a mismatched beam with $M = 0.1$ would have a 10% larger maximum projection.

Finally, we note that another definition of mismatch factor is found in the literature, [18] based on an increase in the effective phase-space area resulting from mismatch. With this definition, Eq. (7.93) is replaced by an equation of the form

$$M' = \frac{\Delta + \sqrt{\Delta(\Delta + 4)}}{2} \tag{7.98}$$

and when $\Delta \ll 1$, $M' \approx \sqrt{\Delta}$.

7.15
Current-Independent Beam Matching

Crandall has discussed the problems associated with beam matching in ion linacs, especially when there are few knobs or adjustable elements [19]. He considered both transverse and longitudinal matching and proposed guidelines for intertank matching that lead to matching solutions that are insensitive to beam current and emittance. These can be achieved by avoiding sharp discontinuities in the focusing systems along the linac. If the linac is designed in this way, it is found that a beam that is matched at the radio-frequency-quadrupole (RFQ) entrance will remain matched throughout the linac approximately independent of beam current and emittance.

This often implies that the initial accelerating gradient of a DTL that follows an RFQ must be reduced to make the longitudinal focusing strength comparable with that of the final RFQ focusing strength. Another consequence is that longitudinal matching between DTL tanks may require changes in the geometry of the cells at the ends. Fortunately, this method of designing linacs can be applied at zero beam current; space-charge forces do not need to be considered. Because the method results in approximate current-independent matching, the results of the zero-current analysis still apply at full beam current.

Figure 7.9 is a schematic depiction of the typical matched beam conditions at the interface between the RFQ and the first DTL tank, where the drift tubes contain quadrupole lenses (FODO), and the distance between the gap centers is $\beta\lambda$. The matched transverse ellipses at location A are similar to those at location C, which is at center of the first gap of the DTL. Figure 7.10 shows longitudinal matching between two DTL tanks.

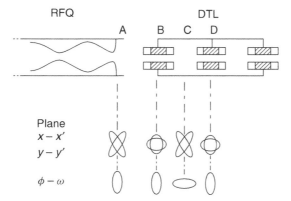

Figure 7.9 Matched phase-space configurations at the end of an RFQ and the beginning of a DTL (courtesy of K. R.Crandall).

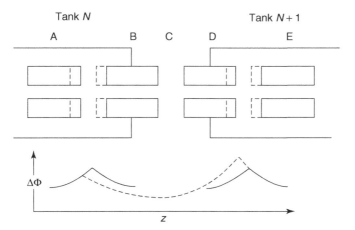

Figure 7.10 Longitudinal matching between two DTL tanks. Solid lines show normal drift tubes and phase profiles, and dashed lines show drift tubes and profiles as modified for matching (courtesy of K. R. Crandall).

7.16
Solenoid Focusing

The longitudinal magnetic field produced by a solenoid has often been used for ion and electron focusing. Comparison of a solenoid lens with a single quadrupole is not valid, since at least two quadrupoles are needed to produce focusing in both planes. Compared with quadrupole doublets or triplets of the same performance, it is usually found that the power dissipation in a solenoid is much higher. Solenoid focusing is a consequence of the interaction between the azimuthal velocity component induced in the entrance fringe-field region, and the longitudinal magnetic field component in the central region [20]. At the ends of the solenoid, the field flares out and there is an interaction between the radial field component and the axial velocity component, producing an azimuthal acceleration. In this way the particle acquires the azimuthal velocity component in the entrance fringe-field region. In the central region, particles traveling parallel to the field are unaffected, but those with an azimuthal velocity component will experience a force causing them to describe an orbit that is helical in space, and circular when viewed from the end of the solenoid.

The radial motion can be approximately described in the smooth approximation as

$$\frac{d^2r}{ds^2} + \left[\left(\frac{qB_{\text{rms}}}{2\gamma mv}\right)^2 - \frac{k_{\ell 0}^2}{2}\right]r = 0 \qquad (7.99)$$

where B_{rms} is the rms value along the axis, and $k_{\ell 0}$ is the wave number for longitudinal motion, defined earlier, which is the equation for a simple harmonic oscillator.

7.17
Smooth Approximation to Linac Periodic Focusing

To simplify the description of the periodic transverse focusing, it is useful to introduce a smooth approximation, which averages over the rapid flutter associated with the individual focusing lenses. There are different versions of the smooth approximation. In this section, an approximate formula for the transverse phase advance per focusing period is obtained by using the matrix method to form the transport matrix over one full period, and then treating both the RF gaps and the quadrupoles as thin lenses. Retaining only the lowest-order terms, we obtain simple approximate formulas. Recall that the thin-lens focal length for the RF gap is given by

$$\frac{1}{f_g} = \frac{\Delta x'}{x} = \frac{\pi q E_0 T \sin(-\phi) L}{mc^2 \lambda (\gamma \beta)^3} \quad (7.100)$$

where L is the length over which the RF impulse is integrated, which is the distance between the gaps. The quadrupole thin-lens focal length is given by

$$\frac{1}{f_Q} = \frac{\Delta x'}{x} = \pm \frac{q G \ell}{mc \gamma \beta} \quad (7.101)$$

For a FODO lattice, where the length L is equal the spacing between lenses, and the period is $2L$; we obtain to lowest order an expression for the phase advance per period

$$\sigma_0^2 \cong \left[\frac{q G \ell L}{mc \gamma \beta}\right]^2 - \frac{\pi q E_0 T \sin(-\phi)(2L)^2}{mc^2 \lambda (\gamma \beta)^3} \quad (7.102)$$

The first term is the quadrupole term, and the second is the RF-defocus impulse, which subtracts from the quadrupole term when ϕ is negative. For stability, the quadrupole term must be larger than the RF-defocusing term. We see that the RF-defocusing term falls off with $\beta \gamma$ faster than the quadrupole term, so that the RF defocusing is relatively more important at low velocities. The phase advance per unit length, $k_0 = \sigma_0/2L$, is an effective wave number for transverse or betatron oscillations. In a smooth approximation, the trajectory is sinusoidal, corresponding to simple harmonic motion, and the local flutter of the trajectories caused by the individual lenses is averaged out. The smoothed wave number is

$$k_0^2 = \left[\frac{\sigma_0}{2L}\right]^2 \cong \left[\frac{q G \ell}{2mc \gamma \beta}\right]^2 - \frac{\pi q E_0 T \sin(-\phi)}{mc^2 \lambda (\gamma \beta)^3} \quad (7.103)$$

Table 7.1 Comparison of lens and lattice properties

Lens	Lattice	1/f	Period	σ_0	k_0
Magnetic quad singlet	FODO	$\dfrac{qG\ell}{\gamma mv}$	$2L$	$\dfrac{qG\ell L}{\gamma mv}$	$\dfrac{\sigma_0}{2L} = \dfrac{qG\ell}{2\gamma mv}$
Magnetic quad singlet	FOFO-DODO	$\dfrac{qG\ell}{\gamma mv}$	$4L$	$\dfrac{2\sqrt{2}qG\ell L}{\gamma mv}$	$\dfrac{\sigma_0}{4L} = \dfrac{\sqrt{2}}{2}\dfrac{qG\ell}{\gamma mv}$
Magnetic solenoid	FO	$\left(\dfrac{qB}{2\gamma mv}\right)^2 \ell$	L	$\dfrac{qB\sqrt{\ell L}}{2\gamma mv}$	$\dfrac{\sigma_0}{L} = \dfrac{qB}{2\gamma mv}\sqrt{\dfrac{\ell}{L}}$
Magnetic quad doublet	FDO	$\dfrac{qG\ell}{\gamma mv}$	$L+D$	$\dfrac{qG\ell\sqrt{LD}}{\gamma mv}$	$\dfrac{\sigma_0}{L+D} = \dfrac{qG\ell}{\gamma mv}\dfrac{\sqrt{LD}}{L+D}$

This result shows that in lowest order the phase advance per unit length is independent of L and of the focusing period. But, as L increases, σ_0 increases and eventually σ_0 reaches the upper limit for stability at $\sigma_0 < \pi$.

In Table 7.1, we present thin-lens approximations for various lens and lattice properties. The phase-advance formulas for the quadrupole term are to be squared and added to the RF-defocus term to obtain $\sigma_0^2 \cong \sigma_Q^2 - k_{\ell 0}^2 P^2/2$. The formulas are generally accurate to within about 10% when $\sigma_0 < \pi/2$. The RF-defocus term in Eq. (7.103) does not change for a different lattice, but in Eq. (7.102), the period $2L$ must be replaced by the appropriate one, denoted by the symbol P in the table. For doublet focusing, the two interlens spacings, L and D, alternate along the lattice, one is larger and the other is smaller. Electric-quadrupole results can be obtained from any of the magnetic-quadrupole results by substituting E/va for the magnetic-quadrupole gradient G, where E is the pole-tip electric field and a is the aperture radius. Table 7.1 is useful for answering questions such as whether, based on overall focusing strength, a singlet or a doublet focusing lattice is preferred. The wave number $k_0 = \sigma_0/P$, which corresponds to the average of the inverse of the Courant–Snyder β function, is a measure of the focusing strength; the average beam size decreases with increasing k_0. Note that if σ_0 is fixed, the largest k_0 corresponds to the lattice with the shortest period.

7.18
Radial Motion for Unfocused Relativistic Beams

The radial RF forces approach zero for a relativistic beam, because in that limit the RF magnetic force cancels the RF electric force. Consequently, transverse focusing is not as important for low-intensity relativistic-electron linacs. For high-intensity electron beams, focusing is required to compensate for space-charge effects at low energies, and for controlling the beam-breakup instability at higher energies. In this section, we consider the radial trajectories for electrons in a linac that experience no RF transverse fields and no transverse focusing. With no transverse fields, the divergence angle of each particle will decrease as the longitudinal momentum increases. Suppose we let $\beta = 1$, and if there is no transverse force, Newton's law gives

$$\frac{d}{dz}\left(\gamma \frac{dx}{dz}\right) = 0 \quad (7.104)$$

or

$$\gamma \frac{dx}{dz} = \gamma x' = \text{constant} \quad (7.105)$$

which is a statement of transverse momentum conservation. We assume that the injected beam at $z = 0$ is relativistic, and we write the initial values as $\gamma = \gamma_0$, and $x' = x'_0$. Therefore transverse momentum conservation implies $\gamma x' = \gamma_0 x'_0$. Assuming the particle energy increases linearly with z, we write

$$\gamma = \gamma' z + \gamma_0 \quad (7.106)$$

Now we integrate Eq. (7.105) and obtain the total transverse displacement at the end of the linac, resulting from the particle divergence,

$$\Delta x = \int_0^z x' \, dz = \gamma_0 x'_0 \int_0^z \frac{dz}{\gamma} = \frac{\gamma_0 x'_0}{\gamma'} \ln\left[\frac{\gamma}{\gamma_0}\right]$$

$$= x'_0 z \frac{W_0}{W - W_0} \ln\left[\frac{W}{W_0}\right] \quad (7.107)$$

Because transverse momentum is conserved, the divergence angle decreases as the beam is accelerated. The trajectory is a logarithmic function of energy, as shown in Fig. 7.11. One finds that for $z = 3$ km (the length of the SLAC accelerator) and for an initial maximum divergence of 0.25 mrad at the energy $W_0 = 40$ MeV, the change in displacement at a final energy of $W = 40$ GeV is only $\Delta x = 5$ mm. This is a remarkably small deflection, and additional insight is obtained by seeing what the accelerator looks like in the moving beam frame. Because of Lorentz contraction, if dz is a longitudinal element of length in the laboratory frame, where the accelerator is at rest, the element in the moving frame dz_m is

$$dz_m = dz/\gamma(z) \quad (7.108)$$

The length of the accelerator ℓ_m in the moving frame is given in terms of the length ℓ in the laboratory frame by

$$\ell_m = \int_0^{\ell_m} dz_m = \int_0^\ell \frac{dz}{\gamma' z + \gamma_0} = \frac{mc^2 \ell}{W - W_0} \ln\left[\frac{W}{W_0}\right] \quad (7.109)$$

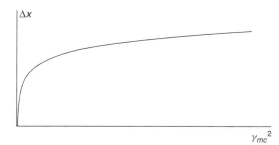

Figure 7.11 Transverse deflection versus energy in relative units for an accelerated relativistic beam.

For the SLAC linac using $\ell = 3$ km, $W_0 = 40$ MeV, and $W = 40$ GeV, we find $\ell_m = 0.27$ m. Because the accelerator appears so short in the moving frame, it is not as surprising that the transverse displacement of the particles is so small.

Problems

7.1. A quadrupole magnet of effective length $\ell = 3.25$ cm is operated at a gradient of 100 T/m, and is used for focusing a 5-MeV proton beam. Calculate the focusing strength K and the thin-lens focal length.

7.2. A 5-MeV proton is focused by the quadrupole magnet of Problem 7.1. The initial coordinates of the proton at the entrance plane of the quadrupole are $x = 3$ mm, $x' = 0$, $y = 0$, and $y' = 0$. Calculate the final divergence angle x' at the exit plane of the quadrupole, using the following assumptions: (a) The quadrupole has constant field over the effective length and is focusing in the x plane. (b) The quadrupole has constant field over the effective length and is defocusing in the x plane. (c) The quadrupole obeys the thin-lens approximation and is focusing in the x plane.

7.3. Consider a FODO lattice composed of the same beam and quadrupoles as in Problems 7.1 and 7.2. Assume the thin-lens approximation is valid, and assume the center-to-center spacing of the quadrupoles is 10 cm. (a) Calculate the phase advance per focusing period σ of the beam particles. (b) Calculate the Twiss parameters $\tilde{\alpha}$, $\tilde{\beta}$, and $\tilde{\gamma}$ of the trajectory ellipses in the middle of a focusing lens. (c) Repeat the Twiss parameter calculation in the middle of a defocusing lens. (d) Repeat the Twiss parameter calculation in the middle of the two drift spaces.

7.4. Assume the emittance in the x plane of the ellipse on which the largest-amplitude particle lies is $\varepsilon_1 = 30$ mm-mrad (3×10^{-5} m-rad). (a) Use the values of the Twiss parameters calculated in Problem 7.3 to tabulate the projections $\sqrt{\varepsilon_1 \tilde{\beta}}$ and $\sqrt{\varepsilon_1 \tilde{\gamma}}$ and the intercepts $\sqrt{\varepsilon_1/\tilde{\beta}}$ and $\sqrt{\varepsilon_1/\tilde{\gamma}}$ at each of the four locations in Problem 7.3, beginning at the middle of the focusing lens. How large must the aperture be in the quadrupoles so that the largest-amplitude particle is transported? **(b)** Use the projections and intercepts calculated in part **(a)** to make a rough sketch of the ellipse at each of the four locations (middle of the lenses and of the drift spaces), beginning with the middle of the thin focusing quadrupole, with the same scales for all plots.

7.5. Suppose the particle with the maximum amplitude of Problem 7.4 has coordinates $x = \sqrt{\varepsilon_1 \tilde{\beta}} = 3.22$ mm and $x' = 0$ at the middle of a thin focusing lens. **(a)** Use the transfer matrix through a period to calculate the particle coordinates x and x' at the middle of the next thin focusing lens. **(b)** Plot these two points on the ellipse at the thin focusing lens sketched for Problem 7.4. The phase advance from one focusing lens to the following one was calculated for Problem 7.3. Do the plotted coordinates look qualitatively consistent with this?

7.6. A FODO array of permanent-magnet quadrupole lenses with gradient $G = 100$ T/m and effective length $\ell = 3$ cm is used to focus a low-current proton beam in a 400-MHz DTL. Assume the array is periodic with lens spacing equal to $\beta\lambda$ (quadrupoles in every drift tube) at the injection energy of 2.5 MeV. **(a)** Use the smooth approximation formula in Section 7.17 for transverse phase advance per focusing period σ to calculate σ for the quadrupoles alone (as if $E_0 T = 0$). **(b)** Calculate σ when $E_0 T = 1$ MV/m and $\phi = -30°$. Are the particles stable transversely? **(c)** For the same $-30°$ phase and the same quadrupole array, what is the maximum accelerating field $E_0 T$ for transverse stability?

References

1 Adlam, J.H., AERE-GP/M-146, Harwell, Berkshire, England, 1953; Good, M.L., *Phys. Rev.* **92**, 538 (1953).
2 McMillan, E.M., *Phys. Rev.* **80**, 493 (1950).
3 Smythe, W.R., *Static and Dynamic Electricity*, 2nd ed., McGraw-Hill, 1950, pp. 13–14.
4 Stratton, J.A., *Electromagnetic Theory*, McGraw-Hill, 1941, p. 116.
5 This section outlines a treatment carried out by Gluckstern, R.L. private communication.
6 Lapostolle, P.M., Meot, F., Valero, S. and Tanke, E., Proc. 1990 Linear Accel. Conf., Albuquerque, NM, September 10–14, 1990, LA-112004-C, p. 315.
7 Carne, A., Lapostolle, P., Prome, M. and Schnizer, B. in *Linear Accelerators*, ed. Lapostolle, P.M. and Septier, A., North Holland, Amsterdam, 1970, pp. 747–783.
8 Prome, M., thesis presented at Universite-Paris-Sud, Centre d'Orsay (1971), English translation in Los Alamos Report LA-R-79-33, 1979.
9 Lapostolle, P.M. *Proton Linear Accelerators: A Theoretical and Historical Introduction*, Los Alamos Report LA-11601-MS, July, 1989.
10 Courant, E.D., Livingston, M.S. and Snyder, H.S., *Phys. Rev.* **88**, 1197 (1952); Courant, E.D. and Snyder, H.S. *Ann. Phys.* **3**, 1–48 (1958).
11 Blewett, J.P., *Phys. Rev.* **88**, 1197 (1952).
12 Teng, L.C., *Rev. Sci. Instrum.* **25**, 264 (1954).
13 Smith, L. and Gluckstern, R.L., *Rev. Sci. Instrum.* **26**, 220 (1955).

14 Brown, K.L. and Servranckx, R.V., AIP Conf. Proc. **127**, 102, 103 (1985).
15 Prome, M., in *Linear Accelerators*, Lapostolle, P.M. and Septier, A.L., North-Holland Publishing and John Wiley & Sons, 1970, pp. 790–794; Hereward, H.G. and Johnsen, K., *CERN Symp. High Energy Accel. Pion Phys.*, 1956, p. 167.
16 Crandall, K.R., *Linear Error Analysis of Beam Transport Systems*, AOT Division Technical Note No. LA-CP-96-16, January 22, 1966.
17 Crandall, K.R., *TRACE 3-D Documentation*, 2nd ed., Los Alamos Report LA-UR-90-4146, December, 1990.
18 Guyard, J. and Weiss, M., Proc. 1976 Proton Linear Accel. Conf., Chalk River, Ontario, Canada.
19 Mills, R.S., Crandall, K.R. and Farrell, J.A., *Design Self-Matching Linacs*, Proc. of 1984 Linear Accel. Conf., Seeheim, Germany, 7-11, 1984, GSI-84-11, pp. 112–114.
20 Reiser, M., *Theory and Design of Charged Particle Beams*, John Wiley & Sons, 1994, p. 98.

8
Radiofrequency Quadrupole Linac

The radiofrequency quadrupole or RFQ is relatively a new type of linear accelerator, and its recent development was a major innovation in the linac field. The RFQ is especially well suited for the acceleration of beams with low velocities in the typical range of about 0.01 to 0.06 times of the speed of light. As such, it is an important accelerator for ions, but not for electrons, which are already emitted from a typical electron-gun source with velocities approaching half the speed of light. All high-energy beams begin at low velocities, and the acceleration technique used at low velocities is an essential contributor to the overall accelerator performance. The RFQ can be used to accelerate high-current proton beams to several megaelectron volts, and can serve as the initial linac structure for a linac system that produces even higher energies. In this chapter, we present the principles of RFQ operation including the vane geometry and the potential function, from which we obtain analytic formulas for the fields that describe the beam dynamics. This is followed by a discussion of the design procedures including adiabatic bunching. We discuss the properties of the four-vane and four-rod structures and introduce an approach on the basis of perturbation theory to understand the effects of field errors.

8.1
Principles of Operation

The principles of operation of the RFQ were first presented by the inventors, Kapchinskiy and Tepliakov (K–T) in their 1969 publication [1]. The RFQ four-vane structure is shown in Fig. 8.1. K–T proposed to modify the shapes of the four electrodes of an RF quadrupole to achieve both acceleration and focusing from RF electric fields. By using a potential function description, K–T showed how to shape the electrodes to produce the fields required by the beam. The achievement of practicable means of applying velocity-independent electric focusing in a low-velocity accelerator gave the RFQ a significant strong-focusing advantage compared with conventional linacs that

RF Linear Accelerators. 2nd, completely revised and enlarged edition.
Thomas P. Wangler
Copyright © 2008 Wiley-VCH Verlag GmbH & Co. KGaA, Weinheim
ISBN: 978-3-527-40680-7

8.1 Principles of Operation | 233

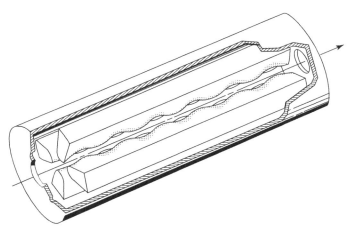

Figure 8.1 Four-vane RFQ accelerator section. The four electrodes are excited with quadrupole-mode RF voltages to focus the beam. The electrodes are modulated to produce longitudinal electric fields to accelerate low-velocity ions.

used velocity-dependent magnetic lenses. This allowed the RFQ to extend the practical range of operation of ion linacs to low velocities, thus eliminating the need for large, high-voltage dc accelerators for injection of the beam into the linac. In a later publication [2], K–T showed how to introduce specific slow variations of the RFQ parameters to bunch the beam adiabatically. This allowed the beam to be injected into the RFQ and to be bunched over many spatial periods, while the beam is contained transversely by the electric-quadrupole forces. Adiabatic bunching allows a large fraction of the beam to be captured, and converted into stable bunches that can be accelerated efficiently to the final energy. Adiabatic bunching result in very compact bunches with minimal tails in longitudinal phase space, and increases the beam-current capacity, because it avoids unnecessary longitudinal compression of the beam at low velocities, which would increase the transverse space-charge effects.

In the RFQ, although no drift tubes or magnetic quadrupole lenses are used, the beam is accelerated by longitudinal RF electric fields, and transversely focused by RF electric-quadrupole fields that are determined by the electrode geometry of the RFQ structure. These electrodes are described as either rods or vanes depending on the type of geometry. The operating principles of the RFQ are most easily explained by first describing how the transverse focusing is accomplished. Suppose we consider four equally spaced conducting electrodes, symmetrically placed about the beam axis, as illustrated in Fig. 8.2. Suppose an ac voltage is applied to each electrode, whose polarity changes sign as we move from one electrode to the next.

Thus, a voltage $\pm V_0 \cos(\omega t)/2$ is applied to the four electrodes in a quadrupolar-symmetric pattern. Off-axis particles experience a transverse

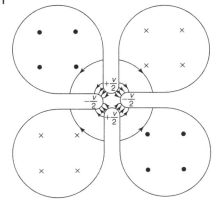

Figure 8.2 Electric-quadrupole cross section with electric-field lines in the transverse plane and magnetic-field lines in the longitudinal direction, perpendicular to the page.

electric field that alternates as a function of time. The particles experience a time-varying electric field with a quadrupole pattern that provides the focusing. An advantage of the RFQ focusing, compared with the more conventional magnetic focusing is that for low-velocity particles, the electric force on a particle is stronger than the magnetic force. Other advantages of the RFQ focusing will be appreciated after the more detailed treatment that follows.

If the electrode geometry is uniform along the axis, there is no axial electric-field component for acceleration of the particles. To see how the geometry can be modified to obtain a nonzero longitudinal electric-field component, consider the geometry shown in Fig. 8.3, where the transverse displacements relative to the axis of the horizontal and vertical electrodes are unequal. The potential on the axis will be nonzero, because it is influenced more by the electrodes that are nearer to the axis than those of the opposite polarity that are further away. If the perturbed geometry of Fig. 8.3 is maintained constant along the axial direction, there will still be no axial electric field, because there is no change of the axial potential. However, if the transverse-electrode displacements are varied along the axis, the axial potential will change as a

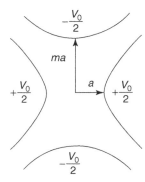

Figure 8.3 Electric-quadrupole geometry with unequal electrode spacing.

function of the longitudinal position, and a corresponding axial electric field will be produced.

Suppose we consider the voltages at the time when they have their maximum value, and suppose the transverse displacement of the electrodes is varied in a sinusoidal-like pattern. If the horizontal and the vertical electrode displacements are out of phase spatially, as shown in Fig. 8.4, one would expect a sinusoidal voltage distribution along the axis, depending on the relative displacements of the horizontal and vertical electrodes. Indeed, the axial potential is maximum at the location where the electrodes have their extreme displacements, and the axial electric field, shown in Fig. 8.5, is maximum at the locations halfway between these extreme-displacement points, where the gradient of the potential is maximum and where the four electrodes also have equal displacements.

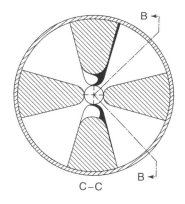

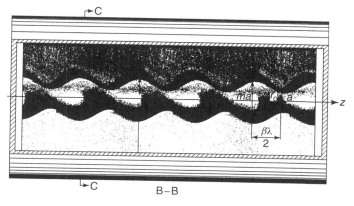

Figure 8.4 Two views of an RFQ accelerator showing the arrangement of the four electrodes, and the unit cell of length $\beta_s \lambda/2$ along with the parameters a, ma, and r_0 that are defined in the text.

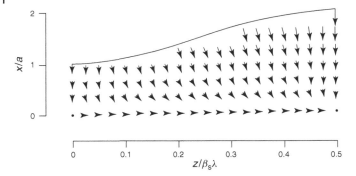

Figure 8.5 Electric-field vectors in the $x - z$ plane over half an electrode modulation cycle. This shows how the pole-tip modulation creates a longitudinal electric-field component.

If the potential on the electrodes did not depend on time, beam particles would be accelerated and decelerated by equal amounts, and the structure would not be a practical accelerator. With time-dependent fields, what is needed to obtain a sustained acceleration of the beam particles is to produce an axial-field pattern that results in synchronism between the accelerating fields and the particle motion. If the pattern of the electrode displacements changes through half a period during the time that the electrode voltages change sign, the axial voltage will maintain the correct sign for sustained acceleration of a synchronous particle. The spatial period of the electrode displacements must match the axial distance traversed by a synchronous particle during one RF period.

8.2
General Potential Function

Next, we analyze the RF quadrupole structure, using the quasistatic approximation. The quasistatic approximation, described in Section 5.14, is valid when the main source of the RF electric field is the electric charge on the four conducting electrodes and the contribution to the RF electric field from the time-changing magnetic field, as given by Faraday's law, can be ignored. It can be shown that the quasistatic approximation is valid, when the electrode displacements from the axis are small compared with the RF wavelength. Because of the particle-velocity increase, the vane geometry is not exactly periodic along the axis. However, it is convenient to make the approximation that the structure is periodic that corresponds to assuming that the rate of acceleration is small, compared with the distance the particle moves in one RF period. Figure 8.6 shows a cut through the $x - z$ plane, and shows the mirror symmetry of the opposite poles about the beam axis. The minimum radius of the electrode tips is a, the maximum radius is ma, where

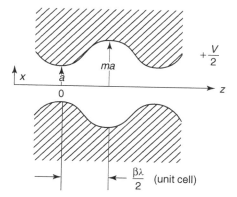

Figure 8.6 RFQ pole-tip geometry.

m is called the *modulation parameter*. The unit cell has a length $\beta\lambda/2$, and the longitudinal electric-field lines flow between the tip of the x electrode near $z = 0$ to the y tip that has a minimum radius at $z = \beta\lambda/2$. There are two unit cells per modulation period, and adjacent unit cells will have oppositely directed longitudinal electric fields. At a given time, every other cell can contain a bunch for acceleration.

Using cylindrical-polar coordinates, the time-dependent scalar potential is written as

$$U(r, \theta, z, t) = V(r, \theta, z) \sin(\omega t + \phi) \tag{8.1}$$

where $\omega/2\pi$ is the RF frequency, ϕ is the initial phase of the potential, and $V(r, \theta, z)$ is a solution of Laplace's equation, given in cylindrical coordinates by

$$\frac{\partial^2 V}{\partial r^2} + \frac{1}{r}\frac{\partial V}{\partial r} + \frac{1}{r^2}\frac{\partial^2 V}{\partial \theta^2} + \frac{\partial^2 V}{\partial z^2} = 0 \tag{8.2}$$

where in Cartesian coordinates $x = r\cos\theta$ and $y = r\sin\theta$. When the electrode displacements vary along the z-axis, we are interested in 3-D solutions that depend on z. Otherwise we are interested in 2-D solutions that are independent of z. Using the method of separation of variables for each case and adding the two solutions, one obtains

$$V(r, \theta, z) = \sum_{s=0}^{\infty} A_s r^{2(2s+1)} \cos(2(2s+1)\theta)$$
$$+ \sum_{n=1}^{\infty} \sum_{s=0}^{\infty} A_{ns} I_{2s}(knr) \cos(2s\theta) \sin(knz) \tag{8.3}$$

Eq. (8.3) is the general K–T potential function, from which the electric field in the vicinity of the beam may be calculated.

8.3
Two-Term Potential Function Description

As a first impression, the RFQ may look like a complicated three-dimensional structure, for which a simple analytical form for the fields would be difficult to obtain. An analytical solution would have great value, because it would allow us to see immediately how the fields depend on the geometry, and this would greatly simplify our task of choosing optimized electrode geometries. Yet, we have written a general solution for the potential, which has an infinite number of terms, and one wonders whether this is of any practical use. An arbitrary electrode geometry will generally require many terms from the general solution of the potential for an accurate description of the potential and the fields. However, if we could choose the electrode geometry, so that only a few terms are large near the beam, the potential description would be greatly simplified. One way to obtain a self-consistent solution is to select only the lowest-order terms from the general solution for the potential, and then construct the electrode shapes that conform to the resulting equipotential surfaces. This is the approach that was used by K–T in their original paper. They chose the pure quadrupole term ($s = 0$) from the first summation of Eq. (8.3) and the monopole term ($s = 0$, $n = 1$) from the second summation. We call this the *two-term potential function*, which is the starting point for our discussion of the RFQ dynamics, and we write it as

$$V(r, \theta, z) = A_0 r^2 \cos(2\theta) + A_{10} I_0(kr) \cos(kz) \tag{8.4}$$

where A_0 and A_{10} are constants that are determined by the electrode geometry, and $k = 2\pi/L$, where L is the period of the electrode modulation. We choose $L = \beta_s \lambda$, where β_s is the velocity of the synchronous particle to satisfy the requirement of synchronism, and we approximate the modified Bessel function as $I_0(v) \cong 1 + v^2/4$. We determine the two constants A_0 and A_{10} as follows. Suppose that at some time the horizontal and vertical vane potentials are $+V_0/2$ and $-V_0/2$, respectively. At $z = 0$, where the modulation term has its maximum value, we express the horizontal ($\theta = 0$) vane displacement as $r = a$, and the vertical vane displacement as $r = ma$, where we assume that $m \geq 1$. At $\theta = 0$ the two-term potential function is evaluated at the horizontal vane tip, which gives

$$\frac{V_0}{2} = A_0 a^2 + A_{10} I_0(ka) \tag{8.5}$$

and at $\theta = \pi/2$ the potential is evaluated at the vertical vane tip, which gives

$$-\frac{V_0}{2} = -A_0(ma)^2 + A_{10} I_0(kma) \tag{8.6}$$

8.3 Two-Term Potential Function Description

Solving Eqs. (8.5) and (8.6) for the constants A_0 and A_{10} yields

$$A_0 = \frac{V_0}{2a^2} \frac{I_0(ka) + I_0(kma)}{m^2 I_0(ka) + I_0(kma)} \quad (8.7)$$

and

$$A_{10} = \frac{V_0}{2} \frac{m^2 - 1}{m^2 I_0(ka) + I_0(kma)} \quad (8.8)$$

It is convenient to define the dimensionless constants X and A as

$$X = \frac{I_0(ka) + I_0(kma)}{m^2 I_0(ka) + I_0(kma)} \quad (8.9)$$

$$A = \frac{m^2 - 1}{m^2 I_0(ka) + I_0(kma)} \quad (8.10)$$

Then $A_0 = XV_0/2a^2$ and $A_{10} = AV_0/2$. The complete time-dependent potential is

$$U(r, \theta, z, t) = \frac{V_0}{2}\left[X\left[\frac{r}{a}\right]^2 \cos(2\theta) + AI_0(kr)\cos(kz)\right]\sin(\omega t + \phi) \quad (8.11)$$

where the time-dependent voltages on the horizontal and vertical electrodes are $+V_0 \sin(\omega t + \phi)/2$ and $-V_0 \sin(\omega t + \phi)/2$, respectively. It is convenient to express Eq. (8.11) in Cartesian coordinates $x = r\cos\theta$ and $y = r\sin\theta$, which results in

$$U(x, y, z, t) = \frac{V_0}{2}\left[\frac{X}{a^2}[x^2 - y^2] + AI_0(kr)\cos(kz)\right]\sin(\omega t + \phi) \quad (8.12)$$

In principle, the geometry of the electrodes is now specified from the $\pm V_0/2$ equipotential surfaces. The transverse cross sections are approximately hyperbolas. At $z = \beta_s\lambda/4$, half way through the unit cell, the RFQ has exact quadrupole symmetry, and the tips of the x and y electrodes have a radius equal to $r_0 = aX^{-1/2}$. In practice, the electrode contours must deviate from the ideal shape to control the peak surface electric field, and to facilitate the machining. However, the electrode tips at $\theta = 0$ and π for the horizontal electrodes, and at $\pi/2$ and $3\pi/2$ for the vertical electrodes, can still be chosen to correspond to the correct equipotential curves of the two-term potential function. From Eq. (8.12), the x tip at $y = 0$ is given by

$$1 = \frac{X}{a^2}x^2 + AI_0(kx)\cos(kz) \quad (8.13)$$

and the y tip at $x = 0$ is given by

$$-1 = -\frac{X}{a^2}y^2 + AI_0(ky)\cos(kz) \quad (8.14)$$

Away from the tips, different transverse profiles can be used, including circular arcs with the same local radius of curvature as at the tip, or with a constant transverse radius of curvature throughout the entire cell.

The peak surface electric field is important, because the probability of electric breakdown increases with increasing surface field. For a circular transverse-electrode geometry with no longitudinal modulations, the peak surface electric field does not occur at the electrode tip, but at the point where the electrodes have minimum separation. The field at the tip is V_0/r_0 and the peak field is $E_s = \alpha V_0/r_0$, where $\alpha = 1.36$. For modulated poles, the value of α is slightly modified by the dependence on the transverse and longitudinal radii of curvature. The spatial dependence of the potential function within a cell is completely determined by the three parameters X, A, and k, which vary from cell to cell along the RFQ, and must be chosen to provide the desired RFQ performance.

For purposes of numerical beam dynamics calculations in the four-vane RFQ structure, eight multipoles are often used to define the fields in an RFQ. The eight-term potential function is:

$$U(r, \theta, z) = \frac{V}{2}\left\{ A_{01}\left(\frac{r}{r_0}\right)^2 \cos 2\theta + A_{03}\left(\frac{r}{r_0}\right)^6 \cos 6\theta \right.$$

$$+ A_{10} I_0(kr) \cos kz + A_{30} I_0(3kr) \cos 3kz$$

$$+ [A_{12} I_4(kr) \cos kz + A_{32} I_4(3kr) \cos 3kz] \cos 4\theta$$

$$\left. + [A_{21} I_2(2kr) \cos 2\theta + A_{23} I_6(2kr) \cos 6\theta] \cos 2kz \right\}$$

(8.15)

In the above equation, $k = \pi/L$, where L is the length of the cell. The A_{mn} coefficients, which depend on m and L/r_0, have been calculated and tabulated for 'normal' cells having several vane-tip geometries. These coefficients vary along the RFQ, and are defined at the end of each cell.

8.4
Electric Fields

The particle dynamics is determined from the electric-field components. The RF magnetic field is small in the vicinity of the beam, and produces only a small force on nonrelativistic particles. The electric field may be obtained from the gradient of the potential function, or $\mathbf{E} = -\nabla U$. By differentiating Eq. (8.12), the components in Cartesian coordinates are

$$E_X = -\frac{XV_0}{a^2}x - \frac{kAV_0}{2}I_1(kr)\frac{x}{r}\cos(kz) \qquad (8.16)$$

$$E_\gamma = \frac{XV_0}{a^2}y - \frac{kAV_0}{2}I_1(kr)\frac{y}{r}\cos(kz) \quad (8.17)$$

$$E_z = \frac{kV_0A}{2}I_0(kr)\sin(kz) \quad (8.18)$$

where each component above is multiplied by $\sin(\omega t + \phi)$. We will be interested in the lowest-order approximation for the modified Bessel function, $I_1(v) \cong v/2$.

The E_z component in Eq. (8.18) provides the accelerating force on the beam. The first term of both Eqs. (8.16) and (8.17) is associated with quadrupole focusing, and the second term leads to the transverse RF defocusing force that acts on the beam, when the phase ϕ is chosen for longitudinal focusing. We call A the acceleration efficiency and X the focusing efficiency. When $m = 1$, $A = 0, X = 1$ and the RFQ becomes a pure quadrupole transport channel with no acceleration. As m increases an accelerating field is produced on axis. The quantity XV_0/a^2 is the quadrupole gradient that is a measure of the quadrupole focusing strength. The most accurate way to calculate the beam dynamics in the RFQ is by numerical integration of the equations of motion through these electric fields. However, considerable insight is obtained by making some simple approximations to examine the effect of these electric fields on the beam.

8.5
Synchronous Acceleration

In the approximation that the radial position and velocity of a particle within a unit cell are constant, the energy gain ΔW for a particle with arbitrary normalized velocity β' can be calculated by replacing $\omega t = 2\pi z/\beta'\lambda$, and integrating the electric field seen by the particle, $E_z \sin(\omega t + \phi)$, over the unit cell. We obtain

$$\Delta W = \frac{qkAV_0I_0(kr)}{2}\int_0^\ell \sin(kz)\sin(k'z + \phi)\,dz \quad (8.19)$$

where $k' = 2\pi/\beta'\lambda$, $k = 2\pi/\beta_s\lambda$, and $\ell = \beta_s\lambda/2$ is the length of the unit cell. Maximum acceleration occurs when the particle travels from the center of one cell to the center of the next in the time that the field reverses its polarity that is exactly a half of an RF period. This is the synchronous condition, and the particle that satisfies this condition is a synchronous particle. Clearly, synchronism requires that $\beta' = \beta_s$, and the energy gain of the synchronous particle is

$$\Delta W = \frac{q\pi AV_0 I_0(kr)\cos\phi}{4} \quad (8.20)$$

The phase ϕ when the particle is at the center of the unit cell is called the *particle phase*; at $\phi = 0$, the electric field is at the peak value. The synchronous-energy gain can be written in a form that corresponds to more conventional linac terminology. First, we calculate the spatial average of the peak axial accelerating field over the RFQ unit cell:

$$E_0 = \frac{1}{\ell} \int_0^\ell E_z \, dz = \frac{2AV_0}{\beta\lambda} \tag{8.21}$$

The resulting average axial field can be interpreted as the ratio of an effective axial voltage AV_0 applied over the length ℓ of the unit cell. The transit-time factor for the synchronous particle is

$$T = \frac{\int_0^\ell E_z \sin(kz) \, dz}{\int_0^\ell E_z \, dz} = \frac{\pi}{4} \tag{8.22}$$

Using these results the energy gain of a synchronous particle per cell of length ℓ has the familiar form

$$\Delta W = q E_0 T I_0(kr) \ell \cos\phi_s \tag{8.23}$$

8.6
Longitudinal Dynamics

The RFQ electrode geometry and fields are chosen to produce a specific velocity gain for a synchronous particle that will continuously gain energy along the RFQ. When the synchronous phase is chosen to correspond to the time when the field is increasing, particles that arrive earlier than the synchronous particle experience a smaller field, and those that arrive later experience a larger field. As was described in Chapter 6, this produces the longitudinal restoring force that provides phase-stable acceleration. The oscillating particles constitute a bunch, and every period in the RFQ contains a single bunch.

It is straightforward to obtain the equations that describe the longitudinal focusing. First, we calculate the energy gain of a particle of phase ϕ, and subtract the energy gain of the synchronous particle, assumed to be on-axis with phase ϕ_s. On expressing this result in differential form as an average rate of change of the relative energy $W - W_s$, we obtain

$$\frac{d(W - W_s)}{dz} = q E_0 T (I_0(kr) \cos\phi - \cos\phi_s) \tag{8.24}$$

where the independent variable z is the axial coordinate. Second, we write the equation that describes the average rate of change of the phase difference between an arbitrary particle and the synchronous particle, as

$$\frac{d(\phi - \phi_s)}{dz} = -\frac{2\pi (W - W_s)}{mc^2 \beta_s^3 \lambda} \tag{8.25}$$

When $-\pi < \phi_s < 0$, one obtains simple harmonic motion at a longitudinal wavenumber

$$k_l^2 = \frac{\pi^2 q A V_0 I_0(kr) \sin(-\phi_s)}{mc^2 \beta_s^4 \lambda^2} \tag{8.26}$$

One finds that Eq. (8.26) can be put in exactly the same form as Eq. (6.43), the expression for longitudinal wavenumber in conventional linacs. By comparison with standard linear accelerator theory, one finds that the longitudinal dynamics of the RFQ are really the same as in conventional linacs.

8.7
Transverse Dynamics

Perhaps the most fundamental impact of the RFQ, as it affects the beam dynamics of low-velocity ions, results from the increased transverse-focusing strength, compared with other methods. The capability of the RFQ to provide this focusing is the reason why the RFQ can be operated at low velocities, and this enables some very attractive features of the RFQ, such as adiabatic bunching, and a high beam-current limit. The focusing or confinement of the off-axis particles in the RFQ is an example of the alternating gradient-focusing principle, already discussed in Chapter 7, and which is the basic beam-focusing principle used in most modern accelerators. The RFQ focusing is different than that described in Chapter 7, only because the polarity of the focusing field varies in time instead of in space.

Now we examine the transverse-focusing properties of the RFQ more quantitatively, using the electric-field components derived in Section 8.4. Suppose we consider the motion in the x plane for a particle with charge q and mass m, and limit ourselves to small displacements from the axis. Nonrelativistically, the equation of motion is

$$\ddot{x} + \left[\frac{qXV_0}{ma^2} + \frac{qk^2 A V_0}{4m} \cos(kz)\right] x \sin(\omega t + \phi) = 0 \tag{8.27}$$

where $\ddot{x} = d^2x/dt^2$. The first term in brackets is the quadrupole term, and the second term is the transverse field associated with the electrode modulation that produces the longitudinal fields, that is, the familiar RF defocus term. Because the time dependence of the axial position of a particle is given by $kz = \omega t$, the second term in brackets is proportional to

$$\cos(\omega t) \sin(\omega t + \phi) = [\sin \phi + \sin(2\omega t + \phi)]/2 \tag{8.28}$$

The term with twice the RF frequency goes through one complete cycle as the synchronous particle moves through a unit cell, and if x is assumed to be constant throughout the cell, this term averages to zero. As a first approximation we ignore the contribution of this term, and we write the equation of motion as

$$\ddot{x} + \left[\frac{qXV_0}{ma^2}\sin(\omega t + \phi) + \frac{qk^2 AV_0}{8m}\sin\phi\right]x = 0 \qquad (8.29)$$

This result has the form of the well-known Mathieu equation. A smooth-approximation solution is obtained by considering a trial solution of the form

$$x = [C_1 \sin\Omega t + C_2 \cos\Omega t][1 + \varepsilon\sin(\omega t + \phi)] \qquad (8.30)$$

where C_1 and C_2 are constants, and Ω and ε are two new parameters, such that $\Omega \ll \omega$, and $\varepsilon \ll 1$. The factor in the first bracket is assumed to vary slowly over the unit cell, and represents the average or smoothed particle trajectory. The angular frequency Ω is the oscillation frequency for the smoothed motion, known as the *betatron frequency*. The factor in the second bracket is a periodic function of time, which varies at the RF frequency, and is called the *flutter factor*, where ε is the flutter amplitude. Values for the two parameters Ω and ε are obtained from the theory. For simplicity we choose $C_1 = 1$ and $C_2 = 0$, differentiate the trial solution twice, and neglecting the smaller terms of order $\varepsilon\Omega/\omega$ and Ω^2/ω^2, we find

$$\ddot{x} \cong -\varepsilon\omega^2 \sin(\Omega t)\sin(\omega t + \phi) \qquad (8.31)$$

An approximate solution of Eq. (8.29) is obtained by choosing the flutter amplitude

$$\varepsilon \cong \frac{qXV_0}{m\omega^2 a^2} = \frac{1}{4\pi^2}\frac{qXV_0\lambda^2}{mc^2 a^2} \qquad (8.32)$$

Next, we look at the smoothed properties of the motion, by substituting Eq. (8.30) into the equation of motion, Eq. (8.29), and averaging over an RF period. We find that

$$\ddot{\bar{x}} \cong -\Omega^2 \sin(\Omega t) \qquad (8.33)$$

Thus, the smooth-approximation solution implies that the average particle displacement satisfies the equation of a simple harmonic oscillator

$$\ddot{\bar{x}} + \Omega^2 \bar{x} \qquad (8.34)$$

where

$$\Omega^2 \cong \frac{1}{2}\left[\frac{qXV_0}{m\omega a^2}\right]^2 + \frac{qk^2 V_0 A\sin\phi}{8m} \qquad (8.35)$$

The first term is always positive and represents the contribution of the quadrupole focusing, and the second term represents the RF defocus term.

If the quadrupole term is large compared with the RF defocus term that depends on the acceleration efficiency A, the transverse motion becomes decoupled from the longitudinal motion, and Ω is approximately the same for particles of all phases. While the magnitude of Ω^2 is a measure of the effective focusing force, it is customary to refer to the betatron phase-advance per focusing period, defined as $\sigma_0 = \Omega\lambda/c$. Substituting the expression for σ_0 into Eq. (8.35), we obtain

$$\sigma_0^2 \cong \frac{1}{8\pi^2}\left[\frac{qXV_0\lambda^2}{mc^2a^2}\right]^2 + \frac{\pi^2 qAV_0\sin(\phi)}{2mc^2\beta^2} \tag{8.36}$$

It is found that the stability criteria depend on σ_0. In the smooth approximation, the beam is stable when $\sigma_0^2 > 0$. When $-\pi/2 \leq \phi \leq 0$, which is the condition for simultaneous acceleration and longitudinal focusing, the second term, representing the RF defocusing effect, is negative, which reduces the net focusing. The magnitude of the second term vanishes at the peak of the accelerating waveform, when $\phi = 0$, and is maximum for $\phi = -\pi/2$. If the second term exceeds the first term, σ_0^2 is negative, and the beam is unstable.

We remarked earlier that Eq. (8.29) has the form of the Mathieu equation. A more accurate treatment, on the basis of the solutions to the Mathieu equation, shows that the beam is stable when $0 < \sigma_0 < \pi$. In standard notation, the Mathieu equation has the form

$$\frac{d^2x}{d\tau^2} + [P + 2Q\sin(2\tau)], \quad x = 0 \tag{8.37}$$

The lower stability limit at $\sigma_0 = 0$ is satisfied when $-P < Q^2/2$, and the upper limit at $\sigma_0 = \pi$ is satisfied when $Q < 1 - P$. The equation of motion is brought into the standard notation by identifying $\tau = \omega t/2$, $P = qAV_0\sin\phi/2mc^2\beta^2$, and $Q = qXV_0\lambda^2/2\pi^2 mc^2 a^2$.

Finally, we note that the transverse focusing in the RFQ is strong for the following reasons: (1) the use of electric rather than magnetic fields for focusing is superior for low-velocity particles, (2) the use of RF rather than dc fields allows higher peak surface electric fields, and larger electrode voltages, (3) the spatially uniform quadrupole focusing increases the fraction of space used by the focusing fields, especially compared with the use of discrete quadrupole lenses in drift-tube linacs at low velocities, and (4) the short focusing period, $\beta\lambda$, helps to reduce σ_0, allowing a larger phase-advance per unit length and stronger focusing without loosing beam stability.

8.8
Adiabatic Bunching in the RFQ

In any RF accelerator the beam must be longitudinally bunched so that all particles will be accelerated. In conventional accelerators such as the drift tube linac (DTL), bunching is accomplished prior to injection into the linac, by the use of one or more RF cavities placed in front of the linac. In buncher cavities the RF electric fields are applied to the unbunched input beam to produce a velocity modulation in which early particles are decelerated and late particles are accelerated. After a suitable drift space, the beam becomes bunched, ready for injection into the linac. The bunching is usually not very efficient and the resulting beam quality is also poor. For high-intensity beams, the bunching process causes an increase in the beam density, which increases the space-charge forces, and often results in a blow up of the transverse beam emittance [3]. These problems were fundamental limitations for the performance of conventional linacs, especially at high beam currents. However, the RFQ can be used to adiabatically bunch the input dc beam, and if the parameters are chosen properly, RFQ bunching nearly eliminates these problems.

The approach to RFQ adiabatic bunching is to inject the beam into the RFQ at low energy, typically 50 to 100 keV, with an initial synchronous phase of about $\phi_s \cong -\pi/2$, where the separatrix has the largest phase width, and the longitudinal acceptance is maximum. After having begun to capture and bunch the beam, the synchronous ϕ_s is gradually increased. While continuing to accelerate the beam, the synchronous phase is moved toward the crest of the accelerating waveform, where the acceleration is more efficient. As the beam is accelerated to higher velocity, the center-to-center bunch spacings get further apart. The phase width of the bunch shrinks, so bunching in phase is accomplished. However, the parameters can be chosen so that, as the velocity increases the geometric length of the bunch remains nearly constant. This keeps the bunch from being spatially compressed, which would result in large space-charge forces within the beam. The electrode modulation and the accelerating field are small initially, so that the focusing is dominated by the quadrupole term, and the longitudinal and transverse dynamics are almost decoupled. As the beam is bunched more in phase, the amplitude of the electrode modulations is increased. Some unique features of the RFQ are useful for adiabatic bunching. (1) The large axial accelerating field at low velocities, which is a result of the inverse proportionality of the field to the synchronous velocity, allows the bunching to be both adiabatic, and to be accomplished in a relatively short length. (2) The velocity-independent electric-quadrupole focusing that is very effective for low-velocity particles, provides the transverse beam confinement even at high currents. (3) The control of the axial field and the synchronous phase is accomplished by the machining of the four vane tips.

Next, we describe the RFQ adiabatic bunching in more detail [4], and define the conditions that lead to a constant geometrical bunch length during the acceleration. It is convenient to change variables from phase and energy differences to position and momentum differences. For small velocity differences relative to the synchronous particle, we can write $W - W_s \cong \beta c \delta p$, where $\delta p \cong p - p_s$ is the axial momentum difference. The phase difference can be converted to an axial position difference by using $\phi - \phi_s \cong -k\delta z$, where $dz = z - z_s$. The solution for small amplitude motion is of the form $z = Z \sin \omega_\ell t$, and the momentum is $p = P \cos \omega_\ell t$, where $P = m\omega_\ell Z$. From Section 5.12 the adiabatic invariant for small oscillations is

$$\oint p \, dz = \pi P Z = \pi m \omega_\ell Z^2 \tag{8.38}$$

Suppose that to control the space-charge effects during acceleration, we want to maintain a constant density for the beam. Constant density can be achieved, if we maintain a constant-amplitude Z for each particle, and we can do that for the particles that undergo small oscillations, because from the constancy of the adiabatic invariant in Eq. (8.38), we see that Z is constant, if ω_ℓ is kept constant. For the RFQ the longitudinal small oscillation frequency is given by

$$\omega_\ell^2 = \frac{\pi^2 q A V_0 \sin(-\phi_s)}{m \beta_s^2 \lambda^2} \tag{8.39}$$

which means, we must choose the RFQ parameters to maintain

$$\frac{A V_0 \sin(-\phi_s)}{\beta_s^2} = \text{cons tan t} \tag{8.40}$$

Equation (8.40) is the first condition for RFQ adiabatic bunching, and assuming V_0 is constant along the RFQ, Eq. (8.40) determines A as a function of ϕ_s and β_s. We can construct the electrodes to produce the desired A by machining the electrodes to have the correct value of the modulation parameter m.

Next we determine a relation for ϕ_s versus β_s, as follows. We would like to require that the constant bunch density extends beyond the region of small oscillations, and includes the particles that undergo large-amplitude longitudinal oscillations. A simple approximate way to accomplish this is to require that the geometric length of the separatrix is constant.

The geometric length Z_ψ of the separatrix is converted from the phase length ψ by using the relation $Z_\psi = \psi \beta_s \lambda / 2\pi$. If the synchronous phase is varied along the RFQ, so that

$$\beta_s \psi = \text{constant} \tag{8.41}$$

the length of the separatrix will also be constant. The angular width ψ has already been given in Section 6.4, and depends only on the synchronous phase. The synchronous phase $|\phi_s|$ is controlled by controlling the center-to-center spacing of the unit cells. Combining Eqs. (8.40) and (8.41) gives a prescription for specifying both $A(\beta_s)$ and $\phi_s(\beta_s)$ to maintain a constant bunch length. This adiabatic bunching approach is the basis of the bunching section of the RFQ, known as the *gentle buncher*. Although the space-charge forces have been neglected in this discussion, numerical simulation studies that include space-charge forces have shown that this procedure leads to an approximately

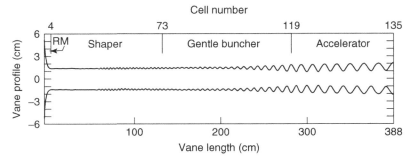

Figure 8.7 A schematic drawing of the pole tips of an RFQ with adiabatic bunching. The transverse dimensions are magnified compared with the longitudinal ones. The beam goes from left to right. Four sections are shown, a radial-matching (RM) section, a shaper section, a gentle-buncher section, and an accelerator section. The bunching is started in the shaper section, and the adiabatic bunching, described in this section, is done in the gentle-buncher section.

constant bunch density, and provides excellent control of space-charge-induced emittance growth. In practice, all of the bunching of an initial dc beam cannot be done adiabatically without making the RFQ too long. The prebunching is usually started in a section called the *shaper*, using a prescription that ramps the phase and the acceleration efficiency linearly with axial distance. A schematic drawing of the pole tips of an RFQ designed for adiabatic bunching is shown in Fig. 8.7.

8.9
Four-Vane Cavity

The RFQ electrodes must be charged periodically at the RF frequency. For typical frequencies ranging from tens to hundreds of megahertz, this is accomplished by placing the electrodes in an RF cavity that is resonant at the desired frequency to charge the vanes in the desired $+--+-$ quadrupole pattern (electric-quadrupole mode). In this section we will discuss the four-vane cavity that is mostly used in the high-frequency range, above about 200 MHz, and is the most common structure for light ions, especially protons. The most popular lower-frequency RFQ structure is the four-rod cavity that will be discussed in Section 8.14.

The four-vane cavity consists of four vanes symmetrically placed within a cavity. The cavity is operated in a TE_{210}-like mode, which is obtained from the natural TE_{211} mode, [5] by tuning specially configured end cells to produce a longitudinally uniform field throughout the interior of the cavity. The transverse electric field is localized near the vane tips, and the magnetic field, which is longitudinal in the interior of the cavity, is localized mostly in four outer quadrants. The efficiency of the four-vane cavity is relatively high, because the vane charging currents are distributed very uniformly along the length of the vanes. There have been many improvements in the mechanical design of the four-vane cavity, since the first cavities were built. The vane-tip patterns are usually produced using a computer-controlled milling machine. The individual vanes can be easily cooled by incorporating suitable cooling channels. Monolithic four-vane RFQ structures with integral vacuum vessels have been fabricated by joining quadrants using both electroforming, and hydrogen-furnace brazing [6].

8.10
Lumped-Circuit Model of Four-Vane Cavity

The separation of the electric- and magnetic-field regions of the four-vane cavity suggests a description based on a simple lumped-circuit model. It is convenient to assume that the cavity has a cloverleaf geometry, shown in Fig. 8.8, consisting of four quadrants, each of which could be analyzed as a resonant cavity. We represent the quadrants with a capacitance C' and an inductance L', shown in Fig. 8.9. (To include dipole modes it would be necessary to include the capacitance between the diametrically opposite vanes.)

Because the four gaps provide separate parallel electrical paths between the vanes of opposite potential, the total capacitance per unit length is $C_\ell = 4C'/\ell_V$. We can derive a number of formulas in terms of the single free parameter C_ℓ. To determine the inductance, we assume that B is approximately uniform over the outer part of the cavity, and write the magnetic flux in each quadrant as $\Phi = BA = \mu_0 AI/\ell_V$, where I is the total transverse current over the vane length ℓ_V, and A is the effective cross-sectional area per quadrant. The inductance of each quadrant is the ratio of the flux to the current, or $L' = \mu_0 A/\ell_V$. To develop the model further, we assume that the effective quadrant area consists of three quarters of a circle of radius r, plus a square with sides of length r (see

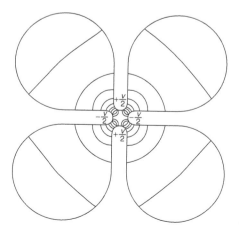

Figure 8.8 Cross section of four-vane RFQ in cloverleaf geometry, used in the lumped-circuit model. The electric-field lines are shown.

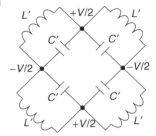

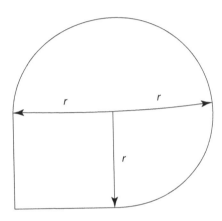

Figure 8.9 Equivalent circuit of the quadrupole mode of a four-vane cavity.

Figure 8.10 Shape of idealized quadrant for inductance calculation.

Fig. 8.10) to represent the area nearest the beam axis, yielding a total quadrant area of $A = (4 + 3\pi)r^2/4$, and a quadrant inductance $L' = \mu_0(4 + 3\pi)r^2/4\ell_V$. The cavity resonant frequency is

$$\omega_0^2 = (L'C')^{-1} = \frac{16}{\mu_0(4 + 3\pi)r^2 C_\ell} \tag{8.42}$$

Equation (8.42) implies that at a fixed frequency, an increase in the capacitive loading reduces the quadrant radius r and the transverse dimensions.

Assuming an $e^{j\omega t}$ time dependence of the currents and voltages, the peak current on the outer wall is given by $I = j\omega_0 C_\ell \ell_V V/4$, and the magnetic field in the quadrants is $B = j\mu_0\omega_0 C_\ell V/4$. We calculate the power loss by assuming the conducting surface area is the perimeter of the quadrant cross section comprised of three quarters of a circle plus a square. Taken over the length ℓ_V, the total surface area for all four quadrants is $S = 2(4 + 3\pi)r\ell_V$. The power dissipation is $P = R_S(B/\mu_0)^2 S/2$, where $R_S = \sqrt{\mu_0\omega_0/2\sigma}$ is the surface resistance, expressed in terms of the conductivity σ. Substituting for B and eliminating r, we obtain the power per unit length for the whole cavity

$$P_\ell = \sqrt{\frac{4 + 3\pi}{32\sigma}}(\omega_0 C_\ell)^{3/2} V^2 \tag{8.43}$$

This shows that for a given vane voltage, the power is proportional to the 3/2 power of the frequency. This is different than the usual $\omega_0^{-1/2}$ scaling of the power dissipation for typical accelerating cavities, because for this case the ratio of surface magnetic field to vane voltage is proportional to the frequency.

The stored energy per unit length is $W_\ell = C_\ell V^2/2$, and the quality factor is

$$Q = \frac{\omega_0 W_\ell}{P_\ell} = \sqrt{\frac{8\sigma}{(4+3\pi)\omega_0 C_\ell}} \quad (8.44)$$

The value of the parameter C_ℓ can be estimated in several ways. As a first approximation, an electrostatic calculation for four rods of circular cross section, whose radius of curvature equals the aperture radius is $C_\ell \cong 90 \times 10^{-12}$ F/m, independent of aperture radius. Using the electromagnetic code SUPERFISH, the value of C_ℓ obtained from the calculations of the resonant frequency of a four-vane cloverleaf cavity, and using the inductance formulas from this section, is found to be a weak function of the vane radius, and for typical values of vane radius to wavelength is approximately $C_\ell = 120 \times 10^{-12} F/m$. Using the latter result for the capacitance per unit length, this model can be used to estimate the properties of a four-vane cavity, and shows the approximate dependence of the cavity properties on the parameters. For an accurate calculation of the cavity properties for any specific geometry, an electromagnetic-field-solver code must be used.

8.11
Four-Vane Cavity Eigenmodes

The ideal quadrupole-mode dispersion curve of the four-vane cavity has the classic hyperbolic shape that is characteristic of a uniform structure. The linac structure is normally operated at the cutoff frequency of the quadrupole-mode passband, where for a given length, the mode separation between the operating mode and the nearest higher longitudinal mode is relatively small. The dispersion curve, showing the longitudinal-mode spectrum for the quadrupole family of modes, is given approximately by the expression

$$\left(\frac{\omega_n}{c}\right)^2 = \left(\frac{\omega_0}{c}\right)^2 + \left(\frac{n\pi}{\ell_V}\right)^2, \quad n = 0, 1, 2, \ldots \quad (8.45)$$

where ω_n is the frequency of the nth mode, ω_0 is the frequency of the operating quadrupole mode, which has a guide wavelength $\lambda_n = 2\ell_V/n$ for an effective vane length ℓ_V. At the vane ends, the magnetic flux from each quadrant splits in half, each half flows around the end of a vane, and returns in the adjacent quadrants. The magnetic field is shown in Fig. 8.11. This produces a continuous path for the flux, and at any cross section the magnetic flux summed over the four quadrants is zero.

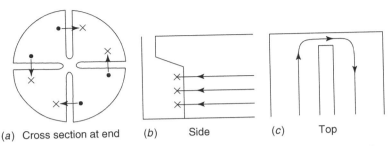

(a) Cross section at end (b) Side (c) Top

Figure 8.11 Magnetic field lines in quadrupole mode at the end: (a) cross section, (b) side view, and (c) top view.

There are three independent azimuthal modes for which the magnetic flux sums to zero, the quadrupole mode and two dipole modes, shown in Fig. 8.12. Suppose that each quadrant has the same cross-sectional area, and that the magnetic field in each quadrant n is determined from a perturbation measurement to have the value B_n, as shown in Fig. 8.13. If the cavity is not properly tuned, the magnitudes of the B_n will not necessarily be equal, and the cavity mode will be an admixture of the quadrupole mode and the two dipole modes. From flux conservation $B_1 + B_2 + B_3 + B_4 = 0$. The quadrupole-mode amplitude is proportional to $A_Q = |B_1 - B_2 + B_3 - B_4|/4$. The amplitudes of the two dipole modes are $A_{D1} = |B_1 - B_3|/2$, and $A_{D2} = |B_2 - B_4|/2$. One of the objectives of the cavity tuning is to minimize the magnitude of the dipole-mode amplitudes, and produce the best approximation to a pure quadrupole operating mode.

An idealized mode spectrum of a four-vane RFQ is shown in Fig. 8.14. The longitudinal-mode spectra for the pair of degenerate dipole modes overlap the quadrupole-mode spectrum. The cutoff frequency of the dipole modes typically

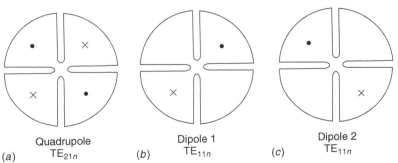

(a) Quadrupole TE$_{21n}$ (b) Dipole 1 TE$_{11n}$ (c) Dipole 2 TE$_{11n}$

Figure 8.12 Azimuthal modes of the RFQ four-vane cavity: (a) quadrupole, (b) dipole 1, and (c) dipole 2. It is convenient to label the RFQ modes by analogy with modes of the pillbox cavity that would exist with open-circuit boundary conditions. The quadrupole modes are labeled as TE$_{21n}$ modes, where n is the longitudinal-mode number, and the dipole modes are TE$_{11n}$ modes.

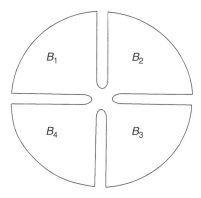

Figure 8.13 General azimuthal magnetic field pattern in four-vane RFQ.

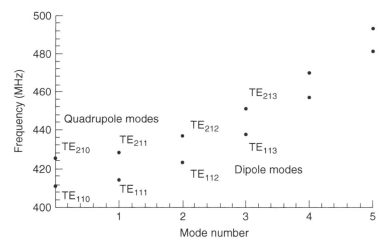

Figure 8.14 Idealized mode spectrum of a four-vane RFQ plotted against a longitudinal-mode number. The RFQ operating mode is labeled the TE_{210} mode because of the appearance of the central field distribution.

lies a few percent lower in frequency than the quadrupole operating mode. This introduces the possibility of accidental degeneracy between the operating quadrupole mode, and a higher longitudinal mode of the dipole family. These effects can make the field distribution of a long, four-vane structure especially sensitive to dimensional construction errors. Field-tilt effects can become important for structures with the length of a few wavelengths. Various methods have been devised to deal with these problems, including vane coupling rings that connect opposing vanes, movable field tuners distributed along the structure, tuning rods that selectively shift the frequency of the dipole-mode spectrum, and resonant coupling of individual RFQ cavities [7].

8.12
Transmission-Line Model of Quadrupole Spectrum

We now construct a model of the four-vane-cavity quadrupole modes that will help us to understand the effects of errors on the field distribution. As shown in Fig. 8.15, we let $L_0(x)$ and $C_0(x)$ be the inductance and capacitance of a shunt resonant circuit that carries the charging current to deliver charge to each vane tip. The variable x is the axial distance along the RFQ. Let $L(x)$ be the inductance associated with longitudinal current that flows along a vane tip, and define $V(x)$ and $I(x)$ as the intervane voltage and the longitudinal current along the vane tip. For a properly tuned RFQ, $I(x) = 0$.

From transmission-line theory

$$\frac{\partial V}{\partial x} = -j\omega L I(x) \tag{8.46}$$

and

$$\frac{\partial I}{\partial x} = -\left(j\omega C_0 + \frac{1}{j\omega L_0}\right) V(x) \tag{8.47}$$

Substituting Eq. (8.47) into Eq. (8.46), we obtain the wave equation

$$\frac{\partial^2 V}{\partial x^2} + k^2 V(x) = 0 \tag{8.48}$$

where the dispersion relation is $\omega^2 = \omega_0^2 + k^2/LC_0$, and $\omega_0^2 = 1/L_0 C_0$ is the cutoff frequency. To represent the known dispersion curve, we must identify $c^2 = 1/LC_0$, where c is the speed of light. Next we choose open-circuit boundary conditions at each end of the line, $x = 0$ and $x = \ell_V$. Thus $I(0) = I(\ell_V) = 0$ and from the transmission-line equation it follows that $dV(0)/dx = dV(\ell_V) \, dx = 0$.

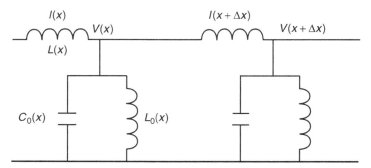

Figure 8.15 Transmission-line model of the RFQ quadrupole mode for the four-vane cavity.

8.12 Transmission-Line Model of Quadrupole Spectrum

The solutions to the wave equation that satisfy the boundary conditions are

$$^0V_n(x) = \sqrt{\frac{2}{\ell_V}} \cos k_n x, \quad n = 0, 1, 2, \ldots \qquad (8.49)$$

where $k_n = 2\pi/\lambda_g = n\pi/\ell_V$, and the normalization is such that $\int_0^{\ell_V} {}^0V_n^2(x)\, dx = 1$. The modes are characterized by the number of half wavelengths of the voltage cosine waves that fit between the boundaries; the frequency of the nth mode is $\omega_n^2 = \omega_0^2 + (n\pi c/\ell_V)^2$. The voltage distribution for the three lowest modes is shown in Fig. 8.16.

We now examine the effects of perturbations on the field distribution, by applying standard perturbation theory for the eigenvalue problem [8]. First we review the theory, recalling that the eigenvalue problem consists in finding the solutions for the eigenvectors V_n and the corresponding eigenvalues γ_n of the operator M that satisfies the equation $MV_n = \gamma_n V_n$. The unperturbed solutions satisfy $M_0\, {}^0V_n = {}^0\gamma_n\, {}^0V_n$, or

$$\left(c^2 \frac{\partial^2}{\partial x^2} - \omega_0^2\right) {}^0V_n = {}^0\gamma_n\, {}^0V_n \qquad (8.50)$$

which is the wave equation, if we identify ${}^0\gamma_n = -\omega_n^2$. The solutions are given by Eq. (8.49). The perturbed problem is described by $M = M_0 + P$, where the perturbation matrix is $P \ll M_0$. The theory tells us that perturbed solutions can be expressed in the lowest order by

$$V_n = {}^0V_n + \sum_{m \neq n} a_{mn}\, {}^0V_n \qquad (8.51)$$

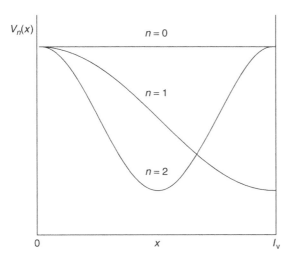

Figure 8.16 Voltage distribution of the three lowest unperturbed modes of the four-vane RFQ.

and

$$a_{mn} = \frac{\int_0^{\ell_V} {}^0V_m P^0 V_n}{{}^0\gamma_n - {}^0\gamma_m} \tag{8.52}$$

We identify the unperturbed operator as

$$M_0 = c^2 \frac{\partial^2}{\partial x^2} - \omega_0^2 \tag{8.53}$$

and the perturbation operator is $P = -\Delta\omega_0^2(x)$, where $\Delta\omega_0^2(x)$ is the squared resonant-frequency error, as a function of x.

Now suppose we have a small resonant-frequency error distribution $\Delta\omega_0(x)$ that depends on x. We assume $\Delta\omega_0^2 \ll \omega_0^2$, where $\Delta\omega_0^2 = 2\omega_0\Delta\omega_0$. In first order the perturbed cavity frequency is

$$^{(1)}\omega_n^2 = \omega_n^2 + \frac{2}{\ell_V} \int_0^{\ell_V} \Delta\omega_0^2(x) \cos^2(k_n x)\, dx \tag{8.54}$$

where $k_n = n\pi/\ell_V$. The perturbed field in the lowest order is

$$^{(1)}V_n(x) = \sqrt{\frac{2}{\ell_V}} \left\{ \cos(k_n x) + \sum_{m \neq n} \frac{a_{mn} \cos(k_m x)}{\omega_n^2 - \omega_m^2} \right\} \tag{8.55}$$

where

$$a_{mn} = \frac{\frac{2}{\ell_V} \int_0^{\ell_V} \cos(k_m x)\cos(k_n x) \Delta\omega_0^2(x)\, dx}{\omega_n^2 - \omega_m^2} \tag{8.56}$$

and

$$\omega_n^2 - \omega_m^2 = (n^2 - m^2)\left(\frac{\pi c}{\ell_V}\right)^2 \tag{8.57}$$

For the RFQ operating mode, $n = 0$ and the perturbed field is

$$^{(1)}V_0(x) = \sqrt{\frac{2}{\ell_V}} \left\{ 1 - \left(\frac{\ell_V}{\pi c}\right)^2 \sum_{m=1}^{\infty} \frac{a_{m0} \cos(k_m x)}{m^2} \right\} \tag{8.58}$$

where

$$a_{m0} = -\frac{2\ell_V}{(m\pi c)^2} \int_0^{\ell_V} \cos(k_m x) \Delta\omega_0^2(x)\, dx \tag{8.59}$$

and Eq. (8.54) gives the perturbed frequency with $k_n = 0$ for the $n = 0$ mode. The perturbed field error is expressed in Eq. (8.58) as a sum over the unperturbed modes. Because of the denominator, the unperturbed modes that are nearest the operating mode will tend to contribute more to the field error. The contribution of any given mode also increases in proportion to the overlap between the squared-frequency-error distribution $\Delta\omega_0^2(x)$, and the unperturbed eigenfunction $\cos k_m x$.

Consider the example of a δ-function frequency error at some point x_0, expressed as $\Delta\omega_0^2(x) = \Gamma\delta(x - x_0)$, where Γ determines the magnitude of the error, and

$$1 = \int_0^{\ell_V} \delta(x - x_0)\, dx \tag{8.60}$$

The frequency perturbation induced by the error is given by

$$^{(1)}\omega_0^2 = \omega_0^2 + \frac{2\Gamma}{\ell_V} \int_0^{\ell_V} \delta(x - x_0)\, dx = \omega_0^2 + \frac{2\Gamma}{\ell_V} \tag{8.61}$$

and the frequency shift is

$$\delta\omega_0 = \frac{\Gamma}{\ell_V \omega_0} \tag{8.62}$$

This means we can also write $\Delta\omega_0^2(x) = \ell_V \omega_0 \delta\omega_0 \delta(x - x_0)$. The perturbed field distribution is

$$^{(1)}V_0(x) = \sqrt{\frac{2}{\ell_V}} \left\{ 1 + \frac{2\Gamma}{\ell_V} \sum_{m=1}^{\infty} \frac{\int_0^{\ell_V} \delta(x - x_0)\cos(k_m x)\, dx}{\omega_0^2 - \omega_m^2} \cos(k_m x) \right\} \tag{8.63}$$

or

$$^{(1)}V_0(x) = \sqrt{\frac{2}{\ell_V}} \left\{ 1 - 8\frac{\delta\omega_0}{\omega_0} \left(\frac{\ell_V}{\lambda}\right)^2 \sum_{m=1}^{\infty} \frac{\cos(m\pi x_0/\ell_V)}{m^2} \cos(m\pi x/\ell_V) \right\} \tag{8.64}$$

Thus,

$$\frac{\delta V_0(x)}{V_0} = -8\frac{\delta\omega_0}{\omega_0} \left(\frac{\ell_V}{\lambda}\right)^2 \sum_{m=1}^{\infty} \frac{\cos(m\pi x_0/\ell_V)}{m^2} \cos(m\pi x/\ell_V) \tag{8.65}$$

is the fractional-field error in terms of the cavity resonant-frequency shift $\delta\omega_0$ caused by the δ-function error at x_0. From Eq. (8.65) we see that the local frequency error adds small amounts of all the other unperturbed-mode fields to the unperturbed mode that corresponds to the error-free state, an effect known as *mode mixing*. We should not think of the error as exciting these other modes, but rather as an error-induced mixing of modes of the ideal system to create a new field distribution for the operating mode. Note that each of the unperturbed modes contributes a term proportional to the field value of each mode at the point of the perturbing error, divided by the mode index squared. Modes with m closest to $m = 0$ tend to contribute more to the field error. An analytic solution exists for the summation in Eq. (8.65), which gives the final result

$$\frac{\delta V_0(x)}{V_0} = -4\pi^2 \frac{\delta\omega_0}{\omega_0} \left(\frac{\ell_V}{\lambda}\right)^2$$
$$\times \begin{cases} \dfrac{1}{3} - \dfrac{x}{\ell_V} + \dfrac{1}{2}\left(\dfrac{x}{\ell_V}\right)^2 + \dfrac{1}{2}\left(\dfrac{x_0}{\ell_V}\right)^2, & x > x_0 \\ \dfrac{1}{3} - \dfrac{x_0}{\ell_V} + \dfrac{1}{2}\left(\dfrac{x}{\ell_V}\right)^2 + \dfrac{1}{2}\left(\dfrac{x_0}{\ell_V}\right)^2, & x_0 > x \end{cases} \quad (8.66)$$

The field error varies as a second-order polynomial in x; it is proportional to $\delta\omega_0$ and to $(\ell_V/\lambda)^2$. An interesting property of the solution is that if an error at some position x_0 causes the local resonant frequency to increase, the field at x_0 decreases. Likewise, if the local frequency decreases, the local field increases. This turns out to be a general rule of thumb, when the modes that are mixed lie higher in frequency than that of the pure mode representing the unperturbed state.

Figures 8.17 and 8.18 show some perturbed field distributions for two examples, (a) a δ-function error at the vane end, where $x_0/\ell_V = 0$ and (b) a δ-function error at middle of the vane, where $x_0/\ell_V = 0.5$. For both examples, we have chosen $\ell_V/\lambda = 2$ and $\delta\omega_0/\omega_0 = 0.01$.

Finally, we consider δ-function frequency errors at each end that are equal and opposite, expressed as $\Delta\omega_0^2(x) = \Gamma\delta(x) - \Gamma\delta(x - \ell_V)$, where again Γ

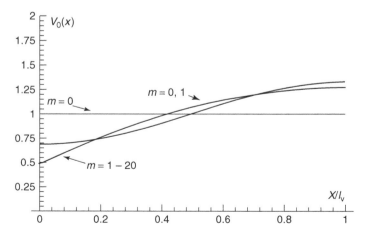

Figure 8.17 Perturbed field distributions for problem with a δ-function error at the vane end, where $x_0/\ell_V = 0$, $\ell_V/\lambda = 2$, and $\delta\omega_0/\omega_0 = 0.01$. (a) unperturbed $m = 0$ mode, (b) adding the $m = 1$ mode, (c) sum of the first 20 modes.

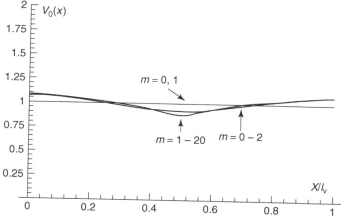

Figure 8.18 Perturbed field distributions for problem with a δ-function error at the middle of the vane, where $x_0/\ell_V = 0.5$, $\ell_V/\lambda = 2$, and $\delta\omega_0/\omega_0 = 0.01$. (a) unperturbed $m = 0$ mode plus $m = 1$ mode, which contributes zero, (b) adding the $m = 2$ mode, (c) sum of the first 20 modes.

determines the magnitude of the error. The induced frequency shift with errors at both ends is

$$^{(1)}\omega_0^2 = \omega_0^2 + \frac{2\Gamma}{\ell_V}\left[\int_0^{\ell_V} \delta(x)\,dx - \int_0^{\ell_V} \delta(x-l_V)\,dx\right]$$

$$= \omega_0^2 + \frac{2\Gamma}{\ell_V}[1-1] = \omega_0^2 \tag{8.67}$$

Thus, the resonant frequency is unchanged. The perturbed field distribution is

$$^{(1)}V_0(x) = \sqrt{\frac{2}{\ell_V}}\left\{1 - 8\frac{\delta\omega_0}{\omega_0}\left(\frac{\ell_V}{\lambda}\right)^2 \sum_{m=1}^{\infty} \frac{1-\cos(m\pi)}{m^2}\cos(m\pi x/\ell_V)\right\} \tag{8.68}$$

There is an analytic solution for the summation that is given by

$$\sum_{m=1}^{\infty} \frac{1-\cos(m\pi)}{m^2}\cos(m\pi x/\ell_V) = \frac{\pi^2}{4}\left(1 - 2\frac{x}{\ell_V}\right) \tag{8.69}$$

Using this result we may express the fractional change in the field distribution as

$$\frac{\delta V_0(x)}{V_0} = -2\pi^2\left(1 - 2\frac{x}{\ell_V}\right)\frac{\delta\omega_0}{\omega_0}\left(\frac{\ell_V}{\lambda}\right)^2 \tag{8.70}$$

The perturbed field distribution of Eq. (8.70) has a linear field tilt. The error term is proportional to $\delta\omega_0$ and to $(\ell_V/\lambda)^2$. We show in the Fig. 8.19 the field distribution for the problem with $\ell_V/\lambda = 2$ and $\delta\omega_0/\omega_0 = 0.01$, summing all the modes.

8.13
Radial-Matching Section

The RFQ can be designed to capture, bunch, and accelerate a continuous unbunched input beam. Matching of the input beam into the RFQ presents a special problem. The matched ellipse parameters in the RFQ vary with the RF phase and are relatively independent of position along the linac. Therefore, the orientation of the acceptance ellipse depends on time. For proper matching into the RFQ, one must provide a transition from a beam having time-independent characteristics to one that has the proper variations with time. At the input, a time-independent set of ellipse parameters is required, which will depend on the beam current. The solution is to taper unmodulated vanes at the input to the RFQ so that the radius decreases and the focusing strength increases from near zero to its full value over a distance of several cells. Quadrupole symmetry is maintained throughout the RM section. The procedure allows a time-independent beam to adapt to the time structure of the focusing system.

Phase-space plots and beam envelopes for a RM section design are shown in Fig. 8.20. In this RM section the focusing strength B is increased linearly as a function of longitudinal position from its value at the input to its full value at the end of the RM section. First, the matched ellipse parameters are found in the interior cells for various phases. Three matched ellipses corresponding to phases 90° apart are shown in the two phase-space plots in the upper right side of the figure for the $x - x'$ and $y - y'$ planes, respectively. This

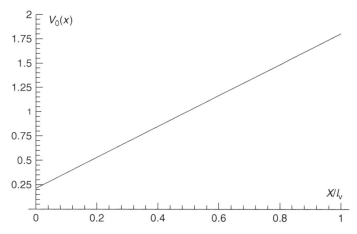

Figure 8.19 Perturbed field distribution for problem with equal and opposite δ-function errors at the ends of the vane, $x_0/\ell_V = 0.0$ and $x_0/\ell_V = 1.0$, $\ell_V/\lambda = 2$, and $\delta\omega_0/\omega_0 = 0.01$.

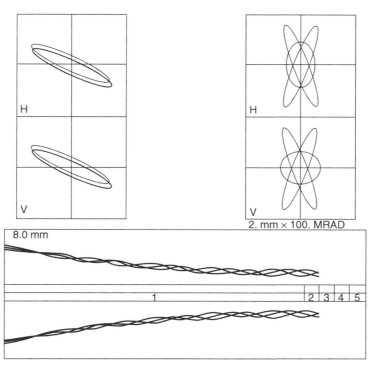

Figure 8.20 Beam envelopes at phases 90° apart in a radial-matching section.

shows graphically how different the matched ellipses are as a function of the phase, and the relatively small area of overlap is common to all the ellipses. The phase-space plots at the upper left are the result of tracking these same three ellipses backward through the tapered RM section, five-periods long (10 cells). These ellipses at different phases are similar and have a high degree of overlap. To obtain a matched beam within the RFQ, the average of these ellipses will be taken as the matched ellipse. The $x - x'$ ellipses and the $y - y'$ ellipses are essentially identical. Thus, the input beam injected into the RM section is axisymmetric. If the beam from the ion source is axisymmetric, the external lenses required for matching the beam would normally be solenoids or quadrupole triplets. The bottom graph shows the horizontal and vertical profiles for the three ellipses that are being tracked through the RM section.

The RFQ can also be designed to accelerate a bunched beam. In this case the RFQ could begin with an $m > 1$ accelerator section. The input beam would generally not be axisymmetric. One or more external cavities would be used to bunch the input beam before injection into the RFQ. With a bunched beam, which has a limited range of input phases, conventional external quadrupole lenses could be used to provide transverse matching that would be suitable for all beam phases. Therefore, for the bunched beam case, a RM section would not be needed. However, if $m > 1$ initially, there would be a time-varying on-axis potential at the input, which is not present with pure quadrupole symmetry when $m = 1$. Consequently, beginning with $m > 1$ would introduce an energy kick at the input, which may be undesirable. However, this could be eliminated by beginning the RFQ with an entrance transition cell, to be discussed in the following section. The transition cell begins with $m = 1$, and makes a smooth transition to the full $m > 1$ value. This transition cell would be immediately followed by the accelerator section.

RM section designs have been improved considerably since the early days when a linear ramp of the focusing strength B was used. A remarkably effective and commonly used RM section is the design proposed by Crandall [9]. This RM section is normally used at the entrance of the RFQ, but can also be used at the exit. It uses a four-term potential function given by

$$U(r, \theta, z) = \frac{V}{2} \sum_{n=0}^{3} A_n T_n(r, z) \cos(2n\theta),$$

$$T_n(r, z) = I_{2n}(kr) \cos(kz) + I_{2n}(3kr) \cos(3kz)/3^{2n+1} \qquad (8.71)$$

where $z = 0$ is at the interface between the RM section and the adjacent normal RFQ cell, and $k = \pi/2L$, where L is the length of the RM section, the distance from the end wall at $z = -L$ to the end of the RM section at $z = 0$. The length L can be chosen by the designer. Crandall considers a range from $L = 2\beta\lambda$ to $12\beta\lambda$, and based on the quality of the match, concludes that there is no reason for choosing L any longer than $3\beta\lambda$. Short matching sections are an advantage since it is found that they require a less convergent injected beam that is easier to achieve.

We treat the four amplitudes A_n for $n = 0, 1, 2,$ and 3 as unknown amplitudes for the RM section geometry. They must be chosen so that the RM section geometry matches the known geometry of the adjacent normal RFQ cell at the $z = 0$ interface. Four equations are needed to determine the four amplitudes. The four equations correspond to requiring at the $z = 0$ interface:

(1) the same x-vane displacements, $\pm x_0$,
(2) the same y-vane displacements, $\pm y_0$,
(3) the same transverse radius of curvature of the x vanes, and
(4) the same transverse radius of curvature of the y vanes.

The properties of Crandall's four-term potential function are:
(1) Each potential term satisfies Laplace's equation.
(2) Each potential term is zero at the end wall at $z = -L$.
(3) Each potential term has a physically reasonable s-shaped z dependence, starting with zero at the end wall, and ending at a maximum at the $z = 0$ interface.
(4) The terms for $n > 0$ vanish at $r = 0$ contributing zero potential on-axis, as would be appropriate for an adjacent

normal RFQ cell with no modulation. A nonzero $n = 0$ term provides for an adjacent normal RFQ cell with modulation $m > 1$.

(5) The longitudinal electric field $-\partial U/\partial z = 0$ at the $z = 0$ interface for all r.
(6) The longitudinal electric field $-\partial U/\partial z \sim 0$ for small kr at $z = -L$ end wall.
(7) A smooth transition is obtained between the field-free region in the end wall and the field within the RFQ.

Four equations must be solved to determine the four unknown amplitudes, A_n. The first requirement which expresses equality of the two horizontal vane displacements at the $z = 0$ interface is

$$U(\pm x_0, 0, 0) = +V/2 \tag{8.72}$$

Likewise, for the vertical vane displacements at $z = 0$,

$$U(\pm y_0, \pi/2, 0) = -V/2 \tag{8.73}$$

We only need to consider the coordinates with the plus sign. Substituting Eq. (8.71) into Eqs. (8.72) and (8.73) gives

$$\sum_{n=0}^{3} A_n T_n(x_0, 0) = 1$$

$$\sum_{n=0}^{3} A_n T_n(y_0, 0) \cos(n\pi) = -1 \tag{8.74}$$

Next we obtain the equations involving the two transverse curvature radii of the vane tips at $z = 0$. By inspection, the transverse slopes at the two horizontal vane tips at $x = \pm x_0$ and $y = 0$ are $dx/dy = 0$, and at the two vertical vane tips at $y = \pm y_0$ and $x = 0$ are $dy/dx = 0$. The transverse radius of curvature for each horizontal vane at $y = 0$ is $1/x_0''$, where $x_0'' = d^2 x_p/dy^2$. The transverse radius of curvature for the y vane at $x = 0$ is $1/y_0''$, where $y_0'' = d^2 y/dx^2$. These transverse radii of curvature are given by their values at the $z = 0$ location of the adjacent normal RFQ cell.

The first equation we obtain that involves a transverse radius of curvature is obtained by differentiating twice both sides of Eq. (8.71) with respect to x and evaluating at $z = 0$. We will be interested in the y vane at $x = 0$ and $y = y_0$. Differentiating the first time gives

$$\sum_{n=0}^{3} A_n \left[\frac{\partial T_n(r, 0)}{\partial r} \cos(2n\theta) \frac{\partial r}{\partial x} - 2n T_n(r, 0) \sin(2n\theta) \frac{\partial \theta}{\partial x} \right] = 0$$

$$\tag{8.75}$$

To avoid a lot of algebra, Crandall notes that the cross section of the four vanes shows that on the y-pole tip at $x = 0$ and $y = y_0$, we have $\partial r/\partial x = 0$. Also, on the y-pole tip at $x = 0$, we have $\theta = \pi/2$ and $\sin(2n\theta) = \sin(n\pi) = 0$. Then, when differentiating Eq. (8.75), any terms multiplying $\partial r/\partial x$ or $\sin(2n\theta)$ will be multiplied by zero. Then, to differentiate Eq. (8.75) the only nonzero terms correspond to differentiating $\partial r/\partial x$ and $\sin(2n\theta)$.

Changing from polar to Cartesian coordinates yields

$$\frac{\partial r}{\partial x} = \frac{yy' + x}{r} \tag{8.76}$$

Differentiating again and evaluating at the vertical pole tip where $y = y_0$ and $x = 0$ results in

$$\frac{\partial^2 r}{\partial x^2} = \frac{1}{y_0} + y_0'' \quad \text{where } y_0'' = d^2y/dx^2 \text{ at } x = 0 \text{ and } r = y_0 \tag{8.77}$$

Also since $\tan(\theta) = y/x$

$$\frac{\partial \theta}{\partial x} = \frac{xy' - y}{r^2} = -\frac{1}{y_0} \tag{8.78}$$

Now, differentiating Eq. (8.75) once more,

$$\sum_{n=0}^{3} A_n \left[\frac{\partial T_n(r, 0)}{\partial r} \cos(2n\theta) \frac{\partial^2 r}{\partial x^2} - 4n^2 \cos(2n\theta) T_n(r, 0) \left(\frac{\partial \theta}{\partial x} \right)^2 \right] = 0 \tag{8.79}$$

Substituting Eqs. (8.77) and (8.78) into Eq. (8.79), and using the results that on the vertical pole tip where $x = y' = 0$, $y = y_0$, and $\theta = \pi/2$, we obtain

$$\sum_{n=0}^{3} A_n \left[\frac{\partial T_n(y_0, 0)}{\partial r} \left(\frac{1}{y_0} + y_0'' \right) - 4n^2 T_n(y_0, 0) \frac{1}{y_0^2} \right] \cos(n\pi) = 0 \tag{8.80}$$

Repeating this procedure for the x vane tip, where $\theta = 0$, $y = 0$, $x' = dx/dy = 0$, $r = x_0$, we obtain

$$\sum_{n=0}^{3} A_n \left[\frac{\partial T_n(x_0, 0)}{\partial r} \left(\frac{1}{x_0} + x_0'' \right) - 4n^2 T_n(x_0, 0) \frac{1}{x_0} \right] = 0 \tag{8.81}$$

We now have four equations, Eqs. (8.72), (8.73), (8.80), and (8.81), that must be solved simultaneously to determine the four unknown amplitudes, A_n, for $n = 0, 1, 2$, and 3. Once these are determined, the potential and fields are determined everywhere in the RM section.

8.14
RFQ Transition Cell

The vane tips in the first RFQs that were built ended abruptly at the end of a full accelerating cell, where one pair of vanes is closer to the axis than the other pair. However, it is not obvious that the two-term potential function, which is valid for a quasiperiodic structure, and therefore a good approximation for the interior cells of the RFQ, is still a good model in the vicinity of the last accelerating cell where the vanes end abruptly. Also, for an RFQ where the vanes end abruptly at the end of the last cell, one would expect physically that the unequal spacing of the vanes at the end of the cell would result in time-varying on-axis potential at the end. In this case, the potential variation at the end would cause an undesirable variation in the energy of the output beam. Furthermore, ending the vanes at the end of a full cell results in an asymmetric output beam, which is strongly converging in one plane and strongly diverging in the other plane, and these characteristics can make output-beam matching difficult. For these reasons it was important to develop a more realistic model for the fields at the ends of the vanes.

Motivated by these arguments, in 1994 Crandall [10] introduced a new type of cell called a *transition cell* to be used after the last full accelerating cell. The transition cell makes a smooth transition from full modulation at the last accelerating cell to no modulation and pure quadrupole symmetry at the end of the transition cell. The transition cell is described using a three-term potential solution of Laplace's equation, and its length is slightly less than that of a normal accelerating cell. The transition-cell parameters are chosen to be continuous with respect to both the potential and the fields from the two-term description used for the final accelerating cell. Crandall pointed out that the RFQ vane tips at the end of the transition cell could also be extended for a short distance with no modulation. The ability to choose the length of the vane extension provides additional flexibility to change the output transverse phase-space ellipses as required for output-beam matching.

In a normal acceleration cell, the fields are derived from a time-varying two-term potential function (Eq. 8.82) that satisfies Laplace's equation:

$$U(r, \theta, z; t) = \frac{V}{2} \left[\left(\frac{r}{r_0} \right)^2 \cos 2\theta + A I_0(kr) \cos kz \right] \sin \omega t \qquad (8.82)$$

where V is the maximum potential difference between vanes, $k = \pi/L$, L is the cell length, and A and r_0 are constants that depend on the boundary conditions. Typical vane-tip profiles are shown in Fig. 8.21. The period of these profiles is $2L$, and the slopes of the profiles (x' and y') are zero at the end of each cell.

If the RFQ linac vanes begin with modulation at the beginning of a full cell ($z = 0$), one pair of vanes, either the x or the y vanes, will begin closer to the axis than the other pair. This means that on-axis beam particles will see a nonzero on-axis field in the entrance fringe-field region, and this means that

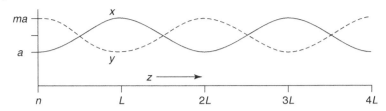

Figure 8.21 Vane-tip profiles from the two-term potential function of Eq. (8.82) (courtesy of K. R. Crandall).

the particles will generally gain or lose energy depending on time or phase. The same argument applies in the exit fringe-field region. The magnitude of the effect can be estimated in the approximation that the two-term potential is valid at the beginning and ends of the vanes. The two-term potential on-axis in the RFQ is

$$U(0, 0, z, t) = \frac{AV}{2} \cos kz \sin \phi \qquad (8.83)$$

where $\phi = \omega t$ is the phase (using the sine convention where zero phase corresponds to the zero crossing of the sine wave at 90° before the crest), V is the intervene voltage, and A is the acceleration efficiency, for which we will assume $A = 0.5$ as a typical value. The synchronous phase, which is the RF phase when the synchronous particle is at the middle of the cell, is typically 30° before the crest or $\phi = 60°$ in the sine convention. When the synchronous particle enters the cell the RF phase is typically 120° before the crest ($\phi = -30°$ in the sine convention) and when it exits the cell the phase is about 60° after the crest ($\phi = 150°$ in the sine convention). At the entrance of the RFQ, using $z = 0$, $A = 0.5$, and $\phi = -30°$ on the sine wave, the on-axis potential is

$$U(0, 0, z = 0, \phi = -30°) = -V/8 \qquad (8.84)$$

Outside the fringe field the on-axis potential is $U = 0$. In a hard-edged approximation the on-axis potential falls from zero to $-V/8$ at $z = 0$, and a synchronous particle of charge q gains kinetic energy $qV/8$ in the entrance fringe field. At the exit of the RFQ, $z = L$, $A = 0.5$, and $\phi = 150°$ on the sine wave, the on-axis potential is

$$U(0, 0, z = L, \phi = 150°) = -V/8 \qquad (8.85)$$

In the hard-edged approximation, the on-axis potential increases from $-V/8$ to 0 at $Z = L$, and a synchronous particle loses kinetic energy $qV/8$ in the exit fringe field.

These kinetic energy changes are only estimates because of the uncertain validity of the two-term potential function at the ends of the RFQ. These energy changes and their uncertainty can be eliminated by using a transition cell.

Crandall's approach for a better analytic model was to find a solution of Laplace's equation for a transition cell, valid after the last accelerating cell, which at the exit makes a smooth transition from full vane-tip modulation at the last accelerating cell to zero modulation with an optional zero-modulation vane segment with pure quadrupole symmetry (equal spacing of the four vane tips from the beam axis). The optional segment with no modulation will be discussed later. To accomplish this, Crandall introduced a three-term potential function to describe the transition cell of length L'. The three-term potential function is given by

$$U(r, \theta, z; t) = \frac{V}{2}\left[\left(\frac{r}{r_0}\right)^2 \cos 2\theta \pm A_{10} I_0(Kr) \cos Kz \right.$$

$$\left. \pm A_{30} I_0(3Kr) \cos 3Kz\right] \sin \omega t \quad (8.86)$$

where $K = \dfrac{\pi}{2L'}$

Periodic vane-tip profiles for the three-term potential function with the period length $4L'$ are shown in Fig. 8.22. The minus sign was used for the second and third terms so that the x vane would start at ma and the y vane would start at a when $z = 0$. Because the second and third terms are zero at $z = L'$, both vane-tip displacements are equal to r_0 at this point. The three unknown parameters are A_{10}, A_{30}, and K (or L'), and these are determined by three conditions. Two conditions are determined at the $z = 0$ interface, where the values of the x- and y-vane-tip potentials in the transition cell must be matched to those at the end of the last accelerating cell. The third condition results from requiring zero slope for both the x- or y-vane-tip profiles at the point of pure quadrupole symmetry at the end in the transition cell, $z = L'$. The horizontal and vertical vane-tip profiles, where $U(r, \theta = 0, z) = V/2$ and $U(r, \theta = \pi/2, z) = -V/2$, respectively, are given by:

$$\left(\frac{x}{r_0}\right)^2 - A_{10} I_0(Kx) \cos Kz - A_{30} I_0(3Kx) \cos 3Kz = 1 \quad (8.87)$$

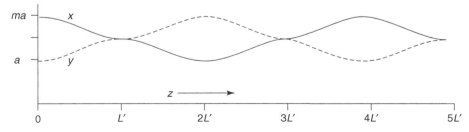

Figure 8.22 Periodic vane-tip profiles satisfying the three-term potential function (Eq. 8.86) (courtesy of K. R. Crandall).

and

$$-\left(\frac{y}{r_0}\right)^2 - A_{10}I_0(Ky)\cos Kz - A_{30}I_0(3Ky)\cos 3Kz = -1 \qquad (8.88)$$

Following Crandall's derivation, we first impose the zero-slope condition at $z = L'$. Differentiating Eqs. (8.87) and (8.88) with respect to z gives expressions for the slopes:

$$\left\{\frac{2x}{r_0^2} - K[A_{10}I_1(Kx)\cos Kz - 3A_{30}I_1(3Kx)\cos 3Kz]\right\}x' =$$
$$- K[A_{10}I_0(Kx)\sin Kz + 3A_{30}I_0(3Kx)\sin 3Kz] \qquad (8.89)$$

$$\left\{\frac{-2y}{r_0^2} - K[A_{10}I_1(Ky)\cos Kz - 3A_{30}I_1(3Ky)\cos 3Kz]\right\}y' =$$
$$- K[A_{10}I_0(Ky)\sin Kz + 3A_{30}I_0(3Ky)\sin 3Kz] \qquad (8.90)$$

Both x' and y' are zero at $z = 0$. For these also to be zero at $z = L'$, the right-hand side of Eqs. (8.89) and (8.90) must be zero there. Since both x and y equal r_0 at $z = L'$, the second terms in both equations are the same and are equal to zero if

$$A_{30}(A_{10}, K) = \frac{\alpha(K)}{3} A_{10} \qquad (8.91)$$

where

$$\alpha(K) = \frac{I_0(Kr_0)}{I_0(3Kr_0)} \qquad (8.92)$$

The vane-tip profiles, showing the final accelerating cell and the transition cell that follows are shown in Fig. 8.23.

Note that because $Kz = \pi/2$ at $z = L'$, $U(r_0, 0, L') = V/2$ and $U(r_0, \pi/2, L') = -V/2$ are satisfied automatically by Eq. (8.86).

Next we return to the conditions at the $z = 0$ interface, where the values of the x- and y-vane-tip potentials of the transition cell must be matched to those at the end of the last accelerating cell. At the $z = 0$ interface between the two cells we must have

$$U(ma, 0, 0) = V/2 \qquad (8.93)$$

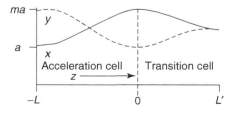

Figure 8.23 Vane-tip profiles in final accelerating cell followed by a transition cell (courtesy of K. R. Crandall).

and
$$U(a, \pi/2, 0) = -V/2 \tag{8.94}$$

Substituting Eqs. (8.93) and (8.94) into the three-term potential function (Eq. 8.86), we find

$$\left(\frac{ma}{r_0}\right)^2 - A_{10}I_0(mKa) - A_{30}I_0(3mKa) = 1 \tag{8.95}$$

and

$$-\left(\frac{a}{r_0}\right)^2 - A_{10}I_0(Ka) - A_{30}I_0(3\ Ka) = -1 \tag{8.96}$$

Multiplying Eq. (8.96) by m^2 and adding the result to Eq. (8.95) and leads to

$$A_{10}[m^2 I_0(Ka) + I_0(mKa)] + A_{30}[m^2 I_0(3\ Ka) + I_0(3mKa)] = m^2 - 1 \tag{8.97}$$

or

$$A_{10}T_{10}(K) + A_{30}T_{30}(K) = m^2 - 1 \tag{8.98}$$

where we identify T_{10} and T_{30} as the terms in brackets multiplying A_{10} and A_{30}, respectively. Substituting Eq. (8.91) to eliminate A_{30}, and solving for A_{10} gives

$$A_{10}(K) = \frac{m^2 - 1}{T_{10}(K) + \alpha(K)T_{30}(K)/3} \tag{8.99}$$

Substituting Eqs. (8.91), (8.92), and (8.99) into Eq. (8.96) yields an expression for K:

$$\left(\frac{a}{r_0}\right)^2 + \left(\frac{m^2 - 1}{T_{10}(K) + \alpha(K)T_{30}(K)/3}\right)(I_0(Ka) + \alpha(K)I_0(3Ka)/3) = 1 \tag{8.100}$$

After K is determined, $A_{10}(K)$ is determined by Eq. (8.99), and $A_{30}(A_{10}, K)$ can be determined from Eqs. (8.91) and (8.92). Also, the length of the transition cell is given by $L' = \pi/2K$.

One can obtain an initial value of K for an iterative solution of Eq. (8.100) in the following way. First, we make the approximation that the modified Bessel functions are given by the leading term in the expansion, so that $I_0 \cong 1$. This allows us to obtain approximate expressions for α, A_{10}, A_{30}, and the acceleration efficiency A of the two-term potential function. The results are $\alpha = 1$, $A_{30} = A_{10}/3$, and $A_{10} = 3(m^2 - 1)/4(m^2 + 1)$. For comparison with the acceleration efficiency parameter A of the two-term potential function, the same approximation $I_0 \cong 1$ gives

$$A = \frac{m^2 - 1}{m^2 I_0(ka) + I_0(mka)} \cong \frac{m^2 - 1}{m^2 + 1} \tag{8.101}$$

Next, the starting value for K can be found by equating the potential functions at the $z = 0$ interface. At the vertical vane tip that is displaced by a, we have

$$-\left(\frac{a}{r_0}\right)^2 - AI_0(ka) = -\left(\frac{a}{r_0}\right)^2 - A_{10}\left[I_0(Ka) + \frac{\alpha}{3}I_0(3Ka)\right] \tag{8.102}$$

The first terms on each side cancel. This time, to obtain an expression for K, we must use the first two terms in the expansion for I_0 because otherwise the expressions for K cancel. Thus,

$$A + A\frac{k^2 a^2}{4} = A_{10}\left(1 + \frac{\alpha}{3}\right) + A_{10}(1 + 3\alpha)\frac{K^2 a^2}{4} \tag{8.103}$$

Using the approximation for A given by Eq. (8.101), using $\alpha = 1$, and $A_{10} = 3A/4$, we obtain

$$K \approx \frac{1}{\sqrt{3}}k \tag{8.104}$$

Eq. (8.104) may be used as an initial value of K for an iterative solution of Eq. (8.100). Also, Eq. (8.104) can be used to estimate the length of the transition cell, since the length, L', of the transition cell and the length $L = \pi/k$ of the accelerating cell are related approximately by

$$\frac{\pi}{2L'} = \frac{1}{\sqrt{3}}\frac{\pi}{L} \tag{8.105}$$

or

$$L' = \sqrt{\frac{3}{4}}L \tag{8.106}$$

By differentiating the slopes x' and y', one can derive the second derivatives, x'' and y''. These can be evaluated for both potential functions at the interface between the accelerating cell and the transition cell, and also for the three-term potential at the end of the transition cell. Crandall shows that the second derivatives agree to first order at the interface, which implies that a smooth transition is made at the interface. Also, the second derivatives of the three-term potential function are zero at the end of the transition cell. Thus, at the end of the transition cell both the first and the second derivatives of the vane profiles are zero, which means that a smooth transition can be made to unmodulated vanes at the end of the transition cell. The potential and fields at the beginning of this transition cell blend smoothly with those at the end of the previous accelerating cell. The potential and fields at the end of the transition cell are the same as for unmodulated vanes. The vane tips at the end of the transition cell have quadrupole symmetry (four vanes equally spaced from the beam axis), so the on-axis potential is zero, and this removes the uncertainty in the output energy of the beam.

Crandall adds that by continuing the vanes with zero modulation for a short distance, the designer can control the point at which the beam exits from the periodic focusing system of the RFQ. This provides more control over the output transverse phase-space ellipses. The beam can be made to exit the RFQ having the characteristics it would normally have in the center of a quadrupole. This would allow an external matching quadrupole to be spaced farther from the end of the RFQ, and the strength requirement for this quadrupole would also be reduced.

Another transition-cell application is to allow a long RFQ to be divided into two shorter ones. The first RFQ would terminate with a transition cell followed by a short zero-modulation section; the second RFQ would begin with a short zero-modulation section followed by a transition cell followed by a normal acceleration cell. A low-current beam would require nothing but a short drift distance between these two RFQs.

An axisymmetric RFQ output beam with identical phase-space ellipses in x and y beam can also be obtained by adding an output RM section at the end of the transition cell. One particular application of an axisymmetric output beam is to allow both positive and negative beams accelerated by the RFQ to be matched to a following magnetic transport system.

8.15
Beam Ellipses in an RFQ

Ellipse configurations in the RFQ depend on time or phase. The intervane voltage is minimum at $-180°$, assuming a cosine wave, and maximum at $0°$. The x and y beam ellipses will be upright at $-180°$ (with, e.g., maximum x projection and minimum y projection), and upright with x and y projections interchanged at $0°$. At $-90°$ and $+90°$, the ellipses are tilted and the beam is round.

Now consider an accelerating cell and a synchronous particle with phase $\phi = -30°$. This means that the synchronous particle is at the middle of the cell when $f = -30°$. It is at the beginning of the cell when $\phi = -120$, and is at the end of the cell when $\phi = +60°$. The phases of the synchronous particle range from $-120°$ to $+60°$. At $30°$ after the beginning of the cell the phase of the synchronous particle is $-90°$ where the phase-space ellipses are tilted and the beam is round. At $30°$ after the particle reaches the center, the phase of the synchronous particle is $0°$, where the ellipses are upright. Thus, the ellipses are almost upright when the synchronous particle is at the middle of the cell. The synchronous particle is at the end of the cell at $30°$ before $+90°$, where the ellipses are tilted and the beam is round. Thus, for a synchronous phase of

$-30°$, the beam is almost round when the synchronous particle is at the end of the cell.

Next, consider a transition cell that follows an accelerating cell. Again we consider a synchronous phase $\phi = -30°$, and we ignore any change of β in the transition cell and treat β as a constant. We use the approximate value for the transition-cell length of $L_T = \sqrt{3}\beta\lambda/4 = 0.433\beta\lambda$, where $\beta\lambda/2$ is the length of the last accelerating cell. The total phase change in the transition cell is $\Delta\phi = 360\, L_T/\beta\lambda = 90° \times \sqrt{3} = 155.9°$. Thus, the synchronous-particle phase changes from $-120°$ at the beginning of the transition cell to $+35.9°$ at the end. At the end of the transition cell, this phase is $35.9°$ away from $0°$ where the ellipses are upright. Thus, the ellipses are almost upright when the synchronous particle is at the end of the transition cell. Suppose we want the beam to be round and have tilted ellipses when the synchronous particle is at the output. We could achieve this if we add an $m = 1$ vane extension, as discussed in the previous section, whose length changes the phase of the synchronous particle from $+35.9°$ to $+90°$. The required phase shift for the $m = 1$ extension is $90 - 35.9 = 54.1°$. The corresponding length for the $m = 1$ extension is $L_1 = 54.1\beta\lambda/360 = 0.15\beta\lambda$. Another round output-beam solution with x and y ellipses interchanged is to add an $m = 1$ extension that gets the synchronous particle from $+35.9°$ to $+270°$. The required phase shift for the $m = 1$ extension is $270 - 35.9 = 234.1°$. The corresponding length for the $m = 1$ extension is $L_1 = 234.1\beta\lambda/360 = 0.65\beta\lambda$. The round output beams are usually good solutions for cases where the beam must be matched into a FODO (+−) quadrupole-focusing system, as shown in Fig. 8.24

Another possibility is the case where a beam must be matched into a FODODOFO (+ − −+) focusing system. In this case good solutions are usually obtained when the output-beam phase is $0°$ or $180°$, when the beam ellipses are upright with a larger size in one plane and smaller size in the other plane. We could achieve the $180°$ solution, if we add an $m = 1$ extension whose length changes the synchronous-particle phase from $+35.9°$ to $+180°$. The required phase shift for the $m = 1$ extension is $180 - 35.9 = 144.1°$.

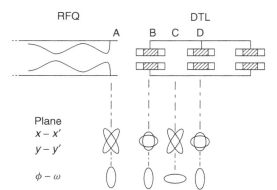

Figure 8.24 Matched phase-space drawings at end of RFQ and beginning of DTL (courtesy of K.R. Crandall).

The corresponding extension length is $L_1 = 144.1\beta\lambda/360 = 0.40\beta\lambda$. The upright-ellipse solution with the $x - x'$ and $y - y'$ ellipses interchanged corresponds to an output phase of $0°$ or $360°$. For this case we can add an $m = 1$ extension whose length changes the synchronous-particle phase from $+35.9°$ to $+360°$. The required phase shift for the $m = 1$ extension is $360 - 35.9 = 324.1°$. The corresponding length for the extension is

$$L_1 = 324.1 \, \beta\lambda/360 = 0.90 \, \beta\lambda \tag{8.107}$$

8.16
Tuning for the Desired Field Distribution in an RFQ

Consider an array of adjustable tuning stubs, also called *slug tuners*, installed in the outer cylindrical walls in each quadrant of an RFQ. Adjusting the radial penetration of these tuning stubs allows us to adjust the magnetic-field distributions. By proper adjustment of these stubs both longitudinal and azimuthal distributions can be adjusted in an RFQ. There would normally be four tuners at each longitudinal position, one in each quadrant. The longitudinal squared magnetic-field distribution near the outer wall of an RFQ can be measured by the standard bead pull method using a metallic bead as described in Section 5.13. The problem to be solved by one who tunes the cavity is to find the mechanical settings of the tuners that produce the desired magnetic-field distributions in the operating quadrupole mode.

The magnetic-field distribution at each longitudinal position is described in terms of the quadrupole and dipole amplitudes of the field, which can be obtained as discussed in Section 8.11. If B_1, B_2, B_3, and B_4 are the measured fields in the quadrants at a given longitudinal position, then the quadrupole amplitude is $A_Q = (B_1 - B_2 + B_3 - B_4)/4$ and the dipole amplitudes are $A_{D1} = (B_1 - B_3)/2$ and $A_{D2} = (B_2 - B_4)/4$. A good check on the measurement is that $B_1 + B_2 + B_3 + B_4 = 0$. The desired values of the dipole amplitudes are $A_{D1} = A_{D2} = 0$. These equations imply that the four fields in each quadrant at each longitudinal position should all be equal for a pure quadrupole mode. The desired longitudinal profile for $A_Q(z)$ is a known function specified by the designer. It is sometimes flat or uniform, but may have a tilt or a more complicated profile depending on the beam dynamics requirements.

Suppose we assume a linear relationship between each tuner setting and the quadrupole or dipole amplitudes at any given longitudinal position, and assume that there are no measurement errors. If the problem is linear and error-free, we could use a familiar method described below to solve a set of simultaneous linear equations to find a solution for the tuner settings that give the desired amplitudes. However, in the real world one may expect some errors in setting the tuner positions and in measuring the fields. Even for these cases, we can still find a solution for the tuner settings by iterating on the simultaneous linear equation solution.

From the linearity assumption, we write

$$V_i - V_{0i} = \sum_{j=1}^{N} \frac{\partial V_i}{\partial T_j}(T_j - T_{0j}), \quad i = 1, P \qquad (8.108)$$

where V_i are the desired A_Q, A_{D1}, or A_{D2} quadrupole or dipole field amplitudes at longitudinal locations i, which are known. The desired amplitudes are $A_Q(z)$, and $A_{D1}(z) = A_{D2}(z) = 0$. These V_{0i} are the actual field amplitudes at locations i, which are known from measurement. The T_j is the unknown desired setting for the tuners at locations j. The T_{0j} is the actual setting for the tuners at locations j. Each $\partial V_i/\partial T_j$ is the measured derivative of ith amplitude with respect to jth tuner.

In matrix form this equation is $V = MT$, where V is the amplitude vector, which is known, T is the tuner vector which is unknown since we do not know the correct tuner settings, and $M_{ij} = \partial V_i/\partial T_j$ is the derivative matrix, which is obtained from measurements. The tuner vector is obtained from the matrix equation above by inverting the M matrix. Then, one obtains the solution for the tuner settings by multiplying by the inverse of the derivative matrix to obtain $T = M^{-1}V$. From the resulting **T** vector, one can then adjust the tuners to new settings, and then remeasure the fields and to obtain a new amplitude distribution. If the new amplitude vector distribution is not close enough to zero, one can use the new V_{0i} and T_{0j} to carry out a new iteration.

8.17
Four-Rod Cavity

The four-rod cavity [11] is used mostly in the lower-frequency range, below about 200 MHz, and is the most commonly used RFQ structure for very low-velocity heavy ions. In principle the four-rod structure, shown in Figs. 8.25 and 8.26, is similar to the Wideröe structure. The four rods are charged from a linear array of conducting support plates or inductive stems. In the ideal

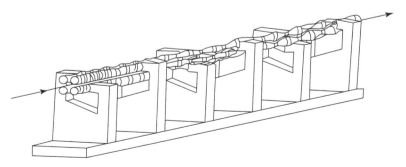

Figure 8.25 Four-rod RFQ (courtesy of A. Schempp).

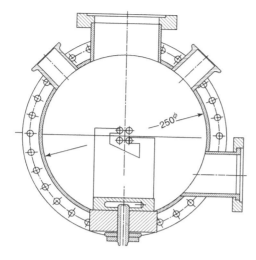

Figure 8.26 Cross section of the four-rod RFQ (courtesy of A. Schempp).

quadrupole geometry, adjacent rods are at opposite potentials, and opposite rods are at the same potential. The opposite pairs of rods are connected to the same plates and the plate connections alternate from one rod pair to the next along the length of the cavity. Because of the low-impedance path between the opposite rods (opposite rods are shorted together), dipole-mode frequencies are much higher than that of the quadrupole mode. Therefore, accidental degeneracy caused by the dipole modes is not a problem for the four-rod cavity, which helps to improve the stability of the fields.

The electric field is concentrated near the rods, and the magnetic fields are concentrated near the inductive stems. The charging currents flow longitudinally along the base from one stem to the next. The current density on the stems is higher than that on the vanes of the four-vane structure, which tends to reduce the efficiency compared with the four-vane cavity. This can be offset by the reduced capacity between the rods, which reduces the required charging current for a given voltage. The outer walls of the cavity carry little current, and the placement of the outer walls does not have a large effect on the resonant frequency. Consequently, the transverse size of the four-rod cavity can be very compact, an advantage for a low-frequency structure. The modulation pattern in the rods can easily be machined on a lathe, but can be machined on a milling machine in the form of short vanes. At higher frequencies, the rods become smaller and can be more difficult to cool than vanes. This can be a disadvantage for high-frequency applications, and is one of the main reasons why the four-vane structure is usually preferred for high frequencies at high duty factor.

8.18
Four Vane with Windows RFQ

At present the most commonly used RFQ structures are the four-vane and the four-rod structures. Usually the four-vane structure is preferred at frequencies above about 200 MHz because of its high RF power efficiency, and the four-rod structure is preferred below 200 MHz because of its compact size at low frequencies. Another type of RFQ structure has attracted interest more recently [12–16], which may be called a *four vane with windows RFQ*. The four vane with windows RFQ may be described as an intermediate resonator configuration between the four-rod and the four-vane RFQ. It may be configured as a four-vane RFQ with holes or windows in the vanes (Fig. 8.27), or as a four-rod RFQ with individual stems supporting each of the four rods, where the stems are oriented 90° apart (Fig. 8.28).

With its close relationship to both the four-vane and the four–rod structures, there would seem to be no reason why this structure could not be used over a broader frequency range that encompasses the range of both the four-vane and the four-rod structures. Most importantly, because of the strong

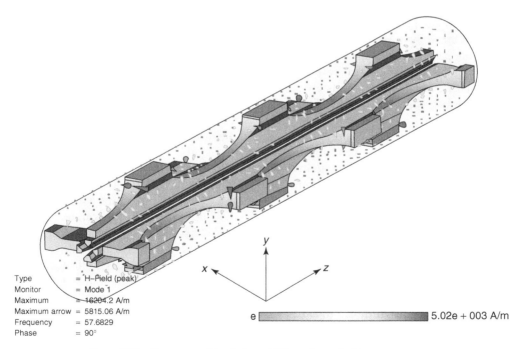

Figure 8.27 Four vane with windows RFQ configured with windows in the vanes and with a fourfold antisymmetric window arrangement where the windows in adjacent vanes are placed longitudinally 180° apart (courtesy of P. Ostroumov).

Figure 8.28 Four vane with windows RFQ configured as a structure with radial stems that support the four vanes or rods. This structure has a fourfold symmetric window arrangement (courtesy of D. Swenson).

magnetic coupling through the windows between neighboring quadrants, the introduction of windows can lower the frequency of the quadrupole operating mode well below that of the dipole modes, reducing the undesirable dipole-mode degeneracy, which complicates the RFQ tuning in a four-vane resonator. At the same frequency, the four vane with windows RFQ is more radially compact than the four-vane RFQ, and is less compact than that the four-rod RFQ. At a given frequency, the four vane with windows RFQ is less efficient than the four-vane RFQ and is more efficient than the four-rod RFQ. The antisymmetric window configuration, shown in Fig. 8.27, where adjacent windows are shifted by 180° is more effective in separating the dipole modes from the quadrupole mode. However, the four vane with windows RFQ can also be built to maintain fourfold symmetry of the windows. This allows the design of an RFQ with zero on-axis potential at the entrance and exit gaps, making matching easier at the entrance and avoiding uncertainty in the output energy at the exit.

Delayen [12] gives quadrupole and dipole-mode frequency results obtained from the electromagnetic-field-solver code MAFIA for different window-geometry parameters.

Delayen has also developed a transmission-line model [12] for the case where the windows extend to the outside diameter of the structure, corresponding to $b = 0$ in Fig. 8.29. The resonant wavelength λ for the quadrupole mode satisfies

$$2\tan\frac{2\pi h}{\lambda}\left\{\tan\left[\frac{2\pi}{\lambda}\left(\frac{L-t}{2}\right)\right] + \frac{\pi t}{\lambda}\right\} = \frac{\varepsilon_0}{C_\ell}\left[\frac{2t}{g} + \alpha\right] \quad (8.109)$$

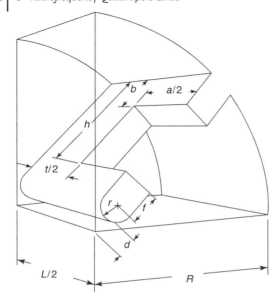

Figure 8.29 Drawing of one quadrant of the periodic four vane with windows RFQ structure with period L (courtesy of J. R. Delayen).

The capacitance per unit length of the four rods, C_ℓ, is given in picofarad per meter, by

$$C_\ell = \frac{39.365}{\cosh^{-1}\left(\frac{d+r}{r\sqrt{2}}\right)} + \frac{31.045}{1 - \sqrt{2} + \frac{d}{r}} + 25.28 \ln\left(1 + \frac{f}{d+r}\right),$$

(8.110)

where a is a constant of order 3, and $g \cong [(2R - h)/2\sqrt{2} - r]$. The right side of Eq. (8.109) is valid when $t < L - \alpha g/2$. When $L - \alpha g/2 < t < L$, the right side of Eq. (8.109) is replaced by $2\varepsilon_0 L/C_\ell g$.

Problems

8.1. Assume an RFQ cell has the following parameters: $\beta_s = 0.025$, $\lambda = 0.75$ m, $a = 1.875$ mm, and $V = 50$ kV. (a) Assume the vane-modulation parameter is $m = 2$. Calculate the focusing efficiency X, the characteristic radius r_0, the acceleration efficiency A, and the average axial field E_0. If the synchronous phase is $\phi = -30°$, what is the energy gain of the synchronous particle in the cell? (b) Repeat part (a) assuming $m = 1$.

8.2. The alternating-gradient focusing principle allows the time-varying electric quadrupole to produce a net focusing effect in both planes. Suppose that the time-dependent electric fields for the electric quadrupole are derivable in the quasistatic approximation from a scalar potential with harmonic time dependence, given by

$$U(x, y) = \frac{V_0}{2} \left[\frac{[x^2 - y^2]}{a^2} \right] \sin \omega t$$

where ω is the RF frequency. We will see that by adding the time-dependent factor, a time-dependent polarity variation is produced, which results in a net focusing effect in both planes. (a) Write the equation of motion for the y plane. (b) Substitute a trial solution in the form of an amplitude-modulated sinusoid, $y = C[1 + \varepsilon \sin \omega t] \sin \Omega t$, where ε is called the *flutter amplitude* of the rapidly oscillating RF factor, C is a constant, and Ω is the frequency of the slowly varying average trajectory. Assume that $\Omega/\omega \ll 1$ and $\varepsilon \ll 1$. By substituting this solution into the equation of motion, identify in lowest order the value of ε that makes the trial solution approximately valid. (c) Show that the averaged trajectory is confined by the time-dependent focusing system, by considering the equation of motion averaged over an RF period, and showing that the averaged trajectory satisfies the equation of simple harmonic motion. Calculate the frequency of this average motion.

8.3. To understand how an RFQ quadrupole-focusing geometry can be modified to function as a linear accelerator, we need to see how to perturb the electrodes to obtain a longitudinal electric field, or equivalently to obtain a spatially varying axial potential. (a) Suppose that the x-pole tips at voltage $+V_0/2$ have radius a, and the y-pole tips at voltage $-V_0/2$ have radius ma with $m \geq 1$. Show that the potential function for this geometry is given by

$$U(x, y) = \frac{V_0}{2} \left(\left[\frac{X}{a^2} [x^2 - y^2] \right] + A \right)$$

where X and A are constants (called *focusing and acceleration efficiencies*), by finding the expressions for X and A as a function of the parameter m that satisfy the boundary conditions. How do X and A change as a function of m, if the radii of the x and y pole tips are interchanged to ma and a respectively, but their voltages remain the same?

8.4. Consider the three-dimensional time-dependent potential function with a spatially oscillating longitudinal term

$$U(x, y, z, t) = \frac{V_0}{2} \left[X \frac{(x^2 - y^2)}{a^2} + A I_0(kr) \cos(kz) \right] \sin(\omega t + \phi)$$

which, by direct substitution, can be shown to be a solution of Laplace's equation. The parameters X and A are constant focusing and acceleration efficiencies, I_0 is the modified Bessel function of order zero, $r = \sqrt{x^2 + y^2}$ and $k = 2\pi/\beta\lambda$, where βc is the velocity of the synchronous beam particle (c is the speed of light), $\lambda = 2\pi c/\omega$ is the RF wavelength and $\beta\lambda$ is equal to the distance traversed by a synchronous particle in one RF period.

The voltage of the horizontal poles is $V_0 \sin(\omega t + \phi)/2$, and that of the vertical poles is $-V_0 \sin(\omega t + \phi)/2$. The geometry in the $y - z$ plane is shifted axially by 180°. At $z = 0$ the horizontal pole-tip coordinates are $x = a$, $y = 0$, and the vertical pole-tip coordinates are $x = 0$, $y = ma$, where $m \geq 1$. (a) Derive the expression for X and A as a function of the geometry parameters a, m, and k, by assuming that the boundary conditions on the pole tips are satisfied at $z = 0$. (b) Derive the expression for the pole tips in the $x - z$ and $y - z$ planes. (c) Obtain the expressions for the three electric-field components. (d) In the approximation that the radial position and the velocity of a particle within a period are constant, calculate the energy gain ΔW for a synchronous particle with velocity βc, by replacing $\omega t = 2\pi z/\beta \lambda$ and integrating the axial electric field seen by the particle over one period.

8.5. An RFQ is designed to operate at a frequency of 400 MHz and has a vane length of 2.25 m. An RF-drive port that is introduced at the midpoint of the RFQ cavity lowers the resonant frequency by 4 MHz. Assuming that this perturbation can be approximated by a δ-function error at the midpoint, and that first-order perturbation theory is valid, calculate the fractional-field error at both ends and at the midpoint of the cavity, and plot the three values versus distance along the cavity. You may use the series summations given in the text material.

8.6. An error in the fabrication of a 400-MHz RFQ, whose vane length is 1.5 m, results in unequal dimensions for the two end cells. Assume that this perturbation can be approximated by equal and opposite δ-function errors at each end of the RFQ, such that the cavity frequency error at one end alone would produce a resonant-frequency error of 2 MHz and that at the opposite end would produce a frequency error of -2 MHz. Using first-order perturbation theory: (a) What is the resonant frequency of the cavity? (b) Calculate the fractional-field error at both ends and at the midpoint of the cavity, and plot the three values versus distance along the cavity. Use these three points versus distance along the RFQ to calculate a slope that characterizes this field tilt. You may use the appropriate series summations as given in the text material.

References

1 Kapchinskiy, I.M. and Tepliakov, V.A., *Prib. Tekh. Eksp.* **2**, 19–22 (1970).
2 Kapchinskiy, I.M. and Tepliakov, V.A., *Prib. Tekh. Eksp.* **4**, 17–19 (1970).
3 The emittance of a beam, which will be defined in Chapter 9, is a measure of its phase space extent, and emittance growth reduces the minimum achievable beam size for a given focusing system.
4 Crandall, K.R., Stokes, R.H. and Wangler, T.P., Proc. 1979 Linac Conf., Brookaven National Laboratory Report BNL-51134, 1979, pp. 205–216.
5 There is no TE_{210} mode of a cylindrical cavity with conducting end walls, because the conducting end-wall boundary condition requiring a zero transverse electric-field component is not satisfied.

6 Schrage, D., *et al.*, Proc. of 1988 Linac Conf., Williamsburg, VA, CEBAF Report 89-001, p. 54; Schrage, D., *et al.*, Proc. of 12th Int. Conf. on Applications of Accelerators in Research and Industry, NIM B79, 1993, p. 372; Schrage, D., *et al.*, Proc. of 1994 Int. Linac Conf., Tsukubu, Japan.

7 Some references to RFQ tuning methods may be found in Stokes, R.H. and Wangler, T.P., Radiofrequency quadrupole accelerators and their applications, *Annu. Rev. Nucl. Part. Sci.* 38, 97–118 (1988); A resonant-coupling method is described in Young, L.M., Proc. 1993 Particle Accel. Conf., May 17–20, 1993, Washington, D.C., p. 3136; Proc. 1994 Linear Accel. Conf., August 21–26, 1994, Tsukubu, Japan, p. 178.

8 Mathews, J. and Walker, R.L., *Mathematical Methods in Physics*, 2nd ed., W. A. Benjamin, 1970, Chapter 10.

9 Crandall, K.R., "*RFQ Radial Matching Sections and Fringe Fields*", Proc. 1984 Linear Accelerator Conf., Seeheim, Germany, May 7–11.

10 Crandall, K.R., "Ending RFQ Vanetips With Quadrupole Symmetry", Proc. of 1994 Linac Conf., Tsukuba, Japan, pp. 227–229.

11 For a review of the four-rod RFQ structure, see Schempp, A., "Radio-Frequency Quadrupoles", CERN Accelerator School, CERN 92-03, Vol II, p. 522.

12 Delayen, *et al.*, "Design and Modeling of Superconducting RFQ Structures", Proc. of the 1992 Linear Accel. Conf., Aug. 24–28, 1992, Ottawa, Ontario, Canada, pp. 692–694.

13 Shepard, *et al.*, "Design for a Superconducting Niobium RFQ Structure", Proc. of the 1992 Linear Accel. Conf., Aug. 24–28, 1992, Ottawa, Ontario, Canada, pp. 441–443.

14 Donald A., Swenson, Wayne, Cornelius and Young, P.E., "Segmented Vane Radio-Frequency Quadrupole Linear Accelerator", United States Patent 5,430, 359, filed Nov. 2, 1992, Date of Patent July 4, 1995.

15 Andreev, V.A. and Parisi, G., "90° Apart Stem RFQ Structure for Wide Range of Frequencies", Proc. of 1993 Part. Accel. Conf., Washington, DC, May 13–18, pp. 3124–3126.

16 Ostroumov, P.N. and Kolomiets, A.A., Design of 57.5 MHz CW RFQ for medium energy heavy ion superconducting linac, *Phys. Rev. Spec. Top. Accel. Beams* 5, 060101 (2002).

9
Multiparticle Dynamics with Space Charge

Particle motion in a linac depends not only on the external or applied fields, but also on the fields from the Coulomb interactions of the particles, and the fields induced by the beam in the walls of the surrounding structure. In Chapters 6 and 7, we treated the longitudinal and transverse dynamics, but we included only the applied forces from the radio frequency (RF) and from static focusing fields. The Coulomb forces play an increasingly important role as the beam-current increases, and that topic is covered in this chapter. The effects caused by the interactions with the structure are the subjects of beam loading and wake fields, which are treated in Chapters 10 and 11. The Coulomb effects in linacs are usually most important in nonrelativistic beams at low velocities, because at low velocities the beam density is larger, and for relativistic beams the self-magnetic forces increase and produce a partial cancellation of the electric Coulomb forces. The net effect of the Coulomb interactions in a multiparticle system can be separated into two contributions. First is the space-charge field, the result of combining the fields from all the particles to produce a smoothed field distribution, which varies appreciably only over distances that are large compared with the average separation of the particles. Second are the contributions arising from the particulate nature of the beam, which includes the short-range fields describing binary, small impact-parameter Coulomb collisions. Typically, the number of particles in a linac bunch exceeds 10^8, and the effects of the collisions are very small compared with the effects of the averaged space-charge field [1]. To describe the space-charge field, we need to understand the properties of an evolving particle distribution, which requires a self-consistent solution for the particles and the associated fields. This is a problem, which has been formulated in terms of the coupled Vlasov–Maxwell equations, for which there are no generally successful analytic solutions in a linac, and computer simulation is the most reliable tool. In this chapter we include an introduction to the most common methods used for computer simulation of the space-charge forces. We also find that among the most important properties of the distribution besides the beam current are the emittances, which are defined from the second moments of the particle distribution and are measures of the phase-space areas occupied by the beam in each of the three projections of position-momentum phase space. The emittances are important measures of the beam quality; they determine the inherent capability of producing, by means of a suitable focusing system, small sizes for the waists, angular divergences, micropulse width, and energy spread. The significance of the space-charge fields is not only that they reduce the effective focusing strength, but also the nonlinear terms, a consequence of the deviations from charge-density uniformity, cause growth of the rms emittances, which degrades the intrinsic beam quality. One consequence of space-charge-induced emittance growth is the formation of a low-density beam halo surrounding the core of the beam, which can be the cause of beam loss, resulting in radioactivation of the accelerating structure.

9.1
Beam Quality, Phase Space, and Emittance

A beam bunch in a linac consists of a collection of particles, moving in approximately the same direction, with approximately the same positions, phases, and energies. The term *beam quality*, although not precise, refers to a coherence property of the beam: the degree to which the beam particles have nearly the same coordinates as the reference particle, which travels on the accelerator axis with a specified energy and phase. The ideal beam with highest beam quality is called the *laminar beam* because it exhibits laminar-like flow. A laminar beam represents the ideal of a highly ordered and coherent beam, which is never exactly realized. As a quantifier of beam quality, it is more convenient to work with a directly measurable quantity known as *beam emittance*, which is introduced in this section.

Every particle in the beam can be described by three position and three momentum coordinates for a total of six coordinates per particle. Then, each particle is represented by a single point in the six-dimensional phase space of coordinates and momenta. In practice, it is more convenient to work with the two-dimensional phase-space projections of the beam, which correspond more directly to what can be measured. For a linac beam, one can refer to a collection of points in the three normalized phase-space projections $x - p_x/mc$, $y - p_y/mc$, and $z - p_z/mc$, where x, y, and z are the coordinates, and p_x, p_y, and p_z are the momentum components. Phase-space areas can be associated with the collection of points inside specified density contours in each of the two-dimensional phase-space projections, as shown in Fig. 9.1. Instead of the transverse momenta, it is convenient to measure the divergence angles, dx/ds and dy/ds. Plots of $x - dx/ds$ and $y - dy/ds$ are known as the *trace-space* or *unnormalized phase-space projections*. In longitudinal phase space, position and momentum relative to the synchronous particle can be used, but more often, these are replaced by the phase and energy variables $\Delta\phi$ and ΔW.

Before defining beam emittance, we need to recognize the special significance of elliptical phase-space distributions. The beam-phase-space

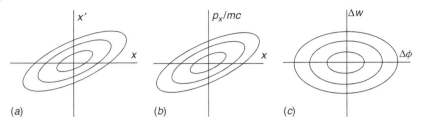

Figure 9.1 Isodensity contours of the beam in (a) unnormalized transverse phase space, (b) normalized transverse phase space, and (c) longitudinal phase space. The emittance associated with any elliptical contour is the area of that contour divided by π.

contours in a linac often have the approximate shape of an ellipse. One reason for this shape is related to the predominance in most accelerators of linear focusing forces. With linear focusing, the trajectory of each particle in phase space lies on an ellipse, which may be called the *trajectory ellipse*. In general, a beam is said to be *matched* when the phase-space isodensity contours of the beam are concentric and geometrically similar to the trajectory ellipses. Therefore matched beams in linear focusing systems have elliptical density contours. Furthermore, with linear focusing, elliptical distributions in phase space, even for unmatched beams, remain elliptical. Because of the tendency of linac beams to exhibit approximately elliptical phase-space distributions, it has become conventional to define for each two-dimensional projection, a quantity called the *emittance*, which is proportional to the area of a chosen beam ellipse. The general equation for an ellipse is written in Chapter 7 as

$$\tilde{\gamma}x^2 + 2\tilde{\alpha}xx' + \tilde{\beta}x'^2 = \varepsilon \tag{9.1}$$

where $\tilde{\gamma}\tilde{\beta} - \tilde{\alpha}^2 = 1$, and where the area of the ellipse is $\pi\varepsilon$. The definition of the emittance ε associated with any two-dimensional elliptical phase-space area is

$$\varepsilon = \frac{\text{Area}}{\pi} \tag{9.2}$$

Having defined the emittance in terms of an elliptical area in 2D phase space, we return to the question of which phase-space area to choose to characterize a real beam. If the beam distribution in phase space had a well-defined boundary, the emittance could be defined simply by the area within that boundary. Generally, real beams do not have well-defined boundaries. One method for assigning an emittance is to choose a specific density contour in phase space, such as one at 50 or 90 or 95% of the maximum density. It can be shown [2] that under certain conditions such emittances are conserved: (1) when Liouville's theorem is satisfied in the six-dimensional phase-space, and (2) when the forces in the three orthogonal directions are uncoupled. As discussed later in this chapter, Liouville's theorem is satisfied when there are no dissipative forces, no particles lost or created, and no small-impact-parameter binary Coulomb collisions between particles. When these conditions hold, the volume of six-dimensional phase space defined by any fixed-density contour of the beam is invariant. While this is an important characteristic, the most useful definition of emittance in a linac is the rms emittance, which is described in the following section.

9.2
RMS Emittance

The presence of nonlinear forces can produce a considerable departure from elliptical trajectories, and can distort the phase-space contours. If the phase-space projections are described by effective phase-space ellipses, such distortions cause an effective-emittance growth, resulting from a filamentation process in six-dimensional phase space, in which outer filaments develop, enclosing regions with rarefied density and resulting in a dilution of the phase-space density. Furthermore, forces that couple the motion between the three directions can cause real growth in the two-dimensional projected areas, even though the six-dimensional volume is constant. Consequently, it becomes important to introduce a definition of an effective emittance. The definition is most conveniently based on mean-square values, or second moments of the coordinates and momenta. An rms emittance definition is more useful because it can easily be defined for an arbitrary particle distribution, being determined only by the rms characteristics of the beam distribution. As shown in Fig. 9.2, the general form of the rms ellipse is defined by Courant–Snyder parameters $\tilde{\alpha}_r$, $\tilde{\beta}_r$, and $\tilde{\gamma}_r$, and an rms emittance ε_r. To be specific we describe the definitions for an unnormalized transverse emittance in x–x' space. The point (x, x') lies on the rms ellipse if

$$\tilde{\gamma}_r x^2 + 2\tilde{\alpha}_r xx' + \tilde{\beta}_r x'^2 = \varepsilon_r \tag{9.3}$$

where $\tilde{\gamma}_r \tilde{\beta}_r - \tilde{\alpha}_r^2 = 1$. We define the ellipse parameters in the following manner. First, we require that the ellipse projections on the x and x' axes are equal to the rms values of the distribution. Thus, we require that

$$\overline{x^2} = \tilde{\beta}_r \varepsilon_r \quad \text{and} \quad \overline{x'^2} = \tilde{\gamma}_r \varepsilon_r \tag{9.4}$$

Next we define $\tilde{\alpha}_r$. For the case of an ideally matched beam with linear focusing, where the isodensity contours of the beam are concentric and similar to the trajectory ellipses, we would like to define $\tilde{\alpha}_r$ so that the rms ellipse also coincides with a trajectory ellipse. Then, because the trajectory ellipses relate $\tilde{\alpha}$ and $\tilde{\beta}$ by $\alpha = -\tilde{\beta}'/2$, we require the same relationship for α_r and β_r. Thus, we define

$$\tilde{\alpha}_r = -\frac{\tilde{\beta}'_r}{2} = -\frac{1}{2\varepsilon_r}\frac{d\overline{x^2}}{ds} = -\frac{\overline{xx'}}{\varepsilon_r} \tag{9.5}$$

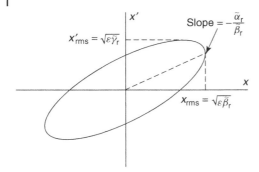

Figure 9.2 The rms ellipse.

where we have assumed that the emittance of the matched beam is constant. The Courant–Snyder parameters for an rms ellipse have been defined in terms of the second moments of the distribution from Eqs. (9.4) and (9.5). We recall that the three ellipse parameters are not independent, but are related by $\tilde{\gamma}_r\tilde{\beta}_r - \tilde{\alpha}_r^2 = 1$. The rms emittance as a function of the second moments is obtained by substituting Eqs. (9.4) and (9.5) into this relationship, which leads to

$$\varepsilon_r = \sqrt{\overline{x^2}\,\overline{x'^2} - \overline{xx'}^2} \qquad (9.6)$$

A simple exercise may be helpful in understanding an important characteristic of the rms emittance. Consider an idealized particle distribution in phase space representing a beam that lies on some line that passes through the origin. This is illustrated by two examples, shown in Fig. 9.3. Assume that, for any x, the divergence x' of the particles is given by

$$x' = Cx^n \qquad (9.7)$$

where n is positive, and C is a constant. The second moments of this distribution are easily calculated using Eq. (9.7), and the squared rms emittance is given by

$$\varepsilon_r^2 = \overline{x^2}\,\overline{x'^2} - \overline{xx'}^2 = C^2\left[\overline{x^2}\,\overline{x^{2n}} - \overline{x^{n+1}}^2\right] \qquad (9.8)$$

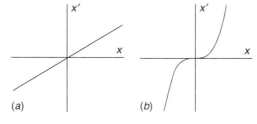

Figure 9.3 The rms emittance ε_r for two distributions with zero area: (a) straight line where $\varepsilon_r = 0$, (b) curved line where $\varepsilon_r \neq 0$.

When $n = 1$, the line is straight and the rms emittance is $\varepsilon_r = 0$. When $n \neq 1$, the relationship is nonlinear, the line in phase space is curved, and the rms emittance is in general not zero. Both distributions have zero area. Therefore, we conclude that, even when the phase space area is zero, if the distribution lies on a curved line, its rms emittance is not zero. The rms emittance depends not only on the true area occupied by the beam in phase space, but also on the distortions produced by nonlinear forces.

9.3
Transverse and Longitudinal Emittance

Transverse emittance can be determined from measurements of the particle distribution as a function of displacement and angular divergence. This distribution determines what is called the *unnormalized emittance*. For example, an upright ellipse, with semiaxes Δx and $\Delta x'$, has an unnormalized emittance

$$\varepsilon = \Delta x \Delta x' \tag{9.9}$$

For an accelerator, it is more convenient to define what is called the *normalized emittance*, for which the dimensionless transverse momentum variable $\Delta p_x / mc$ is used instead of the divergence. The normalized and unnormalized emittances are related by

$$\varepsilon_n = \varepsilon \beta \gamma \tag{9.10}$$

The reason for introducing a normalized emittance is that the transverse momenta of the particles are unaffected by acceleration, but the divergences of the particles are reduced during acceleration because $x' = p_x/p$. Thus, acceleration reduces the unnormalized emittance, but does not affect the normalized emittance. The typical unit of either unnormalized or normalized emittance is meter-radians or millimeter-milliradians and typical normalized rms emittance values range from a few tenths to a few millimeter-milliradians.

The variables used for longitudinal emittance can be chosen in several ways. First, we write a normalized longitudinal emittance for an upright ellipse as

$$\varepsilon_{n\ell} = \frac{\Delta z \Delta p}{mc} = \frac{\Delta t \Delta W}{mc} = \frac{\Delta \phi \Delta W}{\omega mc} \tag{9.11}$$

where we used the relationships $v\Delta p = \Delta W$, $\Delta z = v \Delta t$, and $\Delta \phi = \omega \Delta t$. This emittance is expressed in the same units of length as transverse emittances. However, phase difference $\Delta \phi$ and energy difference ΔW are more convenient longitudinal variables in a linac, and a more common form for longitudinal emittance in a linac is

$$\varepsilon_\ell = \Delta \phi \Delta W \tag{9.12}$$

The units for Eq. (9.12) are usually expressed in deg-MeV. This emittance is proportional to the normalized emittance and does not change because of acceleration; unfortunately, it does depend on the RF frequency. For example, if the frequency of the linac doubles at a certain energy, the emittance ε_ℓ also doubles. For a linac with a frequency transition, an emittance defined by $\Delta t \Delta W$ would not change when the frequency changes.

9.4
Emittance Conventions

Lack of agreement on emittance conventions has led to much confusion. One ambiguity is whether emittance is defined as area/π as we have done, or as area. One convention in the literature is to quote the emittance value as the area, showing an explicit factor of π; for example, $\varepsilon = 1\pi$ mm-mrad. For those who interpret the emittance as area/π, as done in this book, the emittance is really the factor multiplying the π; thus in the example, the emittance is really 1 mm-mrad.

The rms emittance has other areas of confusion. Some people multiply the rms emittance that we have defined, which is the Sacherer [3] convention, by 4 and call this the *rms emittance*. This is the Lapostolle [4] convention, which is useful because, for continuous beams, it gives the total emittance of an equivalent uniform beam. Also, for almost all other real beams, the value of the rms emittance is close to the emittance containing 90% of the beam. Use of the Lapostolle convention becomes confusing, only when it is also called the rms emittance. One way to avoid confusion is to call this the *4-rms emittance*. Some have also found it convenient to quote an emittance of 6 times the rms emittance. This contains 95% of the beam for Gaussian beams. If this is called the *6-rms emittance* instead of the rms emittance, it should lead to no confusion. The only safe solution for these problems is to carefully define the emittance. Because of the lack of a convention for defining and expressing emittance, the reader of accelerator literature must beware, especially if the authors have not carefully defined the emittance they quote.

It is sometimes useful to refer to the total emittance of a distribution that has been generated for computer simulations. A useful result relates the total emittance to the rms emittance for uniform hyper-ellipsoids in n-dimensional space. This was calculated by Weiss [5] and is

$$\varepsilon_r = \frac{\varepsilon_{\text{total}}}{n+2} \tag{9.13}$$

Some examples include $n = 6$, uniform density in 6D space, known as *the 6D Waterbag distribution* where $\varepsilon_r = \varepsilon_{\text{total}}/8$; $n = 4$, uniform density in 4D space, known as the *4D Waterbag distribution* where $\varepsilon_r = \varepsilon_{\text{total}}/6$; $n = 3$, uniform in all three-dimensional projections where $\varepsilon_r = \varepsilon_{\text{total}}/5$; and $n = 2$, uniform density in all two-dimensional projections, known as the *Kapchinsky–Vladimirsky distribution* where $\varepsilon_r = \varepsilon_{\text{total}}/4$.

9.5
Space-Charge Dynamics

Before beginning our discussions of space-charge and other high-intensity effects, we note that in the usual nomenclature for bunched beams in linacs, there are several different definitions of current that are used, which often creates confusion, even among linac experts. For proton linacs there are two definitions of current that are most frequently specified for a pulsed machine. One definition is the value averaged over the RF *macropulse*, given by $I = qNf$, where q is the charge per particle, N is the average number of particles per bunch, and f is the bunch frequency. This is usually called the *peak current*, because it is the current that is present during the macropulse. A second common definition of current in proton linacs is called the *average current*. The average current is the current averaged over a full RF macropulse cycle, including both the macropulse time and the time between the macropulses. For a CW proton linac (100% duty factor), the peak and average currents are equal, while for any pulsed linac, the average current is always less than the peak current. A typical high value for the peak current in a proton linac is about 100 mA, and average currents above about 1 mA are usually considered high values. Generally, these two definitions are sufficient to describe the beam intensity in a proton linac. A more complicated situation occurs when the pulsed-linac beam is chopped, which is done for applications where the linac beam is injected into and stacked in an accumulator ring. An example is an H^- linac injector for a neutron-spallation source, where the RF period is about 1 ns, the chopper period is several hundred nanoseconds, and the RF macropulse period is about 1 ms. For this case, one must distinguish between the peak current defined so as to either exclude or include the chopping cycle. The former current is relevant to the beam dynamics, and the latter is of relevance for the performance of the RF system, which has a longer time constant than the chopping period. In addition, for some of the space-charge models one may refer to a dc current over the time duration of the passage of bunch, which equals the average current while the bunch is passing a given point, given by $I = qN_\ell v$, where N_ℓ is the number of particles per unit length and v is the beam velocity.

For electron linacs, the current is often defined differently than for proton linacs. This difference is probably because for the most common electron-linac applications, such as free-electron lasers and linear colliders, what matters most is the charge per bunch, or the charge divided by the time spread of the bunch. Often the electron linacs bunch the beam using a subharmonic-frequency buncher, such that ratio of filled to total buckets is small. The averaging, which is done for the proton-linac current definitions, is of little interest for some electron linacs. A common definition of electron-linac peak current is the total charge per bunch divided by the full-width half maximum of the micropulse duration. Typical peak currents for high-current electron linacs range from a few tens to a few hundreds of amperes. For some electron-linacs, a current is not quoted, but merely the total charge per bunch. The overall conclusion is that because of the different definitions of current, one must be careful to determine the precise definition that is being used.

Now, we are ready to begin a general discussion of space-charge effects. With increasing beam intensity the interaction between the charged beam particles becomes more important. The charges produce mutually repulsive electric fields that act in opposition to the focusing forces, and also magnetic fields that produce attractive forces. The magnetic forces are smaller than the electric forces, and are unimportant except for relativistic particles. The total Coulomb field experienced by any particle is the sum of the fields due to all the other particles in the beam. For real linac beams with the order of 10^8 or more particles per bunch, the method of directly summing the fields from this many particles does not lend itself to a practical computational approach. One may classify the self-interaction of the particles into two categories, the *collisional regime*, dominated by binary collisions caused by close encounters, and the *collective* or *space-charge regime*, the result of many particles that produce an average or collective field that can be represented by a smooth field as a function of space and time. The space-charge regime occurs when there are enough neighboring particles to shield the effects of density fluctuations, a phenomenon known in plasma physics as *Debye shielding*. Ignoring the effect of the finite-sized beam, the criterion for the collective Debye shielding for a beam with number density n is that the number of particles N_D in a Debye sphere, [6] a sphere with radius equal to the beam-Debye length λ_D, is

$$N_D = n \frac{4\pi \lambda_D^3}{3} \gg 1 \qquad (9.14)$$

where $\lambda_D = \sqrt{\varepsilon_0 k_B T / n q^2}$ and k_B is Boltzmann's constant, and for a spherical bunch with rms projections a, and $k_B T = mc^2 \varepsilon_n / a^2$. For accelerator beams this criterion is usually satisfied and the smoothed space-charge field describes the main effect of the Coulomb interactions.

With collisions neglected, Liouville's theorem, written in the form that expresses continuity of particles in phase space, is

$$\frac{df}{dt} = \frac{\partial f}{\partial t} + \dot{\mathbf{x}} \cdot \frac{\partial f}{\partial \mathbf{x}} + \dot{\mathbf{p}} \cdot \frac{\partial f}{\partial \mathbf{p}} = 0 \qquad (9.15)$$

where $f(\mathbf{x}, \mathbf{p}, t)$ is the particle density in phase space. Expressing $\dot{\mathbf{p}}$ in terms of the sum of the external fields plus the smoothed self fields yields the *Vlasov equation*, [7] also known as the *kinetic equation*, or the *collisionless Boltzmann equation*,

$$\frac{\partial f}{\partial t} + \frac{\mathbf{p}}{\gamma m} \cdot \frac{\partial f}{\partial \mathbf{x}} + q\left(\mathbf{E} + \frac{\mathbf{p} \times \mathbf{B}}{\gamma m}\right) \cdot \frac{\partial f}{\partial \mathbf{p}} = 0 \tag{9.16}$$

Rather than describing the motion of single particles, the object is to follow the evolution of the distribution function that is consistent with the electromagnetic fields. The Vlasov equation and Maxwell's equations form a set of closed equations, which determine the self-consistent dynamics of a distribution of charges satisfying Liouville's theorem. Unfortunately, there is no general method for obtaining analytic solutions of this set of equations for linac beams. In practice, equivalent solutions are usually obtained numerically on the computer from the more direct approach of simulating the particle interactions, typically using 10^4 to 10^5 simulation particles, as discussed in Section 9.6.

As the previous discussion suggests, space-charge-dominated beams behave in some respects like plasmas. Similarities include particle shielding from the effects of density variations and external focusing fields as described above, and density oscillations at the beam-plasma frequency for space-charge-dominated beams. At first, it may seem surprising that a beam with a single charge species can exhibit plasma oscillations, which are characteristic of plasmas composed of two-charge species. But the effect of the external focusing force can play the role of the other charge species to provide the necessary restoring force that is the cause of such oscillations.

It is important to bear in mind that beams in linacs are also different from plasmas in some important respects. One is that beam transit time through a linac is too short for the beam to reach thermal equilibrium. Also, because Liouville's theorem and phase-space conservation are a good approximation, the phase-space volume of the beam is generally a more important parameter than beam temperature. Also, unlike a plasma, the Debye length of the beam may be larger than or comparable to the beam radius, so shielding effects may be incomplete. The space-charge fields can be separated into linear and nonlinear terms as a function of the displacement from the centroid. The linear space-charge term defocuses the beam and leads to an increase in the beam size. The nonlinear space-charge terms also increase the effective emittance by distorting the phase-space distribution, further increasing the spread of the displacements, divergences, and energy.

One must account not only for the electric force between the particles but also for a magnetic force associated with the motion of the charges in the beam. While the electric force causes defocusing, the magnetic force causes attraction. This can be seen by considering the simple example of a long bunch, approximately represented by a cylinder of charge with azimuthal symmetry and charge density $qn(r)$ moving at velocity v. The electric field is directed radially outward and from Gauss's law is

$$E_r = \frac{q}{\varepsilon_0 r} \int^r n(r) r \, dr \qquad (9.17)$$

From Ampere's law, the magnetic field is azimuthally directed, and is

$$B_\theta = \frac{q v \mu_0}{r} \int^r n(r) r \, dr \qquad (9.18)$$

The radial Lorentz force is $F_r = q(E_r - vB_\theta) = qE_r(1-\beta^2) = qE_r/\gamma^2$. Thus the attractive magnetic force, which becomes significant at high velocities, tends to compensate for the repulsive electric force. Therefore space-charge defocusing is primarily a nonrelativistic effect. It is of great concern for intense low-velocity ion beams, and for electron beams at injection into a linac, but not for relativistic electron beams where $\beta \approx 1$. In the paraxial approximation, Newton's second law for transverse motion is

$$F_r = mc\frac{d(\gamma \beta_r)}{dt} = \gamma mc\dot{\beta}_r, \quad \text{or} \quad \frac{d^2 r}{dt^2} = \frac{qE_r}{\gamma^3 m} \qquad (9.19)$$

While the magnetic-self force does not affect the longitudinal motion, the longitudinal motion is affected by a relativistic longitudinal mass $\gamma^3 m$ resulting in an equation of motion

$$\frac{d^2 z}{dt^2} = \frac{qE_z}{\gamma^3 m} \qquad (9.20)$$

In both cases but for different reasons, the space-charge term in the equation of motion has γ^3 in the denominator.

9.6
Practical Methods for Numerical Space-Charge Calculations

Beam-dynamics codes used for simulation of intense linac beams normally include a subroutine for calculating the space-charge forces. In principle the effects of space-charge can be calculated by adding the Coulomb forces between all the particles. However, to represent bunches with a typical number of particles near 10^8 or more, this approach is impractical. Several methods for calculating space-charge forces have been developed [8]. The first method, used in the program TRACE, [9] is based on the assumption of linear space-charge forces. Linear space-charge forces allow the use of analytical and matrix methods resulting in a significant simplification, but imply a uniform density distribution, which is rarely the case for a real beam. However, it was discovered by Lapostolle [4] and Sacherer [3] that for ellipsoidal bunches, where the rms emittance is either constant or specified in advance, the evolution of the rms beam projections is nearly independent of the density profile. This means that for calculation of the rms dynamics, the actual distribution can be replaced by an equivalent uniform beam, which has the same rms values. The resulting space-charge dynamics of the rms sizes can be calculated by the ellipse transformation method, described in Chapter 7; instead of tracing individual particles, the rms beam envelope is traced. The matrix equation,

$$\sigma_2 = R\sigma_1 R^T \tag{9.21}$$

determines the transport of the ellipse parameters $\tilde{\alpha}$, $\tilde{\beta}$, and $\tilde{\gamma}$ between points 1 and 2. These parameters are elements of the σ matrix, and in linear systems Eq. (9.21) can be solved by matrix multiplication if the transfer R matrix is given. To account for linear space-charge forces, the space-charge electric fields, given in Sections 9.8 and 9.9, are used. These fields depend on the rms beam size, which varies in the manner given by the ellipse $\tilde{\beta}$ parameter, an element of the σ matrix. To account for space charge, the code treats each space-charge impulse as a thin lens, which imparts a momentum impulse to each particle given by

$$\Delta p_i = qE_{si}\frac{\Delta s}{v}, \quad i = x, y, z, \tag{9.22}$$

where E_{si} is the space-charge field component, and Δs is the step size over which the field acts [10]. Because of the assumption of linearity, space-charge-induced emittance growth cannot be calculated by this type of subroutine. However, this type of envelope tracking code is very useful for finding the beam-ellipse parameters that are matched to a periodic transport system in a linac, and for determining quadrupole gradients needed to produce a matched beam.

To describe the evolution of the beam distribution and include space-charge emittance growth effects, the space-charge calculation requires particle tracking. The simulation particles are known as *macroparticles*, because as individual sources of the space-charge field, each macroparticle represents the total charge of many (typically 10^4 to 10^5) real beam particles. A frequently used macroparticle-tracking approach is called the *particle-in-cell* (PIC) *method*, in which at each step a mesh is superimposed on the bunch, which allows a smoothing of the fields to reduce the effects of artificially large forces that would otherwise be caused by binary encounters between macroparticles. The number of particles in each cell is counted, and the smoothed space-charge force acting on each particle is obtained by summing the fields from the charges in each cell. Finally, the forces are applied to deliver a momentum impulse to each particle. The time required for the calculation depends both on the number of macroparticles and the number of cells. A three-dimensional calculation needs a large number of cells, and some assumption of symmetry is usually made to speed up the calculation. The most commonly used computer subroutine for space-charge calculations is the SCHEFF subroutine used in the PARMILA codes, and written by K. R. Crandall [11]. SCHEFF, an acronym for space-charge effect, uses a two-dimensional (r–z) PIC method, which calculates the radial and longitudinal space-charge forces for an azimuthally symmetric distribution, or for a distribution with elliptical symmetry by using a transformation to an equivalent azimuthally symmetric distribution. Given either the real or the transformed azimuthally symmetric distribution, each macroparticle is assumed to represent a ring of charge, centered on the axis. In r–z space, a rectangular mesh with N_r radial intervals, each of length Δr, and N_z axial intervals, each of length Δz, is superimposed on the bunch. The total charge in each cell is counted, and the radial and longitudinal electric fields E_r and E_z resulting from the charge in each cell is computed at the nodes, located at the corners of each cell. Then, each macroparticle receives an impulse, using an electric field obtained by a linear interpolation from the nodes to the coordinates of the macroparticle. Typical runs are made with 10^4 to 10^5 particles with a typical mesh given by $N_r = 20$ and $N_z = 40$.

The advantage of allocating the particles to cells is to allow a smoothing method for eliminating the Coulomb singularities that occur when two particles are too close together. In SCHEFF there are two different options for providing this smoothing. The first and simplest option is to assume that, for the purpose of calculating the fields, the total charge contained within each cell is placed at the location of the centroid of the particles within that cell. The second option is to replace the distribution of charged rings in each cell by n charged rings, whose positions are chosen, using a Gaussian quadrature formula, to approximate a uniform charge density within the cell. The number of rings per cell depends on the cell aspect ratio. If the aspect ratio is less than 2, the program chooses $N = 4$; if it is between 2 and 4, $N = 8$, and if it is greater than 4, $N = 12$. To reduce computer time, a Green's function technique is used in which E_r and E_z at every node due to a unit charge are precalculated and saved in a table. Then at each step the space-charge fields are obtained by simply scaling the fields from a unit source charge in proportion to the total charge contained in each cell. The electric field components at position (r, z) due to a circular ring of charge centered at coordinates $(0, z_s)$ with radius r_s is

$$E_r(r, z, r_s, z_s) = \frac{Q}{4\pi^2 \varepsilon_0 r \sqrt{d^2 + 4rr_s}}$$
$$\times \left[K(\alpha) - \frac{(r_s^2 - r^2 + (z - z_s)^2)}{d^2} E(\alpha) \right] \quad (9.23)$$

$$E_z(r, z, r_s, z_s) = \frac{Q(z - z_s) E(\alpha)}{2\pi^2 \varepsilon_0 d^2 \sqrt{d^2 + 4rr_s}} \quad (9.24)$$

where $d^2 = (r - r_s)^2 + (z - z_s)^2$, $\alpha = \sqrt{4rr_s/(d^2 + 4rr_s)}$, and $K(\alpha)$ and $E(\alpha)$ are complete elliptical integrals of the first and second kinds, given by

$$K(\alpha) = \int_0^{\pi/2} \frac{d\theta}{\sqrt{1-\alpha^2 \sin^2\theta}} \tag{9.25}$$

$$E(\alpha) = \int_0^{\pi/2} \sqrt{1-\alpha^2 \sin^2\theta}\, d\theta \tag{9.26}$$

The procedure for handling the elliptical beam is exact only for a uniformly charged elliptical cylinder, or a uniformly charge ellipsoid [12]. First the ellipticity of the beam is calculated from the parameter $\varepsilon^2 = a_x/a_y$, where a_x and a_y are the x and y rms sizes of the beam. The effective radius of each particle is calculated as $\rho = \sqrt{(x/\varepsilon)^2 + (\varepsilon y)^2}$. Each particle is assigned to a cell in this transformed space, and the electric fields E_r and E_z are calculated in the transformed space. Finally the correction, applied to the Cartesian transverse electric-field components to account for the elliptical cross section of the beam, is given by

$$E_x = E_r \left[\frac{2}{1+\varepsilon^2}\right] \frac{x}{r} \tag{9.27}$$

$$E_y = E_r \left[\frac{2\varepsilon^2}{1+\varepsilon^2}\right] \frac{y}{r} \tag{9.28}$$

where E_r is given by Eq. (9.23). Using Eqs. (9.27), (9.28), (9.23), and (9.24), each macroparticle receives an x, y, and z impulse. This feature that allows representation of an elliptical beam, albeit that the procedure is approximate and ignores any departures from transverse ellipticity, converts SCHEFF to a subroutine capable of handling bunched beams in quadrupole focusing systems. The effect of any number of adjacent beam bunches can also be included by assuming periodicity in the bunch train. The calculation for relativistic particles is corrected to include the magnetic effects by Lorentz transforming the particles to the bunch frame, where the space-charge electric fields are applied, followed by a transformation back to the laboratory frame.

Another method for tracking interacting macroparticles is the straightforward *particle-to-particle interaction* (PPI) approach, in which the total three-dimensional space-charge force on a macroparticle is calculated as the vector sum of the individual forces from all the other macroparticles. To reduce the effects of artificially large macroparticle deflections, some method must be used to reduce the effect of the singularity in the Coulomb force, when two macroparticles get too close together. A typical approach [13] for treating the close encounters is to replace the point macroparticles by a spherical volume with uniform charge density, known as a *cloud*, which produces a linear force that falls to zero when the centers of the particles coincide. The choice of cloud radius R_c must be optimized, and a range of values of R_c near the average interparticle spacing can usually be found, where the results are insensitive to the exact value. The main disadvantage of the PPI method is that the computer time required is proportional to the square of the number of particles, making large particle runs prohibitive.

Finally, three-dimensional codes have been written, using analytic expressions that relate the space-charge field to the particle distribution. The first approach [14] of this type, developed at CERN, replaced the macroparticle distribution by an rms-equivalent Gaussian charge density with ellipsoidal symmetry, from which the space-charge fields were calculated by numerical integration. Although this approach leads to rapid three-dimensional computations, the restriction to a Gaussian profile is in principle not compatible with realistic distributions for intense beams. Recently, other approaches have been proposed to generalize this method, the most recent of which is the proposal of Lapostolle, [15] which describes the macroparticle charge density without requiring any assumption of symmetry by using a sum of Hermite polynomials to describe the beam density. Because the lowest-order term is a Gaussian distribution, this method is a generalization of the original CERN method, which promises to provide improved accuracy for a relatively high-speed, truly three-dimensional calculation.

9.7
RMS Envelope Equation with Space Charge

Analytic expressions are presented in Sections 9.8 and 9.9 for the space-charge field of a uniform continuous elliptical beam and a uniform ellipsoidal bunch. These expressions are useful for including space-charge forces in approximate equations of motion, for obtaining current-limit formulas, and for obtaining rms envelope equations. We begin with the latter application, following the early work of Lapostolle [4] and Sacherer [3]. Consider a beam moving in the s direction, where individual particles satisfy the equation of transverse motion

$$x'' + \kappa(s)x - F_s = 0 \tag{9.29}$$

The linear external force is given by $-\kappa(s)x$, and F_s is the space-charge force term, which in general is nonlinear and includes both the self-electric and the self-magnetic forces. The quantity F_s is related to the space-charge electric field E_s by $F_s = qE_s/\gamma^3 mv^2$. For simplicity we assume that the mean displacement and divergence are zero, and we are interested in finding an equation of motion for the rms beam size. First, we write the equations of motion for the second moments of the distribution. Thus,

$$\frac{d\overline{x^2}}{ds} = 2\overline{xx'} \tag{9.30}$$

$$\frac{d\overline{xx'}}{ds} = \overline{x'^2} + \overline{xx''} = \overline{x'^2} - \kappa(s)\overline{x^2} + \overline{xF_s} \tag{9.31}$$

and

$$\frac{\overline{dx'^2}}{ds} = 2\overline{x'x''} = -2\kappa(s)\overline{xx'} + 2\overline{xx'} + 2\overline{x'F_s} \tag{9.32}$$

where the averages are taken over the particle distribution. The first two equations lead to the equation of motion for the rms beam size, $a \equiv \sqrt{\overline{x^2}}$. Using Eq. (9.30), we have

$$aa' = \overline{xx'} \tag{9.33}$$

Differentiating Eq. (9.33), and using Eq. (9.31) gives

$$a'' - \frac{\overline{x^2}\,\overline{x'^2} - \overline{xx'}^2}{a^3} - \frac{\overline{xx''}}{a} = 0 \tag{9.34}$$

The numerator of the second term will be recognized as the square of the rms emittance, defined by Eq. (9.6). Then, substituting Eq. (9.29) to eliminate x'' from Eq. (9.34)

$$a'' + \kappa(s)a - \frac{\varepsilon_r^2}{a^3} - \frac{\overline{xF_s}}{a} = 0 \tag{9.35}$$

Equation (9.35) is the rms envelope equation, and it expresses the equation of motion of the rms beam size. The second term is the focusing term, and the third term is the emittance term. The emittance term is negative and is analogous to a repulsive pressure force acting on the rms beam size. The last term in Eq. (9.35) is the repulsive space-charge term. The equation of motion for the rms beam size is similar to the single-particle equation of motion, except for the presence of the additional emittance term. The third moment equation, Eq. (9.32), has not been used in this derivation. It can be shown that this equation affects the growth of rms emittance, [16] and to obtain a time-dependent solution, one would need to determine independently the evolution of the beam distribution.

9.8
Continuous Elliptical Beams

Sacherer and Lapostolle independently derived the envelope equations for continuous beams with arbitrary density profiles that have elliptical symmetry in x–y space. Although a linac beam is bunched, the continuous beam results are still useful for an approximate description of the transverse fields of a long bunch. The electric-field components for the uniform-density distribution are

$$E_{sx} = \frac{I}{\pi\varepsilon_0 v(r_x + r_y)}\frac{x}{r_x}, \text{ and } E_{sy} = \frac{I}{\pi\varepsilon_0 v(r_x + r_y)}\frac{y}{r_y} \tag{9.36}$$

where r_x and r_y are the semiaxes of the ellipse, related to the rms beam sizes by $r_x = 2a_x$, and $r_y = 2a_y$. Substituting Eq. (9.36) into Eq. (9.35) yields the rms envelope equations for a uniform density beam, which are given by

$$a_x'' + \kappa_x(s)a_x - \frac{\varepsilon_{rx}^2}{a_x^3} - \frac{K}{2(a_x + a_y)} = 0 \tag{9.37}$$

and

$$a_y'' + \kappa_y(s)a_y - \frac{\varepsilon_{ry}^2}{a_y^3} - \frac{K}{2(a_x + a_y)} = 0 \tag{9.38}$$

The quantity K is called the *generalized perveance* given by $K = qI/2\pi\varepsilon_0 m\gamma^3 v^3$, where $I = qN_\ell v$ is the current, expressed in terms of the number of particles per unit length N_ℓ. These equations were first derived by Kapchinsky and Vladimirsky [17] for a stationary uniform beam in a quadrupole-focusing channel, and are known as the K–V envelope equations. However, the remarkable result found by Lapostolle and Sacherer is that Eqs. (9.37) and (9.38) are valid not only for uniform density beams, but for all density distributions with elliptical symmetry. Thus the form of the envelope equation is independent of the density profile of the beam. To calculate the rms beam trajectories, even in the presence of space-charge forces, we can replace the actual beam distribution, which may not be known in advance, with an equivalent uniform beam [18] having the same current and the same second moments as the real beam. It is convenient to work with an equivalent uniform beam because, as we have seen, the space-charge field of a uniform beam, with elliptical or ellipsoidal symmetry, is easily calculated and is linear.

The ratio of the space-charge to the emittance terms in Eqs. (9.37) and (9.38) can be used to determine when space-charge is important compared with emittance in determining the rms beam size. For a round beam, where $a = a_x = a_y$, we have an emittance-dominated beam when $Ka^2/4\varepsilon^2 \ll 1$, and a space-charge dominated beam when $Ka^2/4\varepsilon^2 \gg 1$. A space-charge-dominated beam may be compared with cold plasma, where collective effects are dominant, whereas an emittance-dominated beam is dominated by random or thermal effects. Accelerators that are characterized as high-current machines can often be designed to avoid the space-charge-dominated regime by increasing the focusing force to reduce the rms beam size.

For a round beam in an ideal uniform focusing channel, where $k_0 \equiv \sqrt{\kappa_x} = \sqrt{\kappa_y}$ is the zero-current phase-advance per unit length, a matched solution of Eqs. (9.37) and (9.38) can be obtained corresponding to $a_x'' = a_y'' = 0$ [19]. The phase advance per unit length including space charge for the equivalent uniform beam, is denoted by k, where $k^2 = k_0^2 - K/4a^2$. The matched beam size is given by $a^2 = \varepsilon/k$, so that k plays the role of the inverse of the Courant–Snyder β function. The tune depression ratio is

$$\frac{k}{k_0} = \frac{1}{u + \sqrt{1 + u^2}} \tag{9.39}$$

where $u = K/8\varepsilon k_0$ is a space-charge parameter. An emittance-dominated beam corresponds to $u \ll 1$, so that $k/k_0 \cong 1 - u$, and the space-charge-dominated beam corresponds to $k/k_0 \cong 1/2u$.

The equivalent uniform beam for an arbitrary charge distribution can also be used to characterize an effective overall focusing strength in the presence of space charge. This can be done in a periodic-focusing system by comparing the phase advance per focusing period including the space charge, σ, for an equivalent uniform beam, with the phase advance per focusing period with zero space charge, σ_0. For a uniform density the quantities σ and σ_0 can be obtained from the transfer matrix through one period, with and without space charge. The procedure assumes the beam is matched, so that the beam ellipse is geometrically similar to the trajectory ellipses, including the linear space-charge field. The ratio σ/σ_0, which is always in the range $0 \leq \sigma/\sigma_0 \leq 1$, is called the *tune depression* or the *tune-depression ratio*. When $\sigma/\sigma_0 \approx 1$, the beam is emittance dominated, whereas when $\sigma/\sigma_0 \approx 0$, the beam is space-charge dominated.

9.9
Three-Dimensional Ellipsoidal Bunched Beams

The electric-field components for the three-dimensional uniform ellipsoid are of more interest for describing a typical linac bunch where the three semi-axes are comparable in length. The results are given by [20]

$$E_{sx} = \frac{3I\lambda(1-f)}{4\pi\varepsilon_0 c(r_x + r_y)r_z} \frac{x}{r_x}, \quad E_{sy} = \frac{3I\lambda(1-f)}{4\pi\varepsilon_0 c(r_x + r_y)r_z} \frac{y}{r_y},$$

$$E_{sz} = \frac{3I\lambda f}{4\pi\varepsilon_0 c r_x r_y} \frac{z}{r_z} \tag{9.40}$$

where r_x, r_y, and r_z are the semiaxes of the ellipsoid in the laboratory frame, for N particles per bunch $I = qNc/\lambda$ is the average current over an RF period, and λ is the RF wavelength. The semiaxes r_i are related to the rms beam sizes a_i by $r_i = \sqrt{5}a_i$, $i = x, y, z$. The displacements x, y, and z are to be evaluated relative to the centroid of the bunch. The quantity f is an ellipsoid form factor and is a function of the parameter $p = \gamma r_z/\sqrt{r_x r_y}$. Values of $f(p)$ for $p < 1$ and $f(1/p)$ for $p > 1$ are obtained from Fig. 9.4. For a nearly spherical bunch where $0.8 < p < 5$, a useful approximate form is $f = 1/3p$.

Substituting the field expressions into Eq. (9.35), the rms envelope equations are

$$a''_x + \kappa_x(s)a_x - \frac{\varepsilon_{r,x}^2}{a_x^3} - \frac{3K_3(1-f)}{(a_x + a_y)a_z} = 0 \tag{9.41}$$

$$a''_y + \kappa_y(s)a_y - \frac{\varepsilon_{r,y}^2}{a_y^3} - \frac{3K_3(1-f)}{(a_x + a_y)a_z} = 0 \tag{9.42}$$

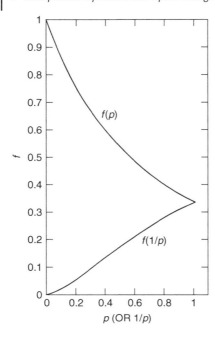

Figure 9.4 Ellipsoid form factor $f(p)$ versus p for $p < 1$ and $f(1/p)$ for $p > 1$.

and

$$a_z'' + \kappa_z(s)a_z - \frac{\varepsilon_{r,z}^2}{a_z^3} - \frac{3K_3 f}{a_x a_y} = 0. \tag{9.43}$$

We have defined a three-dimensional space-charge parameter

$$K_3 = \frac{qI\lambda}{20\sqrt{5}\pi \varepsilon_0 mc^3 \gamma^3 \beta^2} \tag{9.44}$$

Unlike the two-dimensional elliptical geometry discussed in Section 9.8, the space-charge terms in Eqs. (9.41) to (9.43) depend on the density profile, even for different distributions with ellipsoidal symmetry. However the distribution dependence is very weak, and Eqs. (9.41) to (9.43) are still a good approximation for any distribution with ellipsoidal symmetry.

9.10
Beam Dynamics Including Linear Space-Charge Field

A longitudinal equation of motion can be written, assuming that the beam density is described by a uniform three-dimensional ellipsoid with linear space-charge-force components. We begin with the equation of longitudinal

9.10 Beam Dynamics Including Linear Space-Charge Field

motion, derived in Chapter 6, and add a longitudinal space-charge field from Eq. (9.40). The single-particle equation of motion for the phase is

$$\phi'' + k_{\ell 0}^2 \left[(\phi - \phi_s) - \frac{(\phi - \phi_s)^2}{2\tan(-\phi_s)} \right] - \frac{3qI\lambda f(\phi - \phi_s)}{4\pi\varepsilon_0 \gamma^3 \beta^2 mc^2 r_x r_y r_z} = 0 \quad (9.45)$$

where ϕ_s is assumed to be constant and the semiaxes of the uniform ellipsoid are related to the rms sizes by $r_i = \sqrt{5}a_i$, $i = x, y, z$. We define μ_ℓ as the ratio of the longitudinal space-charge force to the linear part of the longitudinal focusing force. Then, Eq. (9.45) can be written as

$$\phi'' + k_{\ell 0}^2 \left[(1 - \mu_\ell)(\phi - \phi_s) - \frac{(\phi - \phi_s)^2}{2\tan(-\phi_s)} \right] = 0 \quad (9.46)$$

where

$$\mu_\ell \equiv \frac{3qI\lambda f}{4\pi\varepsilon_0 \gamma^3 \beta^2 mc^3 r_x r_y r_z k_{\ell 0}^2} \quad (9.47)$$

If we define the phase advance per unit length, including space charge as $k_\ell^2 = k_{\ell 0}^2(1 - \mu_\ell)$, where $k_{\ell 0}$ is given by Eq. (6.43), the linear equation of longitudinal motion, ignoring the quadratic term of Eq. (9.46) is

$$\phi'' + k_\ell^2(\phi - \phi_s) = 0 \quad (9.48)$$

A common figure of merit for evaluating the importance of the space-charge force is the longitudinal tune-depression ratio, defined as $k_\ell/k_{\ell 0}$. This is

$$\left[\frac{k_\ell}{k_{\ell 0}}\right]^2 = 1 - \mu_\ell = 1 - \frac{3qI\lambda f}{4\pi\varepsilon_0 \gamma^3 \beta^2 mc^3 r_x r_y r_z k_{\ell 0}^2} \quad (9.49)$$

An approximate transverse equation of motion can also be derived, again assuming that the beam density is described by a uniform three-dimensional ellipsoid. We use an approximate treatment based on the smooth approximation for a periodic quadrupole focusing array, introduced in Section 7.17. The equation of motion in the smooth approximation has the form

$$\frac{d^2 x}{ds^2} + \left[\frac{\sigma_0}{P}\right]^2 x - \frac{qE_{sx}}{\gamma^3 \beta^2 mc^2} = 0 \quad (9.50)$$

where σ_0 is the zero-current transverse phase advance per focusing period P, and E_{sx} is the x component of the space-charge electric field, given in Eq. (9.40). Identifying σ as the phase-advance per focusing period including space charge, we obtain

$$\left[\frac{\sigma}{\sigma_0}\right]^2 = 1 - \frac{3qI\lambda(1-f)}{4\pi\varepsilon_0 mc^3 \gamma^3 \beta^2 (r_x + r_y) r_x r_z} \left[\frac{P}{\sigma_0}\right]^2 \quad (9.51)$$

9.11
Beam-Current Limits from Space Charge

Using the uniform three-dimensional ellipsoid model, an approximate expression for the longitudinal current limit can be derived [21]. As the beam current increases, the repulsive space-charge forces reduce the size and depth of the longitudinal potential well that defines the bucket. For a given synchronous phase and a given accelerating field, there will be some value of the current at which the space-charge force is just equal to the focusing force and the potential well vanishes. This limiting current defines the longitudinal space-charge limit, or the longitudinal current limit. If we return to the definition for μ_ℓ given in Eq. (9.47), substitute $k_{\ell 0}$ from Eq. (6.43), and solve for the current, we obtain

$$I = \mu_\ell \frac{8\pi^2 \varepsilon_0 c r_x r_y r_z E_0 T \sin(-\phi_s)}{3\beta\lambda^2 f} \qquad (9.52)$$

A current limit can be obtained from Eq. (9.52) that is consistent with numerical simulation results. We use the approximation for the ellipsoid form factor $f = 1/3p$, where p is defined in Section 9.9. We assume that $r_z = |\phi_s|\beta\lambda/2\pi$, which implies that the separatrix width is unaffected by space charge, and that only $\pm|\phi_s|$ of the phase width is available. It is also convenient to introduce a flutter factor for the alternating transverse-focusing system, defined by $\psi \equiv r_{\max}/r_{\min}$. At the point in each transverse focusing period, where r_x is maximum and r_y is minimum, we write $\psi \equiv r_x/r_y = r_x^2/r_x r_y$. The longitudinal current limit is largest, when $r_x = a$, where a is the aperture radius. Substituting all these into Eq. (9.52), we obtain an expression for the longitudinal current limit, which is generally in good agreement with numerical simulation results,

$$I_{\ell,\max} = \frac{2\mu_\ell \beta \gamma a E_0 T \sin(-\phi_s)|\phi_s|^2}{Z_0 \psi^{1/2}} \qquad (9.53)$$

where $Z_0 = 1/\varepsilon_0 c = 376.73\ \Omega$ is the impedance of free space, and $\mu_\ell \approx 1$ for a space-charge dominated beam. Equation (9.53) implies that the longitudinal current limit increases with increasing velocity, increasing beam aperture, increasing accelerating field, and for small angles where $\sin(-\phi_s) \approx -\phi_s$, it increases approximately as the cube of the synchronous phase.

Next we develop an expression for the transverse current limit. Physically, as the beam current increases, the matched beam size grows until it is equal to the aperture radius, which corresponds to the current limit. Any further increase of the current results in beam loss. Returning to Eq. (9.50), we define the ratio of the space-charge force to the focusing force as μ_T, and solving for the current, we obtain for the x motion

$$I = \mu_T \frac{4\pi\varepsilon_0 \gamma^3 \beta^2 mc^3 (r_x + r_y) r_x r_z}{q\lambda(1-f)} \left[\frac{\sigma_0}{P}\right]^2 \qquad (9.54)$$

Again, we use the approximation for the ellipsoid form factor $f \cong 1/3p$, we let $r_z = |\phi_s|\beta\lambda/2\pi$, and we introduce the flutter factor $\psi \equiv r_x/r_y = r_x^2/r_x r_y$. The transverse current limit will be largest, when $r_x = a$. We write the transverse focusing period as $P = N\beta\lambda$, and assume that we can equate the arithmetic and geometric mean of the beam size over a period. Substituting these relations into Eq. (9.54), we obtain an expression for the transverse current limit,

$$I_{T,\max} = \mu_T \frac{4mc^2\gamma^3\beta|\phi_s|\sigma_0^2}{3Z_0 q \psi N^2(1-f)} \left(\frac{a}{\lambda}\right)^2 \tag{9.55}$$

In practice, the phase advance σ_0 is limited to $\sigma_0 < 90°$ to avoid envelope instability, [22] which occurs for high-intensity beams in quadrupole channels. A useful rule of thumb is that if $\sigma_0 > 90°$ and the space-charge force reduces σ so that $\sigma < 90°$, the beam will be unstable. In some cases the value of σ_0 may be limited instead by the quadrupole focusing strength (i.e., corresponding to the Gl product of the quadrupoles). Then, it is more convenient to substitute the expression for σ_0 as a function of the quadrupole parameters. Using the results from Section 7.17, the zero-current phase advance per focusing period in a FODO channel in the smooth approximation, including both the quadrupole and the RF defocusing term, is

$$\sigma_0^2 \cong \left[\frac{qG\ell P}{2mc\gamma\beta}\right]^2 + \frac{\pi q E_0 T \sin(\phi) P^2}{mc^2 \lambda (\gamma\beta)^3} \tag{9.56}$$

Expressing the focusing period as $P = N\beta\lambda$, where N is an integer, and substituting Eq. (9.56) into Eq .(9.55), we obtain

$$I_{T,\max} = \mu_T \frac{4mc^2\gamma^3\beta|\phi_s|a^2}{3Z_0 q \psi (1-f)} \left[\left[\frac{qG\ell}{2mc\gamma}\right]^2 - \frac{\pi q E_0 T \sin(-\phi_s)}{mc^2\gamma^3\beta\lambda}\right] \tag{9.57}$$

For low-velocity beams the RF defocusing term should not be neglected. The maximum ratio of the transverse space-charge force to the focusing force is $\mu_T \approx 1$, which occurs for a space-charge-dominated beam. Typically in a proton linac, the smallest values of both the longitudinal and transverse current limits are found at the end of the gentle buncher section of the RFQ. The current limit formulas can be used as a basis for optimized RFQ-linac designs, by choosing the parameters to yield equal values of the longitudinal and transverse limits, which are chosen to be larger than the design current including a safety margin for errors.

9.12
Overview of Emittance Growth from Space Charge

Beams that are in equilibrium in the focusing channel of a linac experience no emittance growth. Unfortunately, beams observed in linac numerical simulations are rarely in equilibrium, and when they appear to be near equilibrium, any changes that occur in the focusing system produce changes in the beam, usually accompanied by emittance growth. Nonlinear forces that act on a nonequilibrium beam will cause the rms emittance to increase. The space-charge force in high-current beams is typically the major cause of such emittance growth. Four different space-charge mechanisms can be identified, [23] and are described briefly in this section. First, when a high-current, rms-matched beam is injected into the accelerator, the emittance can grow very rapidly as the charges redistribute to provide shielding of the external focusing field. This mechanism, called *charge redistribution*, is the fastest known emittance-growth mechanism, producing growth in only one quarter plasma period, where the beam plasma frequency for a beam with particle density n is $\omega_P = \sqrt{nq^2/\varepsilon_0 m}$. The free energy for the emittance growth comes from the field energy of the initial distribution [21, 24]. Second, if the injected beam is not rms matched, additional energy from the mismatch oscillations of the beam is available for emittance growth. Even for relatively small mismatches, this mechanism can become the largest contributor to emittance growth. Third, for nonsymmetric or anisotropic beams, there can be emittance transfer as a result of space-charge resonances that couple longitudinal and transverse oscillations, where in some cases the kinetic energies in the three planes may approach an approximate equalization, sometimes referred to as *equipartitioning*. In such cases, the emittance grows in a plane that receives energy, and decreases in a plane that looses energy. Finally, the periodic-focusing structure can resonantly excite density oscillations in the beam, the most serious of which is the envelope instability mentioned in Section 9.11. To avoid these resonances in real beams, it is sufficient to keep $\sigma_0 < 90°$. This is a more stringent requirement than for the stability of a zero-current beam in a periodic-focusing channel, for which the stability requirement is $\sigma_0 < 180°$. The four mechanisms and some of their characteristics are summarized in Table 9.1.

Numerical simulations of nonequilibrium linac beams, show that linac beams do evolve, although sometimes slowly, to quasiequilibrium distributions. When the focusing is a linear function of the displacement, but nonlinear space-charge forces are present, the equilibrium spatial distribution consists of a uniform central-density core, and at the edges the density falls to zero over a distance approximately equal to a Debye length [25]. For space-charge dominated beams, the Debye length is small compared with the beam size, and the distribution consists almost entirely of the uniform core, which falls sharply to zero at the edge. For emittance-dominated beams, the Debye length is large compared with the beam size, and the distribution is dominated by a broad Debye edge. The spatial density then has the appearance of a Gaussian-like distribution. It is also observed that the space-charge-induced emittance growth in nonequilibrium beams is often associated with the formation of an outer halo of low-density particles in phase space. The phase-space density of the core is reduced only slightly in the process, but the associated rms

Table 9.1 Properties of space-charge-induced emittance growth.

	Charge redistribution	RMS mismatch	Emittance transfer	Structure resonance
Free-energy source	Nonlinear field energy	Oscillation energy of excited mode	Space-charge coupling resonances	Longitudinal energy
Approximate timescale	$\approx \dfrac{\tau_{plasma}}{4}$	Typically $\geq 10\tau_{plasma}$	Typically $\geq 10\tau_{plasma}$	$\approx 2\tau_{betatron}$
Distribution function sensitivity	Strongly dependent	Weakly dependent	Weakly dependent	Strongly dependent
For minimum growth	Avoid transitions toward stronger tune depression	RMS match	Avoid space-charge coupling resonances	Keep $\sigma_0 < 90°$

emittance growth caused by the halo can be significant. The halo is especially undesirable for high-duty-factor linacs, because it can result in particle loss in the accelerating structure, and radioactivation of the linac.

9.13
Emittance Growth for rms Matched Beams

First, we distinguish between internal matching and rms matching to a periodic-focusing structure. Internal matching corresponds to particle phase-space trajectories that coincide with the isodensity contours of the 6D phase-space density. For an internally matched beam, the distribution will be in equilibrium in the accelerator channel, and no emittance growth will occur, even though nonlinear forces may act on the beam. Such an equilibrium distribution has a constant beam radius if the focusing is uniform along the accelerator, or for a periodic focusing channel has a periodic radial profile with the same period as the focusing channel. Examples of equilibrium distributions have been studied for beam transport channels with linear focusing forces: linear functions of the transverse beam displacement [26]. The most frequently studied is the Kapchinsky–Vladimirsky (K–V) distribution for transverse dynamics in 4D phase space, which is the only equilibrium distribution in a periodic transport channel for which an analytic description has been found [27]. For the K–V distribution the beam is distributed on the surface of a hyperellipsoid in four-dimensional phase space. No particles populate the central core of this 4D space, which is not the case in real beams. Nevertheless, the K–V distribution results in uniform ellipses for all 2D projections, which is convenient since it produces simple linear space-charge forces in the spatial projections.

Given a beam that is not internally matched to a periodic structure, one may wish to transform it into an internally matched equilibrium distribution. We do not know any simple procedure for making such a transformation, and whether in principle such a transformation is possible without increasing the rms emittance in the process. Nevertheless, we do know how to match the rms beam sizes for each degree of freedom. This is accomplished by providing a beam-optics transformation to eliminate the mismatch oscillations of the rms beam sizes. In a uniform focusing channel the resulting rms sizes will be constant; in a periodic channel the rms sizes will undergo a periodic flutter with the same period as that of the focusing lattice. An rms-matched beam is not generally internally matched, so the rms-matched beam is not generally in equilibrium. Therefore, the rms-matched beam can still relax to an equilibrium state with the possibility of irreversible emittance growth. Even so, injection of an rms-matched beam can be considered a first approximation to the desired internally matched beam.

Numerical simulations of nonequilibrium beams in uniform focusing channels show that such beams often evolve to quasiequilibrium distributions, which change only slowly as the beam is accelerated. The evolution of such beams is usually accompanied by rms-emittance change as a result of both nonlinearity and coupling between degrees of freedom, and the velocity distributions are Maxwellian-like. When the focusing is linear, the spatial distribution of a space-charge dominated beam has an approximately uniform density core of density n. The density decreases to zero at the edges of the beam over a distance approximately equal to the Debye length, λ_D, given nonrelativistically by

$$\lambda_D = \sqrt{\varepsilon_0 k_B T / n q^2} \qquad (9.58)$$

where q is the charge per particle and ε_0 is the free-space permittivity. In Eq. (9.58) the thermal energy is given by $k_B T = mc^2 \varepsilon_n^2 / a^2$ where mc^2 is the particle rest energy, a is the rms beam size, and ε_n is the rms-normalized emittance, defined without a factor of 4, but including the relativistic $\beta\gamma$ factor.

For space-charge dominated beams $\lambda_D \ll a$, so the equilibrium spatial distribution is approximately uniform with a sharp falloff at the edges. For emittance dominated beams $\lambda_D \gg a$ and the Debye tail is dominant, resulting in a peaked Gaussian-like charge density. The K–V distribution is an exception because it always has uniform charge density, regardless of the relative importance of emittance and space charge. However, this distribution does not correspond to the final equilibrium state of real beams.

Emittance growth can occur in beams that are internally mismatched even if they are rms matched, through a mechanism referred to as *charge redistribution*. When a high-current rms-matched beam is injected into an accelerator, the charges redistribute rapidly to shield the interior of the beam from the linear external focusing force. Transverse energy conservation is valid in the smooth approximation, and the kinetic energy associated with the emittance growth comes from the difference of the field energy of the initial distribution and that of the final distribution. For linear external focusing fields in the extreme space-charge (zero emittance) limit, this implies a charge rearrangement to a uniform density to provide exact shielding. Finite emittance rms-matched beams in numerical simulations evolve toward an internally matched charge density with a central uniform core and a finite thickness boundary, whose width is approximately equal to the Debye length. The rms-emittance growth results from the nonlinear space-charge fields that are present while the beam has nonuniform density and is undergoing initial plasma oscillations. The emittance growth can also be described as the result of the decoherence of the plasma oscillations of particles with different amplitudes, which causes a phase mixing. In either case this is the fastest-known emittance-growth mechanism, producing full emittance growth in only one-quarter plasma period, where the plasma frequency of the beam with particle mass m and charge q, and with particle density n, is $\omega_p = \sqrt{n q^2 / \varepsilon_0 m}$. The distance a beam particle travels during one plasma period is given by

$$\lambda_p = \sqrt{\frac{8\pi^2 a^2}{K\gamma^2}} \tag{9.59}$$

The initial growth in one-quarter plasma period is followed by damped oscillations for typically 10 or so plasma periods. This mechanism can be important when a beam that is internally matched to a strong-focusing emittance-dominated channel is injected after rms matching into a weaker-focusing space-charge-dominated channel.

For fixed rms beam size, the minimum space-charge field energy corresponds to a uniform-density beam. If the rms-matched input beam has a peaked spatial profile, characteristic of a channel where the space-charge force is small compared with the focusing force, the beam particles in the weaker-focusing channel will rapidly rearrange to produce a more uniform density, and in the process space-charge field energy is converted to kinetic energy. If the field energy difference is known, it can be used to calculate the emittance growth.

To describe these effects more quantitatively, we introduce the concept of nonlinear field energy U, a quantity that depends on the shape of the charge distribution and corresponds to the additional space-charge field energy of beams with nonuniform charge density. The rate of change of unnormalized emittance ε is given by the expression [28]

$$\frac{d\varepsilon^2}{dz} = -2a^2 K \frac{d}{dz}\left(\frac{U}{w_0}\right) \tag{9.60}$$

where a is the rms beam size and K is the generalized perveance, which is proportional to the beam current

$$K \equiv \frac{qI}{2\pi\varepsilon_0 m v^3 \gamma^3} \tag{9.61}$$

The nonlinear field energy is defined as $U = W - W_u$, where W is the self-electric-field energy per unit length of the distribution, and W_u is the self-electric-field energy per unit length of an equivalent uniform beam, which has the same current, emittance, and rms size as the actual beam. The quantity W_u is the field energy per unit length contained within a radius b, which is equal to or larger than the beam radius, and is given by

$$W_u = w_0(1 + 4\ln(b/2a)), \quad b \geq 2a \tag{9.62}$$

where w_0 is the self-electric-field energy per unit length within the boundary of an equivalent uniform beam. The electric field and the field energy outside the beam are independent of the charge density, and the contribution to the difference between W and W_u outside the beam is zero. The special role of the uniform distribution can be explained by its associated linear self-force, which causes no rms emittance growth. Thus, U is the residual self-electric-field energy possessed by beams with nonuniform charge densities. Because nonuniform beams have nonlinear self fields, we call U the nonlinear field energy. Equation (9.60) tells us that a decrease in nonlinear field energy U corresponds to an increase in rms emittance. Both the electric and the magnetic field contributions are contained in the equation by including the factor γ^3 in the definition of K (γ^2 accounts for the self-magnetic field and γ accounts for the relativistic mass). RMS emittance growth induced by space charge is inherently a nonrelativistic effect; it is most important when γ is near unity.

The quantity U/w_0, which affects the emittance growth, is dimensionless. It is zero for a uniform charge distribution, is positive for both peaked and hollow distributions, and increases as the distribution becomes more nonuniform. It is independent of the beam current and the rms beam size, and is a function only of the shape of the distribution. Table 9.2 shows U/w_0 for some common unbunched beam distributions.

If the initial distribution is known, the initial value of U/w_0 can be calculated. In principle Eq. (9.60) could be solved to obtain the emittance growth if we knew the final U/w_0. However, the final distribution is usually not known, so an assumption of some kind is needed.

To proceed further, Eq. (9.60) can be integrated if we assume the beam size of the rms-matched beam is approximately constant. Thus,

$$\Delta \varepsilon^2 = -2a^2 K \Delta (U/w_0) \tag{9.63}$$

and

$$\frac{\varepsilon^2}{\varepsilon_i^2} = 1 - \frac{2Ka^2}{\varepsilon_i^2} \Delta \left(\frac{U}{w_0} \right) \tag{9.64}$$

Next we rewrite the quantity Ka^2/ε_i^2 in a more convenient way in terms of the space-charge tune depression ratio of an equivalent uniform beam as follows. The single-particle equation of motion for an equivalent uniform beam with a linear space-charge term is

$$x'' + k_0^2 x - \frac{Kx}{4a^2} = 0 \tag{9.65}$$

Table 9.2 U/w_0 for some common unbunched beam distributions.

Spatial Distribution	Charge Density $\rho(r)$		U/w_0
Gaussian	$\exp(-r^2/2a^2)$		0.154
Waterbag	$1 - (r/R)^2$	$r \leq R$	0.0224
Uniform	1	$r \leq R$	0.0
Hollow ($n = 2$)	r^2	$r \leq R$	0.0754
Hollow ($n = 10$)	r^{10}	$r \leq R$	0.245

from which we identify the phase advance per unit length, including space-charge, as

$$k^2 = k_0^2 - \frac{K}{4a^2} \tag{9.66}$$

The matched envelope equation for a uniform focusing channel may be written as $a'' = 0$, or

$$k_0^2 a - \frac{K}{4a} - \frac{\varepsilon^2}{16a^3} = 0 \tag{9.67}$$

From these two equations we obtain

$$k^2 = \frac{\varepsilon^2}{16a^3} \tag{9.68}$$

Equation (9.67) can be rewritten as

$$\frac{4Ka^2}{\varepsilon^2} = \frac{k_0^2}{\varepsilon^2/(16a^4)} - 1 = \frac{k_0^2}{k^2} - 1 \tag{9.69}$$

and therefore Eq. (9.64) can be written as

$$\frac{\varepsilon^2}{\varepsilon_i^2} = 1 - \left(\frac{k_0^2}{k^2} - 1\right)\left(\frac{U - U_i}{2w_0}\right) \tag{9.70}$$

If we knew the tune depression and the initial U/w_0, we could calculate the rms emittance growth if the final value of U/w_0 is known, but this would require knowledge of the final distribution. There is one case, where we do know U/w_0. In the extreme space-charge limit we expect the beam to evolve toward a final stationary state with a uniform charge density to obtain complete shielding of the linear applied focusing force. Let us assume that for all tune depression ratios, $U/w_0 = 0$ in the final state, which corresponds to assuming a final uniform beam. This approximation should be very good for space-charge dominated beams, where emittance growth is largest. In the emittance-dominated limit, where the approximation would lead to a larger fractional error in emittance growth, the emittance growth is small anyway. In all cases the approximation will result in an upper limit to the emittance growth. With this approximation the upper limit to the final emittance growth for an rms-matched beam is

$$\frac{\varepsilon^2}{\varepsilon_i^2} = 1 - \left(\frac{k_0^2}{k^2} - 1\right)\left(\frac{U_i}{2w_0}\right) \tag{9.71}$$

Equation (9.71) can also be written as

$$\varepsilon_f^2 = \varepsilon_i^2 + \frac{1}{2}\left(\frac{K}{k_0}\right)^2 \left(\frac{U_i}{w_0}\right) \tag{9.72}$$

9.13 Emittance Growth for rms Matched Beams

This result tells us that as the initial emittance approaches zero, the final emittance approaches a minimum value (the second term) that decreases with increased focusing strength (larger k_0), increases with perveance K, and increases with the initial nonuniformity as measured by U_i/w_0.

Figure 9.5 shows results of a numerical simulation for 10 plasma periods using an initial Gaussian charge density truncated at two standard deviations, and an initial space-charge tune depression of 0.02 (a space-charge-dominated beam). The abscissa is the distance along the beam line in plasma periods. The quantity ρ is a dimensionless radial distribution parameter defined as $\rho = 1 - \bar{r}/\bar{r}_u$ where $\bar{r}$ is the average radius and $\bar{r}_u$ is the average radius of an equivalent (same second moments) uniform beam. For common distributions the parameter ρ is positive for a peaked charge density, zero for a uniform density, and negative for a hollow beam.

The parameter ρ in Fig. 9.5a indicates that the beam density undergoes damped radial plasma oscillations between peaked and hollow configurations. The quantity U_i/w_0 in Fig. 9.5b is maximum at both the extreme peaked and hollow configurations and therefore oscillates at twice the frequency of the parameter ρ. Its minimum value is not exactly zero, expected for a uniform distribution partly because of the formation of some beam halo. The rms emittance-growth ratio, $\varepsilon/\varepsilon_i$ in Fig. 9.5c shows the rapid initial growth during the first quarter plasma oscillation followed by damped oscillations. By 10 plasma periods the beam has evolved to an approximately stationary state.

A second example numerical example is shown in Fig. 9.6 using an initial semi-Gaussian charge density beam (uniform spatially and Gaussian in velocity space) truncated at four standard deviations in velocity space with an initial space-charge tune depression of 0.25 (a moderately space-charge-dominated beam). This example is interesting because the rms emittance decreases. The parameter ρ in Fig. 9.6a rises from its initial value of zero corresponding to uniform density and remains positive throughout indicating a peaked distribution. The radial density profile (not shown) evolves from an initial hard-edged beam to one with a Debye tail. The quantity U_i/w_0 in Fig. 9.6b increases while the rms emittance-growth ratio $\varepsilon/\varepsilon_i$ in Fig. 9.6c decreases in accordance with Eq. (9.71).

Figure 9.7 shows the emittance growth after 100 plasma periods versus the space-charge tune-depression ratio for an initial truncated Gaussian beam density. The agreement of the curve from Eq. (9.71) and the symbols from the second moments taken from the simulations is excellent. Figure 9.7 shows that if one tries to operate at very low tune depression ratio, there can be a risk of very large emittance growth. One might expect that the way to minimize the emittance growth is simply to produce a uniform density beam. However, this may be a little too simple. It may not be easy to maintain the uniformity because the charge redistribution process is so rapid. For moderate or weak tune depressions, a uniform beam will rapidly grow a Debye tail. After that happens, one needs to avoid introducing transitions that result in stronger tune depression, because that will surely lead to field energy conversion to

312 9 Multiparticle Dynamics with Space Charge

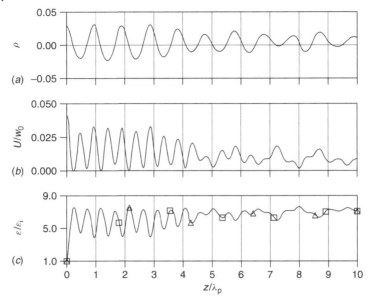

Figure 9.5 Numerical simulation results showing the evolution of three beam parameters for an initial Gaussian beam truncated at two standard deviations for a space-charge tune-depression ratio of $k/k_0 = 0.02$ versus distance along the beam line in plasma periods. The lower curve showing the emittance growth is actually two overlapping curves showing very good agreement, one from Eq. (9.71) and one from the second moments taken from the simulation.

kinetic energy accompanied by emittance growth. As a consequence of the risk of converting nonlinear field energy into thermal energy accompanied by large emittance growth, control of emittance growth in the space-charge dominated regime is a significant challenge.

An experimental test of the theory was carried out at University of Maryland using a 5-keV electron beam with two solenoids used for rms beam matching followed by a 5-m long periodic channel with 36 solenoid lenses [29–31]. In the experiment the initial beam was highly nonuniform, consisting of five separate beamlets created using an aperture plate with five small holes. The beam was then rms matched, and the beam profile was measured at six locations along the beam transport channel. The free energy was entirely due to the nonuniformity of the distribution. The relevant beam parameters were $U/w_0 = 0.2656$, and initial tune depression ratio $k_i/k_0 = \sigma_i/\sigma_0 = 0.31$. The final emittance growth from theory was $\varepsilon_f/\varepsilon_i = 1.56$. The final beam observed on a fluorescent screen was homogenized and showed no evidence for the initial five-beam beamlet structure. The emittance measurements using a slit-pinhole system showed good agreement with the theory.

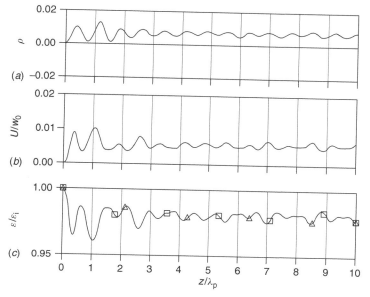

Figure 9.6 Numerical simulation results showing the evolution of three beam parameters for an initial semi-Gaussian beam (uniform spatially and Gaussian in velocity space) truncated at four standard deviations for a space-charge tune-depression ratio of $k/k_0 = 0.25$ versus distance along the beam line in plasma periods. The lower curve showing the emittance growth is actually two overlapping curves showing very good agreement, one from Eq. (9.71) and one from the second moments taken from the simulation.

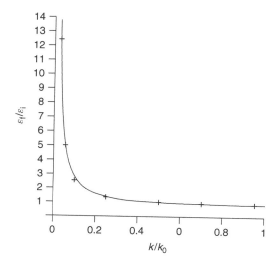

Figure 9.7 Final emittance growth ratio versus initial space-charge tune depression ratio for an initial Gaussian beam truncated at four standard deviations. The curve is from Eq. (9.71) and the symbols are from the simulations after 100 plasma periods.

9.14
Model of Space-Charge-Induced Emittance Growth in a Linac

In this section we construct a nonrelativistic model of transverse emittance growth [32] for the charge-redistribution process, from which the scaling of emittance growth with respect to the beam parameters in a linac can be deduced. As a model for a long bunch, we consider a continuous beam with circular cross section, velocity v, current I, and zero initial emittance. As the beam propagates in a drift space, the space-charge force will change the transverse momentum components of each particle by an amount equal to the product of the space-charge force times the time over which the force acts. To simplify the calculation, we ignore the changes in the transverse positions, so that the spatial density is assumed to remain unchanged as the beam propagates. For simplicity we assume a parabolic density, which will contribute nonlinear space-charge fields that can produce emittance growth. The density distribution as a function of radius r is

$$\rho(r) = \frac{2qN_l}{\pi R^2}\left[1 - \frac{r^2}{R^2}\right] \tag{9.73}$$

where N_l is the number of particles per unit length and R is the beam radius. From Gauss's Law, this corresponds to a nonlinear space-charge field given by

$$E_r(r) = \frac{qN_l}{\pi \varepsilon_0 R^2}\left[r - \frac{r^3}{2R^2}\right] \tag{9.74}$$

Over the drift length z, this field delivers a radial momentum impulse to each particle given by $\Delta p_r(z) = qE_r z/v$. Suppose the initial phase-space distribution is represented by a straight line, corresponding to zero transverse emittance. As described in Section 9.2, the nonlinear field changes an initial straight-line distribution into a line with curvature accompanied by rms emittance growth. Using the definition of rms emittance, it can be shown that the unnormalized rms emittance in x or y as a function of the drift distance z is

$$\varepsilon(z) = \frac{Kz}{12\sqrt{5}} \tag{9.75}$$

where K is the generalized perveance K, given nonrelativistically by $K = qI/2\pi\varepsilon_0 mv^3$. It may seem surprising that the emittance is independent of the radius R. This is because as R increases, although the momentum impulses from space charge are smaller, they extend over a larger radial extent. Because emittance is proportional to phase space area, these two effects cancel.

When the injected beam is traced with numerical simulation codes for a longer distance, sufficient to allow the particle positions to adjust, one finds that the beam approaches a steady-state distribution, and that most of the emittance growth occurs in roughly one quarter of a beam plasma period. To approximate the emittance growth as a function of position, suppose that initially the emittance is assumed to grow linearly, at the rate given by Eq. (9.75), until saturation occurs at the distance $z = \lambda_p/4$, where $\lambda_p = 2\pi R/\sqrt{3K}$ is the beam plasma period. Substituting the saturation distance z into Eq. (9.75), the saturated unnormalized rms emittance becomes

$$\varepsilon = \frac{K\lambda_p}{48\sqrt{5}} = \frac{\pi R}{24}\sqrt{\frac{K}{15}} \qquad (9.76)$$

In contrast to the result of Eq. (9.75) in which the emittance growth rate was independent of R, Eq. (9.76) shows another surprising result that the saturated emittance increases with beam radius R. This result is a consequence of the proportionality of the plasma period to beam radius: for larger R the model predicts that the beam takes longer to relax to a steady state, and during the relaxation process, the emittance continues to grow. Equation (9.76) implies that to minimize rms emittance growth from space-charge forces, the linac should provide strong focusing to make the beam small. This result implies that to control transverse emittance growth, one should not design a high-current, high-brightness linac with weak transverse focusing to produce a large-radius beam, as one might be motivated to do to reduce the transverse space-charge force. Rather, one needs to increase the ratio of the focusing force to the space-charge force, and this requirement produces a reduced beam radius. Indeed, it can be shown that for a round beam in a uniform linear focusing channel, although weaker focusing increases the beam size and decreases the space-charge force, it also decreases the focusing force, such that the ratio of the space-charge force to the focusing force increases (see Problem 9.14).

So far we have not expressed the result of Eq. (9.76) in a way that is suitable for a bunched beam in a linac. To do this, we write the peak current in terms of the number of particles per unit length N_l as $I = qN_l v$. Furthermore, we assume that the linac accelerates a bunch train with bunch frequency f and a number of particles per bunch N. If we assume that a bunch can be represented by a cylinder of length $2b$ with a uniform longitudinal profile, the peak bunch current I is expressed in terms of the average current $\bar{I} = qNf$ over the bunch train, as $I = \bar{I}v/2bf$. Converting the unnormalized emittance of Eq. (9.76) to a normalized transverse rms emittance, and substituting the expression for K and for the current, we find a nonrelativistic result for the saturated normalized emittance of an initially laminar beam, given by

$$\varepsilon_n = \varepsilon\beta = \frac{R}{48}\sqrt{\frac{\pi q\bar{I}}{15\varepsilon_0 mc^2 bf}} \qquad (9.77)$$

Although the magnitude of the expression in Eq. (9.77) is only approximate, it predicts a dependence on the beam parameters that is in good agreement with numerical simulation. To minimize space-charge-induced transverse-emittance growth for fixed average current, one needs to design the linac with strong focusing to produce a small beam radius R, with high frequency f to distribute the charge over more bunches, and to avoid focusing the beam too much longitudinally.

9.15
Emittance Growth for rms Mismatched Beams

A mismatched rms beam envelope oscillates with a period that depends on which envelope mode is excited. There are two linearly independent envelope modes, the antisymmetric mode where the x and y oscillations are out of phase, [33] whose wave number is given by

$$k_{e1} = \sqrt{k_0^2 + 3k^2} \tag{9.78}$$

and the symmetric mode where they are in phase whose wave number is given by

$$k_{e2} = \sqrt{2k_0^2 + 2k^2} \tag{9.79}$$

For an rms-mismatched beam, excess potential energy associated with the mismatch oscillations is available for emittance growth. Emittance growth will occur when the beam, under the influence of nonlinear space-charge forces, relaxes toward an equilibrium state. For a uniform continuous beam with linear external focusing, Reiser has derived an equation for emittance growth of an initially rms-mismatched beam, assuming that the free energy of the initial oscillation is completely converted into thermal energy of the final matched beam [34]. Numerical simulation studies [35] show that the emittance growth is associated with formation of beam halo surrounding the beam core. These studies suggest that rms mismatch may be the source of most beam halo in high-current beams.

To calculate the emittance growth from rms mismatch, Reiser defines a free-energy parameter h as

$$h = \frac{1}{2}\frac{k^2}{k_0^2}\left(\frac{a_m^2}{a_{mm}^2} - 1\right) - \frac{1}{2}\left(1 - \frac{a_{mm}^2}{a_m^2}\right) + \left(1 - \frac{k^2}{k_0^2}\right)\ln\frac{a_m}{a_{mm}} \tag{9.80}$$

where a_{mm} is the initial mismatched beam radius, a_m is the initial beam radius if the beam was matched, and k/k_0 is the initial space-charge tune depression ratio. It is assumed that the two rms beam radii are defined at the location where the beam ellipse is upright. The three terms in Eq. (9.80) represent the contributions of the kinetic energy, the potential energy associated with the external focusing force, and the space-charge field energy, respectively. After calculating h, the ratio a_f/a_m of final beam radius to the initial matched beam radius can be calculated by solving numerically the transcendental equation

$$\frac{a_f^2}{a_m^2} - \left(1 - \frac{k^2}{k_0^2}\right)\ln\frac{a_f}{a_m} = 1 + h \tag{9.81}$$

9.15 Emittance Growth for rms Mismatched Beams

It is assumed that in the final state the entire free energy from the mismatch oscillations is transferred to kinetic energy: there is no residual mismatch oscillation. Finally, the emittance growth may be calculated from

$$\frac{\varepsilon_f}{\varepsilon_i} = \frac{a_f}{a_m}\sqrt{1 + \frac{k_0^2}{k^2}\left(\frac{a_f^2}{a_m^2} - 1\right)} \qquad (9.82)$$

If residual mismatch oscillations are present in the final beam state, the equation for emittance growth is an upper limit to the growth. Figure 9.8 shows the plots of emittance growth versus the free-energy parameter h, which varies between 0 and 1.0, as a function of space-charge tune depression, as calculated from Eq. (9.82). Figure 9.8 shows that for a given free energy, emittance growth from rms mismatch increases as the space-charge tune depression becomes stronger. Control of emittance growth in the space-charge-dominated regime is a significant challenge, when the possibility of rms mismatch is present.

In his book, Reiser presents the above material in a more general form which not only includes the rms mismatch effect, but also the nonlinear field energy effect for an rms-matched beam, as well as the effect of an off-centered beam [36]. The same treatment can be used for all three effects, if the total free energy parameter h is the sum of contribution from all three effects. Thus, $h = h_s + h_m + h_c$ representing the contributions of nonlinear field energy, mismatch, and off-center beams, respectively. Thus, h in Eq. (9.80) becomes h_m, and Eq. (9.81) uses the total h value. Emittance growth for an off-center beam also requires some nonlinearity, since nonlinear space-charge does not affect the coherent oscillations. Otherwise the coherent beam oscillations for an off-center beam will continue indefinitely and the free energy will not be thermalized.

The predictions were tested experimentally at the University of Maryland from the same experiment in a 36-solenoid-lens channel that was discussed previously in Section 9.13 [29,30,37], where in this case the two solenoid matching lenses were set to produce an initial rms mismatched to matched beam-size ratio of $a_{mm}/a_m = 0.5$. The beam was highly nonuniform,

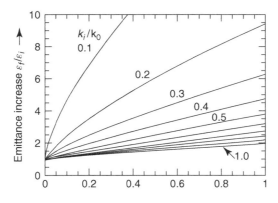

Figure 9.8 Emittance growth from rms mismatch versus the free-energy parameter h as the abscissa as a function of space-charge tune depression k_i/k_0 (courtesy of Martin Reiser).

consisting of five separate beamlets created using an aperture plate with five small holes. However, the main source of free energy was from the mismatch oscillations. The resulting free-energy parameter was $h = 0.396$, and the theoretical emittance growth ratio from mismatch was $\varepsilon_f/\varepsilon_i = 3.72$. The measured emittance growth value was $\varepsilon_f/\varepsilon_i \cong 4.2$. The experiment also showed the formation of beam halo which constituted the majority of the emittance growth. The explanation for the small discrepancy was attributed to greater free energy in the case of the asymmetry produced by the five beamlets.

9.16
Space-Charge Instabilities in RF Linacs from Periodic Focusing: Structure Resonances

An important mechanism that can lead to emittance growth in RF linacs is beam instabilities that occur in periodic focusing channels. The theoretical approach for the stability analysis is known as the *linearized Vlasov theory* in which perturbations in the form of electrostatic charge-density oscillation modes, described by their azimuthal symmetry such as quadrupole ($n = 2$), sextupole ($n = 3$), and so on, are added to known stationary solutions of the Vlasov equation. The theoretical objective is to determine whether the amplitudes for such modes grow or decay. For a beam transport channel comprising a periodic array of quadrupole or solenoid lenses, there is only one known stationary state available for such an analysis, which is the Kapchinsky–Vladimirsky (K–V) distribution. Consequently, the analysis is limited to the study of the K–V distribution. The stability problem for the K–V distribution was first studied in 1970 for the case of a uniform or continuous-focusing channel by Gluckstern, [38] who found that for space-charge tune-depression ratios above 0.4, all the oscillation modes are stable, whereas for tune depressions below 0.4 there existed some unstable modes. In simulation studies it was found that these unstable K–V oscillation modes led to charge-density redistribution with no emittance growth.

In 1983, the stability analysis was carried out for a K–V distribution in a periodic-focusing channel, [39] where it was found that in the presence of periodic focusing some of the oscillation modes became unstable, driven by perturbations in the current distribution. The stability criterion depends on the zero-current phase advance per period σ_0 and the space-charge tune depression ratio, σ/σ_0. The lowest order K–V modes, especially the quadrupole and sextupole modes, generate significant emittance growth in simulations. The quadrupole mode is excited approximately when $\sigma_0 > 90°$, and when σ lies within a stop band slightly under $90°$, whereas the sextupole mode is excited when $\sigma_0 > 60°$.

For simulations with more realistic initial beam distributions, such as the Gaussian distribution, the emittance growth for the sextupole mode is not significant, whereas the quadrupole mode still gives a large emittance growth.

Thus, computer simulations for realistic beams predict that one can safely operate not only with $\sigma_0 < 60°$, but also in the region with $\sigma_0 > 60°$ and $\sigma_0 < 90°$. The region with $\sigma_0 > 90°$ is generally to be avoided.

Next, we describe two experiments that tested the predictions of the theory. A single beam transport channel was constructed at Lawrence Berkeley National Laboratory using 82 electrostatic quadrupole lenses in a FODO configuration, using a cesium beam, as part of the heavy-ion inertial-fusion program. Systematic experiments were conducted by Tiefenback and Keefe, [40] where the beam was matched in both transverse planes, and both σ_0 and σ/σ_0 were varied. The envelope instability predicted by K–V periodic-focusing beam-transport theory for a phase advance per period of $\sigma_0 > 90°$ led to major beam degradation with beam loss. No instability modes predicted by K–V theory, below the $\sigma_0 = 90°$ envelope instability were observed. Similarly, in a systematic experimental study carried out in a solenoid focusing lattice at University of Maryland, [41] the envelope instability was also observed with major beam loss. This was investigated systematically by varying σ/σ_0 and changing σ_0 from below to above 90°. Below 90°, no other instability predicted by the theory, including the third-order (sextupole) mode for $\sigma_0 > 60°$, was found. The conclusion is that for real beams in periodic-focusing channels, the envelope instability predicted by theory for a phase advance per period of $\sigma_0 > 90°$ is the only instability of this theory that leads to emittance growth.

9.17
Longitudinal-Transverse Coupling and Space-Charge Instabilities for Anisotropic Linac Beams

Longitudinal-transverse coupling of the motion, and the behavior of anisotropic beams are subjects of particular importance for RF-linac beam dynamics. The transverse and longitudinal parameters of beams in RF linacs are usually different, which means that the beam bunches are anisotropic. Even without the space-charge forces, the transverse and longitudinal motions in linacs are coupled through nonlinear effects. The sources of the coupling are (1) the dependence of the transverse RF defocusing on the longitudinal phase, and (2) the dependence of the transit-time factor, which determines the longitudinal accelerating force, on the radial displacement. Weiss has calculated the resulting increase in the rms emittance of a bunched beam as it crosses an RF gap [42]. Furthermore, analyses of these nonlinear coupling effects on the transverse motion by Gluckstern [43] showed that substantial amplitude and emittance growth can occur, particularly at low velocities, from resonance when $k_\ell = 2k_t$, where k_ℓ and k_t are the wave numbers of the longitudinal and transverse oscillations, respectively. The analyses showed that other zero-current coupling resonances also may exist, but are rarely significant, with the possible exception of low-velocity resonant amplitude growth, when $k_\ell = k_t$.

When space-charge is included, the physics of anisotropic beams is controlled by collective anisotropy resonances driven by space-charge forces [44–52]. By comparison with these collective effects, collisions between individual particles play no significant role in the physics during the short time that the beam propagates in a linac. A 2D model of the physics by Hofmann [46] is based on analytical results obtained for an anisotropic K–V distribution in a linear uniform-focusing channel. The two dimensions are interpreted as transverse (x), and longitudinal (z) motion, coupled through the space-charge forces. The emittance growth is caused by density perturbations that excite the collective anisotropy resonant modes; some of these modes grow producing the nonlinear fields that cause the rms emittances to change. The resonant regions are conveniently displayed in "stability plots" that show the x space-charge tune-depression ratio k_x/k_{0x} (ordinate), a measure of the importance of space charge, versus the tune ratio k_z/k_x (abscissa), which allow identification of the anisotropy resonances. The "tunes" k_x and k_z are phase advances per transverse focusing period including space charge for an equivalent uniform beam, and the subscript zero refers to zero space charge. Some examples from Hofmann's work are shown in the Fig. 9.9, [50] where in the figures the symbol k is replaced by the symbol ν. Plots are shown for constant values of the emittance ratio $\varepsilon_z/\varepsilon_x$, which is interpreted as the emittance ratio for the initial beam if emittances change.

An energy anisotropy parameter T_a is defined as the ratio of kinetic energies in the two degrees of freedom

$$T_a \equiv \frac{\varepsilon_z k_z}{\varepsilon_x k_x} \tag{9.83}$$

where energy equipartition corresponds to $T_a = 1$. To see this we consider the simple case of an rms-matched beam with uniform focusing in all three planes. Suppose we define a temperature for each degree of freedom by transforming to the center-of-momentum frame of the beam, which we call the *beam frame*, and defining temperatures in the beam frame according to

$$k_B T_x \equiv \frac{\overline{p_{bx}^2}}{m}, \quad k_B T_y \equiv \frac{\overline{p_{by}^2}}{m}, \quad k_B T_z \equiv \frac{\overline{p_{bz}^2}}{m} \tag{9.84}$$

Equipartitioning is defined by the condition $k_B T_x = k_B T_y = k_B T_z$. By performing a Lorentz transformation back to the laboratory frame, it can be shown [53] that equipartitioning implies

$$\frac{\varepsilon_{n,x}^2}{x_{rms}^2} = \frac{\varepsilon_{n,y}^2}{y_{rms}^2} = \frac{\varepsilon_{n,z}^2}{\gamma^2 z_{rms}^2} \tag{9.85}$$

Using the single-particle equations of motion for an equivalent uniform beam, one can also show that the rms beam sizes can be written as

$$x_{rms}^2 = \frac{\varepsilon_{n,x}}{\beta \gamma k_x}, \quad y_{rms}^2 = \frac{\varepsilon_{n,y}}{\beta \gamma k_y}, \quad z_{rms}^2 = \frac{\varepsilon_{n,z}}{\beta \gamma^3 k_z} \tag{9.86}$$

where k_x, k_y, and k_z are the phase advances per unit length of the equivalent uniform beam including space charge, and γ is the relativistic factor for the center of momentum frame. Substituting Eqs. (9.85) into Eqs. (9.86) to eliminate the rms beam sizes, yields a convenient expression for the equipartitioning condition

$$\varepsilon_{n,x}k_x = \varepsilon_{n,y}k_y = \varepsilon_{n,z}k_z. \tag{9.87}$$

The concept of equipartitioning is introduced here, not because linac beams always evolve to stationary states that are equipartitioned, which they do not, but because equipartitioned beams have no free energy available for energy and emittance transfer.

Thus, if $T_a = 1$ there is no free energy available for emittance transfer. If $T_a \neq 1$, the beam is nonequipartitioned and free energy is available for emittance transfer from the degree of freedom that has the larger temperature, given by the larger product $\varepsilon_i k_i$, to the degree of freedom having the smaller temperature. We are interested in the stability with respect to emittance transfer of beams that are rms matched but nonequipartitioned.

The stability charts in Fig. 9.9 show contours for analytically calculated growth rates from the model. These contours identify the calculated stop bands (regions of exponential growth of a mode) that lie in the vicinity of tune ratios with integer tunes, for emittance ratios $\varepsilon_z/\varepsilon_x$ ranging from 0.6 to 5.0. Growth rates in equidistant steps are defined as the e-folding distance per zero-current transverse or betatron period. The most prominent stop bands in the plots are located near tune ratios $v_z/v_x = 1/3$, 1/2, 1, and 2. Note that there is no stop band at $v_z/v_x = 1$ for the emittance ratio $\varepsilon_z/\varepsilon_x = 1$, at $v_z/v_x = 1/2$ for the emittance ratio 2, and at $v_z/v_x = 1/3$ for the emittance ratio 3, all of which are equipartitioned cases. The tune ratios corresponding to equipartitioning at the given emittance ratio are marked as dashed vertical lines in the figures. The typical range for the transverse tune-depression ratio in high-current RF linacs is about $0.5 < v_x/v_{0x} < 0.8$. Energy and emittance transfer are confined to the resonance stop bands, which are relatively narrow if the tune-depression ratio is moderate to weak. Simulations show that the integer stop band, $v_z/v_x = 1$, is usually the only significant mode of concern for emittance transfer. Consequently, nonequipartitioned beams for RF linac designs are safe from emittance transfer, provided that the $v_z/v_x = 1$ stop band is avoided, or if the design trajectory does intersect this stop band that the emittance ratio and the parameter T_a are not far from unity, to limit the available free energy.

As the space-charge tune depression becomes stronger (smaller tune depression ratios), Fig. 9.9 shows that the stop band widths increase and the structure of separated resonances seen at weak to moderate tune depressions, where most linacs are designed, vanishes as the resonance stop bands overlap. The thermodynamic picture that anisotropic beams approach energy equipartitioning for all tune ratios is applicable only very close to the space-charge limit, where the stop bands completely overlap. Then emittance transfer

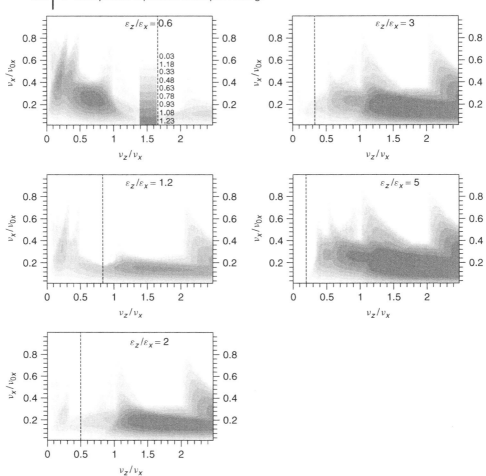

Figure 9.9 Stability charts with analytically calculated stop bands for arbitrary space-charge tune depressions ν_x/ν_{0x} (ordinate), tune ratios ν_z/ν_x (abscissa), and emittance ratios $\varepsilon_z/\varepsilon_x = 0.6, 1.2, 2, 3,$ and 5. Stop bands can be seen in the vicinity of $\nu_z/\nu_x = 1/3, 1/2, 1,$ and 2. The growth rates are defined as the e-folding distance per zero-current betatron period. The dashed vertical lines indicate the tune ratios that correspond to energy equipartitioning, where emittance transfer would not occur (courtesy of I. Hofmann).

approaching equipartition can occur at all tune ratios rather than at only the discrete values; an initially equipartitioned case at very strong tune depressions would show no emittance transfer. While this might appear to avoid emittance transfer, at strong tune depressions emittance growth rather than emittance transfer could still occur in all three degrees of freedom with the free energy coming from nonlinear field energy or from rms mismatch. Maintaining

9.17 Longitudinal-Transverse Coupling and Space-Charge Instabilities for Anisotropic Linac Beams

a bright beam at tune depressions near the space-charge limit remains a significant challenge.

To visualize better the risk of emittance transfer in a real high-current linac design, the design trajectory as a function of beam energy can be superimposed on the stability plot corresponding to the initial emittance ratio. To illustrate the effects discussed above for the parameter regime of a high-current linac, a study was made for three linac designs, the CERN-SPL design, the SNS linac, and the ESS linac [49]. The stability plots are shown in Figs. 9.10 and 9.11. For these two figures the phase advance per period is denoted by k rather than the symbol ν that was used in Fig. 9.9. Figure 9.10 shows the trajectory for the CERN SPL linac design on a stability plot for a nominal emittance ratio $\varepsilon_z/\varepsilon_x = 2$. Three cases are shown. Case 1 is the so-called short design that applies to the superconducting linac from 120 MeV to 2.2 GeV. Its trajectory stays outside the $k_z/k_x = 1$ stop band. Case 2 is a modified design with an increased tune ratio that overlaps with the $k_z/k_x = 1$ stop band. Case 3 corresponds to the full linac. For case 1, where the design trajectory does not enter the $k_z/k_x = 1$ stop band, no emittance exchange was observed. However, for case 2 where the beam trajectory enters the $k_z/k_x = 1$ stop band, significant emittance transfer was observed; the longitudinal emittance decreased from about 0.75 to 0.56 mm-mrad and the transverse normalized emittance increased from about 0.4 to 0.5 mm-mrad. The authors comment that the transferred longitudinal energy is shared equally by both transverse degrees of freedom.

Figure 9.11a shows the design trajectory for the full SNS linac on a stability plot for a normalized emittance ratio $\varepsilon_z/\varepsilon_x = 1.4$. In this design the tune ratio passes through three stopbands located near 1/3, 1/2, and 1. The tune ratio intercepts the $k_z/k_x = 1$ stop band at the low-energy end of the drift-tube linac (DTL) and decreases to about $k_z/k_x = 1/3$ at the end of the DTL. It increases

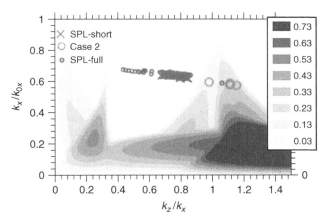

Figure 9.10 Stability plot for $\varepsilon_z/\varepsilon_x = 2$ showing the superimposed CERN SPL design trajectory points as described in the text (courtesy of I. Hofmann).

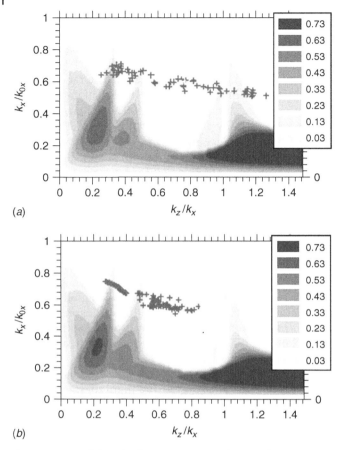

Figure 9.11 Stability plot for $\varepsilon_z/\varepsilon_x = 1.4$ (a) showing the SNS trajectory points, and for $\varepsilon_z/\varepsilon_x = 1.3$ (b) showing the ESS design. The design trajectory points are the + symbols (courtesy of I. Hofmann).

to near $k_z/k_x = 1$ in the superconducting linac. The simulation shows only a small effect of the stop band at $k_z/k_x = 1$, which is mainly in the DTL, and no significant effects for the $k_z/k_x = 1/3$ and $1/2$ stop bands. The net emittance change over the whole linac is $+27\%$ for the transverse case and $+3\%$ for the longitudinal case. The authors attribute some of the growth found in all directions to nonadiabatic behavior at some of the transitions. The design trajectory for the ESS linac design in Fig. 9.11b intersects the $k_z/k_x = 1/3$ and $1/2$ stop bands, but not the $k_z/k_x = 1$ stopband, and no significant emittance transfer was observed.

To summarize, rms-emittance transfer is a small effect in nonequipartitioned beams, if the $k_z/k_x = 1$ stop band is avoided. If this is not possible, the emittance ratio should not be large compared with unity to limit the available free energy for emittance transfer to the transverse degrees of freedom. A good example is the SNS design for which, even though the design trajectory intersects the stop band at $k_z/k_x = 1$, the emittance ratio $\varepsilon_z/\varepsilon_x = 1.4$ was still fairly close to unity. For moderate tune depressions such as in the examples shown, there are large safe regions in the tune depression versus tune-ratio space that are mostly resonance free, which means that nonequipartioned designs, such as that of SNS, can be safe from emittance transfer [54].

9.18
Beam Loss and Beam Halo

Beam halos are of concern for the operation of linacs with high-average beam intensity, because halos increase the risk of beam loss and radioactivation. An important design objective is to restrict beam losses to levels that will allow hands-on maintenance throughout the linac. The maximum tolerable beam-loss rate is conveniently expressed in terms of the lost beam power. If one adopts a hands-on-maintenance criterion that limits the activation level to 80 mrem/h at a distance of 1 ft from stainless steel at 4 h after shutdown of the accelerator, the loss rate for beam energies at 1 GeV must be limited to 1 W/m.

Although not all emittance growth is necessarily associated with the generation of a halo, halo formation is always accompanied by emittance growth. It is known from computer simulations that a halo is produced from space-charge forces, especially from rms-mismatched beams. This is not unexpected physically, because mismatched beams produce time-varying space-charge forces that can transfer energy to some of the particles. Because there is no consensus about its definition, *halo* remains an imprecise term. In any given computer simulation one can unambiguously define an rms beam size, and a maximum particle displacement. Provided that the statistical precision is sufficiently adequate that the results are not sensitive to the motion of a few lone outer particles, the ratio of the maximum displacement to the rms size of the matched beam, which we call the *maximum to rms ratio*, is a useful figure of merit. Qualitatively, one can describe the evolution of the outer regions of the particle distribution for the case of an initial compact particle distribution (excluding the singular K–V distribution). A compact particle distribution might be arbitrarily defined as having initial position and velocity coordinates that are contained within about 3σ. If this beam evolves in an rms-matched state, an equilibrium distribution develops in which the density at the beam edge falls off within about a Debye length. The Debye tail, whose size is a function of both the rms emittance and beam current, is a consequence of the propensity of the charges in a beam to provide shielding within the beam core. For a beam with a given current and emittance, the size of this tail relative to the rms size can be changed by changing the focusing strength. Although there is no consensus about whether to call the Debye tail a halo, values of the maximum to rms ratio larger than about five are generally not observed in simulations for rms-matched beams, and such beams retain a very compact distribution.

The outer region for a beam, with the same initial compact particle distribution in an rms-mismatched state, evolves differently. Many theoretical and numerical studies of halo formation in mismatched beams have been reported, showing larger amplitudes extending well beyond the Debye-tail of a matched beam. For practical estimates of expected mismatch in linacs, values for the maximum to rms ratio as large as 10 to 12 have been observed in simulations, and it is generally agreed that this is called *halo*. The principal halo-formation mechanism is a resonant interaction between the particles in the beam and a beam core that is oscillating because of mismatch. Numerical beam-dynamics simulations show that the largest transverse halo amplitudes are associated with mismatch excitation of the breathing mode of the beam, which is characterized by in-phase oscillations in x and y. The particle-core model of the breathing mode [56] provides understanding of the mechanism that transfers energy to halo particles, and provides a conceptual basis for the aperture choice. Suppose we represent a long bunch with a round, continuous beam in a uniform-focusing channel, and assume a uniform-density core. The core is assumed to be mismatched to excite the radial oscillation of the breathing mode. Halo formation can be studied by introducing single particles that oscillate through the core, driven by linear external focusing fields and the space-charge fields of the core, which are linear inside and nonlinear outside the core. Mathematically, the model may be expressed in dimensionless form as

$$\frac{d^2 X}{ds^2} + X - \frac{\eta^2}{X^3} - \frac{I}{X} = 0 \tag{9.88}$$

and

$$\frac{d^2 x}{ds^2} + x - \begin{cases} Ix/X^2, & x < X \\ I/x, & x \geq X \end{cases} \tag{9.89}$$

where s is the product of the axial distance times the zero-current phase-advance per unit length k_0, X is the ratio of the core radius to the matched core radius, $\eta = k/k_0$ is the tune-depression ratio of the core, k is the single particle transverse phase advance per unit length including space charge, and $I = 1 - \eta^2$ is a dimensionless parameter. Equation (9.88) is the rms envelope equation for the core, and Eq. (9.89) is the equation of motion of a single particle, where the normalized displacement x is defined as the ratio of the particle displacement to the matched rms radius of the core. We note that since I depends on η, the two equations depend on a single dimensionless parameter η. It can be shown that $\eta = 4\varepsilon/k_0 R_0^2$, where ε is the unnormalized rms emittance, and R_0 is the matched core radius. It can also be shown that $1 - \eta^2 = K/k_0^2 R_0^2$, where the K is the generalized perveance, defined earlier.

Although the motion of a particle depends on the normalized core radius X, the dynamics of the core is independent of the motion of the particle, a reasonable approximation, if the halo particles represent only a small fraction of the total beam. The degree of mismatch is measured by the mismatch parameter μ, defined as the ratio of the initial beam radius to the radius of the matched beam; an rms-matched beam has $\mu = 1$.

Equations (9.88) and (9.89) can be studied by systematically varying the initial coordinates of the particle. A stroboscopic phase-space plot, obtained by taking snapshots of many independent particle trajectories, once per core-oscillation cycle at the phase of the core oscillation that gives the minimum core radius, reveals a separatrix defining three regions: (1) core-dominated, (2) resonance-dominated, and (3) focusing-dominated. The stroboscopic plot is shown in Fig. 9.12. The core-dominated trajectories are the innermost trajectories on the stroboscopic plot. The focusing-dominated trajectories are the outermost trajectories, which describe a peanutlike shape in the plot. The resonance-dominated trajectories are those centered about two fixed points on the x axis, and are the trajectories most strongly affected by a parametric resonance, which occurs when the particle frequency is approximately half the core frequency. This resonance is responsible for the growth of the amplitudes that form a halo. For tune ratios below about $\eta = 0.4$, evidence for chaos is observed as the separatrix begins to spread, including the region of phase space very close to the core. This may provide a mechanism for including more

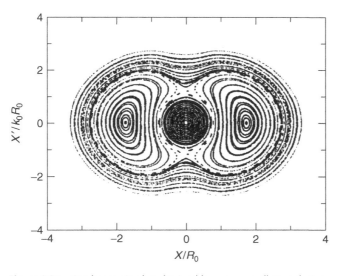

Figure 9.12 Stroboscopic plot obtained by taking snapshots of many independent particle trajectories, once per core-oscillation cycle at the phase of the core oscillation that gives the minimum core radius. Initial particle coordinates were defined on the x and x' axes.

particles in the halo. The observation of chaos is consistent with numerical simulation of space-charge interactions, which often show extreme sensitivity to initial conditions.

The nonlinearity of the resonance, associated with the decrease of the space-charge force outside the core, results in a relationship between frequency and amplitude, which leads to a maximum amplitude for the resonantly driven halo particles, beyond which resonance is not sustained. This important feature is confirmed by numerical simulations using the beam-dynamics codes, from which it is observed that the maximum halo amplitude does not grow without limit as the beam travels through an accelerator with linear external focusing forces. For given values of μ and η, the maximum amplitude for the resonantly driven particles is given by the location of the outermost point of the separatrix. The maximum amplitudes have been calculated from the numerical solution of Eqs. (9.88) and (9.89), as a function of μ between $\eta = 0.5$ and 0.9. The maximum amplitude is very insensitive to η, and can be described by an approximate empirical formula

$$x_{\max}/a = A + B|\ln(\mu)| \tag{9.90}$$

where $x_{\max}$ is the maximum resonant particle amplitude, and a is the matched rms beam size, which is identified in the model with the rms size of the core, and which can be expressed as a function of ε, k_0, and K, using the results in Section 9.8. The constants A and B are weak functions of the tune depression η. In the range $0.500 \leq \mu \leq 0.952$, and $1.05 \leq \mu \leq 2.00$, approximate results are $A \approx B \approx 4$. The ratio of the maximum amplitude to the matched rms beam size increases as a function of the deviation from unity of the mismatch parameter μ. If the main source of beam mismatch is at a single location, usually at a transition point, and if the maximum expected mismatch is known, Eq. (9.90) predicts the maximum halo amplitude. A choice of beam aperture can be made that is larger than the maximum, allowing an appropriate safety margin for a missteered beam centroid, and for trajectory flutter in the quadrupoles. For the more realistic case where the mismatches originate from multiple sources, such as quadrupole gradient errors, the problem must be studied by numerical simulation.

Modeling studies have shown that there are important differences between the transverse and the longitudinal dynamics responsible for beam halo. The transverse halo is caused primarily from particle interactions with the symmetric or breathing envelope mode of the bunch. This mode results in the largest density variations and therefore, the highest frequency, and interacts strongly with particles having high-frequency betatron orbits. The longitudinal dynamics are different in two respects [57]. First, the longitudinal particle frequencies are smaller because of the weaker longitudinal focusing, and second, the halo particles are subjected to relatively stronger nonlinear longitudinal forces. The combination of these effects, shifts the frequency scale for the resonant interaction and changes the primary mode that causes the longitudinal halo from the high-frequency breathing mode to the lower-frequency antisymmetric envelope mode, where the longitudinal rms size oscillates out of phase with the two transverse rms sizes.

9.19
Los Alamos Beam Halo Experiment

A beam-halo experiment was carried out at the Low-Energy Demonstration Accelerator (LEDA) at Los Alamos [58, 59] to test two models of halo formation caused by beam mismatch in high-intensity beams: (1) the free-energy model for mismatched beams, which for a given mismatch strength, determines the maximum emittance growth resulting from the complete transfer of free energy into emittance (see Section 9.15), and (2) the particle-core model discussed in Section 9.18 in which beam mismatch excites a symmetric or antisymmetric (x_{rms} and y_{rms} in or out phase, respectively) oscillation of the core. A high-current, 75-mA, 6.7-MeV proton beam from an RFQ was injected into a 52-quadrupole focusing channel (see Fig. 9.13). The gradients of the first four quadrupoles were independently adjusted to match or mismatch the injected beam. RMS emittances and beam widths were obtained from measured beam profiles for comparison with the theoretical models. The experimental results were also compared with multiparticle simulations. Beam scanners at nine stations, each located midway between pairs of quadrupoles, measured the horizontal and vertical beam distributions. These scanners were unique wire and scraper interceptive beam-profile devices that measured beam profiles including the halo. Secondary electrons were produced in a 33-m carbon wire that was stepped through the beam core, allowing measurement of the beam density in the core. A pair of graphite scraper plates, in which the proton beam stopped, was stepped through the halo, allowing measurement of the density outside the core. Data from the wire and scraper plates were combined in the computer software to produce a single distribution. The scanners were labeled with numbers corresponding to the preceding quadrupole magnet number. The beam was matched, using a least-squares fitting procedure that adjusted the first four quadrupoles to produce equal rms sizes at the last eight scanner locations. Emittances and beam widths were obtained from the measured profiles. The major results are shown in Figs 9.14

Figure 9.13 Block diagram of the LEDA halo experiment showing the 52-quadrupole-magnet lattice and the nine locations of beam-profile scanners.

and 9.15. Figure 9.14 shows the x–y averaged rms-emittance growth results (points with error bars) versus the mismatch parameter μ (ratio of initial rms size to matched rms size) at scanner 20 for a 75-mA breathing-mode mismatch. The maximum emittance-growth curves from the free-energy model are shown for the two tune depression values that bracket the tune-depression values for the debunching beam. It can be seen that the theoretical maximum is insensitive to the tune depression over this range. The breathing-mode data in Fig. 9.14 are consistent at all μ values with the maximum emittance growth predicted by the model. The breathing-mode results at the downstream scanner 45 (not shown) show no significant additional emittance growth, consistent with the upper limits from the model and with complete transfer of free energy. Overall, the data from both the breathing and the quadrupole mode indicate a rapid emittance growth with nearly complete transfer of free energy occurring in less than 10 mismatch oscillations.

The particle-core model, discussed in Section 9.18, predicts the maximum resonant-particle amplitude as a function of the mismatch parameter μ. It was not possible to determine an experimental maximum amplitude for direct comparison with the model because of background halo in the input beam. Instead, the measured amplitudes (x–y averaged half widths of the beam) at three different fractional beam-profile intensity levels (10, 1, and 0.1% of the peak) for breathing-mode mismatches were compared with the maximum amplitude predicted by the particle-core model. This comparison is shown in Fig. 9.15 at scanner 20 for 75 mA. The shapes of all three measured half-width curves are consistent with the shape and magnitude of the maximum amplitude curve from the particle-core model. Similar results are observed at scanner 51. Although the particle-core model based on a single mismatch mode is a simple

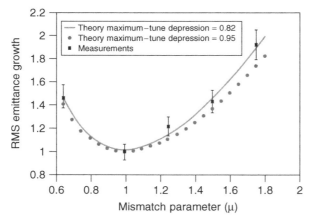

Figure 9.14 Measured rms-emittance growth averaged over x and y for 75 mA at scanner 20 for a breathing-mode mismatch. The curves show maximum growth from the free-energy model.

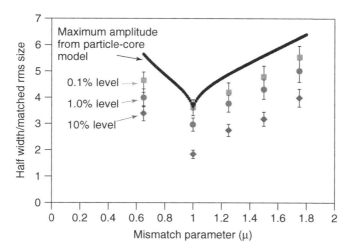

Figure 9.15 Measured beam half widths at scanner 20 (75 mA and a breathing-mode mismatch) at different fractional intensity levels versus mismatch strength μ for comparison with the maximum resonant amplitude of the particle-core model.

description of the beam dynamics, the agreement of the model for the curve shapes and for the consistency of the magnitudes supports the conclusion that the model describes the main physical mechanism responsible for the halo growth.

The 6D-density distribution of the input beam, needed for multiparticle simulations, was not experimentally known. It was found that knowledge of the input Courant–Snyder parameters and emittances alone was not sufficient to obtain agreement between numerical simulations and the observed halo distribution. Agreement between simulations and experiment for the growth rate of the halo will require knowledge of the input distribution, particularly of the tails of the input beam. The results strongly support both models for mismatched beams. This result is important because these models predict upper limits to emittance and halo-amplitude growth in high-current transport channels and linacs, and allow estimation of focusing strength and aperture requirements in new designs.

9.20
Scaling of Emittance Growth and Halo

Assuming that the bunch can be modeled as a long cylinder with a uniform longitudinal profile, we obtain simple scaling formulas for transverse rms emittance growth and maximum particle amplitude for a mismatched beam in a linear focusing channel. The space-charge-induced rms-emittance growth is a function of the tune-depression ratio k/k_0, where k and k_0 are the transverse phase advances per unit length, with and without space charge. In the smooth approximation, where the focusing is represented by an equivalent uniform focusing channel, we may write $k/k_0 = \sqrt{1 + u_t^2} - u_t$, where u_t is a space-charge parameter given by

$$u_t = \left(\frac{qI}{16\pi \varepsilon_0 \ell f \gamma^2}\right)\left(\frac{1}{mc^2 \beta}\right)\left(\frac{1}{\varepsilon_n k_0}\right) \quad (9.91)$$

The parameters appearing in Eq. (9.91) are the charge q, rest energy mc^2, average beam current I, effective bunch length λ, bunch frequency f, relativistic mass factor γ, velocity (relative to that of light) β, and normalized rms emittance ε_n. The average current is related to the number of particles per bunch N and the bunch frequency, by $I = qNf$.

The particle-core model for an rms-mismatched beam predicts that the halo particles created by the resonance have a maximum amplitude for a given core-oscillation amplitude. From Eq. (9.90), the numerical solution predicts that the maximum amplitude is proportional to the matched rms size a of the core, given by

$$a^2 = \frac{\varepsilon_n}{k_0 \beta \gamma}\left[\sqrt{1 + u_t^2} + u_t\right] \quad (9.92)$$

In the space-charge-dominated limit, when $u_t \gg 1$, the rms beam size is

$$a^2 \cong \frac{2\varepsilon_n u_t}{k_0 \beta \gamma} = \left(\frac{qI}{8\pi \varepsilon_0 \ell f \gamma^2}\right)\left(\frac{1}{mc^2 \gamma \beta^2}\right)\frac{1}{k_0^2} \quad (9.93)$$

which is independent of the emittance. The emittance-dominated limit corresponds to $u_t \ll 1$, and we find

$$a^2 \cong \frac{\varepsilon_n}{k_0 \beta \gamma}\left[1 + \left(\frac{qI}{16\pi \varepsilon_0 \ell f \gamma^2}\right)\left(\frac{1}{mc^2 \beta}\right)\left(\frac{1}{\varepsilon_n k_0}\right)\right] \quad (9.94)$$

In Eq. (9.90), the second term is much less than unity, and smaller emittance results in smaller a and smaller maximum amplitude. Higher frequency and longer bunch length produce smaller maximum amplitude. In Eqs. (9.93) and (9.94) stronger focusing (larger k_0) also results in smaller maximum amplitude. These results must be taken with caution, because if the transverse focusing is too strong, the longitudinal space-charge forces will increase and eventually longitudinal halo could become a problem.

9.21
Longitudinal Beam Dynamics Constraint on the Accelerating Gradient

We introduce a simple model to describe a longitudinal envelope instability in a periodic array of cavities that provide longitudinal focusing of the beam.

9.21 Longitudinal Beam Dynamics Constraint on the Accelerating Gradient

The model illustrates the concept of a parametric resonance, which in this case occurs with periodic focusing, and shows how the accelerating gradient is limited by the instability. We write an equation for the longitudinal envelope Z with periodic longitudinal focusing from an array of RF cavities with period L, with space charge treated using the 3D ellipsoid model, and with s as the distance along the accelerator. Assuming a sinusoidal term with amplitude a for the periodic part of the longitudinal focusing, the envelope equation given by Eq. 9.43 is expressed as

$$Z'' + k_{\ell 0}^2 \left[1 + a\, \sin\left(\frac{2\pi s}{L}\right)\right] Z - \frac{\varepsilon^2}{Z^3} - K_\ell = 0 \tag{9.95}$$

where we have introduced k_ℓ as the space-charge parameter.

We write $Z = Z_0$ as the matched solution where $Z_0'' = 0$, and introduce a small mismatch with amplitude φ, where $Z = Z_0 + \varphi$ and assume $\varphi \ll Z_0$. Substituting $Z = Z_0 + \varphi$ into the envelope equation, and using the fact that the matched envelope Z_0 also satisfies the longitudinal envelope equation, we obtain an equation for the mismatch amplitude

$$\varphi'' + k_{\ell 0}^2 \left[1 + a\, \sin\left(\frac{2\pi s}{L}\right)\right] \varphi + \frac{3\varepsilon^2}{Z^4} \phi = 0 \tag{9.96}$$

or

$$\varphi'' + \left(k_{\ell 0}^2 + \frac{3\varepsilon^2}{Z^4}\right)\varphi = -k_{\ell 0}^2 a\, \sin\left(\frac{2\pi s}{L}\right)\varphi = 0 \tag{9.97}$$

This is the equation for an oscillator with a driving term on the right side. Suppose that as a consequence of this driving term, the mismatch amplitude φ is driven sinusoidally. Then the driving term on the right side is the product of two sinusoidal factors, which constitutes the equation for a parametric resonance. The squared wave number of the envelope mismatch oscillation is the factor in the second term on the left:

$$k_{env}^2 = k_{\ell 0}^2 + \frac{3\varepsilon^2}{Z^4} \tag{9.98}$$

Since the driving term is in general the product of two sinusoids with different wave numbers, it can be expressed as the sum of two terms, a first term oscillating at the sum of the two wave numbers and the second term oscillating at the wave number difference. Resonance corresponds to the difference term,

$$k_{env} = \frac{2\pi}{L} - k_{env} \tag{9.99}$$

or $k_{env} L = \pi$. Resonance corresponds to a phase advance of the envelope oscillation equal to π, which means that the period of the envelope oscillation is twice the period of the longitudinal focusing lattice.

Next we search for an approximate expression for the single particle wave numbers k_ℓ and $k_{\ell 0}$, corresponding to the envelope being driven on resonance. These wave numbers will vary with the space-charge tune depression ratio. We begin by writing the single-particle equation of motion in a periodic focusing lattice for a particle whose longitudinal coordinate is z:

$$z'' + k_{\ell 0}^2 \left[1 + a\, \sin\left(\frac{2\pi s}{L}\right)\right] z - \frac{K_\ell z}{Z} = 0 \qquad (9.100)$$

We will replace the second term using the smooth approximation from Section 8.7:

$$z'' + k_{\ell 0}^2 \left[1 + \frac{(ak_{\ell 0}L)^2}{8\pi^2}\right] z - \frac{K_\ell z}{Z} = 0 \qquad (9.101)$$

Assuming $a \ll 1$, we ignore the term that is second order in the amplitude a and obtain

$$z'' + k_{\ell 0}^2 z - \frac{K_\ell z}{Z} \cong 0 \qquad (9.102)$$

The single particle wave number including space charge is

$$k_\ell^2 \cong k_{\ell 0}^2 - \frac{K_\ell}{Z}. \qquad (9.103)$$

Recall that the wave number for the envelope is

$$k_{env}^2 = k_{\ell 0}^2 + \frac{3\varepsilon^2}{Z^4} \qquad (9.104)$$

In smooth approximation the matched envelope Z_0 satisfies $Z_0'' = 0$. Then,

$$k_{\ell 0}^2 Z_0 - \frac{\varepsilon^2}{Z_0^3} - K_\ell = 0 \qquad (9.105)$$

Then,

$$k_\ell^2 = k_{\ell 0}^2 - \frac{K_\ell}{Z_0} = \frac{\varepsilon^2}{Z_0^4} \qquad (9.106)$$

We can eliminate the emittance term and obtain a relationship between the wave number of the envelope mismatch oscillations and the wave numbers of single-particle motion with and without space charge,

$$k_{env}^2 = k_{\ell 0}^2 + \frac{3\varepsilon^2}{Z_0^4} = k_{\ell 0}^2 + 3k_\ell^2 \qquad (9.107)$$

The space-charge tune depression ratio is $\eta = k_\ell / k_{\ell 0}$. Then, $k_{env}^2 = k_{\ell 0}^2 (1 + 3\eta^2)$, or at resonance

$$\sigma_{\ell 0} = k_{\ell 0} L = \frac{\pi}{\sqrt{1 + 3\eta^2}} \qquad (9.108)$$

and

$$\sigma_\ell = k_\ell L = \eta k_{\ell 0} L = \frac{\pi \eta}{\sqrt{1 + 3\eta^2}} \quad (9.109)$$

These two equations give the single particle tunes (phase advance per focusing period) σ_ℓ and $\sigma_{\ell 0}$ as a function of tune depression, when the longitudinal envelope is driven on resonance ($k_{env}/L = \pi$). At the space-charge dominated limit, $\eta = 0$, $k_\ell = 0$ and $\sigma_{\ell 0} = k_{\ell 0} L = \pi$. For the emittance-dominated case, $\eta = 1$, $\sigma_\ell = k_\ell L = \pi/2$, and $k_{\ell 0} L = \pi/2$.

Resonance is always avoided if $\sigma_{\ell 0} = k_{\ell 0} L < \pi/2$, and therefore we will conservatively adopt this as an upper limit to $k_{0\ell}$. Further studies are required to determine whether this criterion is too conservative. Nevertheless, we will see that this constraint will limit the average accelerating gradient, and assuming this constraint and using Eq. 6.43 we write the squared zero-current longitudinal wave number as

$$k_{\ell 0}^2 = \left(\frac{\sigma_{\ell 0}}{L}\right)^2 = \frac{2\pi \langle E_0 T \rangle \sin(-\phi)}{mc^2 \gamma^3 \beta^3 \lambda} \leq \left(\frac{\pi}{2L}\right)^2 \quad (9.110)$$

or

$$\langle E_0 T \rangle \leq \frac{\pi mc^2 \gamma^3 \beta^3 \lambda}{8q \sin(-\phi) L^2} \quad (9.111)$$

The longitudinal envelope instability constraint is important for high charge state, small mass, low velocities (cubic dependence on β), high frequencies, and for long focusing periods (quadratic dependence on L). It is a constraint on the spatial average gradient or real-estate gradient. Reducing the magnitude of the phase below about 30° does not help much because the phase width of the bucket shrinks, causing beam losses. Thus, this longitudinal envelope instability, a parametric resonance in a periodic focusing array, can limit the accelerating gradient at low velocities. It is generally not a problem for relativistic particles and is more of a problem for proton beams than for heavy ions.

Problems

9.1. Given the transfer matrix of a drift space, derive the results for the change in the ellipse parameters. Show that the ellipse parameter $\tilde{\gamma}$ is invariant. What does this mean physically?

9.2. Given the transfer matrix of a thin lens, derive the results for the change in the ellipse parameters. Show that the ellipse parameter $\tilde{\beta}$ is constant. Is this reasonable physically?

9.3. Measurements of transverse beam distribution for a proton beam with velocity $\beta = 0.06$ show an upright ellipse in the $x - x'$ plane, with an rms size $\Delta x_{rms} = 3$ mm and an rms divergence $\Delta x'_{rms} = 0.06$ mrad. (a) What is the unnormalized rms emittance? (b) What is the normalized rms emittance? (c) If the normalized rms emittance is conserved during acceleration to 100 MeV, what is the unnormalized rms emittance at 100 MeV?

9.4. Measurements of the longitudinal beam distribution result in an upright ellipse in the $\Delta\phi$–ΔW plane with rms values $\Delta\phi_{rms} = 1°$ and $\Delta W_{rms} = 100$ keV. (a) What is the longitudinal emittance in deg·MeV? (b) If the RF frequency is 400 MHz, what is the longitudinal emittance in s·MeV? (c) If the beam is now injected into an 800 MHz linac, what is the emittance in deg·MeV and s·MeV?

9.5. Assume a bunched proton beam at 100 keV with an average current of 100 mA and a frequency of 400 MHz, having a uniform spherical density with radius 2 mm. (a) Calculate the charge density and the number of particles per bunch. (b) Calculate the electric field from space charge at the edge of the beam. (c) Calculate the space-charge parameter K_3. (d) Assume that the number of dimensions for which the distribution is uniform is $n = 3$. Calculate the projected rms size on the x axis. (e) If the rms unnormalized emittance in the x plane is $\varepsilon_r \pi = 20\pi$ mm·mrad, what is the ratio of the space-charge term to the emittance term in the envelope equation? Is the beam space-charge or emittance dominated?

9.6. Consider a 100-mA proton beam at 5 MeV in a 200-MHz DTL with parameters $E_0 T = 2$ MV/m, and synchronous phase $\phi_s = -25°$. Assume that the bunch is spherical and has an rms size of 2.24 mm in all three planes. (a) Calculate $k_{\ell 0}$, the zero-current wave number of small longitudinal oscillations. (b) For the equivalent uniform beam, what is the longitudinal tune-depression ratio $k_\ell/k_{\ell 0}$, the wave number of small longitudinal oscillations k_ℓ including space charge, and the ratio of space-charge to focusing force μ_ℓ?

9.7. For the same beam as in Problem 9.6, assume that the transverse focusing is provided by a FODO quadrupole array using quadrupoles with gradient-length product $G\ell = 3$ T placed in every drift tube. (a) What is the period of the focusing lattice at 5 MeV? (b) Using the smooth-approximation formula given in Section 9.11 what is the zero-current phase advance per focusing period σ_0? Does it meet the requirement $\sigma_0 < 90°$ to avoid envelope instability? (c) For the equivalent uniform beam, what is the transverse tune-depression ratio σ/σ_0, the phase advance per focusing period σ (including space charge), and the ratio of space-charge to focusing force μ_T?

9.8. (a) What is the space-charge electric field in megavolt per meter at the edge of the beam bunch in Problem 9.6 assuming a uniform density bunch? (b) The longitudinal focusing force can be written as $F = -mc^2 k_{\ell 0}^2 \beta^2 r$. Calculate the effective longitudinal focusing field at the edge of the beam in megavolt per meter for Problem 9.6 and compare to the axial accelerating field $E_0 T \cos\varphi_s$.

9.9. A 50-mA proton beam is injected into a 400-MHz DTL at a velocity $\beta = 0.05$. Assume the radial distribution is approximately parabolic with radius $R = 4$ mm, and bunch length $2b = 8$ mm. Assume the emittance of the injected beam is essentially zero, and assume all buckets are filled. (a) Use the result of Section 9.14 to calculate the total normalized transverse emittance in units of mm-mrad after the beam is subjected to its own space-charge force over the fixed distance of the first cell. (b) Calculate the beam plasma period in meters at the DTL injection velocity assuming that the bunch parameters are constant during the first beam-plasma oscillation. (c) Calculate the quasisteady state transverse normalized emittance after the beam has gone through one plasma period, using the result of Section 9.14. Express the answer in units of mm-mrad. (d) Suppose that the beam radius is decreased to $R = 2$ mm. How do the results of parts (a), (b), and (c) change as a result of the decreased beam radius.

9.10. (a) Consider a distribution of particles in phase space that lies on a straight line given by $x' = ax$. Use the definition of rms emittance to show that the rms emittance is zero. (b) Suppose the particles lie on a curved line described by $x' = ax^3$. This line still has zero area as in part (a), but is the rms emittance still zero?

9.11. A 100-mA dc proton beam with an initial parabolic charge density, and initial velocity $\beta = 0.01$ is injected from an ion source into an RFQ. Assume the RFQ electric fields will remove neutralizing electrons, so the beam sees the full space-charge force. The beam is matched and the focusing is always adjusted to give an average rms size $R = 7.5$ mm. (a) Calculate the generalized perveance K. (b) The beam charge will rapidly redistribute accompanied by emittance growth, and from the model in section 9.14 the final unnormalized emittance ε for an initially zero emittance beam is given by Eq. 9.76. Assuming that for a beam with initial emittance ε_i, the final emittance is given by $\varepsilon_f = \sqrt{\varepsilon_i^2 + \varepsilon^2}$. Calculate and plot the final normalized rms emittance $\beta\gamma\varepsilon_1$ for initial normalized emittances $\beta\gamma\varepsilon_1 = 0, 0.1, 0.2, 0.3$, and 0.4 mm-mrad. If an ion source breakthrough allows the input rms-normalized emittance to be reduced from 0.2 to 0.1 mm-mrad, would the RFQ be able to preserve this emittance reduction? (c) If the emittance growth and beam relaxation to the final uniform density occurs in one quarter plasma period, what is this relaxation distance?

9.12. Show that the radial electric field and potential from space charge at the edge of a round continuous uniform beam of radius R and velocity β can be written as $E = I_b/2\pi\varepsilon_0 c\beta R = 60I_b/\beta R$ in V/m, and $V = I_b/4\pi\varepsilon_0 c\beta = 30I_b/\beta$ in V, where $I_b = qN_\ell\beta c$ is the average current during the passage of the bunch, expressed in terms of the number of particles per unit length N_ℓ.

9.13. If we choose a round continuous uniform beam of length ℓ as a cylinder model for a bunch, the current I averaged over an RF period can be written as $I = I_b \ell/\beta\lambda$, where I_b is the average current during the passage of the bunch, and λ is the RF wavelength. Use this and the results from Problem 9.12 to calculate the radial electric field and potential from space charge at the edge of the bunch for the parameters $\beta = 0.1$, $R = 0.5$ mm, $I = 0.1$ A, and $\ell/\beta\lambda = 0.2$.

9.14. Resolve the apparent paradox that the most effective approach for controlling space-charge effects is to apply strong focusing, which makes the beam-size small and increases the space-charge force. Assuming an rms-matched, round, continuous beam propagating in a uniform-focusing channel, write an expression for the ratio of the space-charge force to the focusing force. Using the analytic solution of the rms envelope equation, and assuming a fixed generalized-perveance K and fixed emittance ε, show that if the focusing strength k_0 increases, the ratio of the space-charge force to focusing force decreases. Thus, although with increased focusing the space-charge force has increased, the focusing force has increased even more, making the ratio smaller.

References

1 For a more complete discussion of this and other space-charge and emittance topics, the reader is referred to the excellent treatments in the books by Lawson, J.D., *The Physics of Charged Particle Beams*, 2nd ed., Oxford University Press, 1988; Reiser, M., *Theory and Design of Charged Particle Beams*, John Wiley & Sons, New York, 1994.
2 Lawson, J.D., *The Physics of Charged Particle Beams*, 2nd ed., Oxford University Press, 1988, pp. 151–155.
3 Sacherer, F., *IEEE Trans. Nucl. Sci.* **18**, 1105 (1971).
4 Lapostolle, P.M., *IEEE Trans. Nucl. Sci.* **18**, 1101 (1971).
5 Weiss, M., unpublished.
6 Lawson, J.D., *The Physics of Charged Particle Beams*, 2nd ed., Oxford University Press, 1988, p. 197.
7 Ibid. p. 246.
8 For a more detailed treatment of numerical simulation of particle interactions see Hockney, R.W. and Eastwood, J.W., *Computer Simulation Using Particles*, Adam Hilger (IOP Publishing Ltd), 1988.
9 Crandall, K.R., *TRACE, An Interactive Beam Dynamics Code, Linear Accelerator and Beam Optics Codes*, ed. Charles R. Eminhizer, La Jolla Institute, 1988; *AIP Conf. Proc.* **177**, 29 (1988).
10 Crandall, K.R., *TRACE 3-D Documentation*, 2nd ed. Los Alamos Report LA-UR-90-4146, December 1990.
11 Crandall, K.R., unpublished.
12 Crandall, K.R., Los Alamos Group AT-1 Memorandum AT-1-237, September 2, 1980.
13 Guy, F.W., *3-D Space-Charge Subroutine for PARMILA*, Los Alamos Group AT-1 Report, AT-1:85–90, March 6, 1985.
14 Martini M., and Prome, M., Computer studies of beam dynamics in a proton linear accelerator with space charge, *Part. Accel.* **2**, 289 (1971).
15 Lapostolle, P.M., Lombardi, A.M., Nath, S., Tanke, S., E., Valero, and Wangler, T.P., *A New Approach to Space Charge for Linac Beam Dynamics Codes*, Proc. 18th Int. Linac Conf., August 26–30, 1996, CERN, Geneva, 1996, p. 375.
16 Wangler, T.P., Crandall, K.R., Mills, R.S. and Reiser, M., High current, high brightness, and high duty factor

ion injectors, *AIP Conf. Proc.* **139**, 133 (1985).

17 Kapchinsky I. and Vladimirsky, V., 2nd Conf. High Energy Accel., CERN, 1959, p. 274.

18 Weiss, M., Proc. 1979 Linear Accel. Conf., September 10–14, 1979, Montauk, NY, Brookhaven National Laboratory Report, BNL 51134, p. 227.

19 Wangler, T.P., *Some Useful Elementary Relationships for a Continuous Beam with Space Charge, Space-Charge*, Los Alamos Group AT-1 Report AT-1:85–71, February 20, 1985.

20 Lapostolle, P., CERN Report AR/Int SG/65-15, 1965.

21 Wangler, T.P., *Space-Charge Limits in Linear Accelerators*, Los Alamos Report LA-8388, December, 1980.

22 Reiser, M., *Theory and Design of Charged Particle Beams*, John Wiley & Sons, New York, 1994, pp. 240–251.

23 Wangler, T.P., *High-Brightness Injectors for Hadron Colliders*, 1990 Joint US-CERN School on Particle Accelerators, November 7–14, 1990, Hilton Head, SC.

24 Wangler, T.P., Crandall, K.R. and Mills, R.S., Heavy ion inertial fusion, *AIP Conf. Proc.* **152**, 166 (1986).

25 The K–V equilibrium distribution is an exception that always has a uniform spatial density with no Debye tail. However, this distribution bears no resemblance to real beam profiles that have been measured or observed in numerical simulations.

26 Kapchinsky, I.M., *Theory of Resonance Linear Accelerators*, Harwood Academic Publishers (1985), p. 273.

27 Kapchinsky, I.M. and Vladimirsky, V.V., *Limitations of Proton Beam Current in a Strong Focusing Linear Accelerator Associated wit the Beam Space Charge*, Proc. Int. Conf. High-Energy Accel. Instrumentation CERN, Geneva, 1959, p. 274.

28 Wangler, T.P., Crandall, K.R., Mills, R.S., and Wangler, T.P., *Field Energy and RMS Emittance in Intense Particle Beams, High Current, High Brightness, and High Duty Factor Ion Injectors*, AIP Conf. Proc., 139, La Jolla, CA, (1986), pp. 133–151.

29 Reiser, M., Chang, C.R., Kehne, D., Low, K., Shea, T., Rudd, H. and Haber, I., *Phys, Rev. Lett.* **61**, 2933 (1988).

30 Haber, I., Kehne, D., Reiser, M. and Rudd, H., *Phys. Rev. A* **44**, 5194 (1991).

31 Reiser, M., *Theory and Design of Charged Particle Beams*, 1st ed., John Wiley & Sons, 1994, pp. 482–488.

32 Wangler, T.P., Lapostolle, Pierre and Lombardi, A., Proc. 1993 Part. Accel. Conf., Washington D.C., May 17–20, 1993, IEEE Catalog No. 93CH3279-7, p. 3606.

33 Reiser, M., *Theory and Design of Charged Particle Beams*, 1st ed., John Wiley & Sons, 1994, p. 242.

34 Reiser, M., *Theory and Design of Charged Particle Beams*, 1st ed., John Wiley & Sons, 1994, p. 476.

35 Cucchetti, A., Reiser, M. and Wangler, T.P., *Simulation Studies of Emittance Growth in RMS Mismatched Beams*, Proc. 1991 Part. Accel. Conf., San Francisco, CA, May, 1991.

36 Reiser, M., *Theory and Design of Charged Particle Beams*, 1st ed., John Wiley & Sons, 1994, pp. 470–479.

37 Reiser, M., *Theory and Design of Charged Particle Beams*, 1st ed., John Wiley & Sons, 1994, pp. 488–491.

38 Gluckstern, R.L., Proc. 1970 Linear Accel. Conf., Fermi National Accelerator Laboratory, Batavia, IL, 1970, Vol. 2, p. 811.

39 Hofmann, I., Laslett, L.J., Smith, L. and Haber, I., Stability of the Kapchinsky-Vladimirsky (K–V) distributions in long periodic transport systems, *Part. Accel.* **13**, 145 (1983).

40 Tiefenback M.G. and Keefe, D., Measurements of stability limits for a space-charge-dominated ion beam in a long A.G. transport channel, *IEEE Trans. Nucl. Sci.* **NS-32**, 2483–2485 (1985).

41 Brown, N. *Thermal Equilibrium of Charged Particle Beams*, Ph.D. Dissertation, Physics Department, University of Maryland, College Park, MD, 1995.

42 Weiss, M. *Bunching of Intense Proton Beams with Six-Dimensional Matching to the Linac Acceptance*, CERN/MPS/LI report 73-2, Geneva, Switzerland

(1978); and Crandall, K.R. and Rusthoi, D.P., *Appendix F, Emittance Increase in a Gap*, 3rd ed., TRACE 3-D Documentation, Los Alamos Report LA-UR-97-886, May, 1997, pp. 78–80.

43 Gluckstern, R.L., *Transverse Beam Growth Due to Longitudinal Coupling in Linear Accelerators*, Proc. of 1966 Linear Accel. Conf., LA Report LA-3609, Los Alamos, NM, October 3–7, 1966, pp. 207–213; and Gluckstern, R.L., Nonlinear effects, in *Linear Accelerators*, Lapostolle P.M. and Septier, A.L., John Wiley & Sons, New York, (1970) pp. 797–807.

44 Hofmann, I. and Bozsik, I., *Computer Simulation of Longitudinal-Transverse Space-Charge Effects in Bunched Beams*, Proc. 1981 Linear Accel. Conf., Los Alamos National Laboratory Report LA-9234-C 1982, p. 116.

45 Jameson, R.A., *Equipartitioning in Linear Accelerators*, Proc. 1981 Linear Accel. Conf., Los Alamos National Laboratory Report LA-9234-C (1982), p. 125.

46 Hofmann, I., Stability of anisotropic beams with space charge, *Phys. Rev. E* **57**, 4713 (1998).

47 Hofmann, I., Qiang, J., and Ryne, R.D., Collective resonance model of energy exchange in 3D nonequipartitioned beams, *Phys. Rev. Lett.* **86**, 2313 (2001).

48 Hofmann, I., and Boine-Frankenheim, O., Resonant emittance transfer driven by space charge, *Phys. Rev. Lett.* **87**, (034802-1) 2001.

49 Hofmann, I., Franchetti, G., Qiang, J., Ryne, R., Gerigk,F., Jeon, D., and Pichoff, N. *Review of Beam Dynamics and Space Charge Resonances in High Intensity Linacs*, Proc. of EPAC 2002, Paris, France, p. 74.

50 Hofmann, I., Franchetti, G., Boine-Frankenheim, O., Qiang, J. and Ryne, R.D., Space charge resonances in two and three dimensional anisotropic beams, *Phys. Rev. ST-AB*, **6**, 024202 (2003).

51 Reiser M., and Brown, N., Proposed high-current RF linear accelerators with beams in thermal equilibrium, *Phys. Rev. Lett.* **74**, 1111 (1995).

52 Kishek, R.A., O'Shea, P.G. and Reiser, M., Energy transfer in nonequilibrium space-charge dominated beams, *Phys. Rev. Lett.* **85**, 4514 (2000).

53 Reiser, M., *Theory and Design of Charged Particle Beams*, John Wiley & Sons, New York, 1994, p. 573.

54 Hofmann, I., Collective resonance model of energy exchange in 3D nonequipartitioned beams, *Phys. Rev. Lett.* **86**, 2313 (2001).

55 Gluckstern, R.L., *Proceeding 1970 Linear Accelerator Conference*, Fermi National Accelerator Laboratory, Batavia, IL, 1970, Vol. 2, p. 811.

56 Wangler, T.P., Crandall, K.R., Ryne, R., and Wang, T.S., Particle-core model for transverse dynamics of beam halo, *Phys. Rev. ST Accel. Beams* **1**, 084201 (1998); Gluckstern, R.L., Analytic model for halo formation in high current beams, *Phys. Rev. Lett.* **73**, 1247 (1994). See these papers for additional references.

57 Barnard J., and Lund, S., 1997 Part. Accel. Conf., Vancouver, Canada, May 12–16, 1997, to be published.

58 Allen, C.K., Chan, K.C.D., Colestock, P.L., Crandall, K.R., Garnett, R.W., Gilpatrick, J.D., Lysenko, W., Qiang, J., Schneider, J.D., Schulze, M.E., Sheffield, R.L., Smith, H.V. and Wangler, T.P., Beam-halo measurements in high-current proton beams, *Phys. Rev. Lett.* **89**, 214802 (2002).

59 Wangler, T.P., Allen, C.K., Chan, K.C.D., Colestock, P.L., Crandall, K.R., Garnett, R.W., Gilpatrick, J.D., Lysenko, W., Qiang, J., Schneider, J.D., Schulze, M.E., Sheffield, R.L. and Smith, H.V., *Beam Halo in Mismatched Proton Beams*, 2002 Charged Particle Optics Conference, October 22–25, 2003, Greenbelt, MD, published in Nuclear Instruments and Methods in Physics Research A 519, p. 425–431 (2004).

10
Beam Loading

The beam is not simply a medium which absorbs radio frequency (RF) energy and adds an additional resistive load to the cavity, but is really equivalent to a generator, which can either absorb energy from the cavity modes or deliver energy to them. However, in the previous chapters we have assumed that the beam is affected by the cavity, but we have ignored the effects of the beam on the cavity fields. As the beam current increases it becomes important to treat the effects of the interaction between the beam and the cavity more carefully. The effects of the beam on the cavity fields in the accelerating mode are referred to as *beam loading*. They arise physically as a result of the charges induced in the walls of the cavity as the bunches, together with their comoving electromagnetic fields, pass through the cavity. The induced charges produce fields that act back on the particles in the bunch, and this interaction does work on the particles, which radiate energy to the cavity. In this chapter we introduce these effects through the fundamental theorem of beam loading, which relates the energy delivered to the cavity to the charge q of a single point charge passing through it. The proportionality factor between the energy transferred to the cavity and the squared charge is called the *loss factor*. The most important effects of beam-induced fields in modes other than the accelerating mode, are higher-order-mode power losses, and the beam-breakup instability, both of which are discussed in Chapter 11. In this chapter, we emphasize the accelerating mode, a special case in which an external generator is provided to establish the fields needed for acceleration of the beam, and in which the net field is a superposition of the generator-induced plus the beam-induced fields. When the beam-induced field in the accelerating mode becomes comparable to the field induced by the external generator, the net phase and amplitude will only be satisfactory for beam acceleration, if some means of compensation for the effect of the beam is provided. The conventional analysis of beam loading in a standing-wave cavity, including the conventional method used for compensation, is presented in this chapter.

RF Linear Accelerators. 2nd, completely revised and enlarged edition.
Thomas P. Wangler
Copyright © 2008 Wiley-VCH Verlag GmbH & Co. KGaA, Weinheim
ISBN: 978-3-527-40680-7

10.1
Fundamental Beam-Loading Theorem

Consider a point charge q moving through an initially unexcited loss-free cavity. The fields from the moving charge will induce surface charges on the cavity walls and a total beam-induced voltage in the cavity. After the charge has left the cavity, the induced surface charges and voltage will remain, and by energy conservation the moving charge must leave some fraction of its energy there. The energy that the charge looses must be the result of work done on the charge by the induced voltage, and that voltage must decelerate the charge to conserve energy. However, because the voltage is in the process of being established while the charge is in the cavity, it is not obvious what fraction of the final induced voltage is actually seen by the exciting charge. We will apply energy conservation to deduce the answer to this question, using a proof that applies for a single arbitrary cavity mode [1].

We consider two equal charges q, each with kinetic energy W, that are spaced apart by half of a wavelength for the cavity mode of interest, the second charge following the first through the initially unexcited cavity, as shown in Fig. 10.1. We assume that the velocity of each charge is constant while passing through the cavity. The first charge excites the cavity, producing a retarding cavity voltage $V_c = -V_b$, shown in Fig. 10.2. We express the induced voltage seen by particle 1 as $V_{q1} = fV_b$, where f will be the unknown fraction of the total beam-induced voltage that acts on the charge that produces it. Thus the energy loss for charge 1 is $\Delta W_1 = -qfV_b$, and we express the induced stored energy by $U = \alpha V_b^2$, where α is a constant of proportionality. From energy conservation $U = \alpha V_b^2 = -\Delta W_1 = qfV_b$. This implies that $V_b = fq/\alpha$, so the induced cavity voltage is proportional to the charge that produced it. Half a period later, shown in Fig. 10.3, the second charge crosses the cavity. The voltage that was induced by particle 1 has changed its phase by π and, if the cavity is lossless, is now an accelerating voltage $+V_b$. The second charge, being identical to the first, will also induce a retarding voltage $-V_b$. The two induced voltages add out of phase to give a net cavity voltage $V_c = 0$, so after the second particle has crossed the cavity, the cavity stored energy must be $U = 0$.

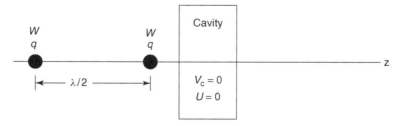

Figure 10.1 Initial configuration showing two identical charges and an unexcited cavity.

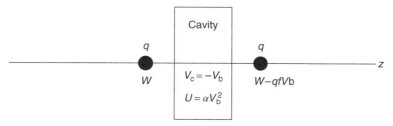

Figure 10.2 Configuration after the first charge has crossed the cavity.

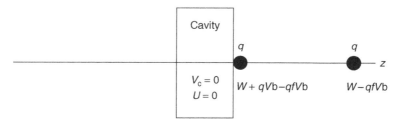

Figure 10.3 Final configuration with an empty cavity. The energy left in the cavity by the first charge is absorbed by the second charge.

From energy conservation, the sum of the energy changes of the two charges must also be $\Delta W_1 + \Delta W_2 = 0$. The energy change of charge 2 is the sum of the energy gain qV_b, from the voltage induced in the cavity by charge 1, which after half an RF period has changed its phase by π, plus the energy loss to the cavity $qV_b f$. Adding the energy gain of charge 2, $\Delta W_2 = qV_b(1-f)$, to the energy loss of charge 1, $\Delta W_1 = -qfV_b$, we obtain $\Delta W_1 + \Delta W_2 = qV_b(1-2f) = 0$, which implies that $f = 1/2$. Therefore, the energy loss of the first charge to the initially unexcited cavity is given by $\Delta W_1 = qV_b/2$. In other words, the induced voltage that the first charge sees is equal to half the induced voltage that the particle leaves in the cavity, or

$$V_q = V_b/2 \qquad (10.1)$$

Equation (10.1) is called the *fundamental theorem of beam loading*. Note that no assumption was made regarding the velocity of the charges, so the result is valid for both relativistic and nonrelativistic charges. We notice also that the mechanism described above produces acceleration of a trailing charge, by using the cavity to transfer energy from a particle to a later particle.

10.2
The Single-Bunch Loss Parameter

It is convenient to define the loss parameter $k = 1/4\alpha$ for each cavity mode, so that the energy loss of a point charge q into an empty cavity mode is

$$U = \alpha V_b^2 = \frac{q^2}{4\alpha} = kq^2 \qquad (10.2)$$

the voltage induced in the cavity mode by the point charge q is

$$V_b = 2kq \qquad (10.3)$$

and the induced voltage in that mode seen by the charge itself is

$$V_q = kq \qquad (10.4)$$

Furthermore, because $k = 1/4\alpha$, we can express the cavity stored energy induced by the point charge as $U = \alpha V_b^2 = V_b^2/4k$, so that the loss parameter is $k = V_b^2/4U$. It is convenient to express k in a form that allows it to be evaluated from the basic properties of the cavity mode, computed from a standard electromagnetic code such as SUPERFISH [2]. We identify the axial voltage V_b as the voltage $V_0 T$ that includes the transit-time factor T. Then we introduce the effective shunt impedance $r = V_b^2/P$ in ohms, where P is the cavity-power dissipation. Introducing the unloaded quality factor $Q_0 = \omega U/P$ for the mode of interest, we express the loss factor, which describes the single-bunch beam-loading characteristics of a mode, as

$$k = \frac{\omega}{4} \frac{r}{Q_0} \qquad (10.5)$$

10.3
Energy Loss to Higher-Order Cavity Modes

When a charge q traverses a cavity, it will induce a voltage $V_{bn} = 2k_n q$ for each mode n. The total induced voltage is the sum of the induced voltages for each mode. We identify the accelerating mode with a subscript 0. The other modes will be called higher-order modes and will be labeled by a subscript $n > 0$. When a charge q crosses the initially unexcited cavity, the total beam-induced voltage V_{bT} is the sum of the fundamental plus the higher-order-mode voltages. Thus

$$V_{bT} = \sum_{n=0} V_{bn} = V_{b0} \sum_{n=0} \frac{V_{bn}}{V_{b0}} = V_{b0} \sum_{n=0} \frac{k_n}{k_0} \qquad (10.6)$$

where $V_{b0} = 2qk_0$ and $V_{bn} = 2qk_n$. We define a higher-order-mode enhancement factor, as

$$B = \sum_{n=0}^{\infty} \frac{k_n}{k_0} \qquad (10.7)$$

For the SLAC linac [3] with a typical RMS bunch length of 1 mm, the higher-order mode enhancement factor is about $B = 4$.

The total energy left by the passage of the charge through the cavity is

$$U_T = q^2 \sum_{n=0} k_n = k_0 q^2 \sum_{n=0} \frac{k_n}{k_0} = k_0 q^2 B \qquad (10.8)$$

A point charge loses energy to all modes of an initially unexcited cavity. The unavoidable excitation of the higher-order modes results in beam-energy loss and cavity-power dissipation. The excess beam-energy loss to higher-order modes is

$$U_{\text{hom}} = U_T - U_0 = k_0 q^2 (B - 1) \qquad (10.9)$$

and the excess power loss is

$$\Delta P_{\text{hom}} = \sum_{n>0} \frac{\omega_n U_n}{Q_{0n}} = q^2 \sum_{n>0} \frac{\omega_n k_n}{Q_{0n}} \qquad (10.10)$$

10.4
Beam Loading in the Accelerating Mode

In this section we describe the effects of beam loading in the accelerating mode of a standing-wave cavity, [4] when an accelerating voltage is already present in the cavity before the bunch arrives. We need to add the generator-induced and beam-induced voltages to obtain the total voltage. The result that the voltages from independent sources can be added together to give the total voltage follows from the principle of linear superposition, which is a consequence of the linearity of Maxwell's equations. For the case of sinusoidally varying fields and voltages, one must add these, taking into account the relative phases. It is convenient to describe the voltages as vector quantities in the complex plane. Thus we write a voltage as $\tilde{V} = Ve^{j(\omega t + \theta)}$ where V is the magnitude and $\omega t + \theta$ is the phase. This is a vector that rotates counterclockwise in the complex plane and is called a *phasor*. It is convenient to choose a frame of reference that is rotating with a frequency ω, so that the phasors remain fixed in time. It is convenient to choose zero phase to correspond to the time of arrival of the center of the bunch at the electrical center of the cavity. Then the component of any voltage that contributes to acceleration of the bunch is the projection of the voltage onto the real axis, which corresponds to the phase of the beam current. In Fig. 10.4 we show a phasor diagram for a cavity that is operating on resonance in a steady state, and is excited both by a generator and a train of bunches. The generator-induced voltage is $\tilde{V}_g$, and it has a phase θ relative to the fundamental Fourier component of the bunch current. The beam-induced voltage, which maximally opposes the beam, is $\tilde{V}_b$. The net cavity voltage $\tilde{V}_c$ is the vector sum of these two voltages, and it has a phase ϕ relative to the beam; ϕ is also the synchronous phase. Figure 10.4 shows that the net cavity voltage deviates from the generator-induced voltage, because of the beam-induced voltage.

Beam-loading compensation is important, because it affects the RF system efficiency. In many high-energy electron linacs the bunches are accelerated at the peak of the accelerating wave to achieve the maximum energy gain. The cavity is driven at its resonant frequency by an RF generator through an input RF power coupler, which acts as a transformer that can be adjusted to match the real part of the cavity impedance to the input transmission line. A schematic drawing of the RF power system is shown in Fig. 10.5. The coupler geometry is ideally adjusted to minimize the reflected power when the beam is present. If the cavity is driven on resonance, the cavity impedance seen by the generator is a pure real number, and the coupler geometry can be chosen to give zero reflected power. The beam-induced and generator-induced voltages are 180° out of phase, and beam-loading compensation requires that the generator current must be increased to correct for the voltage reduction caused by the beam.

If the bunches are injected off the crest of the accelerating wave, the reflected power can be adjusted to zero by employing the two-step approach of (1) compensating for the phase difference by detuning the cavity so that it is driven off resonance, and (2) adjusting the coupler for minimum reflected power. In the following section, we review the theory of a steady-state, beam-loaded cavity.

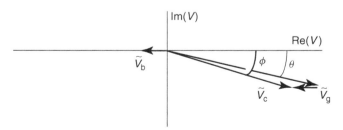

Figure 10.4 Voltage phasors for a steady-state, beam-loaded cavity operating at resonance. The quantities shown in the figure are defined in the text.

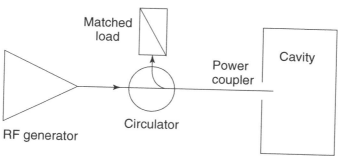

Figure 10.5 Simplified drawing of the RF system including the generator, circulator, RF input drive line or waveguide, power coupler, cavity, and the beam.

10.5
Equations Describing a Beam-Loaded Cavity

In this section, we present the theory by Wilson [5] for a beam-loaded cavity. The equivalent circuit shown in Figure 10.6 is the same as described in Section 5.4, except that an additional generator is added to represent the beam. We see immediately that the beam functions not as an additional load, represented by a resistance, but as a generator, which can excite the cavity. A consequence is that the cavity time constant and the loaded quality factor Q_L do not change when the beam is turned on. A physical argument for representing the beam by an equivalent generator rather than a resistive load is that if the generator is turned off for an operating accelerating structure, the fields do not simply decay to zero, but the structure continues to be excited by the beam to a new steady-state amplitude and phase. Figure 10.7 shows the equivalent circuit for the beam-loaded cavity transformed into the resonator circuit. The quantity R_c/β is the impedance seen looking back toward the generator from the cavity (i.e., the impedance seen by the reflected wave from the coupler or a wave emitted from the cavity).

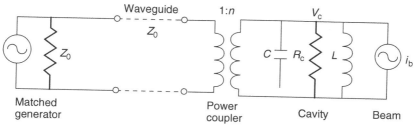

Figure 10.6 Equivalent circuit for the beam-loaded cavity. The cavity is represented by the LRC circuit. The RF generator and the beam are represented by current generators.

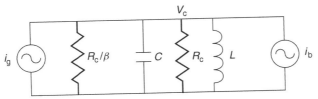

Figure 10.7 Equivalent circuit for the beam-loaded cavity transformed into the resonator circuit.

We introduce the following notation. The generator current is i_g, the average beam current is I, and $i_b \approx 2I$ is the harmonic component of the beam current at the bunch frequency, assuming that the bunches are short compared with their spacing, the beam image current $-i_b$ has opposite sign to the beam current i_b, ϕ is the synchronous phase, which represents the phase of the net cavity voltage V_c relative to the phase of i_b, θ is the phase of the generator current i_g relative to the phase of i_b, $r_s = 2R_c$ is the effective shunt impedance, defined by $r_s = V_c^2/P_c$, so that $V_c = V_0 T$, where T is the transit-time factor, and P_c is the average power dissipation in the cavity walls. The average power delivered to the beam is $P_b = IV_c \cos\phi$. As defined in Chapter 5, the cavity Q or quality factor, unloaded by the external circuit is Q_0, and the quality factor loaded by the external circuit is $Q_L = Q_0/(1+\beta)$, where the waveguide-to-cavity coupling factor is $\beta = Q_0/Q_{ext}$. The external Q, or Q_{ext}, is a function of the geometry of the coupler. The cavity resonant frequency ω_0 is not necessarily exactly equal to the driving frequency of the generator and beam ω. It is convenient to define the detuning angle ψ as

$$\tan\psi = -2Q_L\delta \tag{10.11}$$

where

$$\delta \equiv \frac{\omega - \omega_0}{\omega_0}, \quad |\delta| \ll 1 \tag{10.12}$$

General Results

The currents and voltages are conveniently expressed as phasors, which have a magnitude and phase, and can be represented in complex notation. By convention the beam current i_b is taken along the positive real axis, which places the beam image current and the beam-induced voltage along the negative real axis. If the detuning angle ψ is positive, the generator and the beam-induced voltages lead then respective currents by ψ. The cavity impedance Z_L, loaded by the external resistance, determines the voltage induced by any driving current i according to $V = iZ_L$, and is given by

$$Z_L = \frac{r_s e^{j\psi}}{2(1+\beta)} \cos \psi \qquad (10.13)$$

Then, we write the beam image current, the generator current, the beam-induced voltage, the generator-induced voltage, and the net cavity voltage in phasor notation as

$$\tilde{i}_b = i_b e^{j\pi} = -i_b \qquad (10.14)$$

$$\tilde{i}_g = i_g e^{j\theta} \qquad (10.15)$$

$$\tilde{V}_b = V_b e^{j(\pi+\psi)} = \frac{Ir_s \cos \psi}{1+\beta} e^{j(\pi+\psi)} \qquad (10.16)$$

$$\tilde{V}_g = V_g e^{j(\theta+\psi)} = \frac{2\sqrt{\beta r_s P_g} \cos \psi}{1+\beta} e^{j(\theta+\psi)} \qquad (10.17)$$

and

$$\tilde{V}_c = \tilde{V}_b + \tilde{V}_g \qquad (10.18)$$

Including beam loading and cavity detuning, the general expression has been derived for the required generator power,

$$P_g = P_c \frac{(1+\beta)^2}{4\beta} \frac{1}{\cos^2 \psi} \left\{ \left[\cos \phi + \frac{V_b \cos \psi}{V_c} \right]^2 \right.$$
$$\left. + \left[\sin \phi + \frac{V_b \sin \psi}{V_c} \right]^2 \right\} \qquad (10.19)$$

where V_b is the magnitude of the beam-induced voltage in the cavity, given by

$$V_b = \frac{Ir_s \cos \psi}{1+\beta} \qquad (10.20)$$

From conservation of energy, $P_r = P_g - P_c - P_b$, from which we find the general result for the reflected power

$$P_r = P_c \left[\frac{(1+\beta)^2}{4\beta} \frac{1}{\cos^2 \psi} \left\{ \left[\cos \phi + \frac{V_b \cos \psi}{V_c} \right]^2 \right. \right.$$
$$\left. \left. + \left[\sin \phi + \frac{V_b \sin \psi}{V_c} \right]^2 \right\} - 1 \right]$$
$$- IV_c \cos \phi \qquad (10.21)$$

If the beam current is zero, $V_b = 0$, and the generator power and reflected power reduce to

$$P_g = P_c \frac{(1+\beta)^2}{4\beta \cos^2 \psi} \tag{10.22}$$

and

$$P_r = P_c \left[\frac{(1+\beta)^2}{4\beta \cos^2 \psi} - 1 \right] \tag{10.23}$$

From Section 5.4, we relate the generator current to the generator (forward) power as $P_g = i_g^2 r_s / 16\beta$. The magnitudes of the beam-induced and generator-induced voltages in the cavity are given by

$$V_b = \frac{I r_s \cos \psi}{(1+\beta)} \tag{10.24}$$

and

$$V_g = \frac{2\sqrt{\beta r_s P_g}}{1+\beta} \cos \psi$$

Optimum Detuning

To put the cavity voltage V_c in phase with the generator current i_g, so that the cavity presents a real impedance to the generator, it is necessary that $\theta = \phi$. Then it is found that

$$\frac{V_b}{V_c} = -\frac{\sin \psi}{\sin \phi} \tag{10.25}$$

It is necessary to detune the cavity resonant frequency so that the detuning angle ψ is

$$\tan \psi = -\frac{I r_s \sin \phi}{V_c (1+\beta)} = -\frac{P_b \tan \phi}{P_c (1+\beta)} \tag{10.26}$$

If ϕ is negative, the required choice to obtain longitudinal focusing, then ψ must be positive, and δ must be negative. Negative δ means that the resonant frequency must be tuned higher than the generator and beam frequency.

To minimize the generator power with respect to β, it is necessary to choose β, and therefore Q_{ext}, so that

$$\beta_0 = 1 + \frac{I r_s \cos \phi}{V_c} = 1 + \frac{P_b}{P_c} \tag{10.27}$$

The corresponding generator power is $P_g = P_c + P_b$, and by conservation of energy the reflected power becomes $P_r = P_g - P_c - P_b = 0$, so that this choice of β both minimizes the generator power, and results in zero reflected power. The detuning angle for the optimum coupling factor β_0 is

$$\tan \psi_0 = -\frac{\beta_0 - 1}{\beta_0 + 1} \tan \phi \tag{10.28}$$

Extreme Beam-Loaded Limit

To treat the problem of a very strongly beam-loaded cavity, we consider the extreme beam-loaded case, where $P_b \gg P_c$. Then we obtain $\beta_0 \cong P_b/P_c$, $P_g \cong P_b \cong \beta_0 P_c$, $\tan \psi_0 \cong -\tan \phi$, or $\psi_0 \cong -\phi$, $V_b \cong V_c$, and $\delta \cong \tan \phi / 2Q_L$. The phasor voltages become

$$\tilde{V}_b = \frac{Ir_s}{\beta} \cos \phi \, e^{j(\pi - \phi)} \tag{10.29}$$

$$\tilde{V}_g = \frac{2Ir_s}{\beta} \cos^2 \phi \tag{10.30}$$

$$\tilde{V}_c = \frac{Ir_s}{\beta} \cos \phi \, e^{j\phi} \tag{10.31}$$

If the beam goes away, the generator power is related to the cavity power according to

$$P_g = P_c \frac{(1 + \beta_0)^2}{4\beta_0 \cos^2 \psi} \cong P_c \frac{\beta_0}{4 \cos^2 \psi} \tag{10.32}$$

and the corresponding reflected power is

$$P_r \cong P_c \frac{\beta_0}{4 \cos^2 \psi} \tag{10.33}$$

Consider the case where the cavity is driven on resonance so $\psi = 0$. If the cavity voltage is maintained at a constant level by feedback control, the formulas imply that the generator power is reduced by a factor of 4 when the beam goes off. This would also result in forward and reflected traveling-wave fields in the input guide that are half of the field in the forward wave when the beam is on. The maximum standing-wave electric field in the input guide when the beam is off will be the sum of the fields of the two traveling waves, which gives the same maximum field as when beam is on. Thus, if the cavity field is maintained constant when the beam is off, the peak field in the waveguide is not increased, and the RF window in the waveguide is not subjected to higher fields.

Numerical Example of a Beam-Loaded Cavity

Consider the example of a beam-loaded normal-conducting cavity with bunches injected earlier than the crest of the accelerating wave. Consider typical numerical values: $f = \omega/2\pi = 700$ MHz, $I = 0.1$ A, $r_s = 25 \times 10^6$ Ω, $V_c = 2 \times 10^6$ V, and $\phi = -35°$. We find that $P_c = V_c^2/r_s = 1.6 \times 10^5$ W, and $P_b = IV_c \cos \phi = 1.6 \times 10^5$ W. Then we find that $\beta_0 = 1 + P_b/P_c = 2$, $\tan \psi = \tan 35°/3$, or $\psi = 13.1°$ determines the detuning of the cavity required for minimum reflected power. With beam present, there is zero reflected power, and $P_g = P_c + P_b = 3.2 \times 10^5$ W, $V_b = Ir_s \cos \psi/(1 + \beta) = 0.81 \times 10^6$ V, and $V_g = 2\sqrt{\beta r_s P_g} \cos \psi/(1 + \beta) = 2.6 \times 10^6$ V. If $Q_L = 10^4$, we find $\delta = -\tan \psi/(2Q_L) = -1.2 \times 10^{-5}$, or the resonant frequency must be detuned by $f_0 - f \cong 8.1$ kHz. The resonant frequency must be detuned

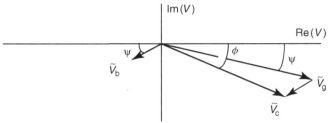

Figure 10.8 Voltage phasors for a steady-state, beam-loaded, normal-conducting cavity, when bunches are injected at a phase ϕ, earlier than the crest of the wave.

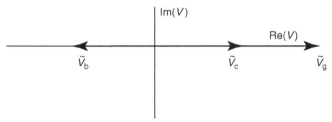

Figure 10.9 Voltage phasors for a steady-state, strongly beam-loaded superconducting cavity, when bunches are injected at the crest of the wave.

higher than the drive frequency. For completeness the cavity time constant is $\tau = 2Q_L/\omega_0 = 4.5$ μs, and the phasor plot is shown in Fig. 10.8.

Example of a Heavily Beam - Loaded Superconducting Cavity with Bunches Injected on the Crest of the Accelerating Wave

Consider a strongly beam-loaded superconducting cavity with the following parameter values; $f = \omega/2\pi = 700$ MHz, $I = 0.1$ A, $r_s = 2.5 \times 10^{12}$ Ω, $V_c = 5 \times 10^6$ V, and $\phi = 0°$. We find that $P_c = V_c^2/r_s = 10$ W, and $P_b = IV_c \cos \phi = 5.0 \times 10^5$ W, so $P_b \gg P_c$. Then we find that $\beta_0 = P_b/P_c = 5 \times 10^4$, $\psi = -\phi = 0°$ (no detuning is required), $P_g = P_b = 500$ kW, $V_b = V_c = 5 \times 10^6$ V, and $V_g = 2V_c \cos^2 \phi = 10 \times 10^6$ V. The magnitude of the beam-induced voltage is equal to the net cavity voltage, and the two voltages have opposite sign, as shown in Fig. 10.9. If the beam goes off and the cavity voltage is kept constant, the generator power would be reduced by a factor of 4.

10.6
Generator Power when the Beam Current is Less than Design Value

When a cavity is matched to the generator, under steady-state conditions all the generator power goes into the beam power plus the cavity wall–power dissipation, and no net power flows back from the cavity.

10.6 Generator Power when the Beam Current is Less than Design Value

If the coupler geometry is not adjustable, it is usually designed for a matched condition at the maximum expected beam current. The coupler provides the matched or reflection-free condition only for this maximum beam current; it will not be matched for any lower beam currents. An alternative approach is to use an adjustable coupler that is capable of providing a matched condition at any beam current of interest. The advantage of an adjustable coupler is obvious, except that adjustable couplers are technically more challenging and generally more expensive than fixed couplers.

To compare the two options of fixed versus variable couplers, we consider the case of an accelerator driven on resonance in the extreme beam-loaded case where the cavity wall–power losses are small compared with the beam power. We assume the bunches are injected on the crest of the accelerating wave, and $\psi = \phi = 0$. We begin with the case of a fixed coupler. From Eq. 10.19 we obtain

$$P_g = P_c \frac{(1+\beta)^2}{4\beta} \left(1 + \frac{V_b}{V_c}\right)^2 \tag{10.34}$$

where from Eq. (10.20) the beam voltage is $V_b = I r_s / (1 + \beta)$, the beam power for an arbitrary beam current I is $P_b = I V_c$, and the cavity power is $P_c = V_c^2 / r_s$. The waveguide-to-cavity coupling factor is given by Eq. (10.27), and in the extreme beam-loaded case, where the beam power is much larger than the cavity power, is given by $\beta \cong P_{b,m}/P_c \gg 1$, where $P_{b,m}$ is the matched beam power corresponding to a beam current I_m chosen for a matched reflection-free condition at a fixed beam power. Using these equations one can obtain $V_b/V_c \cong P_b/P_{b,m}$, and the generator power as a function of beam power P_b is given by

$$P_g \cong \frac{P_{b,m}}{4} \left(1 + \frac{P_b}{P_{b,m}}\right)^2 \tag{10.35}$$

For the matched case $P_b = P_{b,m}$, we obtain $P_g = P_{b,m}$ so all the generator power goes into the beam in this approximation that the cavity power is negligible. When $P_b = 0$ the required generator power is one quarter of the power required for the matched case. The reflected power as a function of the beam power can also be derived in the same beam-loaded limit where $\beta \gg 1$. We obtain

$$P_r = \frac{P_{b,m}}{\beta} \left\{ \frac{\beta}{4} \left[\left(1 + \frac{P_b}{P_{b,m}}\right)^2 + \left(\frac{P_b}{P_{b,m}}\right)^2\right] - 1 \right\} - P_b \tag{10.36}$$

When $P_b = 0$, the equation gives $P_r \cong P_{b,m}/4$, so that in the extreme beam-loaded approximation the generator power all goes into net reflected power. Also, for the matched case where $P_b = P_{b,m}$, the net reflected power is zero, as expected.

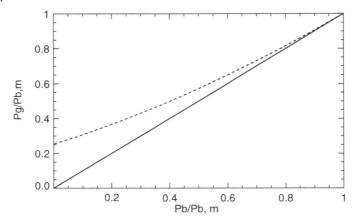

Figure 10.10 Generator power versus beam power at fixed cavity voltage. Both the generator and the beam power are scaled to the matched beam power $P_b = P_{b,m}$. Points on the solid curve ($P_g = P_b$) corresponds to a coupler that is always adjusted for a matched condition. Points on the dashed curve (Eq. 10.35) correspond to a fixed coupler with a matched condition when $P_b = P_{b,m}$.

For the case of an adjustable coupler, the generator power is assumed to be set so that it always equals the beam power; thus $P_g = P_b$ and there is zero net reflected power for all beam currents. The two functions for P_g are plotted in Fig. 10.10. The dashed curve shows that at zero current, the fixed coupling configuration requires the generator to supply one quarter of the beam power of the matched case, nearly all of which is reflected. However, the figure illustrates that not much power is lost with a fixed coupler when operating at beam currents that are lower than the design matched current.

10.7
Transient Turn-On of a Beam-Loaded Cavity

The turn-on of the RF input power and the beam in a standing-wave accelerating cavity should be done in the proper order. It is usually desirable to reach the design field level in a relatively short time to avoid wasting power. With the beam off, the RF generator is turned on at a sufficient input power level that the cavity field approaches an asymptotic value that exceeds the design value. The higher the input RF power level, the shorter the time to reach the design cavity field. When the design cavity field level is reached, the beam can be injected, provided that the field is not still increasing. Different methods can be used to program the RF input power as a function of time. A simple method is based on the result that the cavity time constant and Q_L do not change when beam is turned on. Thus, the responses of the cavity to the RF generator and to the beam are both exponential functions with exactly the same time constant, $\tau = 2Q_L/\omega$, the cavity time constant when the cavity is loaded by the input power coupler. The parameters can be adjusted so that if the beam is turned on at the instant when the cavity reaches the design field level, the field remains constant at that value. Whether the input power from the generator needs to be adjusted when beam is injected is a matter of choice. Usually, the coupling is chosen so that the input waveguide is matched to the cavity with the beam on. With the beam off, some of the input power from the generator will generally be reflected from the input power coupler and some will be transmitted into the cavity.

We describe a simple method, where the input power P_+ from the generator is constant during the entire transient turn-on time. With no beam, the steady-state power dissipated in the cavity is $P_{cn} = P_+(4\beta)/(1+\beta)^2$, the steady-state stored energy is $U_n = Q_0 P_{cn}/\omega = (Q_0 P_+/\omega)[(4\beta)/(1+\beta)^2]$, and the time-dependent cavity voltage, before the beam is injected, is $V(t) = V_n(1 - e^{-t/\tau})$, where $V_n \propto \sqrt{U_n}$. With beam on, and zero reflected power, $P_+ = P_{cw} + P_b = P_{cw}(1 + P_b/P_{cw})$, where P_{cw} is the steady-state power dissipated in the cavity, and P_b is the beam power. From Section 10.5, the coupling factor required for no reflected power with beam, is $\beta = 1 + P_b/P_{cw}$. Thus, the steady-state stored energy is $U_w = Q_0 P_{cw}/\omega = Q_0 P_+/\omega\beta$, and the steady-state cavity voltage with beam is $V_w \propto \sqrt{U_w}$. The time when beam injection should begin is when $V(t)$ reaches the design voltage level of V_w, or

$$t_b = -\tau \ln(1 - V_w/V_n) \tag{10.37}$$

An example of the cavity voltage as a function of time for the case $P_b/P_{cw} = 0.5$ is shown in Fig. 10.11. Problems 10.5 and 10.6 have been chosen to illustrate the method.

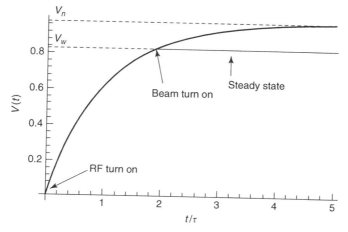

Figure 10.11 Cavity voltage versus time, showing RF power and beam turned on.

10 Beam Loading

Problems [6]

10.1. Consider the following model shown in Fig. P10.1 of the beam-cavity interaction for a relativistic charge that crosses an initially unexcited cavity. The RF cavity is replaced by a plane, parallel-plate capacitor of gap g. The charge q is represented by an infinitesimally thin, uniformly charged disk of radius a, which enters the gap through one capacitor plate at $z = 0$, crosses the gap moving at velocity $v = c$, and exits through the opposite plate. Each plate is a thin metallic foil through which the charge can pass. All electric fields are assumed to be parallel to the axial z direction. If $g \ll \lambda$ the charged disk crosses the gap in a time short compared to the RF period of the fundamental cavity mode, so that the problem is approximately electrostatic.

(a) When the disk enters the gap, it induces images charges in the capacitor plates. To produce a net zero field ahead of the charge q, as is required by causality (when the charge enters the gap its fields cannot move forward any faster than the velocity $v = c$ of the charge itself), we assume that a uniform image charge distribution with total charge $-q$ is induced at the entrance plane $z = 0$. Show from Gauss's law that for any position of the charge q in the gap the induced electric field within the gap is given by: $E = -q/\pi \varepsilon_0 a^2$ (corresponding to a wakefield) behind the charge q and $E = 0$ ahead of the charge. Plot the field as a function of z when the disk is at $z_d = g/2$. (b) Assume the equation of motion for the moving disk is $z = ct$, where $t = 0$ is the time when the disk enters the gap. Calculate the stored electric energy $U_E(t)$ in the cavity as a function of time. Equate the total disk energy loss to the stored electric energy $U_E(t = g/c)$ left in the gap when the charge leaves the gap. From $U_E(t = g/c)$ what is the total effective retarding voltage V_q that has acted on the disk? What do you think the image charges are at the $z = 0$ and $z = g$ conducting planes immediately after the disk leaves

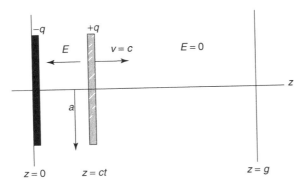

Figure P10.1 For problem 10.1.

the gap. **(c)** Using the induced field behind the charge from part a, calculate the retarding voltage V_b induced in the gap after the disk has left the gap. Show that $V_q/V_b = 1/2$, or the particle sees only half the voltage that it induces. This result is known as the fundamental theorem of beam loading.

10.2. To gain a better understanding for the fundamental beam-loading theorem, we replace the thin relativistic disk of Problem 10.1 with a relativistic, finite-thickness, uniform-density cylinder of charge q, length d, and radius a (see Fig. P10.2). Assume that the image charge $-q$ is induced at the entrance plane as in Problem 10.1, when the cylinder of charge has entered the gap. Suppose the motion begins at $t = -d/c$ with the head of the cylinder at the entrance plane $z = 0$, and describe the motion of the tail by $z_t = ct$, and the head by $z_h = ct + d$. At $t = g/c$ the tail is at $z = g$ and the cylinder has left the gap. Let the variable s, where $0 \leq s \leq d$, represent an arbitrary position within the cylinder relative to the head of the cylinder. Thus $s = 0$ is at the head and $s = d$ is at the tail.

(a) Use Gauss's law to show that for any position of the cylinder, when it is completely inside the gap, the induced electric field is (1) behind the cylinder (wakefield), $E = -q/\pi\varepsilon_0 a^2$, $0 \leq z \leq z_t$, (2) within the cylinder at a point $s = z_h - z$, $E(s) = -qs/\pi\varepsilon_0 a^2 d$, $0 \leq s \leq d$ or $z_t \leq z \leq z_h$, and (3) from causality ahead of the cylinder $E = 0$, $z \geq z_h$. Plot the field as a function of z at $t = 0$, $t = g/2c$, and $t = g/c$. (It can be shown that the field at s within the cylinder is constant for all positions of s within the gap, even when the cylinder is entering through the $z = 0$ plane and leaving through the $z = g$ plane.). **(b)** Using the result of part a, show that after the cylinder has moved completely through the gap, the total voltage seen by a point at s inside the cylinder is $V(s) = -qsg/\pi\varepsilon_0 a^2 d$. **(c)** Multiplying the voltage $V(s)$ by the infinitesimal charge $dq = q\, ds/d$ at s, and integrating over

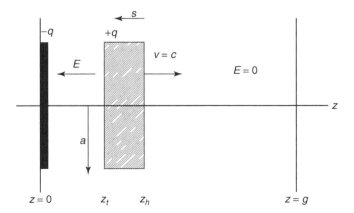

Figure P10.2 For problem 10.2.

s gives the total energy loss of the cylinder. Show that this energy loss of the cylinder to the gap fields is $\Delta W = -q^2 g/2\pi\varepsilon_0 a^2$. Calculate from ΔW the total effective retarding voltage V_q that is experienced by the cylindrical charge. **(d)** Another way to calculate ΔW is to calculate the final stored electric energy U_E in the initially unexcited gap from the field induced by the cylinder of charge. This must equal the energy lost by the cylinder. Show that this does give the same result as part (c). **(e)** Using the induced field behind the charge given in part (a), calculate the retarding voltage $V_b(t)$ induced in the gap over the time interval t. Show that $V_q(t)/V_b(t) = 1/2$ as was found in the previous problem. **(f)** Why are not the voltages V_q and V_b equal? (Hint: Explain by using the result that for a charge at position s in the cylinder, the voltage it sees as it crosses the gap is $V(s) = -qsg/\pi\varepsilon_0 a^2 d$. Examine the behavior as a function of s. The plots from part a may also be helpful to visualize this.) If the cylinder represented a bunch of initially monoenergetic particles, would the gap induce a significant energy spread in the bunch?

10.3. Suppose the capacitor gap of Problem 10.1 is charged with an axial accelerating field E_0 and voltage $V_0 = E_0 g$ before a relativistic charged disk with charge q crosses the $z = 0$ conducting plane. Assume that the image charge is induced on the $z = 0$ plane just as in Problem 10.1, and that the induced fields are superimposed on the preexisting accelerating field.

(a) What is the initial stored electric energy within a cylinder of radius a before the disk enters. **(b)** After the disk crosses the gap, what is the stored energy? From energy conservation what is the energy change for the disk? Express the total voltage seen by the disk as the sum of the applied voltage V_0 plus the induced voltage seen by the disk V_q. Define the beam-loading parameter $\mu = -V_b/V_0$. Will the disk always gain energy and if not, what condition on μ must be satisfied for a net energy gain to the disk? **(c)** Express the total voltage after the disk has crossed the gap as the sum of the applied voltage V_0 plus the induced voltage V_b. Is V_q still equal to half the induced voltage V_b as was true when the cavity was initially unexcited? **(d)** Define an energy extraction efficiency η as the fraction of the initial cavity energy that is extracted by the disk? Show that $\eta = 2\mu - \mu^2$. For what value of μ is the energy extraction maximum and what is η for this case? Can the charged disk remove all the initial gap stored energy? **(e)** If the disk is replaced with a finite-length cylinder, one can show the head of the cylinder sees zero induced voltage and the tail sees the full induced voltage V_b. The total voltage seen by the head and the tail includes the applied voltage V_0. The average voltage seen by the cylinder is the same as for the disk: $V_0 + V_q$. Define the ratio δ as the full energy spread relative to the total energy gain of the cylinder as a function of the beam-loading parameter μ? Plot η and δ as functions of μ in the range $0 < \mu < 2$, where the beam extracts energy from the gap. As μ increases toward 1 to approach 100% transfer of energy from the gap to the beam, what happens to the energy spread parameter δ? (f) For some typical numbers for a bunch in an electron linac, $q = 5 \times 10^{-9}$ C, $a = 3$ cm, and $E_0 = V_0/g = 20$ MV/m, calculate μ, η, and δ. If the cavity length is $g = 3$ m, how much energy per bunch is removed from the cavity in joules? If the beam energy is 1 GeV, what is the induced energy spread as a percent of the beam energy?

10.4. Consider a beam-loaded superconducting cavity in a proton linac with the following parameter values; drive frequency $f = \omega/2\pi = 700$ MHz, beam current $I = 0.1$ A, cavity shunt impedance $r_s = 2.5 \times 10^{12}$ Ω, quality factor loaded by the external circuit $Q_L = 3 \times 10^5$, synchronous phase $\phi = -35°$, and cavity voltage (includes the transit-time factor) $V_c = 5 \times 10^6$ V (see Fig P10.4). (a) Show that the power delivered to the beam is much larger than the cavity wall-power loss. (b) Calculate the detuning angle ψ, and the waveguide-to-cavity coupling factor β_0 required to yield zero reflected power when the beam is on. How much and in what direction must the cavity resonant frequency be detuned with respect the driving frequency? (c) Assuming $\beta = \beta_0$ and the detuning angle ψ from part (b), calculate the voltages induced by the beam, and the generator, and the required generator power. Confirm that the magnitude of the beam-induced voltage is approximately equal to the magnitude of the net cavity voltage, and that the placement of vectors in Fig. P10.4 is qualitatively correct. (d) Suppose that the generator goes off while the beam remains on. Calculate the cavity voltage, the cavity power, and the power emitted from the cavity through the coupler after a steady state is reestablished. (e) Suppose that the voltage amplitude is fixed by feedback control, and that the generator goes off while the beam is on. Calculate the generator power and the reflected power after a steady state is reestablished.

10.5. Consider the transient turn-on of a linac, designed with a ratio of beam power to structure-power loss equal to 0.50. Assume that the generator power is held constant during the entire turn-on. (a) Show that to maintain a constant field value after the beam is injected, the beam should be injected at the approximate time $t = 1.79\tau$, where τ is the

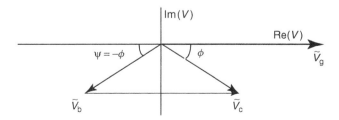

Figure P10.4 For problem 10.4.

cavity time constant (see Fig. 10.11). **(b)** Show that for this case, the steady-state field level with beam on is about 83% of the asymptotic value with beam off.

10.6. A typical operating mode for a standing-wave electron linac is to inject the beam at the crest of the accelerating waveform, to drive the accelerating cavities at the resonant frequency, and to choose the coupling factor β for zero reflected power when the beam is on and at the design current I. For this case, the cavity voltages induced by the generator and by the beam are 180° out of phase. **(a)** Refer to the results of Sections 5.6 and 10.5 to confirm that the cavity voltage during the turn-on transient has the form

$$V(t) = \sqrt{P_+ r_s} \left(\frac{2\beta^{1/2}}{1+\beta} \right) (1 - e^{-t/\tau}) - \frac{Ir_s}{1+\beta}(1 - e^{-(t-t_b)/\tau})$$

where P_+ is the constant value of the forward power from the generator, $t = 0$ is the time that the RF power is turned on, and $t = t_b$ is the time that the beam is injected into the cavity. Thus, $I = 0$ for $t < t_b$, and the current has the constant value I for $t > t_b$. The first term in the expression for the cavity voltage is the generator-induced voltage, and the second term is the beam-current-induced voltage. Both terms have the same time constant $\tau = 2Q_L/\omega$. **(b)** Show that $V(t)$ is constant for $t > t_b$ (see Fig. 10.11), when t_b is

$$t_b = -\tau \ln \left(\frac{Ir_s}{2\beta^{1/2}\sqrt{P_+ r_s}} \right) = -\tau \ln \left(\frac{\beta - 1}{2\beta} \right)$$

where β is chosen to give zero reflected power when the beam is on and at the design-current value.

References

1 Wilson P.B., High energy electron linacs: applications to storage ring RF systems and linear colliders, 1981 summer school on high energy accelerators, *AIP Conf. Proc.* **87**, 450 (1981); Perry Wilson, SLAC-PUB-2884 (Revised) November, 1991.

2 Halbach K. and Holsinger R.F., SUPERFISH-a computer program for evaluation of Rf cavities with cylindrical symmetry, *Part. Accel.* **7**, 213–222 (1976).

3 Wilson P., Wake fields and wake potentials, *AIP Conf. Proc.* **184**, 526–564 (1989).

4 For a discussion of beam loading in traveling-wave linacs see Leiss, J.E., in *Linear Accelerators*, Lapostolle P.M. and Septier A.L., John Wiley & Sons, New York, 1970, pp. 147–172.

5 Wilson P., 1981 summer school on high energy accelerators, *ATP Conf. Proc.* **87**, 450 (1981); P. Wilson, SLAC-PUB-2884 (revised), November 1981.

6 Problems 10.1–10.3 are based on an unpublished wakefield model by J. Lawson.

11
Wakefields

The electric and magnetic fields of the bunch are modified by the conducting walls of the surrounding structure. In general, the fields carried along with the bunch induce surface charges and currents in the walls, and these become the sources of fields that act back on the particles in the bunch. At low velocities, the fields from the beam-induced surface charges are described by equivalent beam-induced image charges, and the effects of these fields on the beam are usually much smaller than the direct space-charge fields. At beam velocities near the speed of light the direct space-charge forces decrease because of the near cancellation of the electric and magnetic self fields. It is well known that the electric-field distribution of a free relativistic charged particle is Lorentz contracted into a disk perpendicular to the direction of motion with a narrow angular spread of the order of $1/\gamma$, and the longitudinal electric-field component approaches zero. The field compression is also present when the moving particles are placed within a smooth perfectly conducting pipe. However, scattered electromagnetic radiation is produced when the fields moving with the relativistic particles encounter geometric variations along the structure, such as radiofrequency (RF) accelerating cavities, vacuum bellows, and beam-diagnostic chambers. In the extreme relativistic limit, the scattered radiation cannot catch up to affect the source particles, but the radiation can and does act on trailing particles in the same or subsequent bunches. As is the convention for relativistic electron beams, the scattered radiation will be called *wakefields*, although strictly speaking for particles with $v < c$, the scattered radiation does not always travel behind the particles that are its source. The longitudinal compression of the fields traveling with the beam is an important effect that tends to increase the energy in the wakefields for short bunches, because of the constructive interference of the radiation from different induced surface charges. At lower velocities the field carried by the beam becomes more isotropic, and when the induced surface charges radiate over a length scale that is comparable to or greater than the wavelength of a particular mode, destructive interference from the different sources of the radiation reduces the energy in the corresponding wakefields.

RF Linear Accelerators. 2nd, completely revised and enlarged edition.
Thomas P. Wangler
Copyright © 2008 Wiley-VCH Verlag GmbH & Co. KGaA, Weinheim
ISBN: 978-3-527-40680-7

Wakefields generally exist as damped, oscillatory electromagnetic disturbances, which can be described equivalently as a sum of all the resonant modes excited within the structure. Wakefields include the higher-order modes and the beam-excited accelerating mode, whose main effect is usually described separately as beam loading (see Chapter 10). The highest frequency components comprising the wakefields do not remain localized, and for frequencies higher than the cutoff frequency of the pipe, propagate away along the beam pipe. The lower frequency modes, below the cut off frequency of the pipe, remain localized near the structures in which they were born, and may linger, depending on the time constants of the individual modes to act on particles in trailing bunches.

The radiated fields are sometimes classified as either short-range or long-range wakefields. For ultrarelativistic bunches, the *short-range wakefields* generated by the particles at the head of the bunch affect trailing particles in the same bunch, causing energy loss, and for off-axis bunches, a transverse deflection of the particles in the tail. The most serious effect of the *long-range wakefields* is caused by the high-Q transverse deflecting modes that induce time-varying transverse deflections in trailing bunches. If the deflecting modes are strongly excited by the beam, they may cause the beam-breakup (BBU) instability, leading to effective emittance growth when averaged over time. Eventually the beam is lost on the walls. The wakefields usually extract only a small fraction of the beam energy, which can be replaced by additional RF power. The wakefield power extracted from the beam ultimately represents additional power dissipation, through Ohmic losses in the structure walls or power delivered to some external load. The wakefields can add significantly to the refrigeration load in a superconducting linac, unless the power is coupled out to a load at higher temperature, by a specially designed higher-order mode coupler. This chapter is intended as an introduction to all those effects caused by the interaction of the beam with the surrounding walls of the accelerating structure. The reader who wishes to delve more deeply into this topic is referred to the book by A. Chao [1]. We begin the chapter with a description of image-charge forces induced by low-velocity beams in a smooth, round beam pipe. Then, we introduce the basic concepts for describing wakefields produced by high-current beams, and the general framework for describing the wakefield interaction with the beam, including wake potentials and beam-coupling impedances. Methods used to mitigate the effects of the wakefields are also discussed. Finally we introduce the topic of the BBU instability in linacs.

11.1
Image Force for Line Charge in Round Pipe

As an approximation for a long bunch, we consider a round, continuous, uniform-density beam with charge per unit length λ and radius R, inside a

11.1 Image Force for Line Charge in Round Pipe

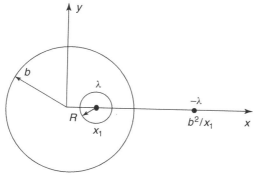

Figure 11.1 Conducting cylindrical pipe of radius b with image charge $-\lambda$ induced by a beam with radius R and line density λ.

perfectly conducting cylindrical pipe of radius b. The geometry is shown in Fig. 11.1. Suppose that the beam centroid is displaced from the x-axis by an amount x_1. Charge will be induced on the surface of the pipe as required to make the pipe an equipotential surface. The electrostatic solution for the fields, valid for $R \ll b$, is obtained placing an equivalent line-charge density $-\lambda$, the image charge, on the x-axis at $x = b^2/x_1$.

The x component of the electric field from the image charge is

$$E_{im,x} = \frac{-\lambda(x - b^2/x_1)}{2\pi\varepsilon_0[(x - b^2/x_1)^2 + y^2]} \cong \frac{\lambda x_1}{2\pi\varepsilon_0 b^2}\left(1 + \frac{x_1^2}{b^2}\right) + \frac{\lambda x_1^2(x - x_1)}{2\pi\varepsilon_0 b^4} \tag{11.1}$$

and the y component is

$$E_{im,y} = \frac{-\lambda y}{2\pi\varepsilon_0[(x - b^2/x_1)^2 + y^2]} \cong \frac{-\lambda x_1^2 y}{2\pi\varepsilon_0 b^4} \tag{11.2}$$

where we have assumed that $(xx_1/b^2)^2 \ll 1$, and $(yx_1/b^2)^2 \ll 1$, and have kept terms up the third order. Equation (11.1) contains the deflecting field with both linear and cubic terms in the displacement x_1, and the defocusing term proportional to $x - x_1$. Equation (11.2) has a focusing term proportional to y. The space-charge electric field inside the beam is

$$E_{sc,x} = \frac{\lambda(x - x_1)}{2\pi\varepsilon_0 R^2} \tag{11.3}$$

and

$$E_{sc,y} = \frac{\lambda y}{2\pi\varepsilon_0 R^2} \tag{11.4}$$

Adding the image field to the space-charge field, and representing the focusing force by a smooth linear restoring force with wave number k_0, we express the equation of motion for the centroid displacement as

$$\frac{d^2 x_1}{dz^2} + k_0^2 x_1 - K \frac{x_1}{b^2}\left(1 + \frac{x_1^2}{b^2}\right) = 0 \tag{11.5}$$

where $K = qI/2\pi\varepsilon_0 \gamma^3 m v^3$ is the generalized perveance, expressed in terms of the current $I = \lambda v$, a factor of γ^2 to account for the magnetic forces, and a factor γ for the relativistic mass factor. Equation (11.5) describes a coherent oscillation that depends on x_1/b. The focusing in both x and y are described by the following equations of motion,

$$\frac{d^2 (x - x_1)}{dz^2} + k_0^2 (x - x_1) - \frac{K}{R^2}\left(1 + \frac{x_1^2 R^2}{b^4}\right)(x - x_1) = 0 \tag{11.6}$$

and

$$\frac{d^2 y}{dz^2} + k_0^2 y - \frac{K}{R^2}\left(1 - \frac{x_1^2 R^2}{b^4}\right) y = 0 \tag{11.7}$$

The effect of the image charge on the focusing, compared with the direct space-charge force, depends on the square of the parameter $x_1 R/b^2$. For a numerical example, suppose $x_1 = b/10$, and $R = b/2$. Then, $x_1 R/b^2 = 1/20$, and $x_1^2 R^2/b^4 = 1/400$, which produces a very small correction compared with the space-charge force. Only if the beam displacement is large enough to put the edge of the beam close to the pipe do the transverse image-charge corrections in a smooth pipe become important in a linac. Generally, image effects decrease the transverse defocusing strength, but increase the longitudinal focusing strength. Image forces have been identified as important for linac beams that debunch in long high-energy beam-transport lines [2]. Additional discussion of transverse image effects and a thorough treatment of longitudinal image effects are given by Reiser. [3]

11.2
Fields from a Relativistic Point Charge and Introduction to Wakefields

We consider the electromagnetic fields in free space from a point charge q with velocity $v = \beta c$ moving along the z-axis [4]. In spherical coordinates r, ψ, where r is the radius from the charge to the field observation point, and ψ is the angle between the unit vector $\hat{r}$ and the z-axis, the electric field is

$$\mathbf{E} = \frac{q\hat{r}}{4\pi\varepsilon_0 r^2 \gamma^2 (1 - \beta^2 \sin^2 \psi)^{3/2}} \tag{11.8}$$

where $\hat{r}$ points from the charge toward the observation point. The field is directed radially, but when $\beta \approx 0$, the magnitude becomes concentrated into the equatorial plane, as shown in Fig. 11.2. When $\psi = 0$ or π, corresponding

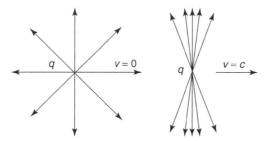

Figure 11.2 Illustration of the concentration of the electric field of a moving charge into the equatorial plane.

to the longitudinal field component, we have $E = q/4\pi\varepsilon_0 r^2 \gamma^2$, and when $\psi = \pi/2$, corresponding to the transverse component, we have $E = q\gamma/4\pi\varepsilon_0 r^2$. Relative to the field of a charge at rest, the longitudinal component is reduced by a factor of γ^{-2}, and the transverse component is increased by a factor γ. In the ultrarelativistic limit [5] where the point charge q moves at the velocity of light c, we approximate $\lambda(z-ct) = q\delta(z-ct)$ where δ is the dirac δ-function. Then the radial electric field, expressed in cylindrical coordinates is [6]

$$E_r = \frac{q\delta(z-ct)}{2\pi\varepsilon_0 r} \qquad (11.9)$$

Ampere's law can be applied, using a circular path centered on the charge at radius r, and writing the current as $I = qc\delta(z-ct)$, we obtain

$$B_\theta = \frac{\mu_0 cq\delta(z-ct)}{2\pi r} \qquad (11.10)$$

Because all field components are zero, ahead and behind the point charge, ultrarelativistic particles with different z coordinates cannot interact. The Lorentz force acting on a test charge q' that is moving with velocity c at radius r with the same z coordinate as the original particle is

$$F_r = q[E_r - cB_\theta] = \frac{q'q\delta(z-ct)}{2\pi\varepsilon_0 r} - \frac{q'q\mu_0 c^2\delta(z-ct)}{2\pi r} = 0 \qquad (11.11)$$

Therefore, in the ultrarelativistic limit there is no direct electromagnetic interaction between the particles in such a beam. The next question is how does the beam pipe affect the field and the interaction of the two charges? If the source charge is centered in the pipe, one finds that the field lines terminate on the pipe, but are otherwise unaffected. If the source charge is not on the axis of the pipe, the field lines are distorted from their free-space configuration, but are still compressed into the flat disk, and the Lorentz force on any test charge is still zero.

If the pipe has a finite conductivity, or if the pipe geometry varies, one finds that the passage of an ultrarelativistic charge is accompanied by the generation of electromagnetic fields behind the charge. The absence of field ahead of the charge is a consequence of causality, that is, no action occurring ahead of the $v = c$ point charge can be causally related to it. For a linac the effect of finite resistivity is usually negligible compared with the effects of geometry variations [7]. It is convenient to think of the incident fields that move with the beam as inducing charges and currents in the conducting walls that radiate electromagnetic energy in directions other than that of the incident fields. The radiation cannot propagate any faster than the ultrarelativistic source charge, and therefore the radiation remain behind a moving ultrarelativistic source. One finds that these electromagnetic wakefields will often remain long after the passage of the source, and can affect trailing particles in the same bunch or in a trailing bunch.

Figure 11.3 illustrates the wakefields excited by a bunch passing through a cavity. A frequency-dependent physical picture has been described by Chao [8]. The lowest-frequency bunch-induced excitations are associated with resonant excitation of the lowest cavity modes by the beam. Wakefields with $\omega < c/a$, where a is the beam-pipe radius, are generated by scattering into each cavity the incident electromagnetic wave from the entrance edge of the cavity. After the bunch leaves each cavity, the wakefields at radius larger than the beam aperture are scraped off by the exit edges and remain behind. An estimate of the radiated electromagnetic energy trapped in the cavity is obtained by calculating the field energy contained within an annular ring of length equal to the bunch length, and of radial thickness extending from the beam-aperture radius a to an effective outer cavity radius, assumed to be limited at $2a$. An equal amount of energy from the high-frequency component of the incident fields with $\omega > c/a$ is diffracted into the beam pipes by the edges of the apertures, and these wakefields propagate along the pipe, because they are above the cutoff frequency of the pipe.

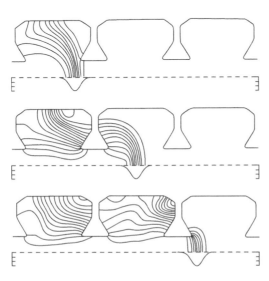

Figure 11.3 Numerically calculated electric wakefields generated by an ultrarelativistic Gaussian line-charge bunch in a PETRA cavity at DESY [9]. The beam moves along the dashed line on an axis of cylindrical symmetry. [From T. Weiland and R. Wanzenberg, Frontiers of Particla Beams: Intensity Limitations, Proc. Joint U.S.-CERN School on Particle Accelerators at Hilton Head Island, *Lecture Notes Phys.* **400**, M. Dienes, M. Month, and S. Turner eds., 39–79 (1992). Copyright 1990 Springer–Verlag.]

11.3
Wake Potential from a Relativistic Point Charge

Consider an ultrarelativistic point charge moving through a linac, which contains typical structure-geometry variations associated with the RF cavities, beam pipes, bellows, diagnostic boxes, and so on. After the charge passes a perturbing element and induces surface charges that radiate electromagnetic energy, a local concentration of the electromagnetic fields may remain that exerts a force on trailing charges. We are interested in the integrated effect of these fields on trailing charges, as they pass through the same perturbing region. It is convenient to define a wake potential that characterizes the net impulse delivered to the trailing charges that are moving at the same velocity along the same or parallel paths. Assume that the source point charge q_1 is moving parallel to a central axis of symmetry, but is displaced by an amount r_1, as shown in Fig. 11.4. Imagine that the source charge moves according to $z_1 = ct$, and imagine a test charge q at a displacement r and at a fixed distance s behind the source charge, that moves according to $z = ct - s$.

We define the δ-function longitudinal and transverse wake potentials [10, 11] per unit source charge as

$$w_z(\mathbf{r}, \mathbf{r}_1, s) = -\frac{1}{q_1} \int_0^L dz \, [E_z(\mathbf{r}, z, t)]_{t=(z+s)/c} \qquad (11.12)$$

$$\mathbf{w}_\perp(\mathbf{r}, \mathbf{r}_1, s) = \frac{1}{q_1} \int_0^L dz \, [\mathbf{E}_\perp + c(\hat{z} \times \mathbf{B})]_{t=(z+s)/c} \qquad (11.13)$$

The interval L must be large enough to include the entire field that a test particle would see. Equations (11.12) and (11.13) are useful as Green's functions from which the wake potentials for any arbitrary charge distribution can be calculated. Each wake potential determines the momentum kick delivered to a trailing test charge. Thus, for the longitudinal momentum

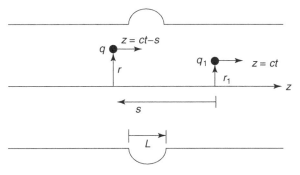

Figure 11.4 An ultrarelativistic source charge q_1 passes through a geometry perturbation and induces wakefields that extend over a length L, followed by an ultrarelativistic test charge q.

change Δp_z or energy change ΔU experienced by the charge q, we have

$$c\Delta p_z(r, r_1, s) = \Delta U(r, r_1, s) = -qq_1 w_z(r, r_1, s) \tag{11.14}$$

and for the transverse momentum change, we have

$$c\Delta p_\perp(r, r_1, s) = qq_1 w_\perp(r, r_1, s) \tag{11.15}$$

The corresponding ith component of the average wake force is

$$\overline{F}_i(\mathbf{r}, \mathbf{r}_1, s) = \Delta p_i(\mathbf{r}, \mathbf{r}_1, s)c/L \tag{11.16}$$

Although the wakefields produced by a given geometric perturbation are complicated functions of position and time, the introduction of wake potentials helps to simplify the problem by providing time-independent functions of position in the bunch, which determine the net momentum impulse at that point. The δ-function wake potential is a characteristic property of the geometric perturbation, and as we will see, can be used as a Green's function to determine the total wake potential of a fixed arbitrary charge distribution.

11.4
Wake Potentials in Cylindrically Symmetric Structures

We now restrict the discussion to an ultrarelativistic beam in a loss-free cylindrically symmetric structure, which we will describe using cylindrical polar coordinates. We assume that the source charge is displaced at radius r_1 along the x-axis where $\theta = 0$, and a test charge is located at the coordinate (r, θ). The radial coordinates will be expressed as a fraction of the aperture radius of the beam tubes, which will be denoted by a. The structure itself will have modes that vary azimuthally as $e^{jm\theta}$, where m is an integer. For example, the $m = 0$, 1, and 2 modes are the monopole, dipole, and quadrupole modes, respectively. The total m-pole component of the longitudinal wake potential can be expressed as a separable function of the coordinates and as a sum over the longitudinal modes n of the particular structure as [12]

$$w_{zm} = \left(\frac{r_1}{a}\right)^m \left(\frac{r}{a}\right)^m \cos m\theta \sum_{n=0}^{\infty} 2k_{mn}(r=a) \cos \frac{\omega_{mn}s}{c} \quad s > 0 \tag{11.17}$$

Likewise, the transverse wake potential, which is nonzero for $m > 0$, is

$$\mathbf{w}_{\perp m} = m \left(\frac{r_1}{a}\right)^m \left(\frac{r}{a}\right)^{m-1}$$

$$\times \left(\hat{\mathbf{r}} \cos m\theta - \hat{\boldsymbol{\theta}} \sin m\theta \sum_{n=0}^{\infty} \frac{2k_{mn}(r=a)}{\omega_{mn}a/c} \sin \frac{\omega_{mn}s}{c}\right) \quad s > 0$$

$$\tag{11.18}$$

These wake potentials are both zero for $s < 0$ from causality. The quantity k_{mn} is the loss parameter for the mode, which was defined in Chapter 10 as half the voltage per unit charge induced in an empty cavity immediately after a charge passes through the cavity. It is expressed in terms of the r/Q for a given mode as $k_{mn} = \omega r/4Q_0$. Thus, the loss parameter for any mode is a property of the structure.

The total wake potentials are obtained by summing over m. If the radial displacements are small compared to a, the dominant terms will be the $m = 0$ monopole term for the longitudinal wake potential, and the $m = 1$ dipole term for the transverse wake potential. Then, the wake potentials are approximated by

$$w_z \cong w_{z0} = \sum_{n=0}^{\infty} 2k_{0n} \cos \frac{\omega_{0n} s}{c}, \quad s > 0 \tag{11.19}$$

and

$$\mathbf{w}_\perp \cong \mathbf{w}_{\perp 1} = \frac{r_1}{a} \hat{\mathbf{x}} \sum_{n=0}^{\infty} \frac{2k_{1n}}{\omega_{1n} a/c} \sin \frac{\omega_{1n} s}{c} \quad s > 0 \tag{11.20}$$

In this approximation, the longitudinal wake potential is independent of the transverse coordinates of both the source and test charges, and the transverse wake potential is linearly proportional to the source-charge radius, is independent of the transverse coordinates of the test charge, and is everywhere directed along the x-axis, the direction of the source displacement. Because each term in w_z varies as $\cos(\omega s/c)$, the wake potential immediately behind the point charge at $s = 0^+$ is positive. From the definition of the wake potential given in Eq. (11.12), this means the wakefield is directed opposite the direction of motion, and test charges immediately behind the source charge are decelerated. The source charge itself at $s = 0$ sees half of the wake potential that is induced behind it (at $s = 0+$), as was discussed in Chapter 10. Each term in the transverse wake potential varies as $\sin(\omega s/c)$. This implies that the transverse wake potential is zero immediately behind the source charge, and unlike the longitudinal case, the source charge does not see any of the transverse wake potential that it generates. As s increases, the sign of both wake potentials can change.

Using the approximations given above, the situation can be summarized as follows. For the longitudinal wake potential

$$w_z(s) = \begin{cases} \sum_{n=0}^{\infty} 2k_{0n} \cos \frac{\omega_{0n} s}{c} & s > 0 \quad \text{(behind source)} & (11.21) \\ \sum_{n=0}^{\infty} k_{0n} & s = 0 \quad \text{(at source)} & (11.22) \\ 0 & s < 0 \quad \text{(ahead of source)} & (11.23) \end{cases}$$

and for the transverse wake potential

$$\mathbf{w}_\perp(s) = \begin{cases} \dfrac{r_1}{a}\hat{\mathbf{x}} \sum_{n=0}^{\infty} \dfrac{2k_{1n}}{\omega_{1n}a/c} \sin\dfrac{\omega_{1n}s}{c} & s > 0 \qquad (11.24) \\ 0 & s = 0 \qquad (11.25) \\ 0 & s < 0 \qquad (11.26) \end{cases}$$

The energy lost by the source charge q_1 to electromagnetic energy in the wake potential is

$$\Delta U = q_1^2 w_z(0) = q_1^2 \sum_{n=0}^{\infty} k_{0n} \qquad (11.27)$$

It is customary to define a total loss parameter k_{tot} as

$$k_{tot} = \sum_{n=0}^{\infty} k_{0n} \qquad (11.28)$$

The energy lost in the structure by the charge q_1 to the wakefields is $\Delta U = q_1^2 k_{tot}$, and the power dissipated from the wakefields can be calculated by introducing the Q of the modes as was discussed in Section 10.3.

As an example, the SLAC electron accelerator is a cylindrically symmetric disk-loaded waveguide, operating at 2856 MHz. The RF wavelength is 10.5 cm, the cell length is $\lambda/3 = 3.5$ cm, a nominal value for the aperture radius is $a = 1.165$ cm, the linac length is 3000 m, and the final energy is 50 GeV. For linear-collider operation the number of particles per bunch is 5×10^{10}, and the RMS bunch length is nominally $\sigma_z = 1$ mm. Figures 11.5 and 11.6 show the computed δ-function wake potentials per cell. By comparing the total wake potentials per cell with the contribution from the accelerating mode alone, one can see where the accelerating mode begins to dominate as a function of s.

11.5
Scaling of Wake Potentials with Frequency

The frequency scaling of the wake potentials can be obtained from the above expressions for the monopole and dipole approximations. Frequency scaling of cavity parameters is usually quoted for the case of a fixed field amplitude and a fixed total length of the structures. All dimensions of the accelerating structures are assumed to be scaled proportional to the wavelength of the accelerating mode, and the ratio of the higher-order mode frequencies to the accelerating mode frequencies are assumed to remain constant. The loss parameters k_{mn} for each mode mn are independent of field level and scale as

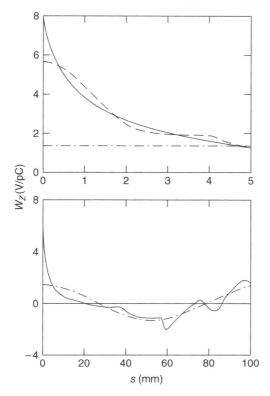

Figure 11.5 Computed longitudinal δ-function wake potential per cell from Ref. [10] for the SLAC linac structure, shown for two distance scales behind the source charge. The solid curve is the total wake potential. The dashed curve shows the contribution from 450 modes, and the dot-dash curve shows the contribution from the accelerating mode. [Reprinted with permission from P. B. Wilson, *AIP Conf. Proc.* **184**, 526–564 (1989). Copyright 1989 American Institute of Physics.]

ω_{mn}^2. If we scale $s \propto \lambda$, $a \propto \lambda$, and assume that r_1 is fixed, we find

$$w_z \propto \omega^2 \tag{11.29}$$

and

$$\mathbf{w}_\perp \propto \omega^3 \tag{11.30}$$

These results show that the wakefield effects increase strongly with increasing frequency.

11.6
Bunch Wake Potentials for an Arbitrary Charge Distribution

In Section 11.3, we defined the δ-function wake potential associated with a geometric perturbation and generated by the passage of an ultrarelativistic point charge. In this section, we discuss approaches for calculation of the wakefields from a bunch of particles. First, we consider a collection of comoving charges, assumed to be ultrarelativistic, and characterized by an

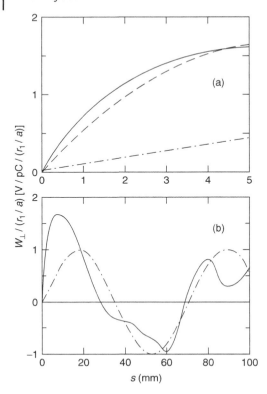

Figure 11.6 Computed transverse δ-function wake potential per r_1/a per cell from Ref. [11] for the SLAC linac structure, shown for two distance scales behind the source charge. The solid curve is the total wake potential. The dashed curve shows the contribution from 495 modes, and the dot-dash curve shows the contribution from the lowest-frequency dipole mode. [Reprinted with permission from P. B. Wilson, *AIP Conf. Proc.* **184**, 526–564 (1989). Copyright 1989 American Institute of Physics.]

arbitrary line-charge density ρ. This will be followed by a discussion of wakefield calculations for real particles with $\beta < 1$. For the ultrarelativistic case, the bunch wake potentials can be obtained in a very straightforward way, using the δ-function wake potentials from Section 11.3 as Green's functions weighted by the charge or current distribution. As shown in Fig. 11.7, we choose a reference plane, which is fixed within the moving bunch. We measure distances backward, so that larger distances correspond to later, more positive times. The coordinate z_1 is an arbitrary source coordinate, z is an arbitrary field coordinate of a test charge, and $s = z - z_1$ is the separation of the charges. Because the δ-function longitudinal wake potential does not depend on the radial coordinates for the dominant monopole mode, and the δ-function transverse wake potential depends only on the displacement of the source charge, we ignore the radial coordinates in this treatment, except for replacing the radial coordinate of the source by the average displacement for the distribution.

An element of charge $dq = \rho(z_1)dz_1$ at z_1 will produce a potential at z, either longitudinal or transverse, given by

$$dV(z) = w(s)\rho(z_1)dz_1 \tag{11.31}$$

11.6 Bunch Wake Potentials for an Arbitrary Charge Distribution

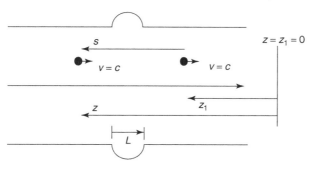

Figure 11.7 Coordinate system relative to a reference plane moving with the bunch. A source point is at z_1, and an arbitrary trailing field point at z. In this figure the perturbing element of length L is the region where the wakefields are generated.

For the ultrarelativistic case, the bunch wake potential at z is defined as the integral overall source charges ahead of z, divided by the total charge in the bunch Q_{tot}. Thus the bunch wake potential for the distribution, in terms of the δ-function wake potential, is

$$W(z) = \left(\frac{1}{Q_{tot}}\right) \int_{-\infty}^{z} w(z-z_1)\rho(z_1)\,dz_1 \tag{11.32}$$

If we change variables to $s = z - z_1$, and integrate over s, we obtain an alternative form

$$W(z) = -\left(\frac{1}{Q_{tot}}\right)\int_{-\infty}^{0} w(s)\rho(z-s)\,ds = \left(\frac{1}{Q_{tot}}\right)\int_{0}^{\infty} w(s)\rho(z-s)\,ds \tag{11.33}$$

Figure 11.8 shows the Gaussian longitudinal wake potential for $\sigma_z = 1$, 5, and 15 mm, where the head of the bunch is toward the left. For the short-bunch case, Wz is positive over the length of the bunch, which means that all the particles lose energy. The energy loss increases almost linearly along most of the bunch, and particles in the tail lose the most energy relative to the head. For the $\sigma_z = 15$ mm case, wz changes sign and the particles in the tail gain energy.

The energy loss of the bunch to wakefields is equal to the work done by the wakefields on the bunch, which can be written in terms of the bunch longitudinal wake potential W(z) [13]. The charge at z experiences an energy loss

$$d\Delta U = \rho(z)\,dz[Q_{tot}W_z(z)] \tag{11.34}$$

The total energy loss is obtained by integrating over the whole charge distribution. Thus

$$\Delta U = (Q_{tot})\int_{-\infty}^{\infty}\rho(z)W_z(z)\,dz \tag{11.35}$$

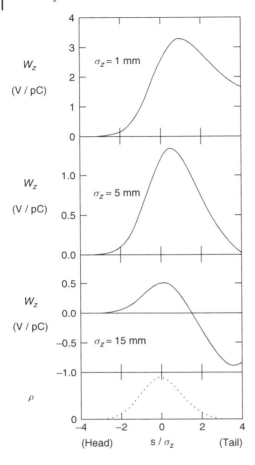

Figure 11.8 Computed Gaussian wake potential per cell from Ref. [10] for the SLAC linac structure for $\sigma_z = 1$, 5, and 15 mm are shown as the solid curves. The dotted curve shows the charge distribution. The head of the bunch is toward the left. [Reprinted with permission from K. L. F. Bane, AIP Conf. Proc. **153**, 972–1014 (1987). Copyright 1987 American Institute of Physics.]

A total bunch loss parameter is defined as

$$k_{tot} = \frac{\Delta U}{Q_{tot}^2} = \left(\frac{1}{Q_{tot}}\right) \int_{-\infty}^{\infty} \rho(z) W_z(z)\, dz \quad (11.36)$$

For the case of a cavity, the beam-induced excitation of the accelerating mode represents beam loading, which is vectorally added to the generator contribution. The contribution to k_{tot} from the higher-order modes [14] of a cavity can be obtained by subtracting the contribution from the accelerating mode; thus

$$k_{HOM} = k_{tot} - \left[\frac{\omega r}{4Q_0}\right]_{accel} \quad (11.37)$$

As discussed by Chao, [15] the wake potential can be computed by numerical solution of Maxwell's equations in the time domain [16, 17], and this is

11.6 Bunch Wake Potentials for an Arbitrary Charge Distribution

usually the most accurate way of computing the short-range wakefields. In numerical computations the restriction to particles with $v = c$ is unnecessary, and causality is automatically satisfied for solutions of Maxwell's equations [18]. Consider an arbitrary position z that is fixed in the coordinate system of a moving rigid bunch, such that z increases as we move from the head to the tail. Suppose that this point has coordinate z' in the linac rest frame, for which the positive direction is the direction of the beam, such that $z' = vt - z$. The longitudinal bunch wake potential is

$$W(r, z) = -\left(\frac{1}{Q_{tot}}\right) \int_0^L dz' [E_z(r, z', t)]_{t=(z'-z)/v} \tag{11.38}$$

where the E_z is the total electric-field z component produced by the wakefields associated with the bunch and the geometric perturbations that are present. For the dominant monopole distribution there is no radial dependence, so the path of integration may be taken at any radius. The energy loss of the bunch to the wakefields and loss parameter are given by Eqs. (11.35) and (11.36), and the average power loss to the wakefields associated with a geometric perturbation is

$$P = (\Delta U) f_b = Q_{tot}^2 k_{tot} f_b \tag{11.39}$$

where f_b is the bunch frequency. For a cavity, the contribution of the accelerating mode should be subtracted as indicated by Eq. (11.37) to avoid double counting.

Numerical calculations, normally carried out on a finite-size section of the accelerating structure, including the beam pipes on each end, require care in specifying the boundary conditions to avoid artificial reflections from artificially abrupt discontinuities at the ends that would modify the wakefields. For the ultrarelativistic problem, an analytic form for the bunch fields, in an infinitely long cylindrical pipe, can be used to prescribe open or infinite-pipe boundary conditions [19] that allow computation to be carried out on a finite section without producing artificial reflections. For particles with velocity $\beta < 1$, one can extend the end beam pipes to a sufficient length, and artificially taper them at the ends so that the waves reflected from the tapered pipes have a longer return path [20]. With this method one hopes to reduce the artificially reflected waves that return to act on the bunch during the computational time for the problem. The tapering also reduces artificial scattering initially, when the bunch is injected, resulting in a more realistic representation of the fields that accompany the beam. The computation should be repeated with the tapered pipe alone, and subtracted as background from the original problem. The subtraction technique is also important for another reason. For $\beta < 1$, space-charge forces act throughout the bunch while it propagates, and even though the bunch is assumed to be rigid, the momentum impulses from space charge will be included in the calculation of the wake potential. It is desirable to subtract the space-charge contribution to avoid including it as a wakefield, and to avoid double counting it, since the space charge is already included in the self-consistent numerical simulation codes. The subtraction approach removes the contributions from the image charges induced in the pipe, but this is usually small enough to be neglected. For a different approach, the short-range wakefields can also be calculated approximately from Eqs. (11.17) and (11.18), using a limited number of modes, and with an analytic extension to include the high-frequency modes [21]. The long-range wake potential is dominated by a few low-frequency modes that continue to execute damped oscillations long after the source charge has passed. For this case, a frequency-domain description, based on the properties of the lowest-frequency modes is useful, and in some cases, Eqs. (11.17) and (11.18) are applicable for long-range wake potentials with only a single dominant mode.

11.7
Loss Parameters for a Particular Charge Distribution

Equation 11.32 expresses the bunch wake potential for an arbitrary charge distribution in terms of the δ-function wake potential. Referring to Fig. 11.7, the integration for an ultrarelativistic problem is overall charges ahead of the point z, so $z_1 < z$. Although some charges have $z_1 > z$, from the causality principle $w_z = 0$ for these charges, because they cannot contribute to the wake at z. Therefore, the integrand is zero when $z_1 > z$ and we can replace the upper limit in the integral by infinity. Thus we write

$$W_z(z) = \left(\frac{1}{Q_{tot}}\right) \int_{-\infty}^{\infty} w_z(z - z_1) \rho(z_1) \, dz_1 \tag{11.40}$$

From Eq. 11.19, the δ-function longitudinal wake potential is

$$w_z(s) = \sum_{n=0}^{\infty} 2k_{0n} \cos \frac{\omega_{0n} s}{c} = \sum_{n=0}^{\infty} 2k_{0n} \, \text{Re} \left[\frac{e^{j\omega_{0n}(z-z_1)}}{c} \right] \tag{11.41}$$

$$= \text{Re} \sum_{n=0}^{\infty} 2k_{0n} \, e^{j\omega_{0n}z/c} \, e^{-j\omega_{0n}z/c_1} \tag{11.42}$$

Substituting Eq. (11.42) into Eq. (11.40), we obtain

$$W_z(z) = \frac{1}{Q_{tot}} \text{Re} \sum_{n=0}^{\infty} 2k_{0n} e^{j\omega_{0n}z/c} \int_{-\infty}^{\infty} e^{-j\omega_{0n}z_1/c} \, dz_1 \tag{11.43}$$

Suppose we choose the origin of the coordinates at the electrical center of the bunch, so

$$\int_{-\infty}^{\infty} \sin\left(\omega_{0n} \frac{z_1}{c}\right) \rho(z_1) \, dz_1 = 0 \tag{11.44}$$

Then, we have

$$W_z(z) = \left(\frac{1}{Q_{tot}}\right) \sum_{n=0}^{\infty} 2k_{0n} \cos(\omega_{0n}z/c) \int_{-\infty}^{\infty} \cos(\omega_{0n}z_1/c)\rho(z_1)dz_1 \quad (11.45)$$

The bunch longitudinal wake potential in Eq. (11.45) for an arbitrary charge distribution has the same form as that of the δ-function longitudinal wake potential of Eq. (11.19), if we identify the bunch loss parameter as

$$k_{0n,distr} = \frac{k_{0n}}{Q_{tot}} \int_{-\infty}^{\infty} \cos(\omega_{0n}z_1/c)\rho(z_1)\,dz_1 \quad (11.46)$$

One can write the same expression for the transverse bunch loss parameter.

11.8
Bunch Loss Parameters for a Gaussian Distribution

The Gaussian distribution is a good approximation for the distribution of a typical relativistic electron beam. Consequently, it is of interest to determine the bunch loss parameters for the Gaussian case. The line-charge density is

$$\rho(z_1) = \frac{Q_{tot}}{\sqrt{2\pi}\sigma} e^{-z_1^2/2\sigma^2} \quad (11.47)$$

This is an even function and using Eq. (11.46), the Gaussian loss parameter is

$$k_{0n,G} = \frac{\sqrt{2}k_{0n}}{\sqrt{\pi}\sigma} \int_0^{\infty} \cos[\omega_{0n}z_1/c] e^{-z_1^2/2\sigma^2} \, dz_1 \quad (11.48)$$

Evaluating the integral, one obtains

$$k_{0n,G} = k_{0n} e^{-\omega_{0n}^2 \sigma^2 / 2c^2} \quad (11.49)$$

The exponential factor is called *the Gaussian form factor*. Equation (11.49) implies that the bunch loss parameter is reduced when the rms bunch length becomes a significant fraction of the wavelength for that mode. Physically, a finite bunch length reduces the loss factor because the wakefields no longer add in phase. The total bunch loss parameter for the Gaussian beam is

$$k_{tot} = \sum_0^n k_{0n} e^{-\omega_{0n}^2 \sigma^2 / 2c^2} \quad (11.50)$$

Figure 11.9 shows k_{tot} for a Gaussian bunch versus the rms length σ_z. For the typical value of $\sigma_z = 1$ mm, k_{tot} is reduced by the Gaussian form factor by nearly a factor of 2 compared with the $\sigma_z = 0$ case.

One can compute the total energy loss of the particles in the bunch to the wakefields from Fig. 11.9 at $\sigma_z = 1$ mm, we have $k_{tot} = 2.1$ V/pC per cell for

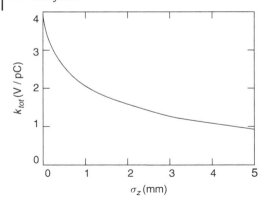

Figure 11.9 Total loss parameter k_{tot} versus rms bunch length from Ref. [10] for one cell of the SLAC linac structure. [Reprinted with permission from K. L. F. Bane, AIP Conf. Proc. **153**, 972–1014 (1987). Copyright 1987 American Institute of Physics.]

the SLAC linac. We assume the number of particles per bunch is $N = 5 \times 10^{10}$, and we calculate the number of cells from the ratio of the total length of 3000 m to the cell length of 3.5 cm. The average energy loss per electron for the whole linac is

$$\frac{\Delta U}{N} = e^2 N k_{tot} N_{cells} \tag{11.51}$$

Substituting the numbers gives $\Delta U/N = 1.44$ GeV per electron, which is approximately 3% of the nominal final energy of 50 GeV. The greater concern associated with the wakefields is not this energy loss, which is not very large, but the loss of beam quality in terms of increased energy spread and transverse emittance growth, as we will discuss later in the chapter.

11.9
Beam-Coupling Impedance

We have described the interaction between a beam and its environment in terms of wakefields and wake potentials, which are functions of distance or equivalently of the time behind a moving charge or distribution of charges. Description in the time domain is convenient for short times, but for longer times the wake potential rings at perhaps one or just a few discrete frequencies. For these problems it is more convenient to describe the problem in the frequency domain. When we studied the problem in the time domain, we found it convenient to treat the basic problem of the response of the beam environment to a δ-function source. We take a similar approach in developing a frequency-domain picture, and consider the response of the system to a single frequency source.

11.9 Beam-Coupling Impedance

The bunch longitudinal wake potential in terms of the δ-function wake potential for the ultrarelativistic case is

$$W_z(z) = \left(\frac{1}{Q_{tot}}\right) \int_0^\infty w_z(z)\rho(z-s)\,ds \qquad (11.52)$$

Because $w_z(s) = 0$ for $s < 0$, we can let the lower limit to the integral extend to minus infinity. Introducing the source current $I = \rho c$, Eq. (11.52) becomes

$$W_z(z) = \left(\frac{1}{Q_{tot}c}\right) \int_{-\infty}^\infty w_z(s) I(z-s)\,ds \qquad (11.53)$$

Now consider a source current consisting of a single frequency ω, so $I(z_1) = I_0 e^{j\omega z_1/c}$, where $z_1 + s = z$. Note that with positions measured backwards, increasing positions correspond to increasing times of arrival, and we can replace position with time. We write

$$I(z-s) = I_0 e^{j\omega(z-s)/c} = I_0\, e^{j\omega z/c} e^{-j\omega s/c} = I(Z) e^{-j\omega s/c} \qquad (11.54)$$

Then

$$W_z(z) = \left(\frac{I(z)}{Q_{tot}c}\right) \int_{-\infty}^\infty w_z(s) e^{-j\omega s/c}\,ds \qquad (11.55)$$

Equation (11.55) may be interpreted as follows. $W_z(z)$ is the wake-potential response to a sinusoidal source current $I(z)$. We now consider frequency as the independent variable and make the following identifications. The source current is $I(\omega) = I_0 e^{j\omega z/c}$. The voltage response is $V(\omega) = Q_{tot} W_z(\omega)$. The remaining factor is called *the beam-coupling impedance* and is given by

$$Z_z(\omega) = \frac{V(\omega)}{I(\omega)} = \frac{1}{c} \int_{-\infty}^\infty w_z(s) e^{-j\omega s/c}\,ds \qquad (11.56)$$

The impedance is the Fourier transform of the δ-function wake potential. Consequently, the impedance function is the frequency spectrum of the δ-function wake potential. A typical impedance spectrum is shown in Fig. 11.10. The beam pipe at the ends of cavity is a waveguide with a cutoff frequency of approximately c/a, where a is the radial aperture. Below the cutoff frequency, the impedance spectrum shows the sharp peaks corresponding to the lower frequency cavity modes. Above the cutoff frequency, the modes can propagate out through the beam pipe, which lowers their Q and creates an overlapping spectrum and a continuous broadband distribution. Recalling that the decay time of the modes is given by $\tau = 2Q/\omega$, we have either low-frequency, high-Q modes with a long decay time τ, or high-frequency, low-Q modes with a short decay time.

11.10
Longitudinal- and Transverse-Impedance Definitions

Just as we have longitudinal and transverse wake potentials to describe the time domain, we define corresponding longitudinal and transverse impedances for the frequency-domain description. The longitudinal relationships are

$$Z_z(x, y, \omega) = \frac{1}{c} \int_{-\infty}^{\infty} w_z(x, y, s)\, e^{-j\omega s/c}\, ds \tag{11.57}$$

$$w_z(x, y, s) = \frac{1}{2\pi} \int_{-\infty}^{\infty} Z_z(x, y, \omega)\, e^{j\omega s/c}\, d\omega \tag{11.58}$$

The transverse relationships are

$$Z_\perp(x, y, \omega) = \frac{1}{jc} \int_{-\infty}^{\infty} w_\perp(x, y, s)\, e^{-j\omega s/c}\, ds \tag{11.59}$$

$$w_\perp(x, y, s) = \frac{j}{2\pi} \int_{-\infty}^{\infty} Z_\perp(x, y, s)\, e^{j\omega s/c}\, d\omega \tag{11.60}$$

Because of causality, the real and imaginary parts of the impedances are not independent. For example, one can show that

$$w_z(s) = \frac{1}{\pi} \int_{-\infty}^{\infty} d\omega\, \text{Re}\{Z_z(\omega)\} \cos(\omega s/c) \tag{11.61}$$

Furthermore, the Panofsky–Wenzel theorem relates the transverse and longitudinal impedances according to

$$\frac{\omega}{c} Z_\perp(x, y, \omega) = \nabla Z_z(x, y, \omega) \tag{11.62}$$

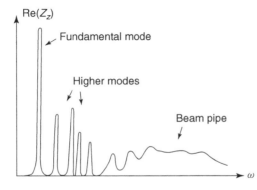

Figure 11.10 Typical frequency spectrum of the real part of an accelerator cavity impedance. [From T. Weiland and R. Wanzenberg, Frontiers of Particle Beams: Intensity Limitations, Proc. Joint U.S.-CERN School on Particle Accelerators at Hilton Head Island, *Lecture Notes Phys.* **400**, M. Dienes, M. Month, and S. Turner (Eds.), 39–79 (1992).

11.11
Impedance and Wake Potential for a Single Cavity Mode

We apply the results of Section 11.10 to calculate the impedance for a single mode of an oscillator, and use the Fourier transform to calculate the wake potential. Consider the shunt resonant circuit of Fig. 11.11, which was discussed in Section 5.1 as a model of an RF cavity. If we identify the driving current with the beam current, and the shunt voltage with the induced axial cavity voltage, we can identify the circuit impedance with the beam-coupling impedance. The circuit impedance for a driving frequency ω is

$$Z(\omega) = \frac{R}{1 + jQ\left[\dfrac{\omega}{\omega_0} - \dfrac{\omega_0}{\omega}\right]} \quad (11.63)$$

where $Q = RC\omega_0$, and $\omega_0 = (LC)^{-1}$. The accelerator shunt impedance r is related to the circuit resistance R by $r = 2R$. The circuit impedance can be written to show the real and imaginary parts more clearly, as

$$Z(\omega) = \frac{R\left[1 - jQ\left\{\dfrac{\omega}{\omega_0} - \dfrac{\omega_0}{\omega}\right\}\right]}{1 + Q^2\left[\dfrac{\omega_0}{\omega} - \dfrac{\omega}{\omega_0}\right]^2} \quad (11.64)$$

The cavity wake potential is the Fourier transform of this circuit impedance, which is

$$w(t) = \frac{\omega_0 R}{Q} e^{-t/\tau} \left\{ \cos\left[\omega_0\sqrt{1 - 1/Q^2}\,t\right] - \frac{1}{2Q\sqrt{1 - 1/4Q^2}} \sin\left[\omega_0\sqrt{1 - 1/Q^2}\,t\right] \right\} \quad (11.65)$$

where the decay time is $\tau = 2Q/\omega_0$. Equation 11.65 is valid for $t > 0$; for $t < 0$, $w(t) = 0$. The real and imaginary parts of the impedance are shown in Fig. 11.12 and the wake potential is shown in Fig. 11.13. The quantity $\omega_0 R/Q$ is equal to $2k$, where $k = \omega_0 r/4Q$ is the loss parameter for the mode. We see

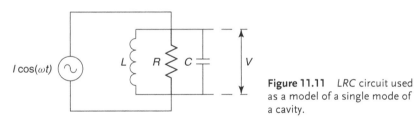

Figure 11.11 LRC circuit used as a model of a single mode of a cavity.

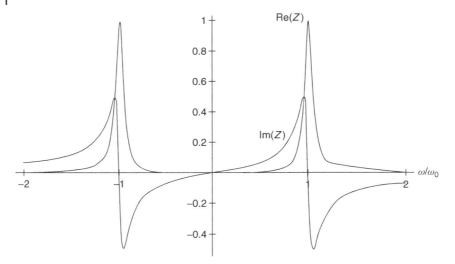

Figure 11.12 The real and imaginary parts of the impedance of a shunt-resonant-circuit cavity model.

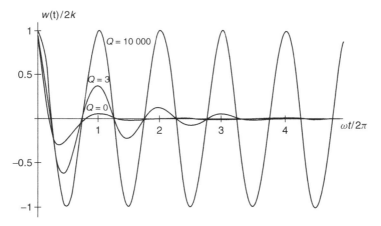

Figure 11.13 Wake potential corresponding to a shunt resonant circuit as a model for the δ-function wake potential of the cavity.

that $w(0) = 2k$. When $Q \gg 1$, the wake potential simplifies to

$$w(t) = 2ke^{-t/\tau} \left\{ \cos[\omega_0 t] - \frac{1}{2Q} \sin[\omega_0 t] \right\} \tag{11.66}$$

For a cavity with no ohmic losses, Q approaches infinity and

$$w(t) = 2k \cos[\omega_0 t] \tag{11.67}$$

This expression for the wake potential will be recognized as the same result as the δ-function longitudinal wake potential in Section 11.4.

11.12
Short-Range Wakefields-Parasitic Losses

The wakefields do work on the beam causing what is called *parasitic energy loss*. The lost beam energy can be restored by increasing the RF applied voltage. Wakefield energy is dissipated through ohmic losses in the walls of the structure. A question is whether the heating effects of the wakefields are sufficient to cause concern. The energy lost to the wakefields per cell from an electron beam with N particles per bunch is $\Delta U = (eN)^2 k_{tot}$, and for a linac section of length L and cell length l_c, the energy loss to wakefields is $\Delta U(L) = (eN)^2 k_{tot} L/l_c$. For a given bunch repetition rate f_R, the average power lost to wakefield is

$$P = \Delta U(L) f_R = \frac{f_R (eN)^2 k_{tot} L}{l_c} \tag{11.68}$$

Consider parameters that are close to those of the SLAC linac; $N = 5 \times 10^{10}$ electrons per bunch, $l_c = 3.5$ cm, $f_R = 120$ Hz, $k_{tot} = 2 \times 10^{12}$ V/C, and $L = 1$ m. We obtain $P = 0.44$ W/m. This is a small parasitic heat load for a copper structure. For a superconducting structure, a heat load of this magnitude is still not large but is more significant. To reduce the load, one would need to reduce the total loss parameter k_{tot}. As might be expected from the result that $k = \omega_0 r/4U$, cavities with nose cones, which have higher shunt impedance r, generate more wakefields, than those without nose cones. At this point we introduce a note of caution. The contributions to the transverse wake potential from geometric perturbations other than the cavities, such as the vacuum bellows, is not always negligible, which means that for an accurate estimate of energy loss to the wakefields, one must include all the components, not only the RF cavities.

11.13
Short-Range Wakefields: Energy Spread

Energy spread must be minimized for applications such as linear colliders or free electron lasers. For a simple analysis suppose we consider a longitudinal wake potential that increases approximately linearly with respect to distance s along the bunch. We approximate the bunch wake potential as $W_z(s) = W'_z s$, and the energy loss per cell for an electron with charge e varies with s as $\Delta W(s) = -eqW'_z s$. For cell length l_c, we find the energy loss over a length L from the approximate expression

$$\Delta W(s, L) = -\frac{eqW'_z s L}{l_c} \tag{11.69}$$

If the particles at the head at $s = 0$ experience a negligible energy loss from the wakefields, the total energy spread across the beam caused by the wakefields generated by the head that act on the tail is

$$\delta W(L) = \frac{eqW'_z s_{tail} L}{l_c} \tag{11.70}$$

where s_{tail} is the bunch length. This energy spread can produce undesirable chromatic aberrations in the linac output beam.

11.14
Short-Range Wakefields: Compensation of Longitudinal Wake Effect

The other effect causing energy spread is the phase dependence of the RF accelerating field. The magnitude and sign of this effect depends on the operating phase of the beam with respect to the RF field. If the phase of the beam centroid is chosen earlier than the peak, then late particles will gain more energy, which is in just the right direction to compensate for the energy spread from the wakefields. To explore this point further, it is convenient to change variables in Eq. 11.70 from s to phase difference along the bunch, using $\Delta\phi = \omega s/c$. The energy loss to the wakefields as a function of phase is

$$\Delta W(\Delta\phi, L) = -\frac{eq[W'_z c/\omega]\Delta\phi L}{l_c} \tag{11.71}$$

The quantity in brackets is the derivative of the longitudinal wake potential with respect to phase. We compare Eq. (11.71) with the energy-gain difference from the RF fields as a function of the phase difference, which is

$$\Delta W_{rf} = \Delta[eEL\cos\phi] \cong -eEL\sin\phi_0 \Delta\phi \tag{11.72}$$

For cancellation we must set $\Delta W_{RF} = \Delta W(\Delta\phi, L)$, which is satisfied for a beam-centroid phase given by

$$\sin\phi_0 = -\frac{q(W'_z c/\omega)}{El_c} \tag{11.73}$$

The larger the derivative of the bunch wake potential, the larger $|\phi_0|$ must be for compensation, and the cost in terms of operating off the peak of the RF waveform increases.

11.15
Short-Range Wakefields: Single-Bunch Beam Breakup

An important effect in high-intensity electron linacs is called *single-bunch BBU*, which is caused by the transverse dipole mode. To illustrate the effect, we use a simple two-particle model for the bunch, shown in Fig. 11.14, in which two

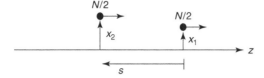

Figure 11.14 Simple two-particle model for single-bunch beam breakup. Each macroparticle contains half the particles in the bunch.

ultrarelativistic macroparticles, each containing $N/2$ particles, are separated by a distance s. As a result of the external focusing forces, an off-axis macroparticle at the head of the bunch will execute transverse (betatron) oscillations about the beam axis. Invoking the smooth approximation, in which the transverse motion is approximated as a simple harmonic oscillation with an angular frequency ω_β, we write for the head macroparticle $x_1 = X \cos \omega_\beta t$. The tail macroparticle would do the same as the head, if it were not for the transverse wakefields induced at the head that affect the tail. The average force F_w caused by the wakefields from a charge $eN/2$ at the head, acting over a cell of length l_c on an electron with charge e in the tail, is obtained from

$$\Delta p_\perp c = F_w l_c = e^2 \frac{N}{2} w_\perp(s) \tag{11.74}$$

We write the wake force as

$$F_w = \frac{e^2 N}{2 l_c} \left[\frac{w_\perp(s)}{x_1/a} \right] \frac{x_1}{a} \tag{11.75}$$

The equation of motion for a particle in the tail in the approximation of a constant-energy beam is

$$m\gamma \ddot{x}_2 = -m\gamma \omega_\beta^2 x_2 + F_w \tag{11.76}$$

or

$$\ddot{x}_2 + \omega_\beta^2 x_2 = \frac{e^2 N}{2 m \gamma l_c a} \left[\frac{w_\perp}{x_1/a} \right] X \cos \omega_\beta t \tag{11.77}$$

This is the familiar equation of motion for an oscillator driven on resonance. Reminding ourselves that γ is assumed to be constant, we write the general solution for a particle in the tail, which has the same initial transverse coordinates as the particle at the head, as

$$\frac{x_2(t)}{X} = \cos \omega_\beta t + \frac{e^2 N t}{4 m \gamma l_c a \omega_\beta} \left[\frac{w_\perp}{x_1/a} \right] \sin \omega_\beta t \tag{11.78}$$

The first term is the free oscillation solution to the homogeneous equation. The second term is a particular solution to the inhomogeneous equation that represents the response to the wake force. Note the factor t in the amplitude of the second term, which implies continuous growth, as shown in Fig. 11.15. If the wakefields were not present we would have $w_\perp = 0$ and would have

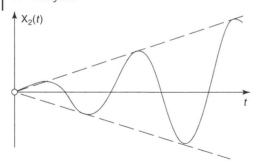

Figure 11.15 Resonant buildup of the displacement of the tail relative to the head of the bunch.

only the first term with unit amplitude. The amplitude of the second term, evaluated at the end of the linac, becomes a measure of the importance of the instability. Over a distance L, the amplitude of the tail displacement relative to that of the head is

$$A(s) = \frac{e^2 N L c}{4(\gamma m c^2) l_c a \omega_\beta} \left[\frac{w_\perp}{x_1/a} \right] \qquad (11.79)$$

The betatron frequency can be written in terms of a betatron wavelength using the relation $\omega_\beta = 2\pi c/\lambda_\beta$. The energy of the beam particle, $\gamma m c^2$, was assumed to be constant. Extending the model over a larger energy range, the energy should be evaluated as an average value over the length L. Figure 11.16 shows the bunch shape, at a particular time during the oscillation, when the tail has swung away from the path of the head. The transverse emittance, averaged over the length of the bunch, grows as the tail amplitude increases. If the amplitude grows large enough, the tail particles will be lost on the walls of the structure. Eq. (11.79) shows how the amplitude depends on the linac and beam parameters, and because of the ω_β factor in the denominator, it shows that the growth of the amplitude can be reduced by stronger focusing.

11.16
Short-Range Wakefields: BNS Damping of Beam Breakup

A method of compensation to inhibit the single-bunch BBU effect, and its associated emittance growth is called *BNS damping* [22], named for Balakin,

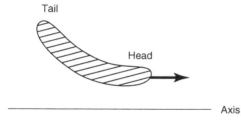

Figure 11.16 Bunch shape caused during beam breakup by the action of the dipole wake potential from the head to the tail.

11.16 Short-Range Wakefields: BNS Damping of Beam Breakup

Novokhatsky, and Smirnov. To introduce this method, we return to the same two-particle model presented in Section 11.15, and consider the case where the betatron frequency of a particle in the tail is not exactly the same as at the head. We will discuss later how this difference can occur. Referring to Section 11.15, the equation of motion of a particle in the tail, located at a distance s from the charge $eN/2$ at the head, and driven by the wakefields from the head, is

$$\ddot{x}_2 + \omega_{\beta 2}^2 x_2 = \frac{e^2 N}{2m\gamma l_c a} \frac{w_\perp(s)}{x_1/a} X \cos \omega_{\beta 1} t \tag{11.80}$$

where $\omega_{\beta 1}$ and $\omega_{\beta 2}$ are the betatron frequencies at the head and tail, respectively, and the trajectory of the charge at the head is $x_1 = X \cos \omega_{\beta 1} t$. Equation (11.80) is the familiar equation of motion of an undamped oscillator, driven off resonance, and if we begin with the condition that at $t = 0$, $x_2(0) = x_1(0)$ and $\dot{x}_2(0) = \dot{x}_1(0)$, and ignore acceleration, the general solution is

$$x_2(t) = X \cos \omega_{\beta 2} t + \frac{F_0(s)}{\omega_{\beta 2}^2 - \omega_{\beta 1}^2} (\cos \omega_{\beta 1} t - \cos \omega_{\beta 2} t) \tag{11.81}$$

where

$$F_0(s) = \frac{e^2 NX}{2(\gamma_2 m)l_c a} \left[\frac{w_\perp(s)}{x_1/a} \right] \tag{11.82}$$

To control the emittance it is necessary to minimize the difference $x_2 - x_1$, which in the two-particle model is

$$x_2(t) - x_1(t) = \left[\frac{F_0(s)}{(\omega_{\beta 2}^2 - \omega_{\beta 1}^2)} - X \right] [\cos \omega_{\beta 1} t - \cos \omega_{\beta 2} t] \tag{11.83}$$

or

$$x_2(t) - x_1(t) = 2X \left[\frac{F_0(s)}{(\omega_{\beta 2}^2 - \omega_{\beta 1}^2)X} - 1 \right]$$
$$\times \left\{ \sin[(\omega_{\beta 2} + \omega_{\beta 1})t/2] \sin[(\omega_{\beta 2} - \omega_{\beta 1})t/2] \right\} \tag{11.84}$$

This difference corresponds to a beat pattern between the driving frequency $\omega_{\beta 1}$, the betatron frequency of the head, and the natural frequency $\omega_{\beta 2}$, which is the betatron frequency of the tail. The frequency difference will generally be small and the resulting beat pattern is shown in Fig. 11.17.

The amplitude of this oscillation is

$$A(s) = 2X \left[\frac{F_0(s)}{(\omega_{\beta 2}^2 - \omega_{\beta 1}^2)X} - 1 \right] \tag{11.85}$$

The amplitude can be made zero if choose $\omega_{\beta 2} > \omega_{\beta 1}$, and

$$\frac{F_0(s)}{(\omega_{\beta 2}^2 - \omega_{\beta 1}^2)X} = 1 \tag{11.86}$$

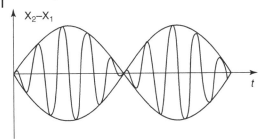

Figure 11.17 Beat pattern describing the difference $x_2(t) - x_1(t)$ between head and the tail in the two-particle model when the betatron frequencies at the head and the tail are not equal.

Assuming that the betatron frequency difference $\Delta\omega_\beta = \omega_{\beta 2} - \omega_{\beta 1}$ is small, we have $\omega_{\beta 2}^2 - \omega_{\beta 1}^2 \cong 2\omega_{\beta 1}\Delta\omega_\beta$. Then, the condition for cancellation implies a betatron frequency difference between tail and head of

$$\Delta\omega_\beta = \frac{F_0(s)}{2\omega_{\beta 1}X} \qquad (11.87)$$

A natural way to produce a betatron-frequency difference is to use the property of magnetic lenses that the focusing force increases at lower momentum, which is true physically because the magnetic rigidity of a particle in a magnetic field is less at lower momentum. One can introduce a correlation between energy and axial position along the bunch, by choosing the synchronous phase later than the crest, so that the particles in the tail see a smaller accelerating field than those at the head. In a relativistic electron linac, the absence of phase oscillations means that the particles in the tail remain in the tail, and have lower momentum throughout the linac. Using the smooth approximation results from Section 7.17 for a quadrupole focusing channel, we obtain the result $\omega_\beta \propto \gamma^{-1}$, where we assume $\beta \cong 1$. Therefore, $\Delta\omega_\beta/\omega_{\beta 1} = -\Delta\gamma/\gamma_1$, and the BNS condition in the lowest-order approximation becomes

$$\Delta W \cong -\frac{e^2 N}{4l_c a(\omega_{\beta 1}/c)^2}\left[\frac{w_\perp(s)}{x_1/a}\right] \qquad (11.88)$$

where $\Delta W = (\gamma_2 - \gamma_1)mc^2$. We can also write this as a fractional energy difference, relative to the average energy $\overline{W}$. Thus

$$\frac{\Delta W}{\overline{W}} \cong -\frac{e^2 N}{4\overline{W}l_c a(\omega_{\beta 1}/c)^2}\left[\frac{w_\perp(s)}{x_1/a}\right] \qquad (11.89)$$

What phase of the accelerating waveform is required to produce this energy difference? Assuming that the energy at any position in the linac $z = L$ is given by $W = eE\cos(\phi_1)z$, the energy difference ΔW relative to the head is $\Delta W = -eE\sin\phi_1 L\Delta\phi = -eE\sin\phi_1 L\omega s/c$. Then

$$\frac{\Delta W}{W} = -\tan\phi_1\frac{\omega s}{c} \qquad (11.90)$$

Equating Eqs. (11.89) and (11.90) gives the result for BNS compensation,

$$\tan \phi_1 = -\frac{\Delta W}{W}\frac{c}{\omega s} \cong \frac{e^2 Nc/\omega s}{4\bar{W}l_c a(\omega_{\beta 1}/c)^2}\left[\frac{w_\perp(s)}{x_1/a}\right] \quad (11.91)$$

The right side of the equation is positive, so the phase ϕ_1 is positive, and the bunch must arrive later than the RF voltage peak, as expected. Recall that for compensation of the energy loss from the longitudinal wake potential at the tail, it was necessary for the bunch to arrive earlier than the peak. The beam-phase requirements for energy-loss compensation and BNS damping are not compatible, so that they cannot be done simultaneously. Fortunately, they can be done sequentially [23]. The BNS compensation of the transverse wake effect for the SLC linac at SLAC is obtained by setting the klystron phases later than the crest at the low-energy end of the linac. At the high-energy end, the klystron phases are shifted to make the bunch arrive early for final correction of the longitudinal energy spread. Typically the phase shifts used are about 15 to 20° relative to the crest. The final energy spread is only about 0.3%.

11.17
Long-Range Wakefields and Multibunch Beam Breakup

The long-range wakefields generally consist of a few low-frequency modes. Of most concern are the deflecting or dipole modes, which can lead to the multibunch BBU instabilities that can limit the intensity in electron linacs. This section presents a brief introduction to the subject, which has been extensively studied by many workers. The classic introduction to the subject is the excellent article by Helm and Loew [24]. The first observations of BBU were made in the late 1950s and early 1960s, and reported from SLAC in 1966. Since then, BBU has been a concern for the design of all high-current linacs, and especially for electron linacs. The most serious deflecting modes are similar to the pillbox-cavity TM_{110} mode. Since the accelerating mode is similar to the TM_{010} mode, one might expect that for a linac structure the ratio of the deflecting-mode frequency to the accelerating-mode frequency is about equal to the ratio of the corresponding zeros of the Bessel functions for these two modes, or $3.8/2.4 = 1.6$. Indeed it is experimentally observed that the ratio of deflecting-mode frequency to the accelerating-mode frequency in linacs is near 3/2. The deflecting modes of real accelerating structures (with beam apertures) are generally neither pure TE nor pure TM, but are linear combinations, called *HEM*, because they are mixed or hybrid electromagnetic modes. Nevertheless, a mode such as the TM_{110} mode of a pillbox cavity without beam holes has many similar properties to the HEM_{110} mode in a cavity with beam holes, and it is useful to refer to the pillbox-cavity mode, which has a simple analytic solution. The nonzero field components of the

TM$_{110}$ mode are

$$E_z = E_0 J_1(k_D r) \cos \phi$$
$$B_r = -j\frac{E_0}{c} \frac{J_1(k_D r)}{k_D r} \sin \phi$$
$$B_\phi = -j\frac{E_0}{c} J_1'(k_D r) \cos \phi \qquad (11.92)$$

where the deflecting-mode wave number is $k_D = \omega_D/c$, and the time dependence of each component is described by the factor $e^{j\omega_D t}$. There are really two modes with orthogonal polarities. The field lines are shown in Fig. 11.18. The magnetic field of the TM$_{110}$ mode is transverse to the cavity axis and produces a transverse beam deflection that is perpendicular to both the beam axis and the magnetic-field direction. The mode has a longitudinal electric field, which vanishes and reverses direction on the cavity axis, as shown in the figure.

Multibunch BBU can be initiated from noise, corresponding to a small transverse modulation of the beam bunches at a deflecting-mode frequency, which can excite a small excitation of the deflecting mode. Once the deflecting mode is excited, it acts on subsequent beam bunches to drive them further off axis, which further reinforces the excitation of the deflecting mode. There are two different mechanisms through which BBU can grow, known as *regenerative* and *cumulative BBU*.

The *regenerative* mechanism occurs in a single multicell linac structure. A standing-wave linac typically has cells that are strongly coupled electromagnetically, and in this case the deflecting mode is a particular normal mode of the multicell array with a phase velocity that is at least approximately synchronous with the beam. The deflecting-mode excitation from the beam-cavity interaction is carried from one cell to the next by the strong electromagnetic coupling of the cells. The strongest excitation can occur if a HEM mode is excited with a phase velocity that is slightly lower than the beam velocity, such that the beam

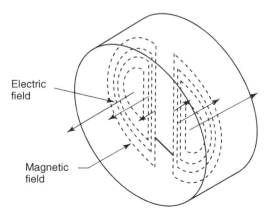

Figure 11.18 Field pattern of a TM$_{110}$ mode of a pillbox cavity.

can slip ahead in the structure by about 180° of the deflecting-mode cycle. To see how the regenerative effect can occur, suppose that in the front part of the structure the phase is such that the beam is deflected by the magnetic field, and acquires a larger transverse displacement. One-quarter period later, the off-axis beam in the back part of the structure can be decelerated by the off-axis longitudinal electric field. In this case the beam is both deflected by and transfers energy to the electromagnetic field of the deflecting mode. The energy increment to the deflecting mode is rapidly distributed throughout the structure by the strong intercell coupling, and this results in an increased amplitude for the particular normal mode causing the deflection. Above a threshold current and pulse length, where the beam-induced growth rate exceeds the ohmic loss rate, the field and beam displacement grow exponentially, and the beam will be eventually deflected so strongly that it is lost in the linac. A starting current for regenerative standing-wave oscillations for beam particles with charge q and rest energy mc^2 can be estimated from [25]

$$I_S \approx \left(\frac{\pi}{L}\right)^2 \frac{\lambda_d \beta \gamma (mc^2/q)}{4Q(r_\perp/Q)} F_a(x) \tag{11.93}$$

where γ and β are the usual relativistic factors corresponding to the final beam energy, λ_D is the free-space wavelength of the deflecting mode, L is the structure length, and $r_\perp$ is a transverse shunt impedance per unit length, defined later in this section. The shunt impedance $r_\perp$ measures the coupling of the beam to the deflecting mode, where a typical value is $r_\perp/Q = 100\ \Omega/\lambda_d$. $F_a(\beta\gamma/\beta_i\gamma_i)$ is a correction factor for acceleration from an initial momentum $mc\beta_i\gamma_i$ to a final momentum $mc\beta\gamma$, given by Wilson [26] as

$$F_a(x) = \frac{(x-1)^3}{6x[(x-1)^2/2 + x - 1 - x\ln(x)]} \tag{11.94}$$

where for large arguments $F_a(x) \approx 1/3$. A starting-current expression for regenerative traveling-wave oscillations is also given by Helm and Loew [27]. Growth rates for the instability are also given for both types of structures in this same reference. The starting current for pulsed beams can be larger than for the CW case, as discussed by Wilson [28].

The *cumulative*-BBU mechanism occurs in a linac consisting of an array of electromagnetically independent cavities, and again may be initiated from a small initial transverse modulation of the beam bunches at a deflecting-mode frequency [29]. The HEM deflecting mode is resonantly excited in each cavity by this modulation, and each cavity in turn amplifies the transverse beam-modulation amplitude. The growing beam displacements, resulting from the deflecting-mode excitation in the earlier cavities, accumulate to produce larger beam displacements in the later cavities, so that the instability is coupled from earlier to later cavities by the beam. The effect is *cumulative* both as a function of the length or number of cavities, and time or number of bunches. Cumulative BBU was important in the early operation of SLAC [30] and was extensively studied there. Cumulative BBU effects are particularly serious in their transient behavior, where the amplitude growth can become very large. Our discussion of cumulative BBU will follow the definitions and results by Gluckstern *et al.* [31] but we have reexpressed the results in the notation used in many earlier references.

An important parameter describing the transverse beam-cavity interaction is a transverse shunt impedance, or transverse coupling impedance, for which several definitions can be found in the literature. From the Panofsky–Wenzel theorem, the transverse momentum change of the particle passing through a cavity excited in a single mode is proportional to a parameter $R_\perp/Q$, the ratio of a transverse shunt impedance in ohms to its quality factor, defined by [32]

$$\frac{R_\perp}{Q} \equiv Z_0 \frac{c^3}{\omega_D^3} \frac{\left| \int \frac{\partial E_z}{\partial x} e^{j\omega_D z/c} \, dz \right|^2}{\int E^2 \, dV} \quad (11.95)$$

where $Z_0 = \sqrt{\mu_0/\varepsilon_0}$ is the impedance of free space and ω_D is the deflecting-mode frequency. By including the complex exponential factor, the transit-time effect is included. Usually $\partial E_z/\partial x$ is nearly an even function of z, and the $\cos(\omega_D z/c)$ term is dominant after the integration, because it is an even function. Also, frequently appearing in the literature is a transverse shunt impedance per unit length, used in the above discussion of regenerative BBU, defined as $r_\perp = R_\perp/\ell$, where ℓ is the active length per cavity for the deflecting mode. [33] Parameter values for the 2856-MHz (accelerating-mode frequency), 3-m SLAC structure with a deflecting-mode frequency about a factor of 1.5 larger than the accelerating mode, are $R_\perp/Q \approx 400\,\Omega$, and $\ell \approx 0.25$ m. The frequency scaling of $R_\perp/Q$ depends on what is held constant. For fixed field amplitude in Eq. (11.92) $\partial E_z/\partial x \propto \omega_D$, and for fixed accelerating structure length, the denominator is proportional to λ_D^2. Then $R_\perp/Q \propto \lambda_D^{-1}$, and useful rules of thumb are $R_\perp/Q \approx 400\,\Omega[0.105m/\lambda_D(m)]$, and $r_\perp/Q \approx 100\,\Omega/\lambda_D$ to $200\,\Omega/\lambda_D$.

Returning to cumulative BBU, the transverse magnetic field of the deflecting mode imparts a transverse momentum impulse to the bunch, and the transverse bunch modulation drives the deflecting mode in each cavity through the interaction with the off-axis axial electric field. Between cavities the momentum impulse is converted to an increased displacement amplitude. When the bunch enters the next cavity, it has a larger displacement and interacts even more strongly with an electric field that is larger at larger radius. The deflecting-mode amplitude in each cavity and the transverse amplitude of the beam are determined by competition between the growth induced by the beam-cavity interaction, and the decay caused by the ohmic power loss in the cavity walls. These amplitudes increase monotonically with length along the linac. As a function of time at a given cavity, the behavior is more complicated. As shown in Fig. 11.19, there is an exponentially growing initial transient for times short or comparable with the deflecting-mode time constant. The growth of the displacement reaches a maximum when the sum of the growth plus decay exponents reach a maximum value, after which the amplitude decreases, a result of both the decay from the finite Q of the deflecting mode, and the driving forces from the bunches which get out of phase because the bunch frequency and the deflecting-mode frequency are not equal. For times much larger than the deflecting-mode time constant, the displacements approach an asymptotic limit, also called *the steady-state regime*. The large increase in displacement with increasing cavity number is evident. The steady-state displacement, observable in Fig. 11.19a, is not obvious in Fig. 11.19b because of the scale change from millimeters to centimeters in the plots.

The net amplitude after the steady state has been reached is proportional to $e^{F_{ss}}$, where ignoring acceleration and focusing, the total growth exponent is

$$F_{ss}(z) = \frac{3^{3/4}}{2^{3/2}} \left[\frac{\pi I R_\perp z^2}{V_0 L \lambda_D} \right]^{1/2} \tag{11.96}$$

Also, $qV_0 = mc^2 \gamma \beta$, L is the center-to-center cavity spacing, z is the position along the linac ($z = N_c L$, where N_c is the cavity number), λ_D is typically 2/3 of the free-space wavelength of the accelerating mode, and I is the average current ($I = qN/\tau$, where N is the number of particles per bunch and τ^{-1} is the bunch frequency). For a relativistic beam, the effect of acceleration can be included by making the substitution, $qV_0 \to mc^2 \gamma' z/4$, where γ' gives the acceleration rate [34]. Thus, for an accelerated beam with no focusing, the asymptotic solution is

$$F_{ss}(z) = \frac{3^{3/4}}{2^{1/2}} \left[\frac{\pi I R_\perp z}{(mc^2/q) \gamma' L \lambda_D} \right]^{1/2} \tag{11.97}$$

In the transient regime, the amplitude growth varies approximately as $e^{F_e(z,t) - \omega_D t/2Q}$ with a growth and a decay term, and which for $t < 2QF_e/\omega_D$ is dominated by the growth term, $F_e(z, t)$. The growth exponent for the transient case, assuming no focusing and no acceleration, is

$$F_e(z, t) = \frac{3^{3/2}}{2^{5/3}} \left[\frac{\pi^2 I (R_\perp/Q) c t z^2}{V_0 L \lambda_D^2} \right]^{1/3} \tag{11.98}$$

Using the prescription above to include the effect of acceleration, the transient solution for the growth exponent for a relativistic beam with no focusing, is

$$F_e(z, t) = \frac{3^{3/2}}{2} \left[\frac{\pi^2 I (R_\perp/Q) c t z}{(mc^2/q) \gamma' L \lambda_D^2} \right]^{1/3} \tag{11.99}$$

Equations (11.96) to (11.99) can be used to estimate the threat of cumulative BBU, by comparing the computed e-folding factors with empirical values of

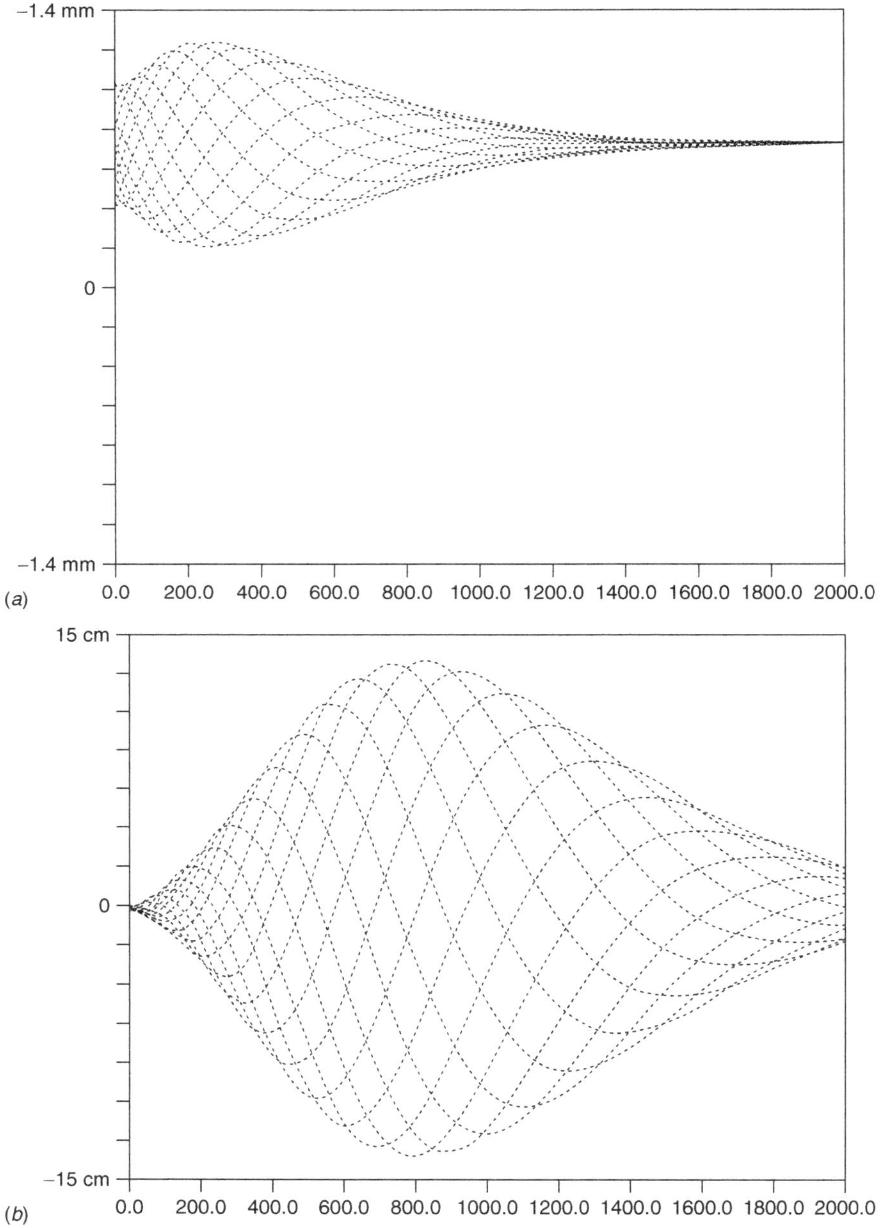

Figure 11.19 Typical bunch-centroid displacement showing cumulative BBU at a fixed cavity versus bunch number M (time is $M\tau$), for the first 2000 bunches, and for a beam in which all the bunches are initially offset at the first cavity by 1 mm. The upper curve (a) is for cavity 15 and the lower curve (b) is for cavity 30. The pattern in the plots is a consequence of the choice of an integer ratio or the deflecting-mode to accelerating-mode frequencies. [From R. L. Gluckstern. R. L. Cooper, and P. Channell, *Part. Accel.* **16**, 125 (1985), Copyright 1985 Gordon and Breach].

about $F = 10$, where one may begin to observe emittance growth, and a value of about 15 to 20 where the beam strikes the walls, [35] the exact values depending on the size of the initial noise-induced transverse amplitude modulation. It was found in the early days at SLAC that the experimental threshold for BBU, when the beam begins to strike the accelerator walls, corresponds to a value of the e-folding factor for transient growth $F_e \approx 20$ [36]. As pointed out by Wilson, with 960 cavities at SLAC, and if all cavities participate, the actual amplification per cavity required to reach the beam-loss threshold was small, since the growth factor for the last cavity, $(1+\varepsilon)^{960} = e^{F_e}$ is satisfied for $F_e \approx 20$, when $\varepsilon \approx 0.021$.

To estimate the value of F_e at the transient peak for any point z in the linac, one needs an expression for the time that the maximum occurs. An approximate result can be obtained from the formulas by finding the value of t that maximizes the total exponent $e_B(z,t) = F_e(z,t) - \omega_D t/2Q$. If we express $F_e = G(z)t^{1/3}$, where $G(z) \equiv F_e(z,t)/t^{1/3}$, we can show that the time for the transient peak is

$$t_{\max} = \left(\frac{2G(z)Q}{3\,\omega_D}\right)^{3/2} \tag{11.100}$$

If $t_{\max}$ is substituted into Eq. (11.98) or (11.99), the growth exponent $F_e(z, t_{\max})$ at the transient maximum is obtained, which may be compared with a value in the range of 15 to 20, above which it is empirically known that beam loss may be expected. These results also predict that at the transient maximum, the decay exponent is one-third the size of the growth exponent.

It has been found empirically that a simple modification of Eqs. (11.96) to (11.99) includes the focusing in a first approximation. In this method one simply replaces z with the average Courant–Snyder beta function [37]. For a ball-park estimate, in a focusing-defocusing (FODO) lattice, the average beta function is approximately equal to the focusing period at a phase advance per period of 60°. A space-charge-corrected beta function can be approximated by dividing the zero-current average β value by the tune depression. More accurate estimates for the effect of focusing in the transient regime of cumulative BBU can be obtained directly from the theory [38]. A dimensionless focusing parameter is defined as

$$\rho = \frac{16}{3\sqrt{3}} \frac{V_0 \sin^2 \sigma_0}{IR_\perp L k_D} \tag{11.101}$$

where σ_0 is the zero-current transverse phase advance per cavity of the focusing lattice [39]. The focusing-dominated regime, where the BBU displacement is a small perturbation, corresponds to $\rho > 1$, whereas $\rho < 1$ is the BBU dominated regime, in which the expressions reduce to the previous results with no focusing when $\sigma_0 = \rho = 0$. Assuming the parameters ρ and σ_0 are constant along the accelerator, the total exponent for the focusing-dominated regime with $\rho > 1$, is

$$e_B(\rho > 1) = F_e(z,t) - \frac{M\omega_D\tau}{2Q} = \left(\frac{I(R_\perp/Q)k_D^2 ctz}{2V_0\sigma_0}\right)^{1/2} - \frac{M\omega_D\tau}{2Q} \quad (11.102)$$

which has a maximum value

$$e_{B,\max}(\rho > 1) = \frac{IR_\perp k_D z}{4V_0\sigma_0} \quad (11.103)$$

The maximum occurs at time

$$t_{\max}(\rho > 1) = \frac{I(R_\perp/Q)k_D^2 c\tau^2 z}{2V_0\sigma_0}\left(\frac{Q}{\omega_D\tau}\right)^2 \quad (11.104)$$

corresponding to bunch number $M_{\max} = t_{\max}/\tau$.

The equations given above assume that the resonant frequency of the deflecting modes is the same for all cavities. Because of fabrication errors, this is generally not true. Although accelerating cavities must be tuned to give nearly the same accelerating-mode frequencies, the same accuracy is not obtained for the other modes. The presence of random frequency errors for the deflecting modes that are comparable to or greater than the cavity bandwidth is a benefit because it inhibits the cooperative behavior of all the cavities in the BBU instability. It can be shown [40] that to evaluate the effect of deflecting-mode frequency errors, the effective Q that should be used in the equations above is reduced to account for fabrication errors. The effective deflecting-mode quality factor for an ensemble of cavities with random fabrication errors is approximately $Q = f_D/\delta f$, where f_D is the deflecting-mode frequency, and δf is the rms resonant-frequency error. For example, if the cavity radius is about $R = 10$ cm, and the rms fabrication error for the radius is $\delta R = 0.1$ mm, then $Q = f_D/\delta f \approx R/\delta R = 1000$. For the case where significant random errors are present, cumulative BBU can only occur on a smaller scale involving a few cavities, which by chance happen to lie close in deflecting-mode frequency. If necessary, this restricted BBU effect can be eliminated by installing higher-order mode couplers to provide additional loading of the deflecting-mode Q values.

In general, there are several effective cures for cumulative BBU: (1) good beam alignment, (2) strong focusing, (3) detuning of the deflecting-mode frequencies from cavity to cavity, a technique which might also result naturally from random construction errors, (4) choosing the accelerating gap to reduce the transit-time factor for the deflecting mode, thereby reducing the transverse shunt impedance, and (5) loading the deflecting modes by installing higher-order mode couplers that selectively propagate the electromagnetic energy to an absorber. Lower frequency also helps to reduce these effects. Finally, the general BBU problem has been studied [41] for a linac with coupled cavities, where the product of coupling strength k and the Q for the deflecting mode determines the BBU regime. Cumulative BBU corresponds to weak coupling, $kQ \ll 1$, and regenerative BBU corresponds to strong coupling between cavities, $kQ \gg 1$. This shows that regenerative and cumulative BBU are fundamentally the same phenomenon, representing two limiting cases of the cavity-coupling parameter kQ.

11.18
Multipass BBU in Recirculating Electron Linacs

In addition to single-pass BBU, which was discussed earlier for conventional single-pass linacs, recirculating electron linacs are susceptible to a regenerative form of a multipass transverse BBU instability. This is an instability caused by excitation of cavity higher-order modes in a recirculating linac, where the beam is reaccelerated in the same linac cavities one or more times. Transverse beam displacements on successive recirculations can excite cavity deflecting modes, which act to further deflect the beam. At sufficiently high currents, and with high deflecting-mode Qs especially for a superconducting cavity linac, the threshold for instability may be exceeded, resulting in undamped transverse beam oscillations. An analytic theory for multipass BBU was developed [42] to calculate the threshold current for instability and was applied to the CW CEBAF superconducting recirculating linac, showing that the instability threshold for that case was more than an order of magnitude greater than the design beam current.

That treatment corresponds to the case where the beam phase is the same on each recirculation and energy is always being added to the beam. Subsequently, the multipass BBU theory was generalized to describe the case where the bunches do not necessarily have the same RF phase at each recirculation [43]. This generalization of the multipass BBU theory allows for the important application to the case of an energy-recovery linac (ERL). After an initial pass where the beam is accelerated and sent through a wiggler to convert some fraction of beam energy into FEL (free-electron laser) radiation, the beam bunches are reinjected into the same linac, but this time with a decelerating phase to convert the large fraction of unused beam energy into RF energy of the accelerating mode for subsequent acceleration of new bunches. Figure 11.20 shows a schematic drawing of the Jefferson Lab ERL system.

A threshold-current formula has been derived for the case of a single higher-order mode (HOM) with angular frequency ω_λ and wave number k_λ, for one recirculation with recirculation time t_r. The result is

$$I_{th} = -\frac{2pc}{e\left(\dfrac{R}{Q}\right) Q_\lambda k_\lambda M_{12} \sin \omega_\lambda t_r} \tag{11.105}$$

where p is the initial beam momentum, c is the speed of light, M_{12} is the transport matrix element for one circulation relating the initial transverse divergence before the cycle to displacement after the cycle, e is the electron

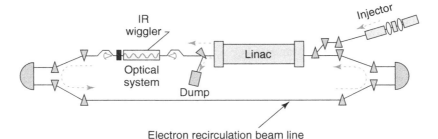

Figure 11.20 Schematic of the original FEL at Jefferson Lab with its photoinjected, superconducting RF linac with energy recovery. After passing through the superconducting linac and the wiggler, the beam is transported back in a recirculation loop and reinjected in the linac at a deceleration phase.

charge, $(R/Q)_\lambda Q_\lambda$ is the shunt impedance and Q_λ is the quality factor of the higher-order deflecting mode. The formula applies for the case where $M_{12} \sin \omega_\lambda t_r < 0$. Several effects can be helpful for increasing the threshold current. Increasing the injection momentum helps, but an efficient ERL requires a low injection energy to reduce the amount of beam energy that goes to the dump and is not recovered. Lowering the quality factor of the deflecting mode, which can be done by use of a HOM-coupler system, is important. Another approach for suppressing the BBU is through appropriate beam-optics design [44]. One such beam-optics solution that can help for regenerative multipass BBU when the deflection per pass is much larger in one transverse plane than the other, is to rotate the beam by 90° so that the large deflection is more equally shared by both transverse planes.

It is observed that the threshold current for regenerative multipass BBU in a recirculating linac can be much lower than the threshold current for regenerative single-pass BBU. One reason for this lower threshold may be the fact that in single-pass regenerative BBU the relatively short length of an accelerating structure converts a deflection at the entrance into a relatively small displacement increase at the exit, contributing a relatively small amplitude increment in the deflecting mode. By contrast, for multipass BBU in a recirculating linac a much larger distance between kicks, nearly the entire orbit of the machine is available for converting a small deflection into a larger displacement, resulting in a larger deflecting mode increase.

However, there is another important difference between single-pass and multipass regenerative BBU. In single-pass BBU the most dangerous modes are those that have a 90° phase slip between the particle and the wave within the structure. This allows a particle to see a magnetic field for deflection at the entrance of the structure, and an electric field producing energy exchange between the beam and the electric field at the exit. In multipass BBU the most dangerous modes are those for which the phase slip between the particle and the wave is zero. This allows the energy exchange that can build up the deflecting-mode amplitude to occur throughout the entire structure [45].

Problems

11.1. The longitudinal wake potential $w_z(s)$ is defined for a given mode as the induced voltage per unit source charge seen by a comoving particle that trails behind the source by a distance s. The loss factor k for the mode is related to the wake potential immediately behind the source charge $w_z(0) = 2k$. (a) Calculate $w_z(0)$, the wake potential behind the source for the fundamental mode of a single cell of length $g = \lambda/2$ using the field $E = -q/\pi\varepsilon_0 a^2$. Note that the charge q in this equation is equal to the source charge q_1. Assume that $\lambda = 10$ cm, $a = 1$ cm and obtain a numerical value in V/pC. (b) Using part (a) calculate the numerical value of the loss parameter k. (c) The loss parameter is a characteristic parameter of a cavity mode even in the absence of an exciting source charge, and is given by $k = \omega r/4Q_0$, where r is the shunt resistance, and Q_0 is the unloaded Q factor or quality factor. Calculate k in V/pC for a 1.3 GHz fundamental mode with $r = 50$ MΩ and $Q_0 = 10,000$.

11.2. Use the method from Section 11.8 to calculate the loss parameter form factor ff(d/λ) for a line-charge density, which is uniform over a distance d. Assume that the line-charge density is $\zeta(z_1) = Q/d$ for $-d/2 \leq z_1 \leq d/2$, and zero elsewhere. Write the line-charge loss parameter as $k_\zeta = k$ ff(d/l). Calculate the form factor for $d/\lambda = 0, 1/4, 1/2, 3/4,$ and 1, and plot ff versus d/λ. Beyond what value of bunch length d/λ does the loss parameter for the uniform beam drop below 90% of the δ-function loss parameter? What is the physical cause of the reduction of the form factor?

11.3. Consider a cylindrically symmetric linac structure containing a wakefield produced by a single longitudinal mode and a single transverse dipole mode. Assume that the cell length is 5 cm, the linac is 10-m long and the final beam energy is 100 MeV. Assume that the longitudinal accelerating-mode frequency is $\omega_0/2\pi = 3$ GHz, and its loss parameter $k_0 = 2$ V/pC per cell. Assume that the dipole-mode frequency is $\omega_1/2\pi = 4.5$ GHz, its loss parameter is $k_1 = 1$ V/pC per cell, and the radial aperture of the iris and beam pipes is $a = 1$ cm. (a) Write the expression for the δ-function longitudinal wake potential per cell as a function of the position s, and plot it for -10 cm $\leq s \leq 20$ cm. (b) Write the expression for the transverse wake potential divided by r_1/a, and plot it versus the position s for the same range as in part (a). (c) Suppose the source charge is $q_1 = 5$ nC, and the test charge is $q = 1.6 \times 10^{-19}$ C, located at a distance of 2 mm behind the source, and with a 1-mm displacement along the x-axis. Calculate the energy loss from the wakefields delivered to the test charge along the whole accelerator in MeV. What fraction of the final energy is this? Calculate the transverse momentum impulse in MeV/c. What fraction of the final momentum is this?

11.4. Beam energy spread from longitudinal wakefields must be corrected in linacs for linear colliders and free electron lasers. Suppose we approximate the bunch wake potential of the 50 GeV SLAC linac with the linear expression $W_z(s) = W'_z s$. Suppose we choose the slope $W'_z = 1$ V/pC/mm. Assume that $s_{tail} = 2$ mm, $N = 5 \times 10^{10}$, $l_c = 3.5$ cm, $L = 3000$ m, and $\omega/2\pi = 3$ GHz. (a) Calculate the final energy spread caused by wakefields that act from head to tail. What percentage of the final energy is this energy spread? (b) By convention, zero phase refers to the point in the RF cycle of maximum energy gain, regardless of the sign of the particle's electric charge. If the magnitude of the accelerating field is 17 MV/m, what nominal phase angle is required for compensation of the energy spread caused by the wakefields to that caused by the sinusoidal accelerating voltage? Does this phase correspond to the beam being earlier or later than the point of peak energy gain?

11.5. Single-bunch BBU can be a serious problem in a high-intensity electron linac unless it is corrected. Use the two-particle model to calculate the ratio of the amplitudes of the tail to the head oscillations at the end of the SLAC linac. Assume that the dipole wake potential for a test charge at $s = 2$ mm behind the source charge at the head is 1.2 V/pC/(r_1/a). Assume the number of particles per bunch is $N = 5 \times 10^{10}$, and assume the betatron wavelength, representing the smoothed focusing force is $\lambda_\beta = 100$ m. For the other parameters, use $L = 3000$ m for the length of the linac, $l_c = 3.5$ cm for the cell length, $a = 1.1$ cm for the radial aperture, and assume that an average electron beam energy of $\gamma mc^2 = 25$ GeV. Is single-bunch BBU a serious problem?

11.6. Apply BNS damping for compensation of the transverse wake effect. Use the same parameter values as in problem 11.5 and two-particle model for BNS damping. (a) Calculate the required fractional energy difference between the tail and the head that provides cancellation of the dipole wake force at the tail. (b) What is the operating phase of the bunch head required to produce this energy difference from the RF accelerating field? Does the bunch come earlier or later than the peak in the energy gain?

11.7. Calculate the starting current for regenerative BBU in a 20-MeV standing-wave electron-linac structure, designed with no transverse focusing, with injection energy of 100 keV, length of 1 m, and accelerating-mode frequency equal to 3 GHz. Assume that the most dangerous deflecting mode lies at a frequency 1.5 times the accelerating-mode frequency, and has $r_\perp/Q = 100$ Ω/λ_d, and $Q = 10,000$. Assume that the acceleration correction factor is $F_a = 1/3$. Based on the result, can this linac be expected to deliver 100 mA output current.

11.8. Consider a 3-GHz CW linac consisting of 36 independent cavities, each separated by $L = 2$ m, that accelerates an electron beam from 5 to 200 MeV with no focusing. Assume the following parameters: $I = 100$ mA, $R_\perp/Q = 400\ \Omega$, and a deflecting mode at $f_D = 4.5$ GHz with $Q = 10^4$. (a) Assuming a uniform accelerating gradient, calculate the cumulative-BBU growth exponent for the steady state in the last linac cavity. (b) Assuming a value for F_{ss} below 15 is required to avoid beam loss, is steady state beam loss from cumulative BBU expected for this linac, and if so where in the linac does it begin to occur? (c) Assuming a value for F_{ss} below 10 is required to avoid emittance growth, is emittance growth from cumulative BBU expected for this linac, and if so where in the linac does it begin to occur?

11.9. Consider the linac of problem 11.8, and assume that a FODO focusing lattice is added with a focusing period $2L$, and an average Courant–Snyder ellipse parameter $\tilde{\beta}$ equal to the focusing period. Calculate the new cumulative-BBU growth exponent at the last linac cavity in the approximation that focusing can be included by replacing $z \to \tilde{\beta}$ in the expression for the exponent. Using the thresholds for emittance growth and beam loss stated in problem 11.8, is emittance growth or steady-state beam loss from cumulative BBU expected when focusing has been added to this linac?

11.10. Writing the total transient cumulative-BBU exponent as $e_B(z, t) = F_e(z, t) - \omega_D t/2Q$, and expressing the growth exponent for transient cumulative BBU as $F_e = G(z) t^{1/3}$, where $G(z)$ is a constant, shows that the value of time that maximizes the total exponent is given by Eq. (11.100). Show that at the time of the transient maximum, the decay exponent is one-third the size of the growth exponent.

11.11. Consider the linac of problem 11.8 with no focusing. (a) Use the result from problem 11.10 to calculate at the last cavity the time for the maximum displacement, which occurs at the transient maximum. (b) Using the result from part (a), calculate the value of the maximum transient cumulative-BBU growth exponent $F_e(z_{\max}, t_{\max})$ in the last cavity. Then calculate the decay term. Which term is larger? (c) Assuming a value for F_e below 15 is required to avoid beam loss, is beam loss expected from transient cumulative BBU?

11.12. Repeat problem 11.11 assuming that the effect of focusing can be approximately included by replacing $z \to \tilde{\beta}$ in all the expressions, and that the average Courant–Snyder ellipse parameter $\tilde{\beta}$ is equal to the focusing period $2L$. Assuming that a value for F_e below 15 is required to avoid beam loss, is beam loss expected from transient cumulative BBU?

11.13. Consider the linac of problem 11.8 with a FODO focusing lattice with a period $2L$, a phase advance per period of $60°$ and a phase advance per cavity of $\sigma_0 = 30°$. Calculate the focusing parameter ρ at 5 and

200 MeV, and decide whether focusing has helped by seeing if the problem is still BBU dominated or has become focusing dominated.

11.14. Consider the linac of problem 11.8 with no focusing and suppose that because of fabrication errors the RMS error in the cavity radius is 0.05 mm. (a) Assuming that the RMS frequency error for the deflecting mode is given by $\delta f \approx f_D \delta R/R$, where $R = 38$ mm, calculate the effective Q for the deflecting mode using the result from the statistical cumulative-BBU theory that the effective $Q = f_D/\delta f$. (b) Using the effective Q from part (a), and assuming a uniform accelerating gradient, calculate the cumulative-BBU growth exponent in the steady state for the last cavity. Is steady-state beam loss from cumulative BBU expected to be a concern for this linac, assuming a value for F_{ss} below 15 is required to avoid beam loss. (c) Using the result of Eq. (11.100), calculate the time for the transient maximum in the last cavity, and calculate the maximum transient cumulative-BBU growth exponent $F_e(z_{\max}, t_{\max})$ in the last cavity. Assuming that a value for F_e below 15 is required to avoid beam loss, is beam loss expected from transient cumulative BBU?

References

1 Chao, A. W., *Physics of Collective Beam Instabilities in High Energy Accelerators*, John Wiley & Sons, 1993.
2 Bongart, K., Pabst, M. and Letchford, A. P., *Space Charge Dominated Beams and Applications of High Brightness Beams*, ed. Lee, S.Y., Bloomington, IN, October 10–13, 1995; AIP Conf. Proc. **377**, 343 (1995).
3 Reiser, M., *Theory and Design of Charged Particle Beams*, John Wiley & Sons, 1994, pp. 252–260, 267–273, 402–428.
4 Jackson, J. D., *Classical Electrodynamics*, 2nd ed., John Wiley & Sons, 1975.
5 An ultrarelativistic particle is an idealized concept of a particle that moves at velocity exactly c.
6 Chao, A. W., *Physics of Collective Beam Instabilities in High Energy Accelerators*, John Wiley & Sons, 1993; and Coherent instabilities of a relativistic bunched beam, SLAC Summer School, AIP Conf. Proc. **105**, 353 (1982).
7 Weiland, T. and Wanzenberg, R., *Wake Fields and Impedances*, Frontiers of Particle Beams: Intensity Limitations, Proceedings of the Joint US-CERN School on Particle Accelerators at Hilton Head Island, Lecture Notes Phys. **400**, 39–79 1992.
8 Chao, A. W., *Physics of Collective Beam Instabilities in High Energy Accelerators*, John Wiley & Sons, 1993, pp. 84–97.
9 Weiland, T., *Part. Accel.* **15**(4), 245–292 (1984).
10 Bane, K. L. F., Wakefield effects in a linear collider, AIP Conf. Proc. **153**, 972–1014 (1987).
11 Wilson, P., Wakefields and wake potentials, AIP Conf. Proc. **184**, 526–564 (1989).
12 Bane, K. L. F., Wilson, P. B. and Weiland, T., Wake fields and wake field acceleration, AIP Conf. Proc. **127**, 875–928 (1985).
13 The energy is eventually dissipated in Ohmic losses.
14 The term higher-order modes to describe other modes than the accelerating mode may be a misnomer. For example, in an iris-coupled N-cell cavity, the π mode is usually the accelerating mode, and it

lies higher than the other N-1 normal modes in that family.
15. Chao, A. W., *Physics of Collective Beam Instabilities in High Energy Accelerators*, John Wiley & Sons, 1993, pp. 110–117.
16. Weiland, T., *Nucl. Inst. Meth.* **212**, 13 (1983).
17. Weiland, T., *Part. Accel.* **15**(4), 245–292 (1984).
18. Strictly speaking, the term wakefields may be misleading for the $\beta < 1$ case, where the scattered fields with v=c are able to advance ahead of the bunch.
19. Weiland, T., *Nucl. Inst. Meth.* **212**, 13 (1983).
20. Chan, K. D. C., Krawczyk, F., private communication.
21. Guignard, G., *Space Charge Dominated Beams and Applications of High Brightness Beams*, ed. S.Y. Lee, Bloomington, IN, October 10–13, 1995; *AIP Conf. Proc.* **377**, 74 (1995).
22. Balakin, V., Novokhatsky, A. and Smirnov, V., Proc. 12th Int. Conf. High Energy Accel., Fermilab 1983, p. 119.
23. Seeman, J. T., *Advances of Accelerator Physics and Technologies*, ed. Schopper, Herwig, World Scientific Publishing, 1993, pp. 219–248.
24. Helm, R. H. and Loew, G. A., Beam breakup, in *Linear Accelerators*, ed. Lapostolle, P. M. and Septier, A.L., John Wiley & Sons, 1970, pp. 173–221.
25. Wilson, P. B., High Energy Electron Linacs: Applications to Storage Rings, RF Systems, and Linear Colliders, 1981 Summer School on High Energy Accelerators, 506.
26. Wilson, P. B., *A Study of Beam Blowup in Electron Linacs*, HEPL-297 (Rev.A), High Energy Physics Laboratory, Stanford University, Stanford, CA, 1963.
27. Helm, R. H. and Loew, G. A., Beam breakup, in *Linear Accelerators*, ed. Lapostolle, P. M. and Septier, A. L., John Wiley & Sons, 1970, p. 187.
28. Wilson, P. B., High energy electron linacs: applications to storage rings, RF systems, and linear colliders, 1981 summer school on high energy accelerators, *AIP Conf. Proc.* **87**, 506 (1981).
29. Wilson, P. B., *A Simple Analysis of Cumulative Beam Breakup for the Steady-State Case*, HEPL-TN-67-8, High Energy Physics Laboratory, Stanford University, Stanford, CA, September, 1967.
30. Panofsky, W. K. H. and Bander, M., *Rev. Sci. Instrum.* **39**, 206 (1968).
31. Gluckstern, R. L., Cooper, R. K. and Channell, P. J., *Part. Accel.* **16**, 125 (1985). On page 142 of this paper the authors provide the substitutions required to compare the formulas in the paper with earlier work. Also provided in this reference are the basic difference equations that will allow the reader to do numerical calculations.
32. This definition is equivalent to that given by Eq.(9) of the previous reference, converted to $R_\perp/Q$ according to the substitution given on page 142 of that reference.
33. Wilson, P. B., High energy electron linacs: applications to storage rings, RF systems, and linear colliders, 1981 summer school on high energy accelerators, *AIP Conf. Proc.* **87**, 507 (1981). SLAC-PUB-2884 (Revised), November 1991, 56. Note that although the cavity end cells are tuned to produce a uniform distribution for the accelerating mode, the field flatness may be poor for other modes, and the deflecting mode may not effectively fill all of the cells with electromagnetic energy. The active length may be chosen to take this into account.
34. For the general case with $\beta < 1$, γ' is replaced by $\beta\gamma' + \beta'\gamma = \gamma^3\beta'$.
35. Wilson, P. B., private communication.
36. Helm, R. and Loew, G., in *Linear Accelerators*, ed. Lapostolle, P. M. and Septier, A. L., John Wiley & Sons, 1970, p. 205.
37. Wilson, P., private communication.
38. Gluckstern, R. L., Neri, F. and Cooper, R. K., *Part. Accel.* **23**, 53 (1988).
39. Be careful here because σ_0 is usually defined as phase advance per focusing period.
40. Gluckstern, R. L., Neri, F. and Cooper, R. K., *Part. Accel.* **23**, 37 (1988).

41 Gluckstern, R. L. and Neri, F. *Part. Accel.* **25**, 11 (1989).
42 Bisognano, J. J. and Gluckstern, R. L., *Multipass Beam Breakup in Recirculating Linacs*, Proc. 1987 Part. Accel. Conf., Washington, DC, (IEEE Catalog No. 87CH2387-9), pp. 1078–1080.
43 Hoffstatter, G. H. and Bazarov, I. V., Beam-breakup instability theory for energy recovery linacs, *Phys. Rev. Special Topics Accel. Beams* **7**, 054401 (2004).
44 Rand, R. and Smith, T., Beam optical control of beam breakup in a recirculating electron accelerator, *Part. Accel.* **11**, 1–13 (1980).
45 Smith, T., private communication.

12
Special Structures and Techniques

The history of linear accelerators is rich in ideas for improving the performance of linac structures and linac systems. In this chapter we will review some of these ideas, which are in various stages of development and some of which have been proposed only recently. We will discuss ideas for increasing the beam current, focusing strength, peak power, and electrical efficiency. It is probable that from some of these ideas, the linac field will advance even further in the years ahead.

12.1
Alternating-Phase Focusing

In Chapter 7, we discussed Earnshaw's theorem from which followed the incompatibility of simultaneous longitudinal and radial focusing from the radio frequency (RF) fields. The conventional solution is to overcome the radial defocusing by the use of external magnetic-focusing lenses. However, the general application of the alternating-focusing principle has led to two other methods that provide a net focusing effect in all three planes from the RF fields alone, without violating Earnshaw's theorem. One method is the use of radiofrequency-quadrupole (RFQ) focusing, as described in Chapter 8. In the RFQ a time-varying alternating-focusing force is applied in the transverse directions x and y, and conventional nonalternating focusing is provided in the longitudinal z direction. In a second method, alternating-phase focusing (APF), [1, 2] the focusing force is alternated in r and z.

Suppose we consider a nonrelativistic beam that is accelerated in a drift-tube linac or DTL. If the synchronous phase is negative, corresponding to particles arriving at the center of an accelerating gap when the sinusoidal accelerating field is increasing, the particles are focused longitudinally, but defocused transversely. If the synchronous phase is positive, corresponding to particles arriving at the center of a gap when the field is decreasing, the particles are focused transversely, but defocused longitudinally. In either case the beam is unstable in at least one direction. However, both the transverse

RF Linear Accelerators. 2nd, completely revised and enlarged edition.
Thomas P. Wangler
Copyright © 2008 Wiley-VCH Verlag GmbH & Co. KGaA, Weinheim
ISBN: 978-3-527-40680-7

and the longitudinal motion can become stable, if the sign of the synchronous phase is changed periodically so that alternating-focusing forces are applied in all directions. Variation of the gap-to-gap synchronous phase in a multicell DTL tank, is obtained by suitable spacing of the gap centers. Instead of a strictly monotonic increase of the drift-tube lengths as in a conventional DTL, a periodic variation is also superimposed, so that the arrival time of the synchronous particle at the gap centers is programmed to provide the desired alternating-focusing pattern. An analysis of the APF linac has been given by Kapchinskiy [3]. Kapchinskiy found that when the synchronous phase alternates symmetrically about the peak, the longitudinal acceptance is small. If the center of the phase modulation is shifted to negative values, the longitudinal acceptance increases, while the transverse focusing strength decreases. APF is most effective at low particle velocities, and at low velocities electrostatic focusing effects, discussed in Chapter 7, become appreciable. Using typical numbers for the low-velocity end of a proton DTL, Kapchinskiy concluded that the optimum center of the phase modulation lies somewhere between $-5°$ to $-10°$.

In principle, the use of APF eliminates the need for installing quadrupoles in the drift tubes, and therefore removes restrictions on the geometry of the drift tubes, enabling higher shunt impedance for the DTL structure. It allows the DTL injection energy to be extended to lower velocities by providing the necessary focusing without quadrupoles in the drift tubes. These potential advantages have stimulated studies [4, 5] during the past two decades to explore the APF parameter space, including a substantial amount of work in the former Soviet Union [6]. Nevertheless, the longitudinal acceptance and the current limit achieved in these studies have generally been too small to be of interest for most high-current linac applications. The region of parameter space that is most advantageous for APF is the same as for the RFQ, and generally the RFQ has been found to give superior performance, providing larger longitudinal acceptance, and stronger longitudinal and transverse focusing. So far, accelerator designers have generally been unable to find applications in which APF could win in the competition with the RFQ.

12.2
Accelerating Structures Using Electric Focusing

The use of electric rather than magnetic focusing is motivated by the need for stronger focusing at low velocities, where the current limit of a linac system is generally determined. The RFQ, the most well-known example of electric focusing, was discussed in Chapter 8. Other methods have also been proposed, but so far none of these has achieved the same popularity as the RFQ. Despite the unmatched advantages of the RFQ as a buncher, it may still be useful to consider the other electric-focusing concepts as possible alternatives for the more inefficient higher energy accelerating section of the RFQ.

An early suggestion [7] for an electric-focusing structure was to construct the drift tubes with four conducting fingers arranged in a quadrupole geometry within the accelerating gap, where two fingers in one transverse plane are attached to the upstream drift tube and two in the orthogonal transverse plane are connected to the downstream drift tube. This arrangement produces an electric-quadrupole field in the accelerating gap. This finger-focusing geometry led to the invention by Tepliakov of a π-*mode electric-focusing structure*, which alternate gaps without fingers, optimized for acceleration, and gaps with fingers, optimized for focusing [8]. The Tepliakov structure was built and operated as an injector linac at the IHEP laboratory in the former Soviet Union [9]. Another proposal is the *Lapostolle matchbox geometry* in which the drift tubes have the form of rectangular boxes with the long axis alternating between the x and y directions from one drift-tube to the next [10]. This geometry has the advantage of avoiding the use of fingers that increase the peak surface electric field.

A related RFQ structure, used for acceleration of low-charge-state heavy ions, is the split coaxial structure [11], which was invented at the GSI laboratory in Darmstadt. The original GSI structure operated at 13 MHz [12]. Instead of vanes or rods, this structure is comprised of many individual small ring electrodes, which are concentric with the beam axis, and to which a pair of diametrically opposite focusing fingers is attached. The rings are mounted on stems that are alternately connected to two pairs of long horizontal spears. Each pair of spears projects from one of the two end plates from which the spears and the attached electrodes are charged. The result is that adjacent electrodes are oppositely charged, and the fingers provide both quadrupole focusing and acceleration in the gap between the rings. The axial potential at the end plates is not zero, which must be taken into account, especially at the low-energy end, where the beam is injected. This RFQ structure can provide efficient acceleration, which is particularly desirable at the high-velocity end of the RFQ, after the beam is bunched.

A recent proposal by Swenson is the RF-focused drift-tube linac or RFD [13]. In the RFD the drift tubes are split into two separate electrodes operating at different electrical potentials. Each electrode supports two fingers which point inward toward the opposite electrode, instead of extending into the accelerating gap, and forms a four-finger quadrupole geometry within each drift-tube structure. These electrodes are electrically isolated from each other by the inductance of their separate water-cooled blade-type stems. During each period the beam traverses two separate regions, an accelerating gap between adjacent drift tubes, and a finger-focusing gap between the two electrodes that comprise each drift tube. As in a conventional linac, the synchronous phase for the accelerating gap corresponds to an accelerating field that is rising in time, as required for longitudinal phase stability. The synchronous phase at the focusing gap corresponds to deceleration, but the longitudinal electric field in this gap is small, since the main electric-field component of this gap is a transverse quadrupole focusing field, produced by the four fingers. The orientation of the fingers in the focusing gaps alternates from gap to gap to produce an alternating focusing-defocusing system. Because of its RF electric focusing, the RFD can provide strong focusing at lower velocities than for conventional magnetically focused DTLs, and consequently, the transition energy between an RFQ that normally bunches the beam and the following RFD can be lower, typically in the 0.5–1 MeV range for protons.

Swenson has also proposed an RF-focused interdigital linac structure (RFI), which is an interdigital Wideroe linac with radiofrequency-quadrupole focusing. Some differences between the RFI and the RFD should be mentioned. For the RFD structure the period length is $\beta\lambda/2$ as compared with $\beta\lambda$ for the Alvarez DTL. Unlike the DTL, the drift tubes of an interdigital structure alternate in potential. The electric fields between the drift tubes also alternate in direction from one gap to the next, and all drift tubes reverse sign every half of an RF period. The particles accelerated in one gap will arrive at the next gap in time for the fields to have reversed sign and produce beam acceleration in every accelerating gap. For the interdigital structure the drift tubes are supported alternately from above and below or from right to left, and the structure is excited in a TE_{110}-like mode that results in excitation of alternate drift tubes with opposite potential (see Fig. 12.1). As is the case for the RFD the drift tubes are split into two electrically independent electrodes, and each electrode has two diametrically opposite fingers that point inward toward the other electrode instead of extending into the accelerating gap, and forms a four-finger quadrupole geometry within each drift-tube structure. However, because of the shorter period of the RFI, the focusing gap must be moved upstream closer to the accelerating gap. This results in an asymmetrical drift tube, with a minor electrode and a major electrode, as shown in Fig. 12.1. The minor electrode is supported on a pair of minor stems and the major electrode is supported on a major stem. To excite the quadrupole focusing fields two minor stems are offset symmetrically on both sides of the major stem, as seen in Fig. 12.2, to couple to the RF magnetic field. As for the RFD, the RFI would be used as the linac structure that follows the RFQ. The main advantages of the RFI over the RFD are reduced size and significantly increased shunt impedance.

An interesting approach for obtaining very high currents of low-velocity ions is to combine both electric focusing and multiple beams. The maximum beam current in an ion-linac channel is usually limited either by the focusing fields that are attainable, or by the envelope instability in a periodic focusing channel. To obtain higher currents than are possible in single linac channels, multiple beams are required. For acceleration of high-current, low-velocity ion beams, Maschke proposed the *multiple-beam electrostatic- quadrupole-array linear accelerator* (MEQALAC) [14], a linac in

12.2 Accelerating Structures Using Electric Focusing

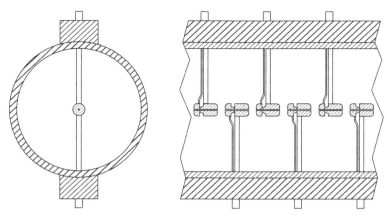

Figure 12.1 Two views of the RFI linac structure. The side view on the right shows asymmetrical split drift tubes comprised of minor and major electrodes which are electrically independent because of inductive isolation. These electrodes contain the focusing gap (courtesy of D. A. Swenson).

Figure 12.2 Drawing of the RFI structure showing the minor stems that are offset symmetrically on both sides of the major stems to provide coupling to the RF magnetic field (courtesy of D. A. Swenson).

which beamlets are accelerated in common RF gaps, and transported within individual channels using electrostatic quadrupoles. This idea was motivated by the desire to produce very bright beams for heavy-ion fusion. Maschke argued from linac current-limit formulas that the brightest linac beams are obtained by filling the available aperture at the lowest practicable velocity and the highest possible RF frequency, which implies using small apertures. The effective emittance, including all beams, is enlarged as a result of the geometrical dilution caused by the finite separation of the beamlets. Maschke argued that brighter beams are still possible by sufficient reduction in the bore sizes and beamlet spacings. He proposed the construction of compact electrostatic-quadrupole arrays that contain thousands of individual channels. The first MEQALAC prototype was built and tested at Brookhaven in 1979 [15]. It accelerated nine beams of Xe^{+1} from 17 to 73 keV, and operated at a very low 4-MHz frequency. The bore diameters were 0.79 cm. The measured average output current for nine beams was 2.8 mA. At the FOM Institute, a MEQALAC was constructed [16] which accelerated four He^+ beams from 40 to 120 keV using an interdigital cavity resonator that operated at 40 MHz (see Fig. 12.3). The bore diameters were 0.6 cm, and a beam current of 2.2 mA per beam was accelerated. Although the MEQALAC can potentially produce high total beam currents, no one has yet demonstrated the design with drastically scaled-down channel dimensions, as was proposed by Maschke.

12.3
Coupled-Cavity Drift-Tube Linac

The *coupled-cavity drift-tube linac* (CCDTL) [17] combines features of both' the Alvarez DTL and the π-mode CCL. The CCDTL resembles the side-coupled linac that was discussed in Chapter 4, except that the accelerating cavity is a short zero-mode DTL. Figure 12.4 shows a CCDTL structure where the accelerating cavity is a two-gap DTL. The chain of cavities operates in a $\pi/2$-like structure mode with adjacent accelerating cavities having oppositely directed electric fields. For large enough bore radius, the CCDTL can have

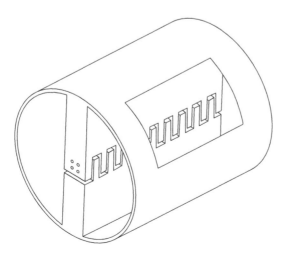

Figure 12.3 The interdigital four-beam MEQALAC. [Reprinted with permission from R. W. Thomae, *et al.*, *AIP Conf. Proc.* **139**, 95–105 (1985). Copyright 1985 American Institute of Physics.]

a larger shunt impedance than the DTL, because of the boundary condition between accelerating cells in a π mode, where it forces the accelerating field to zero, providing longitudinal field concentration to maintain a high transit-time factor (see Section 2.4). Focusing can be provided by replacing an accelerating cavity by a drift space in which an external quadrupole can be installed. If a quadrupole space is inserted, the two end cells on each side of the drift space may still be coupled through a coupling cavity, whose TM_{010}-like mode frequency is equal to that of the accelerating cells. The idea is similar to an excited bridge-coupler cavity, used for the CCL, except that for the CCDTL, the bridge is a nominally unexcited coupling cell. If the distance between the gap centers of the end cells of adjacent accelerating cavities is an odd number times $\beta\lambda/2$, the coupling cavities must be oriented so that the axes of the coupling-cell posts are parallel to the beam axis, as shown in Fig. 12.4. If the distance is an even number times $\beta\lambda/2$, the coupling cavities must be rotated by 90°, so that the axes of the posts are radially directed. This flexibility makes it easy to insert quadrupoles into the chain wherever they are needed, with minimal disruption to the electromagnetic structure mode.

12.4
Beam Funneling

The concept of funneling was proposed at the heavy-ion fusion workshops in the late 1970s [18]. In its simplest form, *funneling* combines two beams with frequency f and current I, by longitudinal interlacing of the bunches, into a single beam of current $2I$, suitable for injection into a linac of frequency $2f$.

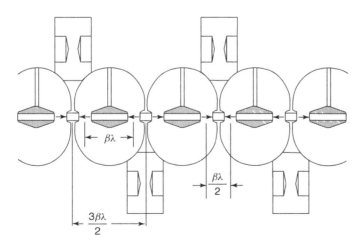

Figure 12.4 A coupled-cavity drift-tube linac (CCDTL) structure with a single drift tube in each accelerating cavity.

After a frequency doubling, the number of available buckets per unit time doubles and all the buckets can be filled. Funneling of high-current beams in RF linacs can be advantageous when the current limits, which increase with increasing energy and decrease with increasing frequency, allow a frequency doubling for the subsequent linac sections. Another motivation for funneling two low-energy beams is that a single ion source may have insufficient output current to supply the required total current for the linac.

A generic beam funnel, shown in Fig.12.5a, contains bending magnets to merge the two beams, and quadrupole lenses, and rebuncher cavities to focus the beam radially and longitudinally. An RF cavity completes the merging and interlacing of the two beams (Fig. 12.5b). If the bunches from the two low-energy injector linacs are 180° out of phase, they can be interlaced by the action of an RF deflecting cavity with frequency f that produces a time-varying transverse electric field. Ideally, funneling would double the beam current per channel with no increase in transverse emittance. However, some emittance growth can be expected because of nonlinear, chromatic, and time-varying effects associated with the funnel-line elements. It should be noted that, because funneling increases the current by interlacing bunches, funneling does not increase the charge per bunch.

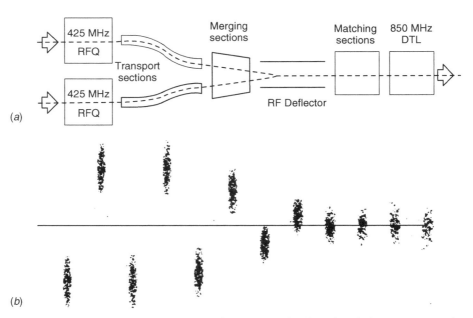

Figure 12.5 (a) Schematic diagram of a beam funnel. The output beams from two RFQs are funneled and the beam is injected into a DTL with twice the frequency. (b) The two bunched beams are shown being merged and interlaced. The transverse scale is magnified compared with the longitudinal scale.

Simulation studies of funneling lines have been reported [19, 20] and an experiment has been performed to test the funnel concept using a beam in one of the two transport lines [21]. Although funneling has been included in some recent proposals for high-power linacs, [22, 23] no complete two-beam funnel demonstration has been reported in the published literature.

A novel proposal for a beam funnel in the form of an RFQ structure for high-current, very low-velocity beams was proposed by Stokes and Minerbo [24]. The structure accepts two bunched ion beams and combines them into a single beam with interlaced microstructure pulses, while providing uninterrupted periodic RF electric-quadrupole focusing.

12.5
RF Pulse Compression

When the SLAC linac was designed it was anticipated that at a later date, the output energy could be increased by adding new klystrons to increase the RF power and the corresponding accelerating field. However, the plan to purchase and install new klystrons for the upgrade would have been very expensive, and an alternative method was devised using the existing klystrons, in which the electromagnetic energy from the klystron pulse would be stored in a high-Q cavity, which could then be discharged into the linac structure. The technique can be generally used for obtaining very short pulses in traveling-wave electron accelerators for linear colliders. The SLAC approach is known as *SLAC energy development* (*SLED*) [25], and the concept is illustrated in a simplified form in Fig. 12.6.

The system consists of a klystron, a fast phase shifter, a circulator, a high-Q storage cavity, and an accelerating structure. The klystron is pulsed on with a pulse length that exceeds the filling time of the accelerating structure. Over most of the klystron pulse, the energy builds up in the storage cavity. No power is reflected back to the klystron because of the circulator, and because

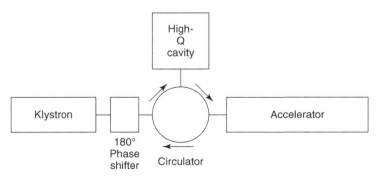

Figure 12.6 Schematic diagram of the SLED pulse compression.

the structure is matched to the waveguide. At a time equal to one structure filling time before the end of the klystron pulse, the phase of the klystron drive signal is rapidly shifted by 180°. The phase reversal allows a discharge of the energy-storage cavity in which the emitted wave from the storage cavity adds in phase with the forward wave coming directly from the klystron. This results in an enhanced RF power delivered to the accelerating structure over the final part of the pulse. At the end of the RF pulse, the accelerating structure is filled with the enhanced electromagnetic stored energy, and beam is then injected into the structure. The implementation of SLED is somewhat different than is indicated by the figure. The major components in the real system are: the fast 180° phase shifter installed on the low-level input side of the klystron, two high-Q cavities used for energy storage, and a 3-dB coupler to replace the circulator. Typical numerical values are 2.5 μs for the full kystron pulse, 0.8 μs for the filling time, and an effective power enhancement of about a factor of 2 [26].

12.6
Superconducting RF Linacs

One of the disadvantages of conventional copper-cavity linacs is the large RF power required because of ohmic power dissipation in the cavity walls. As a result, the RF equipment and ac power for accelerator operations are major if not dominant costs for a copper linac. Also, the design choices for cavity geometry and accelerating fields are dominated by considerations of RF power efficiency. Reducing the fields, to reduce the RF power costs, results in more real estate needed for the linac. For high-duty-factor linacs, the cooling of the copper cavities becomes an important engineering constraint, restricting the fields even further. Because the mean free path for an electron moving in a conductor increases with decreasing temperature, one might hope to improve matters by cooling the copper cavities to cryogenic temperatures. However, because of the anomalous skin effect, which determines the RF surface resistance when the mean free path exceeds the classical skin depth, the surface resistance of cryogenically cooled copper is reduced by less than a factor of 10 compared with room-temperature copper. Unfortunately, the gain in reduced surface resistance is more than offset by the inefficiency of cryogenic refrigeration, so that more total power is required to operate a cryogenically cooled copper linac compared with a room-temperature copper linac. However, the use of superconducting niobium cavities allows a reduction in RF surface resistance by about 10^5, compared with room-temperature copper, which results in a net operating power savings for the superconducting case of one to two orders of magnitude. Thus, the application of superconductivity to linacs has been long recognized as an important step toward better performance and lower costs.

12.6 Superconducting RF Linacs

Throughout this book, we have discussed many issues related to superconducting linacs. The superconducting RF surface resistance was introduced in Chapter 1. In Chapter 2, we presented a model for the transit-time factor for large bore π-mode elliptical cavities, we discussed design issues for superconducting cavities, and we discussed frequency scaling for superconducting linacs. In Chapter 4, we discussed independent-cavity linacs and presented results for π-mode coupled-cavity systems. In Chapter 5, we discussed electron-loading effects, including multipacting and field emission. Then, in Chapter 10 we discussed operation of heavily beam-loaded linac cavities.

In this section, we present general issues regarding superconducting linacs, emphasizing the basic physics. Electric-field limitations arise from field-emission-induced electron loading, and magnetic-field limitations are caused by the presence of normal-conducting impurities on the niobium cavity walls. High performance requires attention to cleanliness during the fabrication and assembly of the cavities and cryomodules. Significant progress has been made, and the applications of superconducting linacs can be expected to increase in the future.

Brief History

Superconductivity was discovered in 1911 by H. Kamerlingh Onnes [27]. Theoretical understanding of the phenomenon of superconductivity developed gradually during the next 50 years, eventually leading to the theory of Bardeen, Cooper, and Schrieffer, known as the *BCS theory* [28]. The application of superconducting technology for RF accelerator cavities began in 1965 with the acceleration of electrons in a lead-plated resonator at Stanford [29]. This initial demonstration was followed in the 1970s by projects using superconducting niobium cavities at Stanford, University of Illinois, CERN and Karlsruhe, Cornell, and Argonne [30]. These early projects were motivated by the promise of high accelerating fields with high-power efficiency. The early applications were for relativistic electron accelerators and storage rings, a particle separator, and a heavy-ion post accelerator for a tandem van de Graaff. For these early projects the accelerating field levels were typically only about 2–3 MV/m.

During the late 1970s and the decade of the 1980s, several important advances led to higher operating field levels in the range of about 5–7 MV/m. The solution of multipacting problems by proper choice of the cavity geometry, combined with improved clean-room techniques and increases in the thermal conductivity of commercially available niobium, made it possible to obtain peak surface electric fields in superconducting cavities of 15–20 MV/m routinely in test facilities. At these levels, field-emission-induced electron loading became an important field limitation. Among the accelerator projects that closely followed these developments were two superconducting linac projects in the United States. First was the ATLAS superconducting linac at Argonne for the acceleration of heavy ions. Second was the superconducting recirculating electron linac at Thomas Jefferson National Accelerator Facility (TJNAF), formerly known as the continuous wave electron-beam accelerator facility or *CEBAF*. Also, in Europe were the electron-positron storage ring facilities, LEP at CERN, and HERA at DESY, while Japan had the TRISTAN facility at KEK. The spallation neutron source (SNS) superconducting linac at Oak Ridge National Laboratory accelerated a negative hydrogen ion beam to 1 GeV in 2007. It contains 11 medium-beta cryomodules with three 6-cell $\beta = 0.61$ cavities per cryomodule (186–379 MeV) followed by 12 high-beta cryomodules with four 6-cell $\beta = 0.81$ cavities per cryomodule (to 1 GeV).

Introduction to the Physics and Technology of RF Superconductivity

Next we summarize the physics and technology of RF superconductivity. A more complete discussion, relevant to accelerator applications, is found in book by Padamsee, *et al.* [31], and the article by Piel [32]. The present theoretical basis for understanding the phenomenon of superconductivity is provided by the BCS theory. Ordinarily, the interaction of the conduction electrons with the crystal-lattice vibrations results in ohmic energy dissipation producing heat. According to the BCS theory, *superconductivity* is the result of an attractive interaction between the conduction electrons through the exchange of virtual phonons, which leads to the formation of correlated electron pairs at temperatures below a critical temperature T_c, and below a critical magnetic field. The correlated electron pairs, known as *Cooper pairs*, occupy the lowest energy state, which is separated from the lowest conduction electron state by a finite energy gap. The energy required to break up a Cooper pair and raise both electrons from the ground state is twice the energy gap, which is about 3 eV. Unless this amount of energy is provided, electrons bound in a Cooper pair cannot be put into a different energy state. Consequently, these electrons are essentially locked into the paired state, and behave like a superfluid, thus producing the phenomenon of superconductivity. As the temperature decreases below the critical temperature, the fraction of electrons that condense into Cooper pairs increases. At $T = T_c$, none of the electrons are paired, while at $T = 0$ all the electrons are paired. Between these temperatures two fluids coexist, the superfluid of Cooper pairs, and the normal fluid of conduction electrons.

Superconductors exhibit zero dc resistance. However, for ac applications a superconductor is not a perfect conductor. A superconductor still experiences ohmic losses for time-dependent fields, because the Cooper pairs that are responsible for the superconducting behavior do not have infinite mobility, and are not able to respond instantly to time-varying fields. Consequently, the shielding is not perfect for time-dependent fields. The fields in the superconductor attenuate with distance from the surface; the characteristic attenuation length is called the *London penetration depth*, which is of order

10^{-8} m in niobium. Then, the unpaired normal electrons that are always present whenever the temperature $T > 0$, are accelerated by the residual electric fields, and dissipate energy through their interaction with the crystal lattice. For the RF surface resistance of superconducting niobium, we use the approximate formula

$$R_s(\Omega) = 9 \times 10^{-5} \frac{f^2(GHz)}{T(K)} \exp\left[-\alpha \frac{T_c}{T}\right] + R_{res} \qquad (12.1)$$

where $\alpha = 1.83$, and $T_c = 9.2$ K is the critical temperature. The first term is the BCS term, although the constant factor in the formula is really a function of many different variables. R_{res} is known as the *residual resistance*; it represents anomalous losses associated with imperfections in the surface, and typically lies in the range $10^{-9} - 10^{-8}$ Ω. The superconducting surface resistance of niobium is of order 10^{-5} lower than that of copper. For accelerating fields in the megavolt per meter range, the power dissipated on the cavity walls is typically only a few Watts per meter. At low frequencies, usually below about 500 MHz, a convenient operating temperature is 4.2 K, the temperature of liquid helium at atmospheric pressure. At higher frequencies, the quadratic dependence of R_s with frequency in the BCS term makes it attractive to lower the operating temperature to typical values of 1.8–2.0 K, which lie below the λ point, where the helium is a superfluid. The most commonly used material for RF superconducting applications has been niobium. Niobium has the highest critical temperature ($T_c = 9.2$ K) of any element, can be machined and deep drawn similar to copper, and is commercially available in bulk and sheet metal forms. Because it getters gases easily, welding needs to be performed under vacuum, and electron-beam welding is primarily used.

Superconducting cavities must be operated within a cryostat, where their cryogenic temperature is maintained, and the small RF power dissipated on the cavity walls is absorbed at low temperatures by a liquid-helium refrigeration system. The ac power required to operate the refrigerator must include the Carnot efficiency for an ideal refrigerator, plus an additional mechanical efficiency factor to represent irreversible effects in the real system. The total required operating power amounts to nearly 1000 times the power absorbed at low temperature. However, even including the refrigeration with its relatively poor efficiency, less total power is required compared with a normal-conducting accelerator, because of the large reduction in the superconducting surface resistance.

The superconducting state in the presence of dc fields will exist if the dc magnetic field is less than the critical field, which at $T = 0$ is $H_c = 2000$ G for niobium. In the presence of RF fields, superconductivity will exist as long as the RF magnetic field is below a critical superheating field, which at $T = 0$ is $H_{sh} = 2400$ G in niobium. This field level has not yet been attained in superconducting cavities. Typical maximum values range from about 500–1000 G. There is no known fundamental reason for the discrepancy, but the practical field limitation is mostly the result of three effects, which are unrelated to the physics of RF superconductivity; electron multipacting, normal-conducting defects, and electron field emission.

Electron multipacting had been an early limiting factor in achieving high fields. Analysis showed that electron multipacting could be eliminated in the TM_{010}-pill-box-like resonators by using a geometry with a rounded shape, sometimes called the *elliptical geometry* [33, 34, 35, 36]. Another limiting factor in achieving higher fields is thermal instabilities produced by normal-conducting surface defects. When such defects are heated in the RF field, the temperature gradient produced across the cavity wall, between the liquid-helium-cooled outer surface and the inner cavity RF surface, can raise the temperature near the defect to a level that exceeds the critical temperature of niobium. Cavity stored energy will be dissipated near the defect. Increasing the thermal conductivity will raise the threshold field for which these thermal instabilities will occur. The thermal conductivity is determined by the mobility of the normal-conducting electrons, which at these temperatures is limited by interstitial gas impurities from O, N, C, and H. Thus, improving the purity of the niobium is the way to increase the thermal conductivity. By the Wiedemann–Franz law the thermal and electrical conductivity of metals are related because the electron mobility affects both. It is customary to measure the purity of the niobium in terms of the residual resistivity ratio (RRR), defined as RRR = electrical resistivity at room temperature divided by the resistance at 4.2 K in the normal-conducting state. The normal-conducting state in 4.2 K niobium can be obtained by applying a dc magnetic field larger than the critical value. Increasing the thermal conductivity of niobium has been achieved through improvements in manufacturing procedures [37, 38]. Refinements in the manufacturing process have produced factors of about 10 increase in RRR of commercially available niobium, corresponding to values today that are typically 250 and higher. These advances have resulted in many cavities that are no longer limited by defect-induced thermal instabilities, but are now predominantly limited by electron field emission. At CERN, a parallel development of plated niobium on copper offers another approach to higher thermal conductivity.

The third limiting factor has been electron field emission, which produces electron loading at high fields. The onset of the field emission occurs at lower surface fields than would be expected even after accounting for metallic protrusions with static electric-field enhancements. Instead, most strong emission sites are believed to be associated with micron-size dust particles on the surface [39]. This means that to achieve high surface fields, it is important to maintain clean working conditions. Furthermore, it has been found that the number of field-emitting sites can be reduced considerably after high-temperature bake-out to temperatures of about 1400 °C. In general, chemical etching, electropolishing, rinsing with pure water, and attention to fabrication and assembly under clean conditions are characteristic requirements for mitigating the effects of field emission and achieving high performance with superconducting cavities.

12.7
Examples of Operating Superconducting Linacs

The first linac to use superconducting cavities was the electron recirculating linear accelerator at Stanford in 1972 [40]. A short time later, RF superconducting booster linacs for heavy-ion nuclear physics applications, were successfully developed [41]. These linacs typically accelerate heavy ions within the velocity range from about $\beta = 0.01-0.20$. Ions with a large range of charge-to-mass ratio are injected into the linac either by a tandem van de Graaff electrostatic accelerator or an ion source on a high-voltage platform. The ion beams have been accelerated up to energies of 25 MeV/nucleon.

Atlas

A well-known example is the ATLAS continuous-wave or CW superconducting linac booster at Argonne National Laboratory, which accelerates heavy ions with mass as large as uranium [42]. The linac consists of independently phased superconducting cavities, and uses superconducting solenoid magnets for focusing. The independent phasing of the cavities provides the flexibility needed to accelerate beams with a wide range of charge-to-mass ratio. The low frequencies (48.5, 72.75, and 97 MHz) provide linear longitudinal focusing across the bunch, allowing the linac to attain the longitudinal beam quality required for precision nuclear physics experiments, producing either very low-energy spread or short subnanosecond pulses for time-of-flight measurements. To keep the cavity sizes manageable at these frequencies, the cavities are loaded capacitively and inductively to resemble lumped circuits. The cavities include quarter-wave resonators loaded with drift tubes for acceleration of the lowest velocity particles, and split-ring resonators for acceleration of the higher-velocity particles. Cavity design velocities include $\beta = 0.06, 0.1$, and 0.16. The split-ring resonators consist of hollow niobium drift tubes mounted onto the ends of a split ring, which is installed in a cylindrical outer housing made of niobium that is explosively bonded to copper. Because the beam current is very low, the loaded Q can be very large, and the RF power requirement is small. The small bandwidth of the high-Q cavities means that the resonators are sensitive to mechanical vibrations. To achieve good phase control the resonators must be mechanically stable, and must use a fast feedback system to compensate for the vibrations. The properties of the ATLAS linac are summarized in Table 12.1.

CEBAF

CEBAF is the CW recirculating linac at the Thomas Jefferson National Accelerator Facility (TJNAF) [44]. The beam is circulated through the linac

Table 12.1 Parameters of ATLAS and CEBAF superconducting Linacs [43]

Facility	ATLAS at Argonne	CEBAF at TJNAF
Application	Heavy-ion linac for nuclear and atomic physics	Recirculating electron linac for nuclear physics
First beam	1978	1994
Species	Lithium through uranium	Electrons
Average beam current	50–500 nA (electrical)	50 μA
Duty factor	CW	CW
Output energy	5–17 MeV/nucleon in normal operation	2 antiparallel 400-MeV linacs used as a recirculating linac to achieve a nominal 4 GeV in 5 passes.
Beam microstructure	12.125 MHz	1497 MHz
Accelerating structures	18 quarter-wave resonators, 48.5 and 72.75 MHz, 2.7-kW RF power; 44 split-ring resonators, 97.0 MHz, 6-kW RF power.	320 5-cell elliptical cavities, each driven by a 5-kW klystron.
Operating temperature	4.2 K	2 K
Length	34.5 m	320 cavities, each with 0.5 m active length.

five times to achieve a nominal output energy of 4 GeV. The recirculation option is attractive because it provides substantial savings in capital and operating costs for RF power. Recirculation is an option for accelerators of relativistic particles, because the beam velocity is essentially fixed at almost the speed of light, and therefore the effective transit-time factor of the multicell cavities is large for every cycle. The subject of recirculating accelerators is not treated in this book. For those who are interested, we recommend the book by R.E. Rand, *Recirculating Electron Accelerators* (Harwood Academic Publishers, 1984). The choice of a CW superconducting linac, compared with the most attractive normal-conducting option consisting of a pulsed linac followed by a storage ring for pulse stretching, enjoys advantages of lower capital and operating costs, higher average currents, fully continuous beams, and much higher beam quality. The injector consists of a 100-keV electron source followed by a 45-MeV superconducting linac, using 18 superconducting cavities. The superconducting cavity design chosen was the proven 5-cell elliptical-geometry design produced by Cornell, and shown in Fig.12.7. The design specifications, 5-MV/m accelerating gradient, and $Q_0 = 2.4 \times 10^9$ at 2 K, were easily met. The average accelerating gradient of the installed cavities is about 8 MV/m (mostly limited by

12.7 Examples of Operating Superconducting Linacs

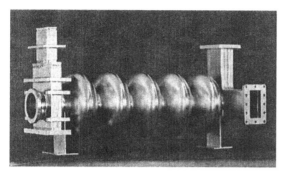

Figure 12.7 Five-cell 1.497-GHz niobium elliptical cavity for the CEBAF recirculating linac at TJNAF showing the waveguide coupling port on the left and two higher-order mode couplers on the right. [Photo Courtesy of P. Kneisel, Thomas Jefferson National Accelerator Facility.]

field emission), and the average $Q_0 = 5 \times 10^9$. Some properties of the CEBAF linac are summarized in Table 12.1. Plans call for an upgrade to 12 GeV, which requires performance improvements in new 7-cell RF cavities [45].

Spallation Neutron Source

The SNS is an accelerator based neutron source that was designed by six national laboratories (Fig. 12.8), and constructed at Oak Ridge National Laboratory for the U.S. Department of Energy. At its full beam power of

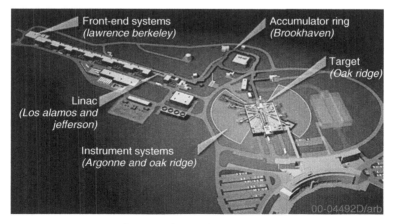

Figure 12.8 Drawing of the SNS facility showing the individual systems and the laboratories that designed them.

1.4 MW, SNS will provide the world's most intense pulsed neutron beams for scientific research and industrial development. In February, 2007 the SNS accelerated beam to its design energy of 1 GeV, a new world energy record for a proton linear accelerator. The SNS accelerator system consists of a negative hydrogen (H$^-$) ion source, a low-energy beam-transport system with a first stage beam chopper, followed by a normal-conducting room-temperature copper linac. The function of the beam chopper is to provide a beam-free gap in the accumulator ring to reduce beam loss during beam extraction from the ring.

The normal-conducting linac consists of a four-vane RFQ (Fig. 12.9) for acceleration to 2.5 MeV, a medium-energy beam-transport system with a second-stage chopper, a 6-tank Alvarez DTL (Fig. 12.10) for acceleration to 87 MeV, and a four-module coupled-cavity linac (Fig. 12.11) for acceleration to 186 MeV. This is followed by the superconducting linac comprised of 11 medium-beta cryomodules with three 6-cell $\beta = 0.61$ cavities per cryomodule (to 379 MeV) and 12 high-beta cryomodules with four 6-cell $\beta = 0.81$ cavities per cryomodule (to 1 GeV) shown in Fig. 12.12. The superconducting niobium linac cavities are cooled with liquid helium to an operating temperature near 2 K (Fig. 12.13).

The SNS linac delivers a 1-ms-long 38-mA peak current chopped beam pulse at 60 Hz to the 248-m diameter accumulator ring injecting the beam in multiple turns. In this way the accumulator ring compresses the 1-ms pulse to ~ 700 ns for delivery by a high-energy beam transport line to the mercury target, where neutrons are produced by spallation. The neutron energies are moderated to useable levels by supercritical hydrogen and water moderators before delivery to 24 beam lines for experiments.

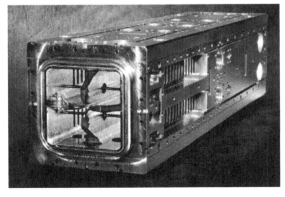

Figure 12.9 402.5 MHz four-vane RFQ that bunches and accelerates the beam from 75 keV to 2.5 MeV.

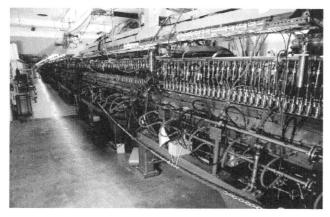

Figure 12.10 402.5-MHz Alvarez drift-tube linac that accelerates the beam from 2.5 MeV to 87 MeV.

Figure 12.11 805-MHz coupled-cavity linac that accelerates the beam from 87 MeV to 186 MeV.

12.8
Future Superconducting Linac Facilities

Superconducting-linac technology has matured sufficiently in recent years that all new RF linac projects are now examined for consideration of a superconducting option. We discuss three applications of superconducting linacs projects currently under study.

International Linear Collider

The International Linear Collider (ILC) is a proposed future international particle-accelerator facility that would produce high-energy particle collisions

Figure 12.12 805-MHz SNS superconducting linac operates near 2 K and accelerates the beam from 186 MeV to 1000 MeV.

Figure 12.13 Six-cell 805-MHz medium-beta ($\beta = 0.61$) and high-beta ($\beta = 0.81$) superconducting niobium cavities designed for the SNS.

between electron and positron beams at center-of-mass energies from 200 to 500 GeV [46]. The machine would also be upgradable in a second phase to 1 TeV. The present design is based on two 11-km-long superconducting linacs operating at 1.3 GHz at an operating temperature near 2 K, and with an average accelerating gradient of 31.5 MV/m (Fig. 12.14). The linacs will be comprised of 9-cell superconducting niobium cavities similar to that shown in Fig. 12.15. The total machine length in the first stage is 31 km long, and the machine would produce a peak luminosity of about $2 \times 10^{34}/\text{cm}^2/\text{s}$ at a center-of-mass energy of 500 GeV. The site length would have to be extended in the second stage to reach 1 TeV in the center of mass. The design achieves this high luminosity through a combination of small emittance beams and

12.8 Future Superconducting Linac Facilities

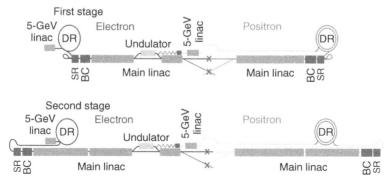

Figure 12.14 Block diagram of the proposed ILC facility showing the two stages of the project.

Figure 12.15 The ILC will use 9-cell superconducting niobium cavities similar to this one. It will operate near 2 K at a 1.3-GHz RF frequency.

high beam power. The beam pulse length is 1.6 ms, the repetition rate is 5 Hz, the duty factor is 0.5%, and the availability goal is 75%.

The electrons are injected from an electron source using a photocathode gun, and the positrons are produced by photoproduction using an undulator-based photon source, driven by a 150-GeV electron beam. The particles will be injected into 5-GeV electron and positron damping rings before being injected into the main linacs. A 4.5-km beam-transport system at the ends of the linacs will bring the two beams into collision at a 14-mrad crossing angle. The R&D effort must focus on achieving high accelerating gradients for the 9-cell superconducting cavities. To ensure an average operating gradient of 31.5 MV/m, the designers wish to achieve 35 MV/m with Q greater than 0.8×10^{10} in acceptance tests during mass production of the cavities. The choice of the operating frequency is a balance between the high cavity cost due to greater cavity size at lower frequency, and a lower sustainable gradient at higher frequency resulting from the increased BCS surface resistance, which causes heating and a global thermal instability that occurs below the RF critical field [47]. In addition, the availability of klystrons limits the frequency choice; 1.3 GHz satisfies all the criteria. The choice of average accelerating gradient of 31.5 MV/m assumes at the present time that further progress will be made in achieving improved cavity performance. The total number of linac components includes 14,460 superconducting cavities and 560 superconducting quadrupoles.

Next-Generation Rare Isotope Facility

A schematic drawing of a next-generation facility for rare isotope beams, previously called the *rare isotope accelerator* (*RIA*) is shown in Fig. 12.16. A superconducting heavy-ion driver linac would accelerate high-intensity beams of stable heavy ions for the production of rare isotopes. The driver linac could deliver beams from protons to uranium at energies of several hundred MeV per amu to one or more targets where the rare isotopes are produced. These rare isotopes would be extracted from the target and could be accelerated in a second superconducting linac called the *post accelerator*. The facility would enable a wide range of studies including measurements of nuclear reactions at astrophysical energies and searches for new heavy elements with long lifetimes.

The driver linac is a CW superconducting linac designed for simultaneous acceleration of stable multiple charge states of heavy-ion elements with masses up to 240 amu. The linac is comprised of independently phased superconducting cavities that provide a wide velocity acceptance as required for acceleration of a wide range of ions with a range of charge states for any ion mass. The beams from an electron cyclotron resonance or ECR ion source are injected into a low frequency RFQ (<100 MHz), followed by the superconducting linac. One or more stripper foils are installed that produce higher charge states to increase the energy gain of the beam and reduce the overall length. In the case of RIA, two stripper foils were used, one near

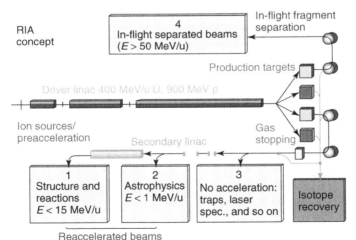

Figure 12.16 The proposed rare isotope accelerator (RIA) shown in this schematic drawing is an example of a next-generation nuclear physics facility for rare isotope beams. Shown are the high-power superconducting driver linacs, which accelerate stable beams from protons to uranium to the target region where the rare isotopes are produced in one or more targets. The rare isotopes are extracted from the target and can be accelerated to higher energies in a second superconducting linac for a variety of nuclear physics and astrophysics experimental studies.

12.8 Future Superconducting Linac Facilities

10 MeV/amu and one near 85 MeV/amu. The net voltage gain of the driver linac was 1.4 GeV. The nominal final beam energy was 400 MeV/amu and the beam power was 400 kW. Different types of resonators available for use in different velocity regions of the driver linac are shown in Fig. 12.17.

Free-Electron Lasers

This topic could have been included in Section 12.7 as an example of operating superconducting linacs, but I have decided to place it in Section 12.8 as future

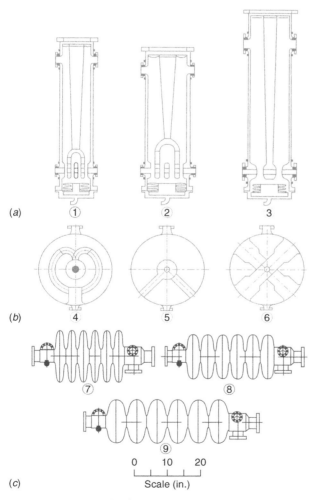

Figure 12.17 Example of superconducting cavities that have been proposed for use in the driver linac including low velocity (*a*), medium velocity (*b*), and high-velocity (*c*) structures.

superconducting linac facilities, because of the great likelihood of further significant expansion of this field. The free-electron laser (FEL) produces intense beams of coherent electromagnetic radiation, and unlike other types of lasers its wavelength is tunable over a wide range. The FEL uses a relativistic electron beam from an accelerator, usually a linac. The electron beam is transported through a periodic alternating-polarity permanent-magnet array called a *wiggler* or *undulator*, which induces a transverse sinusoidal oscillation in the electron beam, causing it to emit synchrotron radiation in a forward cone (Fig. 12.18) at a resonant wavelength that depends on the period and magnetic field of the wiggler, and the electron-beam energy. The photon wavelength of the FEL radiation is given by a resonance condition

$$\lambda_{ph} = \frac{\lambda_u}{2\gamma^2}\left(1 + \frac{K^2}{2}\right) \tag{12.2}$$

where λ_u is the undulator period, γ is the usual relativistic factor, $K = eB_u\lambda_u/2\pi mc$ is the undulator parameter, e and m are the electron charge and mass, c is the speed of light, and B_u is the peak magnetic field of the magnets in the undulator. Equation 12.2 describes the condition that the electrons, while traveling through one full period of the undulator, slip by one wavelength of the radiation relative to the faster electromagnetic wave that is propagating parallel to the undulator axis. Equation 12.2 shows that in principle the resonance condition can be satisfied for arbitrarily small photon wavelength by increasing the electron-beam energy.

The radiation can build up to saturation by inserting the undulator between a pair of normal-incidence mirrors (not shown) to create an optical resonator in which the radiation level builds up from multiple reflections with a properly synchronized electron beam to add more radiation in each pass. This works well over a wide range of wavelengths from infrared to ultraviolet and X-ray

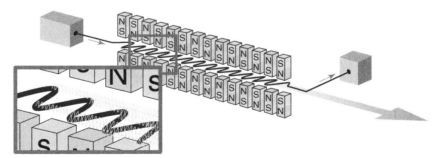

Figure 12.18 A schematic drawing showing an electron beam from an accelerator passing through a wiggler or undulator comprised of an alternating-polarity permanent-magnet array. The wiggler induces an oscillation of the electron beam causing it to emit coherent electromagnetic radiation that propagates in the forward direction and adds to the radiation from previous cycles as shown in the insert.

wavelengths. The electron beam must be produced with a small emittance, and the emittance must be preserved both in the linac and in the transport line to the wiggler. This is because efficient lasing requires high overlap between the electron and photon beams, which is expressed by the condition that the unnormalized emittance ε of the electron beam must satisfy $\varepsilon < \lambda_{ph}/4\pi$.

The two main directions for future FEL development are (1) shorter wavelengths into the vacuum ultraviolet or VUV and X-ray regimes, and (2) higher average power. At shorter wavelengths below about 200 nm, high reflectivity mirrors are not available. In this case a single – pass method is used based on the principle called *self-amplified spontaneous emission (SASE)*. For a SASE FEL a long undulator is used, in which the spontaneous radiation emitted in the first part of the undulator interacts with the electron beam in second part causing amplification by stimulated emission at the resonant wavelength. The SASE method does not require mirrors, does not require an external input signal, and is an attractive solution for the short wavelength VUV and X-ray regimes.

Development of the photocathode electron gun for production of high-brightness electron beams represents a significant technical advance. In this case, the photocathode is placed in an RF cavity with a high electric field and the electrons produced by photoproduction are rapidly accelerated within the cavity to relativistic energies, greatly reducing the effect of the nonlinear space-charge forces on the beam emittance. In addition, a beam-dynamics design approach called *space-charge compensation* [48] is effective in cancellation of much of the remaining space-charge-induced emittance growth. The highest priority for the design of the main accelerator is the control of emittance growth particularly from wakefields, coherent synchrotron radiation, RF input-coupler forces on the beam, and misalignments. The need for high accelerating gradients is of less importance for FELs than for the linear collider application.

At the present time two major SASE FEL projects are (1) the Linac Coherent Light Source (LCLS) at SLAC that uses the final third of the two-mile SLAC linac, and (2) the European XFEL at DESY that will use a 20-GeV superconducting linac. A list of three key design parameters for these two facilities is given in Table 12.2.

For producing higher average radiated FEL power, the most important development has been that of the energy recover linac (ERL) combined with the use of a superconducting linac. This development is motivated by the fact that only a small fraction of the electron-beam energy is converted to

Table 12.2 Some Design Parameters for SASE FEL Projects

Parameter	LCLS	XFEL
Wavelength	1.5 Å	1.0 Å
Electron-beam energy	14.35 GeV	20 GeV
Normalized RMS emittance	1.2 mm-mrad	1.4 mm-mrad

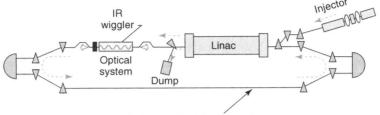

Figure 12.19 Schematic diagram of the original FEL at Jefferson laboratory with its photoinjected, superconducting RF linac with energy recovery. After passing through the superconducting linac and the wiggler, the beam is transported back in a recirculation loop and reinjected in the linac at a deceleration phase.

radiation as the electron beam passes through the wiggler. At high-power levels, dumping the output electron beam after it passes through the wiggler with nearly its full power would result in high levels of radioactivity and severe cooling problems. For a practical FEL device it is desirable to recover as much of the unused electron-beam energy as possible rather than covert all this energy to heat. In the ERL approach the electron beam, after passing through the wiggler, is reinjected into the same linac using an injection phase that decelerates the beam. An example of the FEL at Jefferson laboratory is shown in Fig. 12.19, where most of the electron-beam energy is converted to microwave energy that can be used simultaneously for acceleration of new bunches of electrons [49]. The used electron beam emerges from the linac with only a small fraction of its full energy, and is then transported to the beam dump, where the radioactivity and cooling problems are reduced and are now more manageable. Because of the high beam current, a potential beam-physics problem is the multipass beam breakup or BBU instability caused by higher-order modes that deflect the beam (see Section 11.18). Higher-order mode couplers combined with dampers to absorb the energy in these modes can be used to lower the Q of the modes to raise the instability thresholds. There are also beam optics solutions that can mitigate the BBU effect. The Jefferson laboratory infrared FEL has demonstrated ERL operation in CW mode at high power [50]. In July 2004, 10 kW of radiated power was produced at 6 μm wavelength and in October 2006, 14.2 kW CW was achieved at 1.6 μm.

References

1 Good, M.L., *Phys. Rev.* **92**, 538 (1953).
2 Fainberg, Ya.B., Proc. CERN Symp. High Energy Accel. Pion Phys. Geneva 1956.
3 Kapchinskiy, I.M., *Theory of Resonance Linear Accelerators*, Harwood Academic Publishers, New York, 1985, pp. 184–199.
4 Swenson, D.A., *Part. Accel.* **7**, 61 (1979).
5 Cheng, W.H. et al., *Alternating Phase Focusing Including Space Charge*, 1996

Linac Conf., Chalk River, Ontario, Canada, August, 1992.
6. Wells, N., *Radio Frequency Quadrupole and Alternating Phase Focusing Methods used in Proton Linear Accelerator Technology*, prepared for DARPA, published by Rand Corporation, January, 1985.
7. Vladimirskii, V.V., *Prib. Tekh. Eksp.* **3**, 35 (1956).
8. Tepliakov, V.A., Proc. 1992 Int. Linac Conf., Ottawa, Ontario, Canada, August 24–28, 1992, p. 21.
9. Gorshkov, B.M. et al., *Sov. Phys. Tech. Phys.* **22**(11), 1348 (1977).
10. Lapostolle, P.M., *Compt. Rend.* **256**, 5294 (1963).
11. Muller, R.W., 1979 Linear Accel. Conf., Brookhaven National Laboratory Report BNL-51143 1979, p. 148.
12. Muller, R.W. et al., 1984 Linear Accel. Conf., GSI Report GSI 84-84-11 1984, p. 77.
13. Swenson, D.A., *RF-Focused Drift-Tube Linac Structure*, Proc. 1994 Int. Linac Conf., Tsukuba, Japan.
14. Maschke, A.W., Brookhaven National Laboratory Report BNL 51029, 1 June, 1979.
15. Adams, R. et al., Brookhaven National Laboratory Report BNL 27128, December, 1979.
16. Thomae, R.W. et al., *High Current, High Brightness, and High-Duty Factor Ion Injectors*, Gillespie, G.H., Kuo, Y.Y., Keefe, D. and Wangler, T.P, *AIP Conf. Proc.* **139**, 95 (1986).
17. Billen, J.H., Proc. 1994 Int. Linac Conf., Tsukuba, (1994) p. 341.
18. Young, D.E., Proc. Heavy Ion Fusion Workshop, Brookhaven National Laboratory Report BNL 50769, 1977, p. 17; also Custom R., et al.ibid. 94; also Swenson, D.A., Proc. Heavy Ion Fusion Workshop, Lawrence Berkeley Laboratory Report LBL-10301, 1979, p. 239.
19. Bongardt, K., Proc. 1984 Linac Conf., Gesellschat fur Schwerionforschung report GSI-84-11, 1984, p. 389.
20. Guy, F.W., in *High Current, High Brightness, and High-Duty Factor Ion Injectors*, ed. Gillespie, G.H., Kuo, Y.Y., Keefe, D., and Wangler, T.P; *AIP Conf. Proc.* **139**, 207 (1986); also Guy, F.W. and Wangler, T.P., ibid. 185.
21. Johnson, K.F. et al., Proc. 1990 Int. Linac Conf., Albuquerque, NM, 1990 p. 701.
22. Lawrence, G.P. et al., *Conventional and Superconducting RF Linac Designs for the APT Project*, Proc. 1996 Int. Linac Conf., Geneva, Switzerland, August 26–30, 1996; also Wangler, T.P., et al., Proc. 1990 Int. Linac Conf., Albuquerque, NM, 1990, p. 548.
23. Gardner, I.S.K., Lengeler, H. and Rees, G.H., *Outline Design of the European Spallation Neutron Source*, ESS 95-30-M, September, 1995.
24. Stokes, R.H. and Minero, G.N., *High Current, High Brightness, and High-Duty Factor Ion Injectors*, ed. Gillespie, G.H., Kuo, Y.Y., Keefe, D., and Wangler, T.P; *AIP Conf. Proc.* **139**, 79 (1986).
25. Farkas, Z.D., Hogg, H.A., Loew, G.A., and Wilson, P.E., Proc. 9th Int. Conf. High Energy Accel., SLAC, 1974, p. 576.
26. Loew, G.A., and Talman, R., *Elementary Principles of Linear Accelerators*, AIP Conf. Proc. No. 105, American Institute of Physics, New York, 1983, p. 89.
27. Kamerlingh, H., Onnes, Comm. *Phys. Lab., Univ. Leyden*, Nos. 199, 120, 122 (1911).
28. Bardeen, J., Cooper, L.N., and Schrieffer, J.R., *Phys. Rev.* **108**, 1175 (1957).
29. Pierce, J.M., Schwettman, H.A., Fairbank, W.M., Wilson, P.B., Proc. 9th Int. Conf. on Low Temp. Phys., Part A Plenum Press, New York, 1965, p. 36.
30. For a complete list of references for the application of rf superconductivity to particle accelerators, see the excellent articles by H. Piel and by H. Lengeler in the *CERN Accelerator School for Superconductivity in Particles Accelerators*, ed. S. Turner, CERN 89-04, 10 March, 1989, pp. 149–229.
31. Paadamsee, Hasan, Knobloch, Jens, and Hays, Thomas, *RF*

Superconductivity for Accelerators, John Wiley & Sons, New York, 1975, 1998.

32 Piel, H., *CERN Accelerator School for Superconductivity in Particles Accelerators*, ed. Turner, S., CERN 89-04, 10 March, 1989, pp. 158–165.

33 Lyneis, C.M., Schwettman, H.A., and Turneaure, J.P., *Appl. Phys. Lett.* **31**, 541 (1977).

34 Lagomarsino, V., Manuzio, G., Parodi, R., and Vaccarone, R., *IEEE Trans.* **MAG-15**, 25 (1979).

35 Klein, U., and Proch, D., Proc. Conf. on Future Possibilities for Electron Accel. Charlottesville Va, 1979, McCarthy, J.S., and Whitney, R.R., University of Virginia, Charlottesville, 1979, pp. 1–17.

36 Kneisel, P., Vincon, R., and Halbritter, J., *Nucl. Instr. Meth.* **188**, 669 (1981).

37 Kneisel, P., Amato, J., Kirchgessner, J., Nakajima, K., Padamsee, H., Phillips, H.L., Reece, C., Sundelin, R., and Tigner, M., *IEEE Trans.* **MAG-21**, 1000 (1985), see also p. 149, 1007; and Kneisel, P., Cornell University, SRF 840702 1984.

38 Lengeler,H., Weingarten, W., Muller, G., and Piel, H., *IEEE Trans.* **MAG-21**, 1014 (1985).

39 Piel, H., *CERN Accelerator School for Superconductivity in Particles Accelerators*, CERN 89-04, 10 March, 1989, ed. Turner, S., pp. 177–179.

40 McAshan, M.S. *et al.*, *Appl. Phys. Lett.* **22**, 605 (1973).

41 Bollinger, L., *Nucl. Instum. Meth.* **A244**, 247 (1986); Proc. 1987 Part. Accel. Conf., IEEE Cat. No. 87CH2387-9 (1987).

42 Shepard, K.W., Proc. 4rd Workshop on RF Superconductivity (KEK, Tsukuba, Japan), Y. Kojima, ed. KEK Report o. 89-21 1990.

43 Grunder, H.A., Bisognano, J.J., Diamond, W.I., Hartline, B.K., Leemann, C.W., Mougey, J., Sundelin, R.M., and York, R.C., Proc.1987 Part. Accel. Conf., Washington, D.C., IEEE Cat. No. 87CH2387-9, 1987, p. 13; Dunham, B.M., *Jefferson Lab, A Status Report*, Proc. 18th Int. Linac Conf., Geneva, Switzerland, 26–30 August, 1996, eds. \C. Hill and M. Vretenar, CERN 96-07, November, 15 1996, p. 17.

44 Delayen, J.R., *Upgrade of the CEBAF Acceleration System*, Proc. 1999 Part. Accel.Conf., New York, 1999, pp. 3498–3500.

45 Clendenin, J., Rinolfi, L., Takata, K., and Warner, D.J., *Compendium of Scientific Linacs*, 18th Int. Linac Conf., Geneva, Switzerland, August 26–30, 1996, CERN/PS 96-32 (DI), November, 1996.

46 "International Linear Collider Reference Design Report 2007," ILC-Report 2007-01, April, 2007.

47 Padamsee, Hasan, Knobloch, Jens, and Hays, Tom, *RF Superconductivity for Accelerators*, Wiley, New York, 1998, p. 223.

48 Carlsten, B.E., *Nucl. Instrum. Meth. Phys. Res.*, **A285**, 313 (1989); Carlsten, B.E., *Part. Accel.*, **49**, 27 (1995).

49 Tigner, M., A possible apparatus for electron-clashing experiments, *Nuovo Cimento*, **37**, 1228–1231 (1965).

50 Bogacz, A. *et al.CEBAF Energy Recovery Experiment*, Proc. 2003 Part. Accel. Conf., Portland, Oregon, May 12–16, 2003, pp. 195–197.

Index

a

Accelerating cell
  array of 176
  RFQ transition cell 266–272
Accelerating field
  ion linac, single particle acceleration 175–199
Accelerating gap
  coordinate transformation 208–209
  radial impulse near axis 204–207
  standing-wave electric-field solution 205
Accelerating gradient 34, 401
  constraint on 333
  in multicell cavity 34
  longitudinal beam dynamics constraint 333–335
  two competing effects 87
Accelerating mode
  free-space wavelength 393
  generator-induced, and beam-induced fields 341
Accelerating structure
  higher-frequency Alvarez DTL 89
  principle wave, with largest Fourier amplitude 57
  quarter-wave resonator 24, 88
  resonant coupled 65
  subtraction technique 375
  TEM mode, for superconducting linacs 121
  using electric focusing 406–410
  with periodic geometries, and phase velocity 53
Accelerating structures, heavy-ion 46
Accelerating system
  and accelerators 1–31
Accelerating waveform
  energy difference 388

Acceleration 5
  efficiency of, bunches 4
  space-charge effects 247
  synchronous, radial position and particle velocity 241–242
Acceleration cavity
  power and acceleration efficiency, figures of merit 42–44
Accelerator 2, 5, 14
  current per RF bucket 10
  driven fission-reactor concepts 13
  linear accelerator 1–2
  linear focusing forces, trajectory ellipse 284
  little machine, the 'linac' 1
  normalized emittance 287
  parameters, cavity tuning and RF system 13
  particle accelerator 2
Accelerator cavity
  schematic drawing, RF system 158
Accelerator cavity impedance
  typical frequency spectrum 380
Accelerator production of tritium (APT) 13
Adiabatic bunching
  beam-current capacity 233
Adiabatic phase damping
  phase-space ellipses 187–188
Alternating-phase focusing (APF) 405
Alvarez drift-tube linac 7, 91–96
  DTL cavity 92
  field-free regions 91
  medium-velocity ion, acceleration of 8
Ampére's law 39
Anisotropic beam
  physics of 320

RF Linear Accelerators. 2nd, completely revised and enlarged edition.
Thomas P. Wangler
Copyright © 2008 Wiley-VCH Verlag GmbH & Co. KGaA, Weinheim
ISBN: 978-3-527-40680-7

Index

Anisotropic linac beams
  longitudinal-transverse coupling 319–325
APF *see* Alternating-phase focusing 405
APT *see* Accelerator production of tritium 13
Asymmetrical drift tube
  quadrupole focusing fields 408
Axial electric field 41
  average 93
Axial energy gain 39
Axial field distribution 41
Axial transit-time factor
  aperture-dependent factor 40
  gap, independent 42
  gap-dependent factor 40
Azimuthal magnetic field
  pattern in four-vane RFQ 254

b

BBU *see* Beam-breakup 362
BCS theory
  phenomenon of, superconductivity 416
Beam 7, 23, 24
  acceleration by, RF electromagnetic fields 2
  and emittance 12
  and longitudinal focusing 11
  and vacuum system 6
  beam-loading compensation 13
  energy gain 23
  isodensity contours of 284
  lighter protons and electrons 4
  low velocity, image-charge forces 362
  multiparticle effects 12–13
  net energy gain 4
  phase space 12
Beam axis 251
  and magnetic-field direction 390
  nominal gap-to-gap spacing 90
Beam breakup
  bunch shape 386
Beam bunches 6, 8
Beam cavity
  configuration after first charge 343
  final configuration 343
Beam current
  beam image current, and beam-induced voltage 348
Beam current, acceleration 10
Beam displacement 364
Beam distribution
  proper matching 12

Beam dynamics
  aperture radius 45
  linear space-charge field 300–301
Beam ellipse
  and beam matching 222–223
  in RFQ 272–274
Beam envelope
  phase-space plots 262
Beam frame
  transverse deflection, for accelerated relativistic beam 229
Beam funnel
  schematic diagram 412
  simulation studies, of funneling lines 413
Beam funneling
  concept of 411–413
Beam halos 283
  and beam loss 325–329
Beam holes 44
Beam loading 12, 341–360
  and cavity detuning 349
  and wake fields 282
  fundamental theorem 343
  in accelerating mode 345–346
  standing-wave cavity 341
Beam loss 12
Beam matching
  current-independent beam matching 224
Beam particle
  axial-field pattern 236
  transverse defocusing 12
  transverse focusing 12
Beam pipe
  decoupling of cavities 116–117
  field lines, terminating on pipe 365
Beam velocity
  increased cavity, and cryostat lengths 87
Beam, continuous 3
Beam, ideal
  laminar beam 283
Beam-breakup (BBU) 362
Beam-cavity
  interaction 356–360
  interaction problems 356
Beam-centroid phase 384
Beam-coupling impedance 378–380
  single frequency source 378
Beam-dynamics codes
  numerical space-charge calculations 292–296

Beam-dynamics design
  approach, space-charge compensation 429
Beam-halo experiment
  at Los Alamos 329–331
Beam-induced field 12
Beam-loaded cavity
  equations 347–352
  equivalent circuit 347
  extreme, beam-loaded case 351
  Wilson theory 347
  with bunches 351
Beam-loading compensation 346
Beam-loading ratio 43
Beam-loading theorem
  fundamental 342–343
  initial configuration 342
  surface charge, and beam-induced voltage 342
Beam-particle energy
  accelerator-physics units
    unit conversions, and physical constants 16
Beam-plasma frequency
  space-charge-dominated beams 291
Beat pattern
  describing difference 388
Bessel functions 40, 38, 56
  and modified Bessel function, zero order 38
Bessel's equation 24, 37
Bethe's theory
  coupling of cavities, through apertures 53
BNS damping 387
Boltzmann–Ehrenfest theorem
  adiabatic invariants, of oscillator 135
Boron–neutron capture therapy 13
Bunch
  fields carrying, surface induced charges 361
Bunch loss parameter
  arbitrary charge distribution 377
Bunch wake potential
  arbitrary line-charge density 372
  in arbitrary charge distribution 371–376
  loss parameters, particular charge distribution 376–377
Bunch-centroid displacement
  center-to-center cavity spacing 393
  showing cumulative BBU 394
Buncher cavity
  phase-dependent kick 195

Bunching 4
  longitudinal phase-space ellipse, of incident bunch 195
Bunching effect 194

c
Capital cost
  in accelerator design 47
Cartesian coordinates 209
Cavity
  electrical center 345
  resonance driven, forward and reflected traveling-wave fields 351
Cavity array
  large apertures 86
Cavity design
  cross section of cavity, nose cones and spherical outer wall 44
  issues of 44–46
Cavity field
  and loss factor 341
  beam effect 341
  circular aperture, and aperture radius 70
  transient build up 153
Cavity geometry
  adiabatic invariant of, oscillator 164–165
  and accelerating fields 414
  field-emission-induced electron loading 415
Cavity loading
  with lumped elements 46
Cavity mode 12
  deflecting modes 12
Cavity parameters
  designing linac, and operating frequency 46
  frequency scaling 46–47
Cavity parameters, conventional 154
Cavity power
  dissipation 154
Cavity resonant frequency
  detuning of 350
Cavity resonant-frequency shift 258
Cavity resonator
  circuit model of 136
  coupled, two external circuits 144
  transverse-magnetic mode of, circular cylindrical cavity 24
Cavity shunt impedance 51
Cavity stored energy
  SUPERFISH 344

Index

Cavity system
  wave description, waveguide-to-cavity coupling   148–156
Cavity time constant
  heavily, beam-loaded superconducting cavity with bunches   352
Cavity voltage   342
  versus time   355
Cavity wake potential
  Fourier transform   381
Cavity walls
  ohmic losses   138
Cavity-power dissipation   344
  point charge, loses energy   345
CCDTL see Coupled-cavity drift-tube linac   410
CCL see Coupled-cavity linacs structure   83
CEBAF recirculating linac   419–420
  five-cell 1.497-GHz niobium elliptical cavity   421
CEBAF see Continuous wave electron beam accelerator facility   416
Cell length
  synchronous acceleration   41
Cell resonant-frequency perturbation
  first-order deviations, from nominally flat field amplitude   107
CERN SPL design
  trajectory points   323
Circuit
  with resonant coupling element   64
Circuit impedance
  for driving frequency   381
Circuit model
  dispersion relation   65
  dispersion relation for   64
  of single mode of cavity   381
  periodic array of cavities, with resonant coupling   66
Circular accelerator   10
  relativistic electrons   13
Coaxial half-wave resonator
  and single loading element   123
Coaxial quarter-wave resonator   122
Coaxial resonator   22, 23
  and accelerating cavities   22–24
  coaxial line application   23
  with voltage, and current standing waves   24
Colliding beams   2
Computed transverse δ-function wake potential   372

Conductor
  boundary conditions, and conducting walls   19–20
  resistivity   19
  surface charge density   19
Constant-energy beam   385
Constant-impedance structure
  constant-gradient   74–76
Continuous wave (CW)   5, 50
  ATLAS superconducting linac booster   419–420
  pulsed or continuous operation   10
  reducing, space-charge   10
Continuous wave electron beam accelerator facility (CEBAF)   416
Cooper pairs
  correlated electron pairs   416
Coordinate system
  relative to reference plane   373
Copper-cavity technology
  superconducting niobium cavities   10
Cosine functions
  orthogonality of   37
Coulomb field
  collisional regime, and space-charge regime   290
Coupled cavity
  periodic arrays of   58
Coupled circuits
  electrically, periodic array   62–63
  magnetically, periodic array   63–64
Coupled multicell TM cavities
  Alvarez DTL and CCL   122–125
Coupled oscillators   58
  and time dependence   105
  closest-mode spacing   106
  identical internal oscillators and half-cell   105–108
Coupled three-cavity system
  effects of, power losses   132
Coupled-cavity array   7
Coupled-cavity drift-tube linac (CCDTL)   410–411
  with single drift tube   411
Coupled-cavity linac (CCL)   8, 83
  design of, biperiodic CCL structures   111–114
  dispersion curve of, biperiodic structure   112
  resonant cavities, and multicavity accelerating structure   98–99
  side-coupled structure   111
Coupled-cavity linac (CCL) structure   98
  DYNAC beam-dynamics code   191

Index | 437

Coupled-cavity system
  resonant-frequency errors   132
Coupler geometry
  matched or reflection-free condition   353
Couplers
  fixed versus variable   353
Coupling
  circuit problem, between generator and resonator circuits   142
  electromagnetic energy, to cavities   138–139
Coupling constant k
  coupling between, adjacent oscillators   105
Coupling factor
  backward power, values of   147
Courant–Snyder ellipse parameter
  maximum value   220
Crandall derivation
  zero-slope condition   269
Critical coupling
  corresponding to, outgoing-wave amplitude   153
  round-trip field attenuation factor   153
Cryogenic refrigeration   46
CW neutron sources   13
CW superconducting linac
  driver linac   426
CW see Continuous-wave   5, 419
Cylindrical cavity   6
  transverse-magnetic resonant mode   6
Cylindrical pillbox
  synchronous acceleration problems   49–52
Cylindrical pipe
  image force for line charge   362–364
  radius b, with image charge   363
Cylindrical resonator
  transverse electric modes   28–31

d

Decay time, of modes   379
Design parameters
  for SASE FEL Projects   429
Design particle
  equation of motion   191
Disk-loaded traveling-wave structure
  electric field acceleration   8
Disk-loaded waveguide   5
Dispersion curve
  cutoff frequency   255
  for uniform waveguide   21
  measurement of (a) π mode, resonant frequencies   67
  measurement, in periodic structures   65–67
Dispersion curve (Brillouin diagram)
  for uniform waveguide   54
Dispersion relation   27
Drift space
  transfer matrix of   335–338
Drift tube   3, 4, 8, 9
  and accelerating gaps   176
  axial bore holes   39
  charges and currents, in DTL   92
  compact   90
  constructing, with nose cones   35
  eddy currents, and power dissipation   92
  field penetration   39
  first accelerating system   1
  gap between   35
  loading   89
  longitudinal phase space   198
  transverse focusing, with magnetic quadrupole lenses   94
Drift tube linac (DTL)
  adiabatic bunching, in RFQ   246
  electric-field pattern   93
  for accelerating protons and heavier ions   96
Drift-kick-drift method   97
Drift-kick-drift treatment   185
Drift-tube linac
  with post couplers for, field stabilization   96
Drift-tube linac (DTL)   4, 83, 210, 324
  accelerating gaps   8
  cooling difficulties   45
  design of   96–98
  nonrelativistic beam   405
  sinusoidal accelerating field   405
Drift-tube support stems   210
Drift–kick–drift
  momentum kick   204
Drive cell
  intercell coupling strength k 107
  power-flow phase shift   107
Drive line
  wave-interference effects   135
Driven oscillator   138
DTL design code, PARMILA   97
DTL tank
  longitudinal matching   225
  matched phase-space configurations   224

DTL *see* Drift-tube linac 4, 210, 324
DYNAC beam-dynamics code 191

e
Earnshaw's theorem 202
ECR *see* Electron cyclotron resonance 426
Eigenvectors
  first-order corrections 102
Electric field 4, 6–8, 10, 11, 17, 22, 23, 25, 28, 30
  accelerating gap, and energy gain 33–36
  and charged-particle beam 2
  and energy gain in each cell 97
  average axial electric-field amplitude 34
  beam, maximal energy transfer 5
  determination of, particle dynamics 240–241
  electric-quadrupole, cross section 234
  localized into, bunches 3
  longitudinal and radial, in RF gap 206
  moving charge into, equatorial plane 365
  pattern of disk, and washer modes 120
  peak surface, pulsed operation 10
  square profile of 36
  tngential 22
Electric focusing 9
Electric wakefields
  ultrarelativistic, Gaussian line-charge bunch 366
Electric- and magnetic-field patterns 7
Electric-field line
  in accelerating gap, longitudinal focusing 11
  in RF gap 202
  projections of DTL cells 94
  radial RF electric, and magnetic forces 202
  transverse RF focusing and defocusing 201–202
Electric-field vectors
  $x - z$ plane 236
Electric-focusing structure
  quadrupole geometry 407
Electrical cell
  periodic chain of 59
Electrical center 35
  calculating, transit-time factor 52
  gap, geometric center 35
  standard electromagnetic field solver code
    SUPERFISH 191

Electrode
  potential of 4
Electrode geometry 238
  electric-quadrupole, with unequal electrode spacing 234
  two-term potential function 238–240
Electrode, isopotential 4
Electromagnetic field
  deflecting-mode cycle 391
  fields, from relativistic point charge 364–366
  time-dependent, quasistatic approximation 167–168
Electromagnetic field–solver codes 44
Electromagnetic mode 3
Electromagnetic standing wave
  in RF cavity 32
Electromagnetic stored energy 23
Electromagnetic traveling wave
  and electromagnetically coupled cells 7
Electromagnetic wave
  phase velocity 7
  uniform waveguide 7
Electromagnetic-cavity structure
  beams accelerated with 3
Electromagnetically coupled cells 7
Electron
  multipacting 159–160
Electron cyclotron resonance (ECR) 426
Electron field emission
  and superconducting cavities 162–163
  limiting factor 418
Electron linac 5
  and beam injection 11
  applications of, modern RF linacs 13–15
  derivation, pillbox cavity 24
  drift tubes and supporting stems 4
  high-energy 346
  in cancer therapy 13
  resonant antennas, with high power losses 4
Electron linear accelerator
  and quarks 1
Electron loading 125
Electron multipacting
  limiting factor 418
Electron traverse gap
  two-point multipacting 159
Electron-linac application
  free-electron lasers, and linear colliders 289
Electron-linac peak current 290
Electron-positron colliders 13

Electrostatic accelerator 2
  electric breakdown 2
  limitation of 2
Electrostatic field
  electron volt (eV) 2
Electrostatic focusing
  gap, effects of position and velocity changes 207–208
Ellipse
  general trajectory ellipse, and its parameters 214
Ellipse parameters
  time-independent set 261
Ellipse transformation
  transporting beam ellipses, between two locations 221–222
Ellipsoid form factor 300
Elliptical beam
  continuous 297–299
  RMS envelope equation, with space charge 296–297
Emittance conventions
  rms emittance 288–289
Emittance growth
  for rms mismatched beams 316
  from rms mismatch 317
  initial semi-Gaussian beam 313
Energy conservation
  and energy loss 343
Energy gain 10, 29
  common trigonometric identity 34
  limited by, fixed potential drop 3
  maximum, for relativistic electrons 34
  radial dependence, and synchronous wave 39
Energy loss
  higher-order cavity modes 344–345
  higher-order-mode enhancement factor 345
Energy recover linac (ERL)
  and superconducting linac 429
Energy spread
  short-range wakefields 383–384
Energy velocity
  velocity of, electromagnetic energy flow 22
Energy-gain process 32
Energy-gain, result
  standing-wave field pattern 39
Energy-recovery linac (ERL) 397
Envelope mode
  emittance growth, for mismatched rms beam 316–318
Equivalent circuit

containing generator 145
generator circuit 141
into resonator circuit 143
Equivalent circuit model
  for periodic structures 59–61
  resonator-to-guide coupling factors 145
ERL see Energy-recovery linac 397, 429
ESS see European spallation source 13
Europe
  electron-positron storage ring facilities 416
European spallation source (ESS) 13

f
FEL see Free-electron laser 397
Field distribution
  equal and opposite δ-function errors 261
  periodic iris-loaded structure 55
Field distribution, perturbed, —δ-function error 260
  (a) a δ-function error 259
Field emission 45
Field errors
  shift in, final energy 182
Field pattern
  TM mode of, pillbox cavity 390
Field tilt
  and coupling 108
  fabrication errors, and distortions 97
Field-free drift spaces 212
Fields and currents
  skin depth 19
Finger-focusing geometry
  Tepliakov, π-mode electric-focusing structure 407
First accelerator
  and ground potential 4
Floquet theorem 53
  and space harmonics 54–57
  loss-free periodic structure 55
  stopbands, intervals of 55
Focusing
  thin drift tubes, without quadrupoles 89
Focusing-defocusing (FODO) 395
FODO lattice structure
  thin-lens 215
FODO quadrupole lattice
  with accelerating gaps 211
FODO see Focusing-defocusing 395
FODO-lattice structure
  thin-lens focusing structure 215–217

Four vane, periodic
  with windows RFQ structure 279
Four-rod RFQ 276
Four-vane cavity 248
  cloverleaf geometry 249
  vane-tip patterns 248
Four-vane cloverleaf cavity 252
Four-vane RFQ
  field-tilt effects 254
  idealized mode spectrum 253
  plotted against, longitudinal-mode
    number 254
Four-vane RFQ structure 240
Fourier integral
  accelerating gap 36
  and particle velocity 37
  longitudinal electric field 36–39
  nonzero field region 37
Fourier integrals
  in exponential form 38
Fractional-field error 258
Free-electron laser (FEL) 397
Free-electron lasers
  pulsed neutron sources 13
Free-energy model
  maximum emittance-growth curves
    330
Full cell
  RFQ linac vanes 266

g

Gap factor
  for finite chamber radius 41
  gap and particle velocity 86
Gap geometry
  and field distribution 33
Gaussian
  longitudinal wake potential 373
Gaussian beam
  final emittance growth ratio 313
  three beam parameters, for initial 312
  total bunch loss parameter 377
Gaussian bunch
  ktot versus the rms length 377
Gaussian distribution
  bunch loss parameters 377–378
Gaussian form factor 377
Generator power
  beam current, less than design value
    352–354
  versus beam power, fixed cavity voltage
    354
  zero reflected power 350
Generator-induced voltage 346

Geometry factor QR_s
  versus design velocity β 127
Golf club see Separatrix, shape 182
Group velocity 20
  stored energy and fields 69
  total attenuation parameter 73
  velocity of, amplitude-modulation
    envelope 21
Gustav Ising
  first accelerator 3

h

H-mode accelerating structures
  transverse-electric (TE) 89–91
H-mode linac structure
  longitudinal beam dynamics 196–198
H-mode structure
  high shunt impedance, in high RF power
    efficiency 90
H-mode structures
  IH structure
    crossbar H-mode 90
Half-wave resonator 23
Halo 12
Halo see Beam halo 304
Hamiltonian
  and Liouville's theorem 182
Hamiltonian theorem
  motion ignoring acceleration 182
Heavy-ion acceleration
  quasi-Alvarez DTL, and quasi-Alvarez
    approach 94
Heavy-ion beam
  buncher for 30
Heavy-ion inertial-fusion program
  Lawrence Berkeley National Laboratory
    319
Heavy-ion linacs 13
  flexibility of 190
HEM deflecting mode
  transverse beam-modulation amplitude
    391
HEM see Hybrid electromagnetic modes
  389
High-energy linac
  CW spallation neutron sources 13
  nuclear wastes, transmutation of 13
Hill's equation
  linear second-order differential equation
    212
  solution, phase-amplitude form 213
  transfer-matrix solution 211–213
Hybrid electromagnetic modes (HEM)
  389

i

ILC *see* International linear collider 423
Impedance function
  frequency spectrum, δ-function wake potential 379
Independent-cavity ion linac
  longitudinal dynamics 190–191
Independent-cavity linacs
  and longitudinal beam dynamics 87
  computational approach 191
Inductance calculation
  idealized quadrant shape 251
Input coupler 44
Input guide
  reflection coefficient 145
Input waveguide
  and traveling waves 148
  or transmission line, and power losses 44
Intercell coupling constant
  electric and magnetic dipoles 114–116
Interdigital four-beam MEQALAC 410
Interdigital H-mode (IH) structure 89
  sequence of electrodes, for acceleration 90
International linear collider (ILC) 423–425
  9-cell superconducting niobium cavities 425
International linear collider (ILC) facility
  showing two stages 425
Ion implantation
  semiconductor fabrication 13
Ion linac
  beam matching, current-independent 224
  independent cavities 11
  longitudinal dynamics, in coupled-cavity linacs 189–190
  longitudinal stability 201
  repulsive space-charge forces 12
  synchronous traveling wave 204
Ion-linac channel
  maximum beam current 408
Iris(disk)-loaded
  traveling-wave structure 68
Iris-loaded
  traveling-wave accelerating structure 70
Iris-loaded structure
  periodic, analysis of 69–72
  traveling-wave accelerator, for relativistic electrons 80–81

Iris-loaded waveguide *see* Disk-loaded waveguide 5
Isotope beams
  next-generation rare isotope facility 426–427

j

Jacobian
  transformation increases, phase-space area 184
Jacobian determinant 182
Japan
  TRISTAN facility at KEK 416
Jefferson laboratory
  infrared FEL, ERL operation in CW mode 430

k

K–T potential function 237
Kapchinsky and Vladimirsky
  quantity K, generalized perveance 298
Kilpatrick criterion 45
  choosing, design field level 163
Kilpatrick formula
  and Kilpatrick criterion 163
KONUS scheme
  ten drift-tube, IH drift-tube array 198

l

LANSCE DTL 94
LANSCE linac
  highest average proton beam 14
  or LAMPF 14
LANSCE proton accelerator
  DTL parameters 95
LANSE *see* Los Alamos Neutron Science Center 9
Lapostolle convention
  and rms emittance 288
Lapostolle matchbox geometry 407
LCLS *see* Linac coherent light source 429
LEDA halo experiment
  52-quadrupole-magnet lattice 329
LEDA *see* Low-energy demonstration accelerator 329
Linac 4–7, 14
  800-MeV LANSCE proton linac 15
  acceleration method 7
  and linac structure 6–10
  application 135
  BBU instability 362
  beam bunch, beam quality
    phase space, and emittance 283–285

Linac (continued)
  beams, paraxial approximation   38
  betatron frequency, betatron wavelength   386
  better beam quality   2
  biperiodic structures   108–111
  block diagram   6
  bunched beams, space-charged dynamics   289–292
  Coulomb effects   282
  designed waveguides, or high-Q resonant cavities   5
  emittance growth, space-charge-induced   314–316
  high-power vacuum-tube amplifiers   156
  independent-cavity   83–87
  length, and ohmic power consumption   10
  linear accelerator   1
  modular array of, accelerating structures   5
  multiparticle dynamics with space charge   282–338
  parameters of, SLC and LANSCE linacs   15
  pulsed machine   289
  quadrupole focusing   209–211
  risk of, emittance transfer   323
  rms emittance   285
  single-pass device   10
  strong focusing   5
  superconducting cavities, accelerator in Stanford   419–422
  superconducting linacs   6
  synchronous acceleration   182
  synchronous-velocity profile   181
  tandem van de Graaff electrostatic accelerator   419
  transverse stability plot   217–218
  wake potential, from relativistic point charge   367–368
Linac accelerating structures   83–132
Linac beam   283
  dynamics, multicell ion linac   10–12
  equipartitioning   321
  stable phase   10
  synchronous phase   10
Linac bunch
  three-dimensional, ellipsoidal bunched beams   299–300
Linac coherent light source (LCLS)   429
  European XFEL at DESY   429
Linac cost model   47
Linac economics
  impact of, design choices   47–49
  resistive losses   47
Linac injector
  neutron-spallation source   289
Linac particle dynamics
  first approximation for, computer simulations   184
Linac periodic focusing
  smooth approximation   226–227
Linac structure
  and linac systems   405
  bunches   6
  longitudinal dynamics, of low-energy electron beam   193
  mode concept   107
  traveling-wave, disk-loaded or iris-loaded waveguide structure   68–69
Linac technology
  group velocity, and energy velocity   20
Linac terminology
  synchronous-energy gain   241
Linac see Accelerator
  little machine, the 'linac'   1
Linac, efficient
  construction of   7
Linac-cavity field
  maintaining phase and amplitude   157
Lincac
  radio frequency (RF) acceleration   32–52
Linear accelerator   5
  alternating-phase focusing (APF)   405–406
  and high-power beams   2
  beams, high energy   1
  historical perspective   2–5
  new type   232
  special structures and techniques   405–430
  spin-off, nuclear physics research   1
  straight-line trajectory, power loss avoidance   2
Linear colliders
  and circular colliders   13
Linear focusing
  ideally matched beam with   285
Linearized Vlasov theory
  theoretical approach, for stability analysis   318
Liouville's theorem   183, 284
  and dashed parallelogram   186
  phase and position impulses   209

Long cavity
  figure of merit, field level and cavity length 43
Long-range wakefields
  and multibunch beam breakup 389–397
Longitudinal beam dynamics 175
Longitudinal mode 27
Longitudinal motion
  accelerating cells 176
  differential equations 178
  equations of, Hamiltonian and Liouville's Theorem 182–186
  small, acceleration rate 178–182
Longitudinal phase space
  and longitudinal potential well 180
Longitudinal phase-space ellipse
  at phase focus 196
  buncher cavity, after kick 195
Lorentz condition 18
Lorentz force 29
Los Alamos beam halo experiment 329
Los Alamos Neutron Science Center (LANSCE) 9
Low-energy beams
  longitudinal dynamics of 192–194
Low-energy demonstration accelerator (LEDA) 329
Low-frequency structures 46
Low-pass filter
  periodic array 61–62
  periodic, basic cell of 61
Low-pass filter, periodic
  dispersion curve of 61
Lumped-circuit model
  four-vane cavity 249–252
  four-vane RFQ, in cloverleaf geometry 249
  lumped-circuit picture 93
  numerical, electromagnetic-field-solver code, SUPERFISH 93

m

402.5-MHz Alvarez drift-tube linac 423
805-MHz SNS superconducting linac 424
805-MHz coupled-cavity linac 423
Macropulses 9
Magnet
  quadrupole 89
Magnetic field
  calculating, effective loop area 171–174
  longitudinal, and solenoid focusing 225–226
  predominant RF longitudinal 89
  standard perturbation techniques 91
  typical(upper curve) and electric field (lower curve) 93
Magnetic field line
  in quadrupole mode at end 253
Magnetic field, axial
  two-cavity klystron 156
Magnetic flux
  voltage divider, behavior of 92
Magnetic mode
  cylindrical resonator
    transverse-magnetic modes 27–28
Magnetic quadrupole lens 233
Magnetic-quadrupole 9
Matched phase-space drawings 273
Mathieu equation 244
  betatron frequency 244
Maximum acceleration 241
Maxwell's equations 17–19, 39
  accelerating cavity, approach of Condon 137
  electric and magnetic fields 17
  quasistatic approximation 135
MEQALAC see Multiple-beam electrostatic-quadrupole-array linear accelerator 409
MHz four-vane RFQ 422
Micropulses 9
Microwave
  and microwave systems 135–174
Microwave power system
  for linacs 156–158
Mismatched beams
  halo-formation mechanism 326
Mode 12, 23, 30
  beam-breakup, or instability 12
  higher-order mode 12
π-mode 41
π-mode boundary conditions 41
Mode, electromagnetic 3
Modulation parameter 237
Multicell, superconducting elliptical cavity 41
Multigigaelectron volt linac
  for heavy-ion-driven inertial-confinement fusion 13
Multipacting
  characteristics of 160–162
  lower electric field levels 160
Multiparticle system
  Coulomb interactions 282
Multiple-beam electrostatic-quadrupole-array linear accelerator (MEQALAC) 409

n

Net cavity voltage 346
Nonequipartitioned beam
  rms-emittance transfer, having small effect 325
Nonlinear space-charge field 314
Normal-mode spectrum
  of coupled oscillator system 106
Nose cone
  capacitance 44
  elliptical cavity shape 45
Nose cones 35
nth space harmonic
  wavenumber, and phase velocity 56

o

Ohmic power
  dissipation, effects of 103–104
Ohmic power loss 5
Operating temperatures 45
Optimum gap geometry
  risk of, RF electric breakdown 36
Oscillation
  small amplitude, and synchronous particle 186–187
Oscillator
  impedance and wake potential, single cavity mode 381–383
Oscillator error 106
Outgoing wave
  reflected wave, and emitted wave 154

p

Panofsky equation 35
Panofsky–Wenzel theorem 392
  beam-breakup instability 168–171
Particle
  peak energy gain 43
Particle acceleration
  normal mode characteristics 76–79
Particle interaction
  high-frequency betatron orbits 328
Particle velocity
  electrode spacing, increase of 89
  transit-time factor, and energy gain 37
Particle-in-cell (PIC)
  macroparticle-tracking approach 293
Particle-to-particle interaction 295
Peak electric stored energy
  and electromagnetic stored energy 25
Peak magnetic field 45
Peak surface electric field 125, 240
  accelerating-gradient ratio
    versus design velocity 126
Peak surface electric-field
  normal-conducting cavities 45
Peak surface field 44, 45
Peak surface magnetic field 45
  versus design velocity 126
Periodic accelerating structures 53–81
  dispersion curve 59
  general description 57–58
Periodic focusing channels
  space-charge instabilities 318
Periodic structure
  phase velocity 54
Periodic structures
  designing of 67
Permanent magnet quadrupoles, compact 91
Perturbation theory
  effects of resonant-frequency errors 101–103
Perturbation-theory, general 107
Phase damping
  longitudinal beam ellipse 188
Phase oscillation 11
  and phase damping 188
  final, synchronous energy 11
  synchronous particle 11
Phase-space area
  trace-space, and unnormalized phase-space 283
Phase-space ellipse
  rms emittance 285–287
  transformation, between two locations 222
Phase-space trajectory
  in coupled-cavity linac structure 190
Phase-volume 12
Phasor
  currents and voltages 348
Physical constants 16
Pillbox-cavity 5, 24
  array of 53
Pillbox cavity, unperturbed 71
Pipe geometry
  finite conductivity 365
Potential 18, 29
Power densities 45
Power loss 46
PPI see Particle-to-particle interaction 295
Program TRACE
  linear space-charge forces 292
Proton drift-tube linac
  parameters of 198–199

Proton linac
 and modern applications 13
 cross section, elliptical cavity 45
 superconducting 83
Proton linear accelerator
 spallation neutron source (SNS) 422
Pulsed linac
 micropulses and macropulses 9

q
Quadrupole
 misaligned, centroid oscillation
   amplitude 221
Quadrupole focusing system
 dynamics of 210
Quadrupole lens
 periodic lattice of 201
Quadrupole magnet
 focusing 5-MeV proton beam,
   229–230
 showing four poles, coils, and
   magnetic-field pattern 210
Quadrupole spectrum
 transmission-line model 255–261
Quadrupole-mode dispersion curve
 four-vane cavity eigenmodes 252–254
Quality factor 23
 waveguide-to-cavity coupling, through
   iris 154

r
Radial field distribution
 mode, resonant frequency of 25
Radial impulse
 RF-defocusing impulse 203
Radial momentum 203
Radial motion
 unfocused relativistic beams 227–229
Radiated field
 short-range or long-range wakefields
   362
Radiation
 synchrotron 5
Radio frequency (RF) 2, 282
Radio-frequency quadrupole (RFQ) 2
 electric field 9
 transverse RF electric-restoring force 9
Radiofrequency (RFQ)
 transverse dynamics 243–245
Radiofrequency quadrupole (RFQ)
 radial-matching 261–265
Radiofrequency quadrupole linac (RFQ)
 new linear accelerator 232–281
 principles of operation 232–236
Radiofrequency–Quadrupole (RFQ) 90
Random error
 effects, for periodic or quasiperiodic
   accelerator systems 218
Random quadrupole misalignment
 effects of errors 218–221
Rare isotope accelerator (RIA) 426
Recirculating electron linacs
 multipass BBU 397–399
Reference particle
 maximum phase excursion 189
Regions, kinematic conditions
 two parallel plates 160
Relativistic electron beams 361
Relativistic electron linac 12
Relativistic mechanics
 basic formulas 16
 Lorentz force, particle with charge and
   velocity 17
 useful relationships 16–17
Residual resistance 20
Residual resistivity ratio (RRR) 418
Resonance accelerator 5
 linac, cyclotron, and synchrotron 3
Resonance-detection electronics
 RF pickup signal 158
Resonant buildup
 of displacement of tail 386
Resonant cavities 5
Resonant cavity
 theory of 137–138
Resonant coupling
 coupling elements 117–121
 post-coupled drift-tube linac structure
   119
Resonant coupling element
 cavities, periodic array 64–65
 shunt resonant admittance 64
Resonant mode 24
Resonant-cavity system
 equivalent circuit 139–144
 methods of, coupling to cavities 140
 transient behavior of 146–148
Resonator
 quality factor of 42
 time dependence of 148
Resonator circuit
 equivalent circuit 348
Resonator field
 turn-on transient 146
RF accelerating field
 compensation of, longitudinal wake
   effect 384

RF accelerator
    adiabatic bunching   245–248
RF accelerator cavity
    application of, superconducting technology   415
RF accelerators   2, 3, 13
RF bunching
    RF linac input beam, and bunches   194–196
RF cavity
    periodic or quasiperiodic, array of cells   32
    pillbox cavity   33
RF cycle   9
    bucket, stable region   9
    electric field, longitudinal   9
RF efficiency   10, 51
RF electric breakdown
    Kilpatrick criterion   163–164
Rf electric breakdown
    risk of   10
RF electric field   9
    beam, sustained energy transfer   32
RF electrode
    four-vane cavity   248–249
RF field
    energy-gain difference   384
RF fields
    alternating-phase focusing approach   201
    focusing difficulties   201
RF generator efficiency   44
RF input power
    transient turn-on, beam-loaded cavity   354–355
RF linacs
    beam bunches of   6
    funneling of, high-current beams   412
    space-charge instabilities   318–319
    stability analysis, and linearized Vlasov theory   318
RF linear accelerator   3, 5
    and resonance accelerator   5
RF pickup probes
    for vacuum pumping   30
RF power   2, 30
RF power costs
    and operational costs   49
RF power losses
    RF power dissipation scales   46
RF power system
    and coupler geometry   346
RF pulse compression
    klystron pulse   413

RF quadrupole structure
    general potential function   236–237
    quasistatic approximation   236
RF superconductivity   2
    Padamsee, accelerator applications   416
    physics and technology   416–418
RF surface resistance   19
    superconducting niobium cavities   414
RF system   13
    block diagram, and equivalent circuit   141
    circulator, RF input drive line   347
RF *see* Radio frequency   2, 282
RF-focused drift-tube linac (RFD)   407
RF-focused interdigital linac structure (RFI)   408
RF-power   6
RF-power tubes
    klystrons   5
RFD *see* RF-focused drift-tube linac   407
RFI linac structure, views of   409
RFI structure
    showing minor stems   409
RFI *see* RF-focused interdigital linac structure   408
RFQ
    and bunched beam   262
RFQ accelerator
    four electrode arrangement   235
RFQ adiabatic bunching   246
    radial-matching (RM)   248
RFQ bunches   9
RFQ cell
    parameters   279–281
RFQ electric field
    transverse electric focusing   9
RFQ electrode geometry
    and longitudinal dynamics   242–243
RFQ four-vane cavity
    azimuthal modes   253
RFQ operating mode   257
    squared-frequency-error distribution   258
RFQ operation
    four-vane RFQ accelerator   233
    principles   232
RFQ quadrupole mode
    four-vane cavity   255
RFQ quadrupole-focusing geometry   280
RFQ structure
    electrode geometry   233
    four-rod cavity   275–276
    windows RFQ, four vane   277–279

RFQ tuning
  in four-vane resonator   278
RFQ see Radio-frequency quadrupole   2
RIA see Rare isotope accelerator   426
RM section designs   263
RM section geometry
  Crandall, four-term potential function   263
RMS ellipse   286
  Courant–Snyder parameters   285
RMS emittance
  transverse and longitudinal   287–288
RMS emittance growth   315
RMS matched beam
  emittance growth   306–312
RMS-emittance growth
  measured   330
  measured, beam half widths   331
RMS-mismatched beam   316
  particle-core model   332
  Reiser's, free-energy parameter   316
r over Q   43
Rolf Wider
  concept of, Wideröe drift-tube linac   3
Rolf Wideröe
  first experimental test   2
RRR see Residual resistivity ratio   418

S

SASE FEL projects
  three key design parameters   429
SASE see Self-amplified spontaneous emission   429
Scalar potential   18
Scaling formulas
  for $\lambda/4$, superconducting structures   131
  stored energy, for TM   130
Scattering matrix
  for n-port device   149
  reciprocity   150
Schematic diagram
  SLED pulse compression   413
SCL see Side-coupled linac   9
Self-amplified spontaneous emission (SASE)   429
Separatrix
  fish   181
  golf club, longitudinal phase-space trajectories   183
Separatrix see Golf club   179
Short-range field
  binary, small impact-parameter Coulomb collisions   282

Short-range wakefield
  single-bunch beam breakup   384–386
Short-range wakefields   362, 376
Shunt impedance   5
  and figure of merit   42
  electromagnetic-field-solver codes, SUPERFISH   97
  normal-conducting cavities   43
  per unit length   43
  TM and $\lambda/2$ superconducting structures   128–130
  unit, megohms per meter   43
Shunt impedance, transverse
  transverse beam-cavity interaction   392
Shunt impedance   43
Shunt resonant circuit model
  parallel resonant circuit   135–137
Shunt-resonant-circuit cavity model
  impedance of   382
Side-coupled linac (SCL)   9
Simple two-particle model
  for single-bunch beam breakup   385
Single gap
  multicell transit-time factor   85
  multigap periodic structure   40
Single wave
  same phase velocity, and net energy   39
Single-bunch BBU effect
  BNS damping of beam breakup   386–389
Single-bunch loss parameter   344
Single-point multipacting   161
  limiting, pillbox-like superconducting cavities   162
Sinusoidal voltage distribution   235
Skin depth see Fields and currents
  skin depth   19
Skin effect
  RF electric and magnetic fields   19
Skin effect see Fields and currents
  skin effect   19
SLAC   5
SLAC electron accelerator
  disk-loaded waveguide   370
SLAC linac
  RF pulse compression   413–414
  typical RMS bunch   345
SLAC linac structure   371
  total loss parameter k_{tot} versus rms bunch length   378
SLAC see Stanford Linear Accelerating Center   5
Slater perturbation method
  field-measuring apparatus   167

Slater perturbation theorem 53
  resonant frequency shift 165–167
  resonant-frequency change 69
SLC electron-positron linear collider 14
SLC linac
  BNS compensation, of transverse wake effect 389
Smooth-approximation solution 244
SNS facility
  individual systems 421
SNS see Spallation neutron source 13, 416, 421
Space charge
  beam current limits 302–303
Space harmonics
  traveling waves, infinite number 56
Space-charge calculation 294
  SCHEFF subroutine, in PARMILA codes 293
Space-charge field
  charge-density uniformity 283
  coupled Vlasov–Maxwell equations 282
  image field 363
  smoothed field distribution 282
Space-charge force 282
Space-charge forces 12
  extended halo, of beam 12
Space-charge mechanism
  emittance growth 303–306
  rms-matched beam, charge redistribution 304
Space-charge-induced emittance growth 283, 305
Space-charge-induced rms-emittance growth
  scaling of 332
Spallation neutron source (SNS) 13, 416, 421–422
Split-ring resonator
  concept for 115-MHz $\beta = 0.13$ resonator, 122
Square waves
  transit-time factor 35
Square-wave
  transit-time factor 36
Stability chart
  for transverse motion, Smith and Gluckstern 219
  Smith and Gluckstern, for DTL 218
  stop bands 322
Stability plot
  showing ESS design 324
Stable operating point 10

Standing waves
  forward and backward traveling waves 105
Standing-wave linacs
  difference equations, of longitudinal motion 177
Stanford Linear Accelerating Center (SLAC) 5
Stroboscopic plot
  independent particle trajectories 327
Structure efficiency 51
Superconducting cavities
  quench, critical magnetic field 45
  RF power efficiency 44
Superconducting linac
  at Oak Ridge National Laboratory 416
Superconducting linac facilities
  free-electron lasers 427–429
Superconducting linacs 83
  electric-field limitations 415
Superconducting niobium cavities 424
Superconducting RF linac
  schematic drawing, Jefferson Lab ERL system 398
Superconducting RF Linacs
  large RF power 414–418
Superconducting structures
  half-wave resonators 124
Superconducting structures, $\lambda/2$
  half-wave resonators 121–124
Superconducting surface resistance 20
Superconducting-linac technology
  future facilities 423–430
Superconductivity
  discovery of 1
  phenomenon of 415–416
Superconductors
  London penetration depth 416
  zero dc resistance 416
SUPERFISH see Electrical center standard electromagnetic field solver code
  SUPERFISH 191
Surface-resistance
  and power loss scaling, in superconducting RF resistance 46
Synchronism factor 86
Synchronous acceleration 49
  and periodic structures 53
Synchronous particle
  accelerated reference particle 189
  and synchronous phase 32
  energy gain 242

longitudinal focusing   175–176
  stable phase   176
Synchrotron radiation   2, 13

t

TE mode   28
TEM *see* Transverse electromagnetic   84
Thin-lens approximation
  and lattice properties, comparison   227
Thin-lens quadrupole
  random transverse misalignments   219
Thomas Jefferson National Accelerator Facility (TJNAF)   416
Three coupled oscillators   99–100
TJNAF *see* Thomas Jefferson National Accelerator Facility   416
TM and TEM structures
  RF properties and scaling laws   125–127
TM cell array
  dispersion relation, for electrically coupled periodic   62
TM circuits
  basic cell of, periodic array   63
TM circuits, electrically coupled
  basic cell of, periodic array   62
TM mode
  cylindrical cavity resonator   25
TM superconducting structures
  or elliptical cavities   122
TM_{011} cavity mode
  node, and energy gain   36
Transfer matrix
  transfer-matrix elements, through one period   214
Transfer-matrix formalism   221
Transit-time effect
  and transit-time factor   35
Transit-time factor   35
  and synchronous velocity   40
Transit-time-factor models
  penetration of field   39–42
Transition cell, Crandall   266
Transition-cell application   272
Transmission lines   3, 5, 20
Transverse coupling impedance   392
Transverse electromagnetic (TEM)   84
  resonant standing-wave mode   22
Transverse emittance
  unnormalized emittance   287
Transverse magnetic wave   53
Transverse particle dynamics   201–230
  off-axis particles   201
Transverse-magnetic mode   27

Traveling
  and standing-wave structures, physics regimes   79–80
Traveling wave
  constant-impedance structure   72–74
  frequencies and wave numbers   20
Traveling wave, synchronous
  radial impulse   203–204
Traveling-wave accelerator
  principles of operation   68
Traveling-wave power   72, 147
Traveling-wave structure
  50-GeV electron linac   7
Tune-depression ratio
  space-charge-induced rms-emittance growth   332
Tuning stubs
  tuning for, the desired field distribution in RFQ   274–275
Two-conductor
  two-point multipacting   159
Two-gap cavities
  schematic drawing   85

u

Ultrarelativistic beam
  wake potentials, cylindrically symmetric structures   368–372
Ultrarelativistic bunches   362
Ultrarelativistic source charge   367
Uniform waveguide
  cutoff wavenumber   54

v

Vane geometry
  potential function   232
Vane voltage
  ratio of surface magnetic field   252
Vane-tip profile
  in accelerating cell   269
Vane-tip profiles, periodic   268
Vector potential   18
Vectors, transformed   52
Velocity change
  and transit-time factor   35
Velocity modulation   4
Voltage
  axial radio frequency   34
Voltage distribution   256
Voltage phasor
  beam-loaded superconducting cavity   352
  for beam-loaded cavity   346, 352

W

Wake force
  free oscillation solution   385
Wake potential
  arbitrary charge distribution, and Green's functions   367
  computed Gaussian wake potential per cell   374
  corresponding to, shunt resonant circuit   382
  frequency scaling of   370–371
  longitudinal   399–402
    and transverse-impedance definitions   380
Wake potential δ-function wake potential   368
Wakefields
  bunch, electric and magnetic fields   361–402
  damped, oscillatory electromagnetic disturbances   362
  equivalent beam-induced image charges   361
  higher-order modes, and beam-excited accelerating mode   362
  parasitic energy loss, short-range   383
  scattered radiation   12
Wave
  phase velocity, matching beam velocity   32
Wave attenuation
  and decibels (db)   31
Wave equations   18
  in cylindrical coordinates   37
Wave group
  and wave packet   21
Wave packet   21
Wave power   22
Waveguide
  and cavity section, conducting plate with iris   150
  cavities, coupling of   135
  equivalent circuit, for cavity coupled to   144–146
  filling time   76
  particle acceleration, in RF Field   32–33
Waveguide mode   30
Waveguide-to-cavity coupling parameter   155
Wideröe
  linac, forerunner to, modern RF accelerators   3
Wideröe Linac   4
  acceleration of, single particle species   87–89
Wideröe-type linac
  and drift tubes   4

Lightning Source UK Ltd.
Milton Keynes UK
UKOW07f1551260116

267118UK00001B/1/P